Environmental Science

Series Editors: R. Allan · U. Förstner · W. Salomons

Springer

Berlin
Heidelberg
New York
Barcelona
Hong Kong
London
Milan
Paris
Singapore
Tokyo

R. Ebinghaus · R. R. Turner · L. D. de Lacerda
O. Vasiliev · W. Salomons (Eds.)

Mercury Contaminated Sites

Characterization, Risk Assessment and Remediation

With 171 Figures and 95 Tables

Springer

Editors

Dr. Ralf Ebinghaus
GKSS Research Centre, Geesthacht, Germany

Dr. Ralph R. Turner
Frontier Geosciences, Seattle, USA

Prof. Dr. Luiz D. de Lacerda
Universidade Federal Fluminense, Rio de Janeiro, Brazil

Prof. Dr. O. Vasiliev
Institute for Water and Environmental Problems (IWEP)
Novosibirsk, Russia

Prof. Dr. Wim Salomons
GKSS Research Centre, Geesthacht, Germany

ISBN 3-540-63731-1 Springer-Verlag Berlin Heidelberg New York

Library of Congress Cataloging-in-Publication Data

Mercury contaminated sites : characterization, risk assessment, and remediation /
R. Ebinghaus ... [et al.]. p. cm. -- (Environmental science)
Includes bibliographical references and index.
ISBN 3-540-63731-1 (hardcover)
1. Mercury--Environmental aspects. 2. Hazardous waste site remediation.
I. Ebinghaus, Ralf. II. Series: Environmental science (Berlin, Germany)
TD196.M38M464 1999 98-27745

Production: ProduServ GmbH Verlagsservice, Berlin
Typesetting: SPS Madras, India
Cover layout: Struve & Partner, Heidelberg
SPIN: 10551370 32/3020-5 4 3 2 1 0 - Printed on acid -free paper

Preface

Mercury is outstanding among the global environmental pollutants of continuing concern. Especially in the last decade of the 20th century, environmental scientists, legislators, politicians, and the public have become more aware of mercury pollution in the global environment. It has often been suggested that anthropogenic emissions are leading to a general increase in mercury on local, regional, and global scales. Numerous industrial activities, including the mining of gold, silver, and mercury itself, have caused mercury contamination of terrestrial and aquatic ecosystems. Mercury-contaminated sites are abundant worldwide. Mercury has been accumulated as an "industrial legacy" in the rocks, soils, and sediments at, and adjacent to, industrial sites and now may pose significant risks to human and ecological health. Direct inputs of mercury into the environment by industrial activities have generally decreased significantly in the Western world over the past three decades. However, sources such as coal combustion and the use of mercury in gold and silver mining continue to be of importance on a global scale. The same is true for diffuse sources such as terrestrial landscapes that have been impacted by mercury inputs from the atmosphere in the past. Soils (and vegetation) located close to large atmospheric mercury sources have clearly been sinks while these sources were active in the past. In the long run, however, these soils may have become important diffuse area sources after the original emissions were discontinued. Unlike other metals, which are generally not very volatile, mercury from contaminated sites can have a significant impact on remote ecosystems via the atmospheric pathway. Thus, mercury contamination is not only just a local issue but also has global dimensions.

This book summarizes, for the first time, information on the characterization, risk assessment, and remediation of mercury-contaminated sites on the European, Asian, and American continents. Review chapters are supplemented by detailed inernational case studies. Included are papers which were initially presented at the 4th International Conference on Mercury as a Global Pollutant held in Hamburg, Germany, August 4 - 8, 1996. The conference was organized jointly by GKSS Research Centre Geesthacht, Germany, and Oak Ridge National Laboratory, USA, and was attended by over 400 participants from 35 nations.

It was the decision of the technical advisory team to publish the Hamburg conference papers in the peer reviewed open literature. This has now been achieved through the preparation of five special journal issues and this book covering all topics from the conference. These publications contain about one-third of the original presentations from the conference. Copies of the special issues can be obtained directly from the publishers as follows:

- Analytical Developments in the *Fresenius Journal of Analytical Chemistry*
- Atmospheric Cycling in the journal *Atmospheric Environment*
- Biogeochemical Cycling in the journal *Biogeochemistry*
- General Topics on Mercury in the journal *Science of the Total Environment*
- Human Health Issues in the journal *Water, Air and Soil Pollution*

The sixth special issue in this series is the present book that contains a number of submitted technical papers and invited overview papers.

In Chapter 1 Ebinghaus et al. review what is known about natural and anthropogenic emissions of mercury to the atmosphere and evaluate this information in the context of regional and global budgets. Ferrara (Chap. 2) describes mercury mining in Almaden (Spain), Idrija (Slovenia), and Mt. Amiata (Italy), contrasting environmental impacts at both active and inactive sites. Lacerda and Salomons (Chap. 3) consider the amalgamation technique for extraction of gold, noting that environmental impacts of this activity can be serious, but that little information is available on the long-term risks or on measures available to remediate contaminated sites. In Chapter 4, Turner and Southworth provide an overview of mercury-contaminated sites in North America and offer some "lessons learned" from experiences at these sites. Hempel and Thoeming review a number of remediation techniques for soil at contaminated sites, noting that wet classification remains the most common technique (Chap. 5). Mukherjee examines nine technologies for the removal of mercury in gases from metallurgical industries (Chap. 6). A brief overview and update on the situation at Minamata Bay in Japan are presented by Kudo and Turner (Chap. 7). The broad review chapters conclude with a discourse (Chap. 8) by da Costa on the behavior of mercury species in biological systems, and more specifically, on the surface chemistry of microbial cell walls.

The second portion of the book is devoted to international case studies and includes papers from the 1996 Hamburg conference under the general topics: Industrial Sites, Mining, Emissions and Atmospheric Dispersion, Remediation, and Mercury Contamination in Aquatic Systems. Under *Industrial Sites* the various chapter authors provide an example of a site where mercury compounds were used to treat wood (Schöndorf et al.) and three examples (1) where mercury contamination of a marine system originated from an industrial area in Brazil without major point sources of mercury (Marins et al.), (2) where the contribution of electrical lamp (fluorescent) manufacturing plants to total emissions in the CIS was evaluated (Yanin et al.) and (3) where mercury in oil and gas deposits of the former Soviet Union has been characterized with respect to geological origin, geographic distribution, and relationship to other constituents (Ozerova et al.).

Under *Mining* are five chapters related to mercury mining and two chapters covering the use or association of mercury to gold and silver mining. Ferrara et al. summarize mercury emission estimates and ambient air concentrations for one of the closed mercury mines near Mt. Amiata, Italy, using LIDAR remote sensing and point measurements. Miklavčič reports mercury concentrations in air, lichens, carrots, and beans in the town of Idrija (Slovenia), while Kobal et al. examine whether urinary mercury concentration in Idrija miners is a valid indicator of individual internal doses received during intermittent exposure to elemental mercury. Gnamus and Horvat evaluate the degree of contamination with mercury and its transfer in terrestrial food webs in the active mercury mining area of Idrija. Banasova describes changes in plant communities around an Hg mine and smelter in Slovakia, noting that observed changes are due to the combined emissions of sulfur dioxide, mercury, and copper. Lechler found that mercury still generally occurs in the same elemental form in which it was used in the 19th century to mine for gold in two areas in the western United States. The last chapter (Laperdina et al.) in this section summarizes mercury concentrations in environmental media in the gold mining districts of Siberia and points out the irreversible ecological degradation which has accompanied placer mining in these districts.

The section on *Emissions and Atmospheric Dispersion* begins with a chapter by Krüger et al. which describes determination of mercury emissions from a major industrial site in Germany occupied by both closed and still-operating facilities (chloralkali and acetaldehyde). Petzoldt et al. describe the use of modified zeolites to remove merucry from various industrial gas streams. Špirič and Hraste share practical experiences with a sulfur-impregnated activated carbon system designed to remove mercury from natural gas. Analytical techniques for the speciation of mercury emissions from a Municipal Solid Waste Incinerator (MSWI) are described by Wang et al.

Under the topic of *Remediation* the book includes papers by Matsuyama, who describes a novel low temperature thermal process for the treatment of Hg-contaminated soils involving addition of iron chloride (Matsuyama), and by Thoeming et al., who describe a hydrometallurgical technique (electroleaching of both mercury and gold) applicable to mercury-contaminated soils from the Brazilian gold mining areas. Finally, Meschede and Vogelsberger describe the demolition and decontamination of a chloralkali plant in Alexandria, Egypt.

The book concludes with four chapters describing *Mercury Contamination in Aquatic Systems*. Glass et al. examined mercury concentrations as a function of depth in sediments in six reservoirs on the lower St Louis River, a major tributary to Lake Superior. Patel et al. found mercury and other metals in sediments to be largely derived from mineral- and coal-rich regions in India. Probst et al. describe studies of mercury partitioning among water, suspended matter, and bottom sediments of the Ill-Thur river system in France, noting the presence of temporal and spatial gradients. Finally, Pandit et al. describe results of sampling and analysis for total and methylmercury in a tidal creek ecosystem near Bombay, India.

The various chapters clearly show that each contaminated site has its own history of pollution and, furthermore, that the risks associated with mercury depend on site-specific biogeochemical conditions. No universal treatment procedure is available and the selection of the most effective one should be made on a site by site evaluation.

Acknowledgment. The joint Editors appreciate the contributions of the many authors and scientific reviewers who made this book a reality. In addition, Karin Rahn provided administrative services and other valuable assistance. Thanks are due to Stefan Schmolke for the preparation of the picture for the cover of the book. Lastly, the Editors thank the staff at Springer-Verlag for their patience and their skill in copyediting and final production of this Volume.

Ralf Ebinghaus, Ralph R. Turner, Luiz D. de Lacerda,
Oleg F. Vasiliev and Wim Salomons

List of Contributors

S. G. Aggarwal
School of Studies in Chemistry
Pt. Ravishankar Shukla University
Raipur-492 010 (M.P.), India

Dr. H. Akagi
National Institute for Minamata Disease
4058-18 Hama, Minamata
Kumamoto 867, Japan
E-mail: hiroakagi@web.nimd.go.jp

Dr. M. Balcer
Lake Superior Research Institute
University of Wisconsin, Superior
1800 Grand Ave.
Superior, WI 54880, USA
E-mail: mbalcer@staff.uwsuper.edu

Dr. V. Banásová
Institute of Botany
Slovak Academy of Sciences
Dúbravská 14
842 23 Bratislava, Slovakia
E-mail: banasova@bou.savba.sk

Dr. Ph. Behra
Institut de Mécanique des Fluides
de l'Université Louis Pasteur
URA CNRS 854
2 rue Boussingault
67000 Strasbourg, France
E-mail: behra@imf.u.strasbg.fr

Dr. H. Biester
Institut für Umwelt-Geochemie
Universität Heidelberg
69120 Heidelberg, Germany
E-mail: biester@classic.min.uni-heidelberg.de

Dr. H. W. Bräuer
Lurgi Bamag GmbH
Wetzlarer Straße 136
35510 Butzbach, Germany

Dr. C. K. Chandrawanshi
School of Studies in Chemistry
Pt. Ravishankar Shukla University
Raipur-492 010 (M.P.), India

Dr. A. E. Cherrnova
Institute of Geology of Ore Deposits,
Petrology, Mineralogy and Geochemistry
Russian Academy of Sciences (IGEM RAS)
35 Staromonetny per
109017 Moscow, Russia
E-mail: ozerova@igem.msk.su

Dr. S. Chikhalikar
School of Studies in Chemistry
Pt. Ravishankar Shukla University
Raipur-492 010 (M.P.), India

Prof. Dr. A. C. A. da Costa
Universidade do Estado do Rio de Janeiro
Instituto de Qulmica
Departamento de Tecnologia
de Processos Bioqulmicos
Rua São Francisco Xavier 542
Pavilhão Haroldo Lisboa da Cunha
Sala 312-A
20550-013, Rio de Janeiro, Brazil
E-mail: acosta@centroin.com.br

Prof. Dr. L. D. de Lacerda
Dept. de Geoqulmica
Universidade Federal Fluminense
Niteroi, 24020-007, RJ, Brazil
E-mail: geodrud@vm.uff.br

Dr. F. Di Francesco
C.N.R. - Instituto di Biofisica
Via S. Lorenzo, 26
56100 Pisa, Italy

Dr. L. A Dobryansky
Ukrainian Overseas Oil and Gas
34 Palladina pr.
252142 Kiev, Ukraine

Dr. R. Ebinghaus
GKSS Research Centre
Institute of Physical
and Chemical Analytics
Max-Planck-Straße
21502 Geesthacht, Germany
E-mail: ralf.ebinghaus@gkss.de

Dr. H. Edner
Department of Physics
Lund Institute of Technology
Box 118
22100 Lund, Sweden
E-mail: hansed@fysik.lth.se

Dr. M. Egli
Universität Zürich
Geographisches Institut
Winterthurerstraße 190
CH - 8057 Zürich, Switzerland

Dipl.-Ing. H.-J. Fell
Lurgi Bamag GmbH
Wetzlarer Straße 136
35510 Butzbach, Germany
E-mail: hans-jochen_fell@lurgi.de

Dr. R. Ferrara
CNR - Istituto di Biofisica
Via S. Lorenzo, 26
56127 Pisa, Italy
E-mail: romafe@ib.pi.cnr.it

Dr. Sci. A. A. Ganeev
St. Petersburg State University
Institute of Chemistry
7/9 Universitetskaya nab.
199034 St. Petersburg, Russia
E-mail: root@aga.chri.pu.ru

Dr. Ch. Geilhufe
Buna Sow Leuna
Olefinverbund GmbH
06258 Schkopau, Germany

Dr. G. E. Glass
National Health Environ
Effects Res Lab
USEPA
6201 Congdon Blvd.
Duluth, MN 55804, USA
E-mail: gglass@d.umn.edu

Dr. A. Gnamuš
Department of Biology
University of Ljubljana
Večna pot 111
1000 Ljubljana, Slovenia
E-mail: ales.gnamus@ijs.si

Dr. M. Hempel
GALAB
Technology Centre GKSS
Max-Planck-Straße 1
21502 Geesthacht, Germany
E-mail: maximilian.hempel@gkss.de

K. Higaki
Taisei Corporation
Technology Division
3-25-1 Sanken Bld
Hyakunin-cho Shinjyuku-ku
Tokyo 169, Japan
E-mail: kanji.higaki@sakura.taisei.co.jp

Dr. M. Horvat
Head, Department of
Environmental Sciences
Josef Stefan Institute
Jamova 39
1111 Ljubljana, Slovenia

Prof. Dr. M. Hraste
University of Zagreb
Faculty of Chemical Engineering
and Technology
Maruličev trg 19
10 000 Zagreb, Croatia
E-mail: marin.hraste@pierre.fkit.hr

H. Iwasaki
Taisei Corporation
Engineering Division
1-25-1 Nishi Shinjyuku
Tokyo 163-0606, Japan
E-mail: iwasaki@kiku.taisei.co.jp

Dr. S. K. Jha
Scientific Officer
Environmental Assessment Division
Bhabha Atomic Research Centre
Trombay, Mumbai 400085, India
E-mail: ead@magnum.barc1.ernet.in

A. Kamavisdar
School of Studies in Chemistry
Pt. Ravishankar Shukla University
Raipur- 492 010 (M.P.), India

T. E. Khvostova
Chita Institute of Natural Resources
Siberian Branch of Russian Academy of Sciences
P. O. Box 147
672090 Chita, Russia
E-mail: ecogeo@cinr.chita.su

Dr. A. Kobal
MD, M. Sc., Ph. D.,
Prim. Occupational Physican
Dept. of Occupational Medicine
Idrija Mercury Mine
Arkova 43
5280 Idrija, Slovenia
E-mail: svd@s5.net

H. H. Kock
GKSS Research Centre
Institute of Physical and Chemical Analytics
Max-Planck-Straße
21502 Geesthacht, Germany
E-mail: Hans-Kock@gkss.de

G. Krempp
CNRS, Centre de Géochimie de la Surface
1 rue Blessig
67084 Strasbourg Cedex, France
E-mail: gkrempp@illite.u-strasbg.fr

Dr. T. M. Krishnamoorthy
Scientific Officer
Environmental Assessment Division
Bhabha Atomic Research Centre
Trombay, Mumbai 400085, India
E-mail: ead@magnum.barct1.ernet.in

Dr. O. Krüger
GKSS Research Centre
Institute of Hydrophysics
Max-Planck-Straße
21502 Geesthacht, Germany
E-mail: olaf.krüger@gkss.de

Prof. Dr. A. Kudo
Research Reactor Institute
Kyoto University
Kumatori-cho
Osaka-fu, Japan

Dr. T. G. Laperdina
Chita Institute of Natural Resources
Siberian Branch of Russian Academy of Sciences
P.O. Box 147
672090 Chita, Russia
E-mail: ecogeo@cinr.chita.su

Dr. P. J. Lechler
Chief Geochemist
Nevada Bureau of Mines and Geology
University of Nevada (MS178)
Reno, Nevada 89557, USA
E-mail: plechler@nbmg.unr.edu

Prof. Dr. O. Lindqvist
Department of Inorganic Chemistry
Chalmers University of Technology and University of Göteborg,
41296 Göteborg, Sweden

Dr. W. Mailahn
Institut für Wasser-, Boden-, Lufthygiene
Umweltbundesamt
Corrensplatz 1
14195 Berlin, Germany

Dr. R. V. Marins
Dept. de Geoquìmica
Universidade Federal Fluminense
Niteroi, 24020-007, RJ, Brazil
E-mail: rmarins@centroin.com.br

Dr. B. E. Maserti
CNR - Istituto di Biofisica
Via S. Lorenzo, 26
56127 Pisa, Italy
E-mail: elima@ib.pi.cnr.it

Dr. N. R. Mashyanov
St. Petersburg State University
Institute of the Earth Crust
7/9 Universitetskaya nab.
199034 St. Petersburg, Russia
E-mail: nrm@nrm.usr.pu.ru

A. Matsuyama
Taisei Corporation
Engineering Division
1-25-1 Nishi Shinjuku
Tokyo 163-0606, Japan
E-mail: matuyama@kiku.taisei.co.jp

Dr. B. Mazzolai
CNR - Istituto di Biofisica
Via S. Lorenzo, 26
I - 56127 Pisa, Italy
E-mail: mazzolai@ib.pi.cnr.it

M. V. Melnikova
Chita Institute of Natural Resources
Siberian Branch of Russian
Academy of Sciences
P. O. Box 147
672090 Chita, Russia
E-mail: ecogeo@cinr.chita.su

Dr. Th. Meschede
Gerling Consulting Gruppe
Frankfurter Straße 720-726
51145 Köln, Germany
E-mail: thomas.meschede@gerling.de

Dr. A. Messaïtfa
CNRS, Centre de Géochimie
de la Surface
1 rue Blessig
67084 Strasbourg Cedex, France

Dr. V. Miklavčič
Head of Ecological Laboratory
Idrija Mercury Mine
Arkova 43
5280 Idrija, Slovenia

Dr. N. N. Moskalenko
Institute of Mineralogy, Geochemistry
and Chrystal Chemistry (IMGRE)
of the Ministry of Natural Resources
of Russian Federation and Rus. Acad.
of Science
15 Veresayeva St.
121357 Moscow, Russia
E-mail: nmoskal@imgre.iitp.ru

Dr. A. B. Mukherjee
Department of Limnology
and Environmental Protection
University of Helsinki
P.O. Box 27
00014 Helsinki, Finland
E-mail: arun.mukherjee@helsinki.fi

E. Nanut
Idrija Mercury Mine
Arkova 43
4580 Idrija, Slovenia

Prof. Dr. N. A. Ozerova
Institute of Geology of Ore Deposits,
Petrology, Mineralogy and Geochemistry
Russian Academy of Sciences (IGEM/RAS)
35, Staromonetny per.
109017 Moscow, Russia
E-mail: ozerov@rinet.ru

P. Panday
School of Studies in Chemistry
Pt. Ravishankar Shukla University
Raipur-492 010 (M.P.), India

Dr. G.G. Pandit
Scientific Officer E
Environmental Assessment Division
Bhabha Atomic Research Centre
Trombay, Mumbai 400085, India
E-mail: ead@magnum.barct1.ernet.in

Dr. K.S. Patel
School of Studies in Chemistry
Pt. Ravishankar Shukla University
Raipur-492 010 (M.P.), India

R. M. Patel
School of Studies in Chemistry
Pt. Ravishankar Shukla University
Raipur-492 010 (M.P.), India

Dr. Dipl.-Chem. O. Petzold
Lurgi Bamag GmbH
Wetzlarer Straße 136
35510 Butzbach, Germany

Y. I. Pikovsky
Lomonosov State University
Moscow, Russia

Dr. J.-L. Probst
CNRS, Centre de Géochimie
de la Surface
1 rue Blessig
67084 Strasbourg Cedex, France
E-mail: jlprobst@illite.u-strasbg.fr

Dr. G. R. Rapp Jr.
Archaeometry Laboratory
University of Minnesota, Duluth
10 University Drive
Duluth, MN 55812-2496, USA
E-mail: grapp@d.umn.edu

Dr. I. Richter-Politz
Buna Sow Leuna
Olefinverbund GmbH
06258 Schkopau, Germany

Dr. W. Rotard
Institut für Wasser-, Boden-, Lufthygiene
Umweltbundesamt
Corrensplatz 1
14195 Berlin, Germany

V. V. Ryzhov
St. Petersburg State University
Institute of Earth Crust
7/9 Universitetskaya nab.
199034 St. Petersburg, Russia
E-mail: vova@ryv.usr.pu.ru

Prof. A. Sabadin
B. Sc., M. Sc. Ph. D. Psychologist
Head Dept. of Work Psychology
Faculty of Arts and Sciences
University of Ljubljana
Aškerčeva 12
1000 Ljubljana, Slovenia

Prof. Dr. W. Salomons
GKSS Research Center
Max-Planck-Straße
21502 Geesthacht, Germany
E-mail: salomons@gkss.de

T. Sano
Taisei Corporation
Technology Division
3-25-1 Sanken Bld
Hyakuninn-cho Shinjyuku-ku
Tokyo 169, Japan
E-mail: sano@kiku.taisei.co.jp

Dr. Th. Schöndorf
HPC Harres Pickel Consult GmbH
Am Wäldele 16 b
79112 Freiburg-Tiengen, Germany
E-mail: tschoend@hpc-ensr.de

L. Schwarzkopf
Fond Du Lac Indian Reservation
Natural Resources Program
105 University Road
Cloquet, MN 55720, USA

S. E. Sholupov
St. Petersburg State University
Institute of Earth Crust
7/9 Universitetskaya nab.
199034 St. Petersburg, Russia
E-mail: sergey@ses.usr.pu.ru

Dr. L. Sobral
Centro de Tecnologia Mineral
21941-590 Rio de Janeiro RJ, Brazil

Dr. G. R. Southworth
Environmental Sciences Division
Oak Ridge National Laboratory
P.O. Box 2008
Oak Ridge, TN 37831, USA
E-mail: sgr@ornl.gov

J. A. Sorensen
Archaeometry Laboratory
University of Minnesota, Duluth
10 University Drive
Duluth, MN 55812-2496, USA
E-mail: jsorense@d.umn.edu

Z. Špirič, M. Sc.
INA-NAFTAPLIN
Šubičeva 29
10000 Zagreb, Croatia
E-mail: zdravko.spiric@ina-np-sir.tel.hr

Prof. P. Stegnar
Biochemistry & Environmental Research
Dept. of Environmental Sciences
Laboratory for Radiochemistry
Jožef Stefan Institute
Jamova 39
1001 Ljubljana, Slovenia

Dr. S. Svanberg
Department of Physics
Lund Institute of Technology
Box 118
22100 Lund, Sweden
E-mail: sunes@fysik.lth.se

Dr.-Ing. J. Thöming
TU Hamburg-Harburg
Eißendorfer Straße 40
21073 Hamburg, Germany
E-mail: thoeming@tu-harburg.de

A. Tripathi
School of Studies in Chemistry
Pt. Ravishankar Shukla University
Raipur-492 010 (M.P.), India

Dr. R. R. Turner
Frontier Geosciences
414 Pontius Ave. North
Seattle, Washington 98109, USA
E-mail: ralpht@frontier.wa.com.

Dr. R. C. Villas Boas
Centro de Tecnologia Mineral
Rio de Janeiro, 21941-900, RJ, Brazil

R. Vogelsberger
Dames & Moore Group
Goernestraße 32
20249 Hamburg, Germany
E-mail: dames.moore.ham@t-online.de

Dr. E. Wallinder
Lighten AB
Ideon Research Park
22370 Lund, Sweden
E-mail: eva@lighten.se

Dr. D. Wallschläger
Frontier Geosciences Inc.
414 Pontius North, Suite B
Seattle, WA 98109, USA

Dr. J. Wang
Astra
Jungfrustigen 1
24441 Kavlinge, Sweden

Prof. Dr. R.-D. Wilken
ESWE Institute for Water Research
and Water Technology GmbH
Söhnleinstraße 158
65201 Wiesbaden-Schierstein, Germany
E-mail: wilken@goofy.zdv.uni-mainz.de

Dr. Z. Xiao
Astra
Jungfrustigen 1
24441 Kavlinge, Sweden

H. Yabuta
Taisei Corporation
Technology Division
3-25-1 Sanken Bld
Hyakunin-cho Shinjyuku-ku
Tokyo 169, Japan

Dr. E. P. Yanin
Institute of Mineralogy, Geochemistry
and Chrystal Chemistry (IMGRE)
of the Ministry of Natural Resources
of Russian Federation and Rus. Acad.
of Science
15 Veresayeva St.
121357 Moscow, Russia
E-mail: imgregol@centro.ru

Contents

I Reviews

II International Case Studies

II.a Industrial Sites

II. b Mining

II. c Emissions and Atmospheric Dispersion

II. d Remediation

II. e Mercury Contamination in Aquatic Systems

I Reviews

Natural and Anthropogenic Mercury Sources and Their Impact on the Air-Surface Exchange of Mercury on Regional and Global Scales

R. Ebinghaus, R.M. Tripathi, D. Wallschläger, and S.E. Lindberg

1 Introduction

Mercury is outstanding among the global environmental pollutants of continuing concern. Especially in the last decade of the 20th century, environmental scientists, legislators, politicians and the public have become aware of mercury pollution in the global environment. It has often been suggested that anthropogenic emissions are leading to a general increase in mercury on local, regional, and global scales (Lindqvist et al. 1991; Expert Panel 1994).

Mercury is emitted into the atmosphere from a number of natural as well as anthropogenic sources. In contrast with most of the other heavy metals, mercury and many of its compounds behave exceptionally in the environment due to their volatility and capability for methylation.

Long-range atmospheric transport of mercury, its transformation to more toxic methylmercury compounds, and their bioaccumulation in the aquatic foodchain have motivated intensive research on mercury as a pollutant of global concern. Mercury takes part in a number of complex environmental cycles, and special interest is focused on the aquatic-biological and the atmospheric cycles. Environmental cycling of mercury can be described as a series of processes where chemical and physical transformations are the governing factors for the distribution of mercury in and between different compartments of the environment. Mercury can exist in a large number of different physical and chemical forms with a wide range of physical, chemical, and ecotoxicological properties and consequently with fundamental importance for the environmental behavior. The three most important chemical forms known to occur in the environment are: elemental mercury [Hg(0)], which has a high vapor pressure and a relatively low solubility in water; divalent inorganic mercury (Hg^{2+}), which can be far more soluble and has a strong affinity for many inorganic and organic ligands, especially those containing sulphur; and methylmercury (CH_3Hg^+), which is strongly accumulated by living organisms. Conversions between these different forms provide the basis of mercury's complex distribution pattern on local, regional, and global scales.

This chapter was prepared while R.M.T. and S.E.L. were visiting scientists at the Institute of Physical and Chemical Analysis at GKSS Research Centre, Geesthacht, Germany. Publication number 4745, Environ. Sci. Div. Oak Ridge National Lab.

Environmental Science
Mercury Contaminated Sites (ed. by R. Ebinghaus et al.)

Extensive information exists on environmental and health effects of mercury and its behavior in the environment (e.g., Wheatley and Wyzga 1997). Much less information is available on the fluxes of the element and its compounds to the air, water, and soils. This chapter summarizes present knowledge on natural and anthropogenic mercury fluxes to the atmosphere from various sources and their relative importance on regional and global scales.

2
The Atmospheric Mercury Cycle

The atmospheric cycle of mercury is determined by natural and anthropogenic emissions, a complex atmospheric chemistry, and wet and dry deposition processes. Atmospheric chemistry and especially the deposition of mercury are strongly linked to the speciation of mercury released into the atmosphere by different types of sources.

The linkage between the atmospheric and biological cycles is manifested in the deposition of atmospheric mercury species. Schroeder and Lane (1988) illustrated (Fig. 1) the most important processes in the emission and deposition cycle of atmospheric mercury. The deposition pathway is dominated by the flux of emitted Hg(II) compounds (formal reactive gaseous mercury or RGM), the oxidation of elemental mercury vapor to Hg(II), and subsequent wet and/or dry deposition. Mercury species attached to particles can be removed from the atmosphere by precipitation or dry deposition (Expert Panel 1994), but these fluxes are generally less important.

Once deposited, the formation of volatile gaseous mercury, especially the formation of highly toxic methylmercury, its enrichment in organisms and nutritional chains, and finally destruction (demethylation) of methylmercury are the main features of the biological cycle of mercury.

2.1
Speciation of Emissions

On a global scale, the atmospheric mercury cycle is dominated by elemental mercury vapor (generally > 95% of total airborne Hg). However, the emission speciation of mercury is determined by the source characteristics and consequently shows large regional variability. To characterize the main emission pathways, the Expert Panel on Mercury Atmospheric Processes (1994) defined three different terms for mercury emissions:

1. Anthropogenic mercury emissions: the mobilization or release of geologically bound mercury by man's activities, with mass transfer of mercury to the atmosphere.

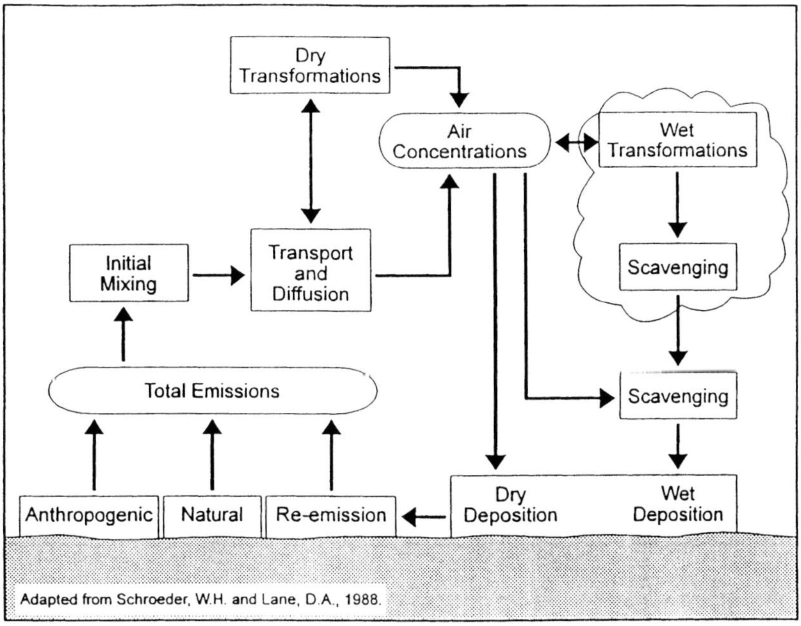

Fig. 1. Mercury emissions-to-deposition cycle, (After Schroeder and Lane 1988)

2. Natural mercury emissions: the mobilization or release of geologically bound mercury by natural biotic and abiotic processes, with mass transfer of mercury to the atmosphere.
3. Reemission of mercury: the mass transfer of mercury to the atmosphere by biotic or abiotic processes drawing on a pool of mercury that was deposited to the Earth's surface after initial mobilization by either anthropogenic or natural activities. Together, these last two pathways are also designated mercury emissions from natural surfaces, and they represent large uncontrolled area source emissions which must be taken into account by global models.

Speciation of atmospheric mercury originated from these source types is discussed in the following sections.

2.1.1
Anthropogenic Mercury Emission Speciation

Fuel combustion, waste incineration, industrial processes, and metal ore roasting, refining, and processing are the most important point source categories for anthropogenic mercury emissions into the atmosphere on a worldwide basis. Besides elemental mercury an important and variable fraction can be emitted as reactive gaseous mercury (RGM) or particulate Hg(II) (Expert Panel 1994). For Europe, Pacyna (1993) estimated the anthropogenic mercury emissions for the above mentioned species summarized in Table 1. It is evident from the table that the major species is Hg(o). The relative distribution between elemental mercury vapor and gaseous or particulate Hg(II) varies from country to country, however.

The major portion of mercury emissions from combustion of fuels is in the gaseous phase. In the combustion zone, mercury present in coal or other fossil fuels evaporates in elemental form and then most likely a portion of it is oxidized in the flue gases (Prestbo et al. 1995). Emitted mercury species into the environment depend upon the nature of the source of emission. Mercury emitted from high temperature processes such as coal combustion and pyrite roasters will probably be converted to the elemental form, Hg(o). However, in flue gases, where the temperature drops, Hg(o) may be oxidized by HCl and O_2 in presence of soot or other surfaces (Hall et al. 1991; Prestbo et al. 1995). In modern combustion plants equipped with flue gas cleaning facilities such as wet scrubbers, the oxidized and particulate forms should be removed easily. Hence, the primary emission will be Hg(o). Table 2 summarizes the speciation of mercury emissions from flue gases and other industrial emissions.

2.1.2
Speciation of Natural Mercury Emission and Re-Emission

The mercury that evades from natural sources is generally entirely in the elemental form (S.E. Lindberg et al. 1979, 1995, 1998). Natural sources are, for example, the evasion from surface waters, from soils, from minerals, and from vegetation located in terrestrial and wetland systems. Volcanism, erosion, and exhalation from natural geothermal and other geological crevices also mainly emit elemental mercury.

Global volcanic emissions were estimated from the Hg/S ratio and account for approximately 20 to 90 t year^{-1}, which is about 1–5% of the annual emissions from human activities (Fitzgerald 1996).

Generally speaking, the distinction between natural and "quasinatural reemissions" of mercury (that which was formerly deposited from the

Table 1. Anthropogenic Emission (t year^{-1}) of mercury and its species in Europe. (After Pacyna 1993)

Country	Hg(o) gas	Hg(II) (gas)	Hg (particles)	Hg (total)
Belgium	5.3	2.2	1.4	8.9
Czechoslovakia	7.8	4.5	2.7	15.0
Denmark	2.1	1.9	0.8	4.8
Finland	3.1	0.8	0.3	4.1
France	15.3	9.0	5.6	29.9
GDR	203.0	99.0	28.0	330.0
FR Germany	38.0	20.0	7.0	65.0
Netherlands	3.0	3.8	1.4	8.2
Norway	1.4	0.4	0.2	2.0
Poland	23.3	13.1	8.3	44.7
Sweden	5.6	1.4	0.5	7.5
Soviet Union	45.0	25.7	17.0	87.7
United Kingdom	21.0	14.0	5.0	40.0

Table 2. Speciation of mercury in flue gases and other industrial emissions. (After Munthe 1993)

Process	% Hg(o)	% Hg(II) (g)	% Hg(II)(s)	Reference
Chlorine alkali	50–90	10–50	–	Leavander (1987)
Coal combustion	50	30	20	Brosset (1983)
				Bergstrom (1983)
Roasting of sulphide ores	80–90		10–20	Leavander (1987)
Pyrite burning	100			Leavander (1987)
Waste incineration	20	60	20	Bergstrom (1986)
				Vogg et al. (1987)
				Lindqvist (1986)

atmosphere to surfaces) is difficult to discern and may be regarded as an unresolved problem. In addition, the so-called natural emissions also include mercury that was previously deposited from natural sources.

Another example of a quasi-natural emission source are frequently occurring forest fires. Forest biomass is known to be a large pool of temporarily stored mercury. These emissions are of interest with respect to the emission speciation since some partially oxidized mercury species may be emitted as particulate or gas-phase species in addition to elemental mercury (Porcella 1995). An extensive Amazon monitoring network showed that the annual atmospheric mercury emissions from slash and burn deforestation in the Amazon was 2–9 t $year^{-1}$. This is relatively small compared to mercury emissions from gold mining in the Amazon, which are estimated to be about 140 t $year^{-1}$ (Lacerda et al. 1995).

Photochemical and photobiological processes play an important role in the reemission of mercury from aquatic systems. Nriagu estimated that an important fraction (10 to 50%) of the dissolved mercury in lakes is in the elemental form (Nriagu 1994). However, recent measurements in surface waters place this fraction closer to 5–10% (Amyot et al. 1995; Fitzgerald and Mason 1996; Schmolke et al. 1997) and new flux measurements in Sweden indicate that the emitted mercury is also in the elemental form (Lindberg et al. 1996).

Globally, it has been suggested that there exists an important reemission of elemental mercury from marine surfaces (Mason et al. 1994a). Recent data from forested areas suggest first evidence that a similar process of Hg(o) emission may be globally important for terrestrial plants as well (Lindberg 1996).

2.2 Speciation of Mercury in the Atmosphere

Vapor-phase mercury is the predominant physical state in relatively clean ambient air, where both vapor-phase and particulate-phase mercury generally coexist. When speciating the vapor-phase fraction, elemental mercury nearly always constitutes almost all of the mass with only minor amounts of other volatile species normally being detected (Brosset and Lord 1991; Stratton and Lindberg 1995). The speciation of mercury emitted to the atmosphere is of great importance for the atmospheric fate of mercury. Hg(o) will add to the global background and will be significantly deposited only after atmospheric transformation processes or through plant surface interactions such as stomatal uptake at elevated concentrations (Hanson et al. 1995). However, particulate-phase mercury (Hg_{part}) and reactive gaseous mercury (RGM) will deposit more rapidly on a local or regional scale (Lindberg et al. 1992). Estimates made by Slemr et al. (1985) based on the relations between vapor pressure and the ratio of particulate concentrations to gaseous concentrations of different species in the atmosphere (Junge 1977) suggest that inorganic gaseous mercury species will be predominantly found on aerosols and thus will share their atmospheric fate, i.e., will have a residence time of about 5 days (Prospero et al. 1983) and cannot be transported over large distances. According to Junge, another consequence is that concentrations of Hg(II) should not be too different from particulate phase concentrations of mercury because the particulates encompass Hg(II). However, recent measurements of RGM using newly designed techniques including treated filters, denuders, and refluxing mist chambers all show that RGM generally exceeds Hg_{part} at a variety of sites (Stratton and Lindberg 1995; Xiao et al. 1997; S.E. Lindberg and Stratton 1998; Ebinghaus et al. 1998)

Measurements of operationally defined Total Gaseous Mercury (TGM) are being made on a routine basis at a number of sites in Europe and North America. A recently conducted field intercomparison of measurements of atmospheric mercury species shows good comparability of the commonly accepted methods for TGM, but less so for RGM and HgP (Ebinghaus et al. 1998). Long-term studies suggest that atmospheric TGM concentrations were increasing on a hemispheric or a global scale until the early 1990s; a decrease has been suggested for the period 1990–1994 (Slemr and Langer 1992; Fitzgerald 1995; Slemr 1996).

Fewer data are available for particulate-phase mercury or inorganic gaseous mercury species. Table 3 summarizes background concentrations of these species at a remote marine location at the Irish west coast. The measurements had been performed by a number of different laboratories taking part in the above field intercomparison study (Ebinghaus et al. 1998). Table 4 summarizes the most extensive data published to date on vapor-phase Hg speciation in ambient air, at two sites in the US. (Lindberg and Stratton 1998) It is readily seen that Hg(o) dominates in all cases. In addition to the inorganic species, both methylmercury and dimethylmercury have been detected in ambient air at background and urban sites (Bloom et al. 1996b). However, the concentrations are far below those of the inorganic species.

Table 3. Background concentrations of atmospheric mercury species detected at Mace Head, Ireland, in September 1995 (R. Ebinghaus et al., in press)

TGM (ng m^{-3})	Hg_{part} (ng m^{-3})	Inorg. gaseous Hg (ng)
1.3–3.8[a]	0.005–0.026[c]	0.013–0.023[e]
1.2–2.1[b]	0.028–0.115[d]	0.041–0.094[f]

[a] Measured with manual methods. [b] Measured with automated analyzers. [c] Collected on disk filters. [d] Collected on quartz wool plugs or coated glass beads. [e] Defined as reactive gaseous mercury. [f] Defined as divalent gaseous mercury.

Table 4. Extensive measurements of the speciation of vapor-phase Hg in mid-continental ambient air using a refluxing mist chamber method. (Lindberg and Stratton, 1998)

		TGM ($ng\ m^{-3}$)			RGM ($ng\ m^{-3}$)		
Site	Location	Mean	SD	N	Mean	SD	N
Walker Branch[a]	Tennessee	2.19	0.64	68	0.065	0.040	75
Earlham College	Indiana	4.02	1.78	153	0.104	0.057	177

[a] Hg_{part} concentrations at this site are generally in the range of 0.01–0.03 $ng\ m^{-3}$ (Lindberg et al. 1992).

Naturally occurring species distribution of atmospheric mercury as it can be detected at remote locations is significantly influenced by the presence of emission sources. The emitted species are dependent on the source characteristics. Depending on the atmospheric residence time and the deposition properties of the individual species, the relative proportions of the atmospheric concentrations are related to the distance of the sources. For central Europe it has become clear that several "hot spots" of anthropogenic Hg(0) emissions are located at industrial sites in the former German Democratic Republic. Annual atmospheric emissions from this relatively small but highly industrialized area were estimated at 330 tons for 1988. A comparison with the emissions of entire Europe, estimated at 726 tons for the same year, shows the significance of this locally limited but relatively strong source of Hg(0) for central Europe (Helwig and Neske 1990).

2.3 Speciation of Deposited Mercury

Mercury in all of its forms can be readily deposited on the Earth's surface. However, because of its reactivity and solubility, deposition of the RGM forms of mercury vapor are generally favored. Although Hg(0) can be directly absorbed by soils and plants (e.g. Lindberg et al. 1979), these processes are most important near local sources (e.g. Hanson et al. 1995).

The major atmospheric deposition process for elemental mercury vapor in background air is the aqueous oxidation by ozone followed by an in-droplet adsorption primarily onto soot particles (Munthe 1992; Petersen et al. 1995). It is important to be aware of the fact that the relation between emissions and depositions may be nonlinear. Consequently, deposition rates cannot be derived directly from emission rates because they are strongly influenced by secondary pollutants such as ozone and soot (Iverfeldt 1995).

Because a small but significant fraction of the atmospheric mercury consists of vapor-phase RGM or oxidized Hg(II), gaseous methylmercury, and particulate phase mercury species, these forms must also be included in regional deposition models because of their shorter atmospheric residence time relative to Hg(0). Of the three forms, oxidized mercury species are of great importance for the total deposition of mercury from the atmosphere according to an Eulerian model

(Petersen et al. 1995). Although these species are typically present at less than 10% of the total atmospheric mercury, they appear to influence the deposition flux significantly (Petersen et al. 1996; Pai et al. 1997).

Any oxidized Hg species will be subject to wet and dry deposition because of its significantly higher deposition velocity compared with elemental mercury (Lindberg et al. 1992; Lindberg and Stratton 1998).

A major fraction of the total mercury in precipitation is associated with particles, at least in urban and industrialized areas (Ferrara et al. 1988; Iverfeldt 1991), although it is not clear if this association occurs before or after rain formation. The particulate fraction in precipitation varies between less than 10% at remote sites to more than 90% at polluted sites (Iverfeldt 1991; Ebinghaus et al. 1995).

Gaseous methylmercury will not influence the total deposition of mercury to any great extent (Petersen et al. 1995). It may, however, constitute an important contribution to the overall loadings of methylmercury in terrestrial and aquatic ecosystems (Hultberg et al. 1995). Methylmercury is present in precipitation at concentrations generally corresponding to around 5% of the total Hg (Bloom and Watras 1988; Hultberg et al. 1995; Munthe et al. 1995; Petersen et al. 1996); however, the sources of this methylmercury are still not known. Bloom et al. (1996a) and Carpi et al. (1998) have each presented evidence for atmospheric methylmercury emissions from marine and continental sources, respectively.

3
Sources and Magnitudes of Anthropogenic Emissions

Mining, smelting, and refining of metals and their myriad applications have resulted in the mobilization and dispersion of large amounts of heavy metals into the environment. Emissions into the atmosphere during extraction, production, fabrication, application, and end-use have resulted in the widespread distribution of their compounds, including mercury. While the emissions in the developed countries are relatively well quantified, those in the developing world are poorly known.

The main types of anthropogenic emission sources maybe categorized as follows (Porcella et al. 1996):

1. Combustion
   Fossil fuels (coal, oil, gas) and wood
   Wastes (municipal, medical, hazardous wastes)
   Sewage sludge
   Crematories
2. High temperature processes
   Smelting
   Coking
   Ore roasting
   Cement and lime production

3. Manufacturing/commercial
   Chlor-alkali plants
   Metal processing
   Chemical and instruments industry (mercury chemicals, paints, batteries, thermometers, process reactants and catalysts).
4. Gold extraction
5. Other sources
   Fluorescent lamps
   Hazardous and municipal waste sites
   Mine spoils
   Land disturbance (e.g., deforestation, reservoir construction)

When they are produced during combustion processes, oxidized forms of mercury can be retained in modern flue gas cleaning systems (Pacyna 1996). Mercury retained in flyash as well as in bottom ash is disposed of on land where it may be released by volatilization or find its way in a form of direct releases to the aquatic environment. The mercury content of coal from different parts of the world are summarized in Table 5. Apart from the coal and fossil fuels, a significant proportion of mercury emissions is attributed to mercury in the various raw materials used for industrial purposes. The concentrations of mercury in various raw materials are summarized in Table 5. Though the concentrations of mercury in these raw materials are small, their contribution to the total emission of mercury is significant, as they are used in extremely large quantities. Historically, and today in developing countries, the chlor alkali industry is a significant source of atmospheric mercury emission. As outdated Hg-consuming processes are replaced by other methods, active mercury emissions are eliminated. However, wastes from earlier operations stored nearby have been shown to emit Hg to the atmosphere at rates which may exceeed those allowed for modern mercury process alkali plants (Lindberg and Turner 1977).

Mercury is used throughout the world for the production of gold using the amalgamation technique (Cleary 1996). Starting in Latin America in the 1980s, a

Table 5. Mercury content in different types of raw materials. (After Mukherjee 1996)

Source categeory	Origin	Hg content ($\mu g\ g^{-1}$)	Reference
Coal	Various	0.04–3.3	Meij (1991) Airey (1982)
Coal		0.02–1.0	Swaine (1990)
Coal	Australia	0.01–0.25	Sloss (1995)
Oil	Former USSR	0.005	Mukherjee et al. (1995)
Peat	Finland	0.01	Mukherjee et al. (1995)
Wood	Finland	0.01	Mukherjee et al. (1995)
Cu-ores	USA	0.5 (average)	Jaisinski (1995)
Zn-concentrate	Finland	100–500	Kuivala (1984)
Cu-concentrate	Finland (domestic)	13	Rantalahti (1996)
	Finland (imported)	110	Rantalahti (1996)
Cu-concentrate	Sweden	200	Dyvik (1995)
Pb-concentrate	Sweden	50	Dyvik (1995)

revival of the use of amalgamation in gold panning has spread rapidly in tropical areas and roughly 10 million people were estimated to be engaged in this activity (Ramel 1996). The gold/mercury amalgam is usually caught in a sluice box and then heated in a pan where the mercury evaporates into the air and the raw gold remains. In Brazil alone, gold production comprises about 2000 sites ("garimpos") of which about 80% are located in the Amazonas. It is mostly small-scale activity in remote villages and the amount of mercury released to the environment is about 1.3 kg mercury for each kg of gold or an overall estimated emission of 180 t $year^{-1}$ (Ramel 1996). It has been estimated that about 2 million people in Brazil alone are directly or indirectly dependent on gold extraction (Branches et al. 1993). In Tanzania, 250 000 people live in gold-mining villages along riverbeds (Ikingura and Mutakyahwa 1995). Official statistics concerning gold production and use of mercury are often underreported because of illegal use and widespread smuggling of both mercury and gold. The current and total emissions of mercury to the atmosphere from different gold mines is summarized in Table 6. It is evident from Table 6 that mercury emissions from mining activities are quite significant.

3.1 Anthropogenic Mercury Emissions in North America

The total anthropogenic emissions of mercury from all anthropogenic sources for North America was estimated by Porcella et al. (1996). The anthropogenic emissions for Canada, Mexico, and USA are summarized in Table 7. The basic data for mercury emissions from Canada was provided by Environment Canada (Deslauriers and Schroeder 1995). It can be seen from Table 7 that the

Table 6. Estimates of current (t $year^{-1}$) and total (tons emitted to date of original reference) emissions of mercury to the environment from gold mining sites. (After Lacerda 1996)

Site	Period of operation	Annual emissions	Total emissions	Reference
Amazon, Brazil	Since 1979	180	3000	Pfeiffer and Lacerda (1988)
Hindanao Is., Phillipines Cramer (1990)	Since 1985	26	200	Torres (1992), Cramer (1990)
Rio de Janeiro, Brazil	Since 1985	0.5	3.0	Lacerda et al. (1995)
Puyango River, Peru	Since 1987	2.9	14	CIMELCO (1991)
Narino, Colombia	Since 1987	0.5	3.5	Priester (1993)
Victoria L fields, Tanzania	Since 1991	6.0	24	Ikingura (1994)
Pando Department, Bolivia	Since 1979	7–30	300	Zapata (1994)
Dia pi Valley, China	Since 1938	2.4	130	Ming (1994)
Dixing region, China	Since 1992	120	360	Yshuan (1994)
Guyana Shield, Venezuela	Since 1989	40–50	300	Nico and Taphorn (1994)
Total	Since 1938	385.3–418.3	4336	

Table 7. Estimated north American anthropogenic sources of mercury emissions as of base year 1990 (t $year^{-1}$) to the atmosphere. (Porcella et al. 1996)

Source categeory	Canada	Mexico	USA
Power generation	6.0	2.7[a]	42.0
Industrial/commercial			
Coal	6.8	1.2[a]	17.9
Residential coal	–[a]	0.2[a]	3.2
Municipal waste Combustion	1.4	15.5[a]	45.5
Medical Waste Combustion	0.3'	3.8[b]	11.1
Manufacturing/smelting	32.0	55.9[a]	27.7
Miscellaneous[b]	–[a]	3.9[a]	7.7
Total	46.5	83.2[a]	154.1
Percentage	16.4%	29.3%	54.3%

[a] Extrapolated.
[b] Some sources remain unaccounted for.

contribution of Canadian mercury emission to North American emission is <20%; the USA has the maximum contribution of just over 50%.

Based on these estimates, Porcella et al. (1996) calculated that North America contributes about 8% of the global anthropogenic estimates of mercury emissions into the atmosphere. It is evident from Table 7 that the contributions from power generation and municipal waste combustion are the most significant sources in North America. However, the overall pattern of the relative importance of the individual source categories varies from country to country.

3.2 Anthropogenic Mercury Emissions in Europe

A summary of atmospheric emissions of mercury from anthropogenic sources in Europe is presented in Table 8 (Pacyna 1996). The most recent European emission inventory for mercury also used the 1990 statistical data (Pacyna 1994). The inventory was prepared on the basis of emission factors and the results of emission estimates for 1987 (Axenfeld et al. 1991). These regional estimates were compared with the available national emission data. The percentage emission of mercury from different sources has also been summarized in Table 8. It is clear from the table that the contribution of combustion sources is about 30%, while industrial emissions make the maximum contribution (>60%).

The dominance of the main source categories varies countrywise. However, it appears that there are four different categories or types in which the countries may be grouped as follows.

Category 1: mainly industry: ex-GDR, ex-Czechoslovakia, Belgium, The Netherlands
Category 2: mainly combustion: Russia, Yugoslavia, Poland, Romania
Category 3: combustion + industry: Bulgaria, France, Italy, Portugal, Spain, UK
Category 4: balance of all four sources: ex-FRG, Austria, Scandinavian countries

Table 8. Anthropogenic emission of Hg to the atmosphere in Europe as of base year 1990 (in t year^{-1}). (Data from Pacyna 1994)

Country	Combustion of fuels	Waste incineration	Industry	Other sources	Total
Albania	0.06	–	–	0.2	0.8
Austria	1.0	0.7	2.0	0.6	4.3
Belgium	2.6	0.2	5.2	–	8.0
Bulgaria	6.0	0.2	4.7	–	10.9
Czechoslovakia	1.9	–	12.4	0.6	14.9
Denmark	0.4	2.0	0.1	1.0	3.5
Finland	1.0	0.2	0.4	–	1.6
France	12.5	1.9	8.9	–	23.3
Germany (FRG)	10.0	11.0	14.0	11.0	46.0
Germany (GDR)[a]	17.2	0.3	278.5	–	296.0
Greece	1.4	–	–	–	1.4
Hungary	2.1	–	–	0.1	2.2
Iceland	v.l.	–	–	–	v.l.
Ireland	0.6	–	–	–	0.6
Italy	4.0	1.0	5.4	–	10.4
Luxembourg	v.l.	–	–	–	v.l.[b]
Netherlands	0.9	0.1	3.2	–	4.2
Norway	0.3	0.3	0.1	0.1	0.8
Poland	25.1	–	8.2	–	33.3
Portugal	2.9	–	2.3	–	5.2
Romania	12.5	–	2.4	–	14.9
Spain	5.5	–	4.4	–	9.9
Sweden	0.3	0.3	0.6	0.3	1.5
Switzerland	0.2	–	–	–	0.2
UK	19.2	9.4	22.2	0.4	51.2
USSR (Europe)	61.5	–	13.9	–	75.4
Yugoslavia	5.4	–	1.3	–	6.7
Total	195.1	27.6	390.2	14.3	627.2
Percentage	31.1%	4.4%	62.2%	2.3%	100%

[a] The ex-GDR emissions were significantly reduced after 1990.
[b] Value very little.

3.2.1
Anthropogenic Mercury Emissions in Finland

Finely detailed mercury emission data are not widely available. However, a recent study in Finland is worth summarizing because it is one of the more complete surveys available, including emissions to air, water, and land. In addition, these estimates, determined independently, illustrate that the uncertainty involved is at least a factor of 2 (Tables 8 and 9).

Finland is an interesting example for a category 2 country because the anthropogenic emissions are dominated by combustion of fuels; however, the total amount of emitted mercury is comparably small (about 3.5 t year^{-1}; Table 9). The total natural mercury emission in Finland has been reported to be 0.4 t year^{-1}, of which mercury emissions from land and water are 0.3 t year^{-1} and 0.4 t year^{-1}, respectively (Mukherjee et al. 1995). In Finland, anthropogenic sources of mercury belong to both, i.e., point and diffuse categories. The mercury discharged into the air, water, and land has been summarized in Table 9. It is evident from the table that metallurgical plants still represent the largest fraction of anthropogenic emission in the Finnish mercury inventory, followed by power plants. It is also evident from the table that atmospheric mercury emissions from industrial sources have decreased from 3.5 t $year^{-1}$ in 1987 to 2 t $year^{-1}$ in 1992 due to decreased atmospheric emission from copper smelter at Harjavalta, decreased chlorine production, and reduced discharges during jerosite production (Mukherjee et al. 1995).

3.3 Anthropogenic Mercury Emissions in Siberia

Detailed mercury emission data also became recently available for one of the Eastern countries (Yagolnitser et al. 1996). In Siberia, industries emitting mercury and mercury-containing compounds can be divided into three main groups. The first group contains plants of the nonferrous metallurgical industry which produce metallic mercury directly from mercury ores and ore concentrates and from recycled mercury-containing products. The second group comprises mainly plants from chemical, electronic, and electrical industries where mercury is one of the elements in the manufacturing process. The major mercury consumers in the chemical industry are chlorine and caustic soda plants and manufacturers of various mercury compounds for other industries such as mercury oxides and fungicides for agriculture. The third group contains the enterprises involved in the extraction and processing of various metal ores and cement producers. In these operations, mercury is a contaminant but its release can be high. Such plants include ferrous and nonferrous metallurgy, oil and gas

Table 9. Total release of mercury (t $year^{-1}$) into the Finnish environment, 1992. (After Mukherjee et al. 1995)

Categeory	Air	Water	Land
Zn plant	0.9	0.002	1.5
Copper smelter	0.08	0.02	–
Power plant coal, oil, peat, and wood	0.689	0	0.09
Chlor alkali plant	0.162	0.025	0.008
Refuse incineration	0.004	0	0.02
Iron and steel industry	0.03	0	–
Cement industry	0.02	0	0
Sludges	0.077	0.15	0.442
Domestic waste and batteries	0	0	1.6
Crematories	0.008	0	0.008
Total	2.0	0.20	3.67
Hg discharged in 1987[a]	3.5	0.15	20.0

- Indicates that the quantity discharged is not known.
[a] Mukherjee (1989)

Table 10. Mercury emissions into the environment from different industries in Siberia. (After Yagolnitser et al. 1996)

	Mercury emissions (t $year^{-1}$)	Mercury emissions (t $year^{-1}$)	Mercury emissions (t $year^{-1}$)	Mercury emissions (t $year^{-1}$)	
Industry	Atmosphere	Waters	Soil[a]	Total	%
Chemical Industry	58.5	109.0	66.7	234.2	78.9
Nonferrous metallurgy and Hg production	10.6	0.2	4.8	15.6	5.3
Ferrous metallurgy	1.1	–	–	1.1	0.4
Terminal electric power generation	10.8	–	–	10.8	3.6
Electrical and electronic engineering	0.11	0.06	0.67	0.8	0.3
Gold mining	10.6	10.6	13.2	34.4	11.5
Total	91.7	119.9	85.4	296.9	100
Percentage	(30.9%)	(40.3%)	(28.8%)	(100%)	

[a] Soil (including dumps, solid waste, etc.).

processing plants, coking plants, and hydrocarbon-fired thermal electric power plants.

The estimates of anthropogenic mercury emissions to the environment in Siberia are summarized in Table 10. The total mercury emission from all anthropogenic sources in Siberia is estimated to be approximately 300 t $year^{-1}$. The contribution of chemical industries to mercury emission in the environment is quite significant. Chlorine and caustic soda plants alone account for nearly 80% of the total mercury emitted. These industries together with gold mining account for more than 90% of total mercury emissions in this region. The emissions of mercury from anthropogenic sources to air, water and soil in Siberia are also given in Table 10. It can be noted from the table that mercury emissions to air, water and soil are each of the same magnitude. Yagolnitser et al. (1996) calculated an annual average release into the atmosphere of 3.8 g of mercury per capita for Siberia and compared with the global annual average of 0.9 g (Lindqvist et al. 1991)

3.4 Anthropopgenic Mercury Emissions in South America

Detailed mercury emission data for South America are available for only one country, Brazil. The industrial emission of mercury from major sources into the atmosphere of Brazil has been summarized in Table 11 (Lacerda et al. 1995). Sources such as the pharmaceutical industry, dentistry, chemical industry, and oil and gas combustion, although emitting mercury, have not been considered due to their low emission to the atmosphere in this region. Practices such as refuse and waste incineration, which are particularly important due to high emission factors, are not common in Brazil, and have also not been included. The chlorine and alkali industry was the major consumer of mercury in Brazil up to

Table 11. Mercury emissions to the atmosphere from industrial sources in Brazil. (After Lacerda et al. 1995)

Source	Emission (t $year^{-1}$)	Percentage
Chlorine industry	11.65	7.2
KOH production	0.37	0.2
Paint and dye industry	0.34	0.2
Electrical industry	0.02	–
Coal combustion	0.01	–
Pyrometallurgy	4.54	2.8
Slash and burn agriculture	8.7	5.4
Gold mining	136.16	84.2
Total	161.79	100

1979 and was responsible for nearly 60% of total mercury consumed in the country (Ferreira and Appel 1991). However, from 1980 onwards strict emission control policies and improved technologies were adopted, which resulted in the decrease of mercury consumption as well as emission. At present, chlorine and alkali production emits 12 t $year^{-1}$ of mercury to the atmosphere in Brazil, which is 7% of the total atmospheric emission. The contribution of paint and dye together with electro-electronics industries to the direct atmospheric emissions is very small and accounts for less than 2% of the total emission. Because of the climate in Brazil, the mercury emission to the atmosphere from energy production (coal, oil, and gas combustion) is very low and contributes less than 0.1% of the total emission. (>95% of electricity is produced by hydroelectric power plants); (MME 1992). The burning of natural vegetation to produce agriculture lands or pastures is a major practice in Brazil, in particular in the Amazon region, where an estimated area of 11 000 km^2 is annually burned for this purpose (INPE 1992). Although mercury content in tropical vegetation is low, the mercury emission is significant due to the large forest areas involved and high biomass of the forest. The contribution of these sources accounts for 5% of the total Hg emission in Brazil. Gold mining using mercury amalgamation is the most important source of mercury to the atmosphere with a total annual emission of nearly 140 t, or >80% of the total mercury emissions to the atmosphere in Brazil. To alleviate these pollutant impacts, a closed retort was developed for use by the Brazilian gold miners (Cleary 1996) which would substantially reduce pollution while creating a purer gold product. However, most Brazilian miners have refused to use the retort, since the gold burning in the retort takes place where miners cannot physically see the burning, and therefore did not trust the device. This illustrates that improved technology and better economics do not necessarily become adopted in real world situations, particularly in the developing countries (Cleary 1996).

3.5
Summary of Anthropogenic Emissions

The emission studies described above may be used to make a crude estimate of regional scale mercury emissions to the air for vast areas of the globe. The values given in Section 3 for North and parts of South America, Western Europe, and parts of Asia add up to a total atmospheric mercury emission of just over 730 t $year^{-1}$. However, vast areas of the globe have not been included here, especially China and many developing countries, and this value of 730 t $year^{-1}$ is thought to represent only 15% of the total anthropogenic emissions, which is further discussed in Section 6.

4 Natural Emissions of Mercury

The assessment of truely "natural" mercury sources and their relative importance compared to direct anthropogenic emissions and indirect (re)emissions is a fundamental problem in studying the global balance and cycling of mercury in the environment. This knowledge is critical to our understanding of the fate of mercury, since so-called natural emissions cannot be reduced or controlled in most cases. Unfortunately, as discussed below, it is very difficult to differentiate between natural and indirect anthropogenic mercury emissions into the environment. Natural mercury emissions in this context are taken to include mercury transport phenomena that would take place in the absence of human existence and activity. However, once any pool of mercury has formed in an environmental compartment, it is no longer important where it originated, since all mercury is subject to the same transport and transformation processes, although it is possible that mercury from different origins may maintain a different speciation in the same compartment and thereby react differently. This section will demonstrate the problems associated not only with assigning relative importance of anthropogenic and natural contributions to the same flux phenomena, but also with separating and comparing different controlling processes leading to these flux phenomena.

There are numerous environmental pathways exchanging Hg with the atmosphere that come to our mind when we think of "natural" processes (Table 12). Among these, however, only the first three are clearly and unambiguously natural and undisturbed by anthropogenic influence. They could be classified as geological sources (meaning they have their origin in the deeper regions of the Earth's crust) and are also the only ones that are exclusively

Table 12. Natural sources and exchange processes of atmospheric Hg

Sources
1. Wind erosion and degassing from Hg mineralized soil and rock formation
2. Volcanic eruptions and other geothermal activity
3. Evasion of Hg from the Earth's subsurface crust
Exchange processes
4. Atmospheric interactions with terrestrial compartments
5. Interactions with the oceans
6. Interactions with freshwater ecosystems

sources. The other three processes are all (more or less) influenced by human activities and are both sources and sinks of atmospheric mercury. In contrast to the former, they could be characterized as "surface-linked" processes for they occur only at the very surface of the Earth's crust. The big question for those interactions is: are they net sources or sinks? In the following sections, we will attempt to quantify the first three processes and to evaluate the flux direction and magnitude for the three natural air-surface exchange processes. Their absolute and relative importance for regional and global Hg budgets will then be discussed in Section 6.

It seems to be the general opinion that whenever elevated mercury concentrations are encountered in any ecosystem compartment in the absence of obvious local or direct anthropogenic sources, this can be interpreted as proof of an anthropogenic influence via atmospheric long-range transport and deposition. While this may be true in many cases, some of these anomalies may be caused, at least partially, by natural local emissions due to underlying geologic anomalies. For example, in the case of elevated Hg concentrations in plants, foliar atmospheric uptake of crustal mercury emissions or root uptake from geologically enriched soils are possible alternative explanations to long-range transport of pollutants (Lindberg et al. 1979). Also, surface enrichment mercury in depth profiles measured in sediment cores may be caused by upward migration due to diagenetical processes or groundwater movement rather than solely by increased discharges or atmospheric deposition to the aquatic systems (Rasmussen 1994). However, the direct evidence for such sediment redistribution of mercury has yet to be published. Since these features may be overlooked in ecosystem mass balances, they will be explored in more depth hereafter. The flux estimates presented in the following sections are based on limited direct measurements, simple concentration measurements, or modeling results.

4.1
Erosion and Degasing of Mercury from Mineralized Surface Soils

Various estimates assume mercury volatilization rates between 1 and 5 ng m^{-2} h^{-1} for background soils in the mercuriferous belts and mineralized areas compared to below 1 ng m^{-2} h^{-1} in totally unimpacted areas (e.g. Lindqvist et al. 1991). Far higher emission rates (10–50 ng m^{-2} h^{-1}) have actually been recently measured over temperate forest soils impacted by atmospheric deposition (Lindberg et al. 1992; Kim et al. 1995) and over seismic zones (Varekamp and Buseck 1986). Flux chamber measurements over cinnabar rich soils in Almadén, Spain, yielded much higher fluxes, on the order of 330 ng m^{-2} h^{-1} (Lindberg et al. 1979), and fluxes on the order of 50–1000 ng m^{-2} h^{-1} were recently measured over geothermal zones in Nevada (Gustin and Lindberg 1997), so, locally, these evaporation rates can be of increased importance. Estimates of global natural fluxes from continents based on lower flux estimates suggest a total of 700 t $year^{-1}$ degassing from soils, with 500 t $year^{-1}$ being contributed from the mercuriferous belts (Lindqvist et al. 1991). Clearly, the newer measurements may suggest larger natural emissions.

Studies in Siberia reveal that the average content of Hg in ores varies from 0.06–1.2% (Obolensky 1996), while the mercury content in minerals of those ore deposits ranges from 0.01 to 2.700 ppm (Obolensky 1996). From these concentration data, it has been estimated that natural emission due to surface degassing of mercury in Siberia alone (approximate surface area 10^7 km^2) amounts to about 40 t $year^{-1}$ (Obolensky 1996). If volcanoes and forest fires are included, the estimate rises to 1700 t $year^{-1}$ (Nriagu 1989), while 1290 t $year^{-1}$ were calculated as the sum of soil vapor flux, volcanic and geothermal activities (Varekamp and Buseck 1986). To put these numbers into context, the mercury pool in a 1-m-deep layer of the continents (total area 1.5×10^8 km^2, assumed density 2 g cm^{-3}, background Hg concentration 50 ng g^{-1}) is 1.5×10^7 tons.

These estimates demonstrate that degasing from mineralisation zones (ore bodies, geothermal areas, primary and secondary geochemical haloes) is a major contribution to the total continental Hg emissions, and seems to be comparable in magnitude to emissions from volcanic and geothermal activity. Other sources like forest fires or vapor emissions from background soils also contribute significantly to natural continental mercury emissions, but it seems that particulate export in the form of dust is not one of them. The reason for this seems to be that soil particles are fairly coarse (10 μm) and although they might be resuspended by wind, they will settle quickly and generally not be transported very far in the atmosphere. Typical concentrations of particles of this size in background atmosphere are 25 μg m^{-3} (Finlayson-Pitts and Pitts 1986) and assuming they have an Hg content similar to background soils (50 ng g^{-1}), this kind of particulate Hg would have a global average concentration of 2.5 pg m^{-3} in ambient air, which is 3 orders of magnitude lower than total gaseous Hg in background air (Ebinghaus et al. 1995). This also indicates that particulate emission of Hg from mineralized soils to the atmosphere is not likely to be a major source of atmospheric Hg and is much less important than gaseous emission.

4.2 Volcanic Eruptions and Other Geothermal Activities

Though volcanic eruptions may contribute significantly to the natural emission of mercury, there are only a few studies to report the contribution from this source. Global estimates for Hg release from volcanoes total 830 t $year^{-1}$, with the vast majority being contributed by active, erupting volcanoes (800 t $year^{-1}$) and only small amounts (30 t $year^{-1}$) coming from passive degassing of volcanoes (Varekamp and Buseck 1986). The fluxes of Hg in the gases from eruptive plumes at Mt. Etna, Italy, and Kuala, Hawaii, and in fumarolic gases at Kilauea, Hawaii, and at White Island, New Zealand, have been estimated by measuring the Hg/S ratios and then correlating those to the sulfur flux from these sources (Fitzgerald 1996). These results were scaled up globally and it was concluded that the annual mercury flux from volcanic activity is between 20 and 90 t $year^{-1}$ and represents only a small fraction (<3%) of the annual mercury emissions from anthropogenic sources (Fitzgerald 1996). This result agrees well with the above-mentioned emission from passive degassing volcanoes, but active volcanic eruptions appear to be responsible for the large difference between the two global estimates; thus, their contribution constitutes a large uncertainty factor and should be reevaluated. Studies at Solfatara volcano, Italy, estimated the Hg emission at 0.9–4.5 g day^{-1} (0.3–1.6 kg $year^{-1}$); (Ferrara et al. 1994), which is comparable to estimates for the Kilauea volcano, Hawaii (1.5 kg $year^{-1}$), but much smaller than estimates for more active volcanoes, e.g., the Colima volcano, Mexico (440 kg $year^{-1}$), or the Etna volcano, Italy (2.7 t $year^{-1}$) (Varekamp and Buseck 1986). Data for Hg emissions from other geothermal sources are even rarer than for volcanoes, but it has been measured that one volcanic geyser on Iceland alone emits 8 kg $year^{-1}$ Hg to the atmosphere (Edner et al. 1991). Also, atmospheric Hg concentrations over a geothermal area were elevated by about a factor of 10 compared to background regions (Gustin et al. 1996), but there is a need for measurements of the contribution of geothermal sources to the global Hg emissions. Although few data are published, a recent study reported mercury emission rates in the order of 10–1000 ng m^{-2} h^{-1} in geothermal areas of the western US (Gustin and Lindberg 1997). Global Hg emission from geothermal sources has been estimated at 60 t $year^{-1}$ (Varekamp and Buseck 1986).

4.3 Evasion of Mercury from the Earth's Subsurface Crust

It is reasonable to assume that anthropogenic influences on the Earth's geochemistry does not go deeper than the immediate surface layer (probably much less than 0.1 km). However, mercury is permanently released from deeper regions of the Earth's crust and permeates as mercury vapor to the surface through faults and fractures in bedrock. This phenomenon is being explored as a potential indicator of earthquakes, since it has been observed that Hg concentrations in soil air increase dramatically directly before earthquakes occur. Crustal Hg emissions probably display extreme spatial and temporal variation. This has been demonstrated by measurements of mercury concentrations in a subterranean vault in Precambrian Shield bedrock (Klusman and Webster 1981) where pronounced diurnal and seasonal cycles were observed. Air and soil temperature, barometric pressure, and relative humidity were identified as the most important controlling factors. It has been argued that crustal mercury emission on the global scale is mainly driven by crustal heat flow rather than by bedrock mercury content (Varekamp and Buseck 1986). From this approach, continental crustal mercury evasion from geologic sources alone has been calculated by one author as 3000 to 6000 t $year^{-1}$ (Rasmussen 1994). This does not include volcanic or geothermal zones, which have to be treated separately due to their much higher heat flow. Overall, this would give a generally larger natural geological mercury flux from the continents to the atmosphere than generally assumed in other mass balances (e.g., see Sect. 4.1). A review of earlier estimates of natural mercury sources made between 1970 to 1982 summarizes published values in the range 2500 to 30 000 t $year^{-1}$ (Lindqvist et al. 1984). A more recent work has ended up with an estimate of 3000 t $year^{-1}$ (Nriagu and Pacyna 1988).

Results from the geological heat flux approach suggest that mercury transfer through mid-ocean ridges could be on the order of 1900 to 3800 t $year^{-1}$, while the whole oceanic crust would emit 7300 to 14 700 t $year^{-1}$ Hg (Rasmussen 1994). These results, in combination with other sources of submarine mercury like hydrothermal vents, seismic activity, and erosion of ocean ridges, as well as submerged parts of the continents, could give the impression that atmospheric deposition might be negligible in the total oceanic mercury budget (Camargo 1993). However, other authors assume that deposition is the major source of mercury (re)emitted to the atmosphere from ocean surfaces (Fitzgerald 1993). These findings contradict each other to such a degree that two main questions have to be asked: (1) is crustal degassing really the major source of oceanic Hg and (2) if so, then what percentage of that Hg actually reaches the atmosphere? These pathways (and their terrestrial counterparts) definitely have to be investigated much more thoroughly (if possible, by actual flux measurements) before the large discrepancies in global Hg budgets can be resolved. It should, however, be noted that the scientific community is beginning to develop programs for expansion of direct measurement campaigns in international collaborative studies (e.g., Gustin and Lindberg 1997)

4.4
Natural Air/Surface Exchange Processes with Soils, Oceans, Freshwaters, and Vegetation

This section will attempt to generalize the common features of mercury exchange between the atmosphere and terrestrial or aquatic compartments and general parameters determining direction and magnitude of the observed fluxes. It has to be kept in mind, though, that only gaseous dry deposition and volatilization are directly comparable in this way. The net emission that is usually observed as a sum of these two processes has to be balanced against particulate dry deposition (generally small) and wet deposition (generally important) in any study to finally evaluate a compartment as a source or a sink of atmospheric mercury.

The exchange of gaseous mercury at any interface is driven by a concentration gradient. If one compartment has a higher Hg(0) concentration than another, a net emission will take place, while net deposition will occur if the competing compartment is clean compared to the overlying atmosphere. Since many investigated waterbodies seem to be supersaturated with gaseous Hg compounds compared to the atmosphere, mostly evasion fluxes from waters to the air have been reported. However, under some conditions (at night and following periods of very high winds) downward fluxes to waters in subtropical Florida have been measured (Lindberg et al. 1997). This is also mostly true for soils, but there are some measurements over background soils suggesting net deposition under limited conditions, also mostly at night or over very wet soils (Kim et al. 1995). In earlier work, it has been shown that soils adsorb Hg vapor if exposed to elevated concentrations (Klusman and Matoske 1983). Therefore, it seems appropriate to propose the existence of a "compensation point", meaning that soils below a certain Hg(0) concentration (or probably: soil gas concentration) tend to absorb

Hg, while soils above that concentration emit mercury depending on the Hg(0) concentration in the overlying air (Kim et al. 1995); even if the mechanisms of emission and absorption may not be the same). A similar compensation point concept has also been reported for vegetation (Hanson et al. 1995).

Other properties whose gradients influence mercury transfer across the air-surface boundary include temperature, pressure and moisture. Therefore, soil temperature correlates well with mercury emission (Kim et al. 1995), higher fluxes are observed at low barometric pressure (McNerney and Buseck 1973) (the soil air is "sucked out" of the soil), and soils emit more mercury in the moist (not flooded) state compared to the dry state (Wallschläger et al. 1997b). In fact, one study of forest soils demonstrated that fluxes could change from net emission to net deposition depending on the soils moisture status (Advokaat and Lindberg 1996). In addition, any condition that favors conversion of oxidized Hg(II) to volatile mercury species in the investigated compartment, e.g., sunlight (Amyot et al. 1994; Carpi and Lindberg 1997) or bacterial activity, increases Hg fluxes to the atmosphere. Also, processes that accelerate the transport across the boundary layer increase the Hg flux. For example, high turbulence or windspeed conditions (Kim et al. 1995) that transport the liberated Hg away from the surface produce an elevated volatilization rate. Similarly, wave breaking increases the surface area and thereby the overall Hg flux from oceans to the atmosphere (Baeyens et al. 1991).

Mercury emissions from natural surfaces at various locations have been measured with different techniques or calculated by model simulations. Table 13 summarizes measured and calculated emission rates of mercury.

The role of vegetation in the overall picture seems to be ambiguous and needs further investigation, since all processes are apparently very specific to both the site and plant species. While some plants do not accumulate mercury from soils, others reflect a contaminated growth environment with elevated mercury content probably as a result of atmospheric exposure (Lindberg et al. 1979). Among the latter, some keep the Hg in the plant structure and release it back to the terrestrial environment after decay, while others transform the mercury into volatile compounds or simply transport already reduced mercury and release it

Table 13. Measured and calculated emission rates of mercury from natural surfaces

Location	Method	Emission rate ($\mu g\ m^{-2}\ year^{-1}$)	Reference
Lakes, SW Sweden	Chamber	18–180	a, b
Forest soil, SW Sweden	Chamber	<2	a, b
Forest, Tennessee, USA	Model and measurements	70–400	c, d
Equatorial Pacific	Meas./model	4–80	d
Almaden, Spain (soils near Hg mine)	Chamber	1140–2890	e
Lakes, Wisconsin, USA	Meas./model	0.7–1.5	f, g

[a] Schroeder et al. (1989). [b] Xiao et al. (1991). [c] Lindberg et al. (1992).
[d] Kim and Fitzgerald (1986) [e] Lindberg et al. (1979). [f] Fitzgerald et al. (1991).
[g] Vandal et al. (1991).

Table 14. Measured mercury fluxes over forests and annual emission estimates. (After S.E. Lindberg et al. in press)

Site	Canopy	Measured net emission Median ($ng\ m^{-2}\ h^{-1}$)	Mean ($ng\ m^{-2}\ h$)	Estimated net emission Mean daily (summer) (ng/m^2)	Annual mean ($\mu g/m^2$)
Walker Branch Watershed, Tennessee, USA	Mature Hardwood	39	37	350	70
Wartburg, Tennessee, USA	Young pines	17	18	150	30
Gardsjon, Sweden	Forest floor	1	1	9	2

to the atmosphere (Varekamp and Buseck 1983; Hanson et al. 1995). Both types of plants have been employed for phytoremediation of contaminated sites, but at least in the second case, this redistribution approach is very questionable. It has been shown that Hg in lichen originates predominantly from degassing from local soils (Bargagli et al. 1989). Vegetation surveys even reflect mercury sources at depths greater than 200 m (Kovalevsky 1986). Trees are a special case in that they continuously take up Hg with their foliage, which is then deposited on the soil as litterfall, while simultaneously taking up Hg via the transpiration stream through the roots and releasing it back to the atmosphere from the foliage. The overall budget seems to indicate that forests could be a net source of gaseous mercury, with the soils emitting more Hg than the foliage assimilates (Lindberg et al. 1992). Some fluxes from forests to the atmosphere have been calculated (Table 14), but litterfall and wet deposition have to be taken into account before characterizing forests as global sinks or sources of atmospheric mercury. The source of mercury in litterfall is at present not completely understood and could include both atmospheric (external) and soil (internal) sources. However, analysis of temporal trends in foliar mercury suggest an atmospheric source (Lindberg 1996; Rea et al. 1996).

The measured values in this field study included both emission and deposition fluxes, and the median was used to indicate a true central tendency weighted by the frequency of fluxes in both directions. Short-term fluxes were scaled using long term data on evapotranspiration which correlate well with mercury emissions (Lindberg et al. 1998).

5
Reemissions of Mercury

The discussion about the relative significances of natural and anthropogenic emissions is complicated (if not rendered impossible) by processes that we classify here as indirect anthropogenic reemissions. This term is chosen to describe secondary reemission of mercury from anthropogenic sources following partial initial deposition of primary anthropogenic Hg emissions. Unfortunately (for our understanding of global atmospheric Hg exchange), these processes usually take place in compartments that also exhibit a natural mercury surface exchange like waters and soils, thereby making it impossible to distinguish between the two parallel ongoing natural and anthropogenically induced processes and to separate them quantitatively. Examples of these indirect anthropogenic emissions are discussed in this section and include emissions from soils as a result of mercury spills, reemissions of formerly deposited mercury in the proximity of large point sources, and emissions from anthropogenically contaminated waterbodies and their adjacent wetlands.

5.1 Reemissions from Rivers

Little is known about mercury emissions from the waterbodies of rivers. It is well established that rivers transport mercury as a result of anthropogenic discharges and/or natural surface runoff, and both transport and deposition behavior in the river and especially in the estuaries with respect to discharges into the oceans have been studied. The few studies that investigate Hg exchange between rivers and the atmosphere all deal with the estuarine section of the rivers and are discussed in the following section on oceans. One recent study (Bahlmann 1997), however, deals with the occurrence of volatile Hg species along the transsect of the highly contaminated Elbe River in Germany. This investigation shows that volatile Hg compounds (defined as purgeable within 1 h; the exact speciation was not investigated, but we designate them here as dissolved gaseous mercury, or DGM) in the waterbody can be detected along the course of the river. It is shown that the highest concentrations of "free" volatile Hg compounds are found in the estuarine region. However, continuous purging of the waters revealed that a much larger amount of Hg could be volatilized from the limnic samples, and this may represent in-situ production of Hg(0) by reduction of Hg^{2+}. The levels of DGM found in this study are illustrated in Table 15, which compares DGM levels in several waters.

This process is quite likely, since the suspended particles in the waterbody of the Elbe are rich in both Hg (Wilken and Hintelmann 1991) and bacteria which have been shown to be resistant to the high Hg levels. These bacteria were shown to demethylate MeHg compounds as a detoxification mechanism and might also be able to reduce divalent Hg compounds (Ebinghaus and Wilken 1993). On the base of the determined concentrations and volatilization potentials, it has been calculated that the Elbe River could emit between 100 and 500 kg of Hg to the atmosphere anually (Bahlmann 1997). As in the case of oceans, there is a severe lack of experimental flux measurements for rivers, so this estimate has to be taken with great caution. One investigative experiment failed to detect a significant Hg concentration gradient right above the water surface of the Elbe, indicating that emission fluxes (if they occur at all) might be rather small

Table 15. Range of DGM concentrations measured in rivers, marine systems, and lakes around the globe

Location	DGM concentrations (pg/l)	Reference
Elbe River	47–152	Bahlmann (1997)
Elbe River Estuary	54–122	Coquery (1995)
German Bight (North Sea)	18–284	Bahlmann (1997)
German Bight (North Sea)	17–87	Reich (1995)
North Sea	<20–90	Coquery (1995)
Baltic Sea	14–22	Schmolke et al. (1997)
Equatorial Pacific	6–46	Fitzgerald (1986)
Equatorial Pacific	8–65	Mason and Fitzgerald (1991)
Northern Wisconsin Lakes	7–70	Vandal et al. (1991)
Scheldt Estuary	190–500	Baeyens et al. (1991)
Scheldt Estuary	45–80	Baeyens and Leermakers (1996)
St. Lawrence Estuary	72	Cossa and Gobeil (1996)
Seine Estuary	<10–90	Coquery (1995)
Long Island Sound	38–254	Vandal et al. (1996)

(Wallschläger 1996). However, gradient methods exhibit a rather high detection limit for fluxes (Lindberg et al. 1996). The only direct flux measurements over a river surface to date showed Hg emission from the St. Laurent River in southern Quebec and came to the conclusion that fluxes were fairly small, on the order of 1 ng m^{-2} h^{-1} (Poissant and Casimir 1996). These studies indicated that the Hg emission was photoinduced by reduction of divalent Hg under the influence of sunlight. Consequently, it will be of fundamental importance for understanding air-water exchange of Hg in rivers to determine to what extent the transported Hg is bound by suspended particles, for this parameter may have a significant effect on the availability for conversion to volatile species. The Elbe seems to represent one extreme in that respect for more than 99% of the transported mercury has been shown to be particulate (Wilken and Hintelmann 1991), while other contaminated rivers and streams seem to exhibit quite different behavior (Turner and Lindberg 1978). On a global base, it has been estimated that only 10% of the rivers' annual Hg loads originate from direct anthropogenic sources, while the remainder results from natural and indirect anthropogenic emissions (Cossa et al. 1996); however, on a regional base, there may be vast differences in these proportions.

5.2 Reemissions from Lakes

The air-surface exchange of mercury over lakes has been the focus of some intensive studies over the past few years. Only two studies have performed experimental flux measurements over lake surfaces, and they were both conducted in Sweden. However, several studies have reported measurements of DGM in lakes, some of which have been used to model evasion. Earlier flux chamber measurements (Schroeder et al. 1992) over five Swedish lakes found fluxes between 3 and 20 ng m^{-2} h^{-1} from the lakes to the overlying atmosphere. On one occasion, a net deposition was observed, but all of these values must be questioned because those measurements were hampered by blank problems (Kim and Lindberg 1995). It was also found that fluxes during daytime were larger than at night, indicating that sunlight, biological activity, and temperature might play an important role in the Hg volatilization process.The authors were also able to demonstrate pronounced seasonal differences in the Hg volatilization rate, with fluxes in May and June being much larger than in November. This phenomenon could be due either to temperature alone or to other seasonal cycles associated with it. More recently, methods have been developed which allow Hg fluxes to be measured directly over water surfaces with the micrometeorological modified Bowen ratio gradient approach (Lindberg et al. 1996). During June, 1994, the first measurements of Hg vapor fluxes over a boreal forest lake were made at Lake Gårdsjön, Sweden. Using highly accurate methods with multiple replicate samplers, the authors measured concentration gradients of Hg vapor, CO_2, and H_2O over the lake surface. Mercury was found to be readily emitted from the lake surface, and there was no evidence of Hg(0) dry deposition to the lake surface. Emission rates over the lake averaged 8.5 ng m^{-2} h^{-1}, and appeared to be weakly influenced by water temperature and solar radiation. Overall, the fluxes ranged from −2 to 18 ng m^{-2} h^{-1}, comparable to fluxes previously measured using surface chambers as discussed above. Overall, the surface water of the lake appears to be a more active zone for Hg exchange than the surrounding soils based on two independent studies (Xiao et al. 1991; Lindberg et al. 1998).

All other studies on Hg emission from lakes use calculations and models to estimate Hg fluxes from measured concentrations of atmospheric mercury and DGM in the water phase. It is then assumed that all gaseous Hg in both the waterphase and the atmosphere is Hg(0) – which is a reasonable initial assumption because other volatile Hg species in the ambient atmosphere or in water occur only at very low concentrations (e.g. Stratton and Lindberg 1995; Bloom et al. 1996 a,b) – and that this Hg(0) partitions between water and atmosphere according to Henry's law. With regard to this, most surface lake waters seem to be supersaturated in Hg(0), which is assumed to give rise to an evasional flux. The difference between the concentrations of Hg(0) measured in the lake water and back calculated from the atmospheric concentration is fit into a thin film model and combined with wind speed and other meteorological parameters to estimate the Hg flux from the lake to the atmosphere. It is evident that this calculation is based on pure phase thermodynamic properties and therefore unlikely to reflect reality in all natural ecosystems. Although, it is undetermined for many systems to what extent theoretical calculations and experimentally determined fluxes disagree, one study in a contaminated lake experimentally measured both DGM and fluxes simultaneously (Lindberg et al. 1997). This study found that measured fluxes exceeded those modeled from DGM using reported values of the exchange coefficients in other lakes.

The majority of studies on mercury exchange between lakes and the atmosphere have been conducted using modeling approaches in a set of seven lakes in Northern Wisconsin as part of the MTL (mercury in temperate lakes) project. All investigated lakes are seepage lakes (meaning they have no permanent surface water in- or outflow) and are not directly impacted by anthropogenic Hg sources. Consequently, the general finding was that atmospheric deposition was the major source of Hg for these lakes, which represent net sinks for atmospheric mercury since the modeled volatilization rates were always significantly smaller than the total Hg deposition from the atmosphere (consisting of measured wet deposition through rain and snow and estimated particulate dry deposition). Even though two out of three parameters in this mass balance are not determined experimentally, it still seems reasonable to conclude from these studies that lakes are net mercury sinks on an annual basis. In contrast, studies in anthropogenically contaminated lakes like the Davis Creek Reservoir in northern California show the opposite behavior, with the evasion flux being at least twice as big as the atmospheric Hg deposition (Porcella 1994). This is reflected in the fact that dissolved Hg(o) concentrations in a contaminated lake (Mason et al. 1995a) are five times higher than in remote lakes (Vandal et al. 1995). Some studies suggest that DMM may be directly formed from inorganic Hg under suboxic/oxic conditions and could, by decomposition, be the actual source of MMM observed in freshwaters (Mason et al. 1995a).

The ratio between atmospheric deposition and volatilization varied greatly between the individual MTL lakes, apparently due to differences in the water chemistry between the lakes (Watras et al. 1994). Between 10 and 50% of the mercury deposited from the atmosphere were reemitted, and the revolatilization was correlated positively with pH, DOC, and residence time in the upper water column. Calculated flux rates range from 1 to 5 $\mu g\ m^{-2}\ year^{-1}$ (0.11 to 0.57 $ng\ m^{-2}\ h^{-1}$). All mercury not reemitted was usually removed from the water column via sedimentation. It was also observed that MeHg was wet deposited from the atmosphere (about 1% of the total Hg deposition), and all deposited MeHg was removed by sedimentation. It has to be noted that the calculated flux rates on an hourly basis are much smaller than those measured at the Swedish lakes mentioned above. Since both the American and Swedish study lakes should be fairly comparable and also receive comparable atmospheric Hg inputs, it cannot be ruled out that the model calculations underestimate Hg volatilization considerably, which could make them less important sinks or even – in the worst case – alter the overall role of lakes in regional Hg budgets from sinks to sources. In the Lake Gårdsjön study, it was estimated that the lake could be a net source of Hg(o) during warm dry summer months, but is a net sink for the rest of the year (Lindberg et al. 1996).

The mechanism of Hg volatilization from lakes remains unclear. In contrast to oceans, information on DMM in lake waters is not known to the authors. Thus, Hg(o) is the only volatile Hg species contributing to the quantification of the emission flux. Since it is known that Hg(II) reduction can occur through both biotic and abiotic reactions, attempts have been made to further elucidate the formation of Hg(o). It has been found that mercury reduction is lower in heat-sterilized lake water, suggesting that the reduction process might be biologically mediated (Vandal et al. 1995). This is in agreement with studies in Upper Mystic

Lake, Massachusetts (Mason et al. 1995a), which found that both the concentration of Hg(o) and its production rate are highest in the surface waters and then decrease sharply down the water column. These authors were able to identify heterotrophic bacteria as the primary source of Hg(II)-reduction, while minor contributions were made by abiotic processes and possibly by phytoplankton and cyanobacteria. On the other hand, there is strong evidence that Hg volatilization from lakes is induced by sunlight (Amyot et al. 1994), but this could be due either to photochemical reactions or light-induced biochemical processes, so it does not necessarily contradict findings that favor biotic reduction of reactive Hg(II).

Another way of obtaining information on the processes that form Hg(o) in the water column is by looking at its depth concentration profiles and seasonal variations in the water columns of lakes. During studies in Onondaga Lake, an industrially polluted lake in near Syracuse, New York, USA, maximum concentrations were observed in April and August near the surface, while the lowest concentrations were found in the hypolimnion during periods of anoxia (Jacobs et al. 1995). This matches with observations in Pallette Lake, Wisconsin, where it was also found that Hg(o) concentrations are highest in the epilimnion and decrease with depth through the thermocline and the hypolimnion (Vandal et al. 1995); this observation was taken as an indication that Hg reduction occurs only in the mixed layer (Porcella 1994). Finally, in wind-sheltered parts of the lake, the Hg(o) concentrations in the epilimnion were elevated compared to wind-exposed regions. This finding strongly suggests that wind is an important driving force behind gas exchange at the lake surface; thus, Hg fluxes to the atmosphere are temporarily increased by windy conditions and waves.

Reactive Hg(II), the most likely precursor for Hg(o), showed variable seasonal concentration depth profiles. Its concentration was highest in the epilimnion in the early summer with low concentrations in the thermocline, but exactly the opposite in late summer. Since this parameter seems to be influenced by numerous and more complex processes, it is not surprising that two relevant sources of reactive Hg(II) were identified in these studies, one being atmospheric deposition and the other being remobilization from the sediments. Rain events increase reactive Hg(II) in the epilimnion, either by direct wet deposition of reactive Hg(II) or by deposition of particulate Hg that becomes transformed to reactive Hg(II) in the epilimnion. Since it has been shown that both reactive Hg(II) and particulate Hg are major fractions in wet deposition (Fitzgerald et al. 1994), a mixture of both processes seems to be the most likely source of reactive Hg(II) in the water column. Although Hg(o) concentrations in sediment porewaters are comparable to those in the epilimnion, they are lower than in the overlying lake water and it was concluded that Hg(o) export from the sediments is not an important source of Hg(o) in the water column. On the other hand, Hg(II) concentrations are higher in porewaters than in lake water and it also exceeds Hg(o) concentrations in porewater by at least 1 order of magnitude. Thus, it was concluded that export of Hg(II) from the sediments is a significant source of Hg(II) in the water column and, consequently, of Hg(o) evasion to the atmosphere (Fitzgerald et al. 1994).

5.3 Reemissions from Marine Systems

The distribution, transport, and fate of mercury at the Earth's surface is critically dependent on the biogeochemical cycling and atmospheric exchange of mercury in the marine environment since, by its shear size, the ocean is both a significant source and sink for atmospheric mercury (Mason et al. 1994a; Mason and Fitzgerald 1996). A number of studies have measured volatile Hg species well above saturation levels in ocean waters primarily as DGM. Hence, the emission of mercury from the oceans to the atmosphere is certainly occurring. However, since deposition is also occurring at the same time, a major question is whether or not emissions exceed deposition. As even the direction of the net Hg flux between oceans and atmosphere has not been determined conclusively, it is only natural that the size of that flux is even more doubtful. As a matter of fact, to date no field experiments have been attempted to actually measure air-surface exchange of mercury over oceanic waters, despite the existence of proven methods for smaller water bodies. All we know is that elemental Hg seems to occur commonly in surficial ocean waters in supersaturated concentrations. From this observation, it is frequently derived that there is a net mercury emission (Vandal et al. 1991; Fitzgerald et al. 1991, 1994), but we have to keep in mind that this assumption is based on physical chemical theories whose applicabilities to the water-air system in oceans have not been proven for mercury beyond doubt. All available flux data were obtained by this model (see sect. 5.2) and must therefore be seen as rough estimates rather than exact values. Estimated net Hg emissions range from average hourly fluxes of 0.4 to 16 ng m^{-2} h^{-1}. The concentrations of elemental mercury and its modeled evasional flux to the atmosphere in various oceanic regions are given in Table 16. From these data, the global mercury emission from oceanic sources has been estimated at 2000 t $year^{-1}$ (Mason et al. 1994a). Another important factor is that all these flux calculations are based on calm sea and moderate wind conditions (under which also all of the sampling campaigns were probably performed). It has been modeled, though, that fluxes may increase by a factor of up to 25 during storms and rough sea conditions (Baeyens et al. 1991), so this estimate may be systematically low. However, such fluxes would be very short-

Table 16. Concentrations of elemental mercury measured in various ocean regions and the estimated associated evasional flux to the atmosphere. (After Mason and Fitzgerald 1996)

Region	Conc. range (pg/l)	% Hg(0)/react. Hg	Potential flux (ng m^{-2} h^{-1})
EPO[a] 1990	10–72	14 ± 10	1.3–9
EPO 1984	6–46	5 ± 3	0.5–9.2
150° W, 10° N–12° S	8–18	–	0.8–2.8
N Pacific	6	–	0.2
N Atlantic	50–250	15–100	15.2 ± 10.4
Mediteranean	<26	<20	

[a] Equatorial Pacific Ocean.

lived while the system reequilibrated with ambient Hg(0), and the overall effect of these pulses on annual fluxes has not been assessed.

Both elemental Hg and DMM are found in oceanic waters in the DGM fraction. A number of studies (Fitzgerald et al. 1994; Mason et al. 1994a) suggest that direct reduction of ionic mercury is predominant in low oxygen waters (Mason and Fitzgerald 1993). Generally, both volatile species are found in comparable concentration ranges, although there always seems to be more Hg(0) than DMM.

In contrast to terrestrial systems, dry deposition of mercury does not seem to be an important process in marine ecosystems although no attempts have been made to measure it. Thus, deposition estimates are derived from precipitation measurements only. Reported results range from 0.04 to over 800 ng m^{-2} h^{-1}. The consequence of this enormous range seems to be that while on a global basis emission and deposition over oceans seem to equalize approximately, there are large regional differences depending on both the regional emission activity and the regional deposition as a result of precipitation rates and degree of local contamination. In the Arctic, for example, deposition was estimated to be five times larger than emission, while the ratio is exactly opposite in the Equatorial Pacific, leading to the conclusion that air-sea exchange of mercury actively contributes to the global Hg transport from midlatitude to Arctic shelves. On the basis of the existing data, it would, however, be more than daring to characterize either the oceans or the oceanic margins as either absolute sinks or sources of atmospheric mercury until more detailed and larger-scale studies have been done.

Studies in the northern Pacific Ocean show that Hg volatilization and deposition in the tropical Pacific Ocean are pretty much equal, but while the evasional flux was fairly constant at different latitudes, deposition varied significantly, thus making the equatorial Pacific Ocean a source of atmospheric Hg while the higher latitude regions turn out to be a sink (Mason et al. 1994b). Since the equatorial part of the Pacific Ocean is an upwelling area, it was suggested on the base of model calculations that reactive Hg(II) supplied by the upwelling process is the source of the observed increased Hg(0) concentrations and thus leads to the inferred atmospheric emission flux. Studies in the northern Atlantic Ocean also identify this region as an atmospheric Hg source, with emission fluxes averaging 139 ± 95 μg m^{-2} $year^{-1}$ (Mason and Fitzgerald 1996). The same trend has also been observed for coastal regions. While Hg evasion in remote arctic shelf areas was very small (3.7–5.8 μg m^{-2} $year^{-1}$), it was shown to increase in more polluted oceanic boundary regions (7.3–22 μg m^{-2} $year^{-1}$). The authors conclude from these data that net atmospheric transfer of Hg from lower latitudes to arctic regions is going on despite a net global balance for emissions and depositions. Somewhat higher atmospheric mercury evasion fluxes have been reported for estuaries (3.7–51 μg m^{-2} $year^{-1}$), and the emission seems to be correlated to the rivers' Hg content (Cossa et al. 1996). In a very detailed review of studies on Hg biogeochemistry in coastal regions, Cossa et al. (1996) propose the following semiquantitative budget for coastal areas. Overall, imports and exports of Hg to ocean margins seem to neutralize within the range of error. The major source of Hg is particulate matter transported in rivers. However, this amount seems to be deposited in the estuaries via sedimentation under the influence of the salinity gradient. Exchange between ocean margins and their sediments, as well as with the open ocean, also seem to balance. Atmospheric exchange in general seems to be a net emission process. They also attempted a mass balance for methylmercury in ocean margins and came to the conclusion that no boundary exchange process had significant impact on the MeHg pool in coastal zones, except for upwelling processes that imported about half of the annual biomass uptake of MeHg into ocean margin areas, with the other half being produced in situ.

Considering the enormous part of the Earth's surface covered by oceans, it is of vital importance to understand if oceans only reemit mercury that is transported by rivers or deposited from the atmosphere, or if they actually contribute actively to the global Hg budget as a net source. There is evidence that Hg from oceanic sediments is released to the overlying water during diagenetic processes (Cossa et al. 1996). These investigations suggest that Mn oxides bind most of the Hg in oceanic sediments, and as these oxides become reduced and dissolved during bacterial degredation of organic matter, Hg concentrations in sediment porewaters are elevated by 1 order of magnitude compared to the overlying waters. Fluxes calculated from these concentrations based on molecular diffusion equal 26 ng m^{-2} day^{-1} (9.5 μg m^{-2} $year^{-1}$). Although this flux apparently accounts for only 3% of the total input into these systems (Cossa et al. 1996), when scaled to the total ocean sediment surface area of 3.6×10^{11} km^2 (=71% of the Earth's total surface area) (Fairbridge 1966), it yields over 3×10^6 t $year^{-1}$. It is therefore vital to investigate if this flux estimate is accurate and globally representative, how it compares to direct sedimentation (i.e., if there is a net release of Hg from oceanic sediments to the water column) and to what extent this flux is carried through the water column to lead to atmospheric emission.

5.4 Reemissions from Terrestrial Systems and Wetland Areas

Compared to the seas, there are more data available regarding mercury exchange between the atmosphere and several types of soils with respect to both the origin and the level of their Hg contamination. Beginning with impacted sites, studies in the vicinity of a strong industrial point source of mercury in eastern Germany revealed wet deposition fluxes of 1200 μg m^{-2} $year^{-1}$ to the soils surrounding the factory premises. For comparison, the combined source strength of the four individual production sites (one active, three shut down) including their highly contaminated vicinity was backmodeled to be 800 to 1600 kg $year^{-1}$ and the model indicates that as much as 30% of the total annual emission is deposited within 5 km from the spot of emission (Ebinghaus and Krüger 1996). Thus, soils located close to large atmospheric Hg sources are clearly sinks while these sources are active. In the long run, however, these soils may turn into important sources when the original emissions are discontinued and direct flux measurements will be important at such sites.

The first direct flux measurements of mercury at an industrially impacted site used a laboratory flux chamber (Lindberg and Turner 1977). Emissions were measured over waste deposited in holding ponds from a former chlor-alkali plant in Virginia. The mercury concentrations in the surface waste deposits averaged 160 $\mu g\ g^{-1}$. The authors reported the highest fluxes yet published, on the order of 120 000 $ng\ m^{-2}\ h^{-1}$ to 170 000 $ng\ m^{-2}\ h^{-1}$ at 35 °C and estimated an annual atmospheric emission from this one area alone of about 40 kg $year^{-1}$. Fluxes increased exponentially with surface temperature, and resulted in ambient air concentrations near the area of over 950 $ng\ m^{-3}$ in summer. Globally, hundreds of such sites still exist and their contribution to regional mercury emissions could be significant.

Elevated mercury emissions have also been reported for soils near the Almadén mercury mine in Spain (Lindberg et al. 1979). Emission rates over naturally Hg-enriched soils (concentrations averaged 97 $\mu g\ g^{-1}$) ranged from 600 to 750 $ng\ m^{-2}\ h^{-1}$ at 35 °C and also increased exponentially with soil temperature. For the first time direct plant uptake of this emitted airborne mercury was measured, leading to elevated levels in leaves, although root uptake from the soil pool was limited.

More recently, mercury fluxes from highly contaminated mill tailings from former gold and mercury mines in the western US have been crudely estimated from atmospheric concentration gradients combined with a simple one-dimensional diffusion model. Fluxes were estimated to be 5–120 $\mu g\ m^{-2}\ h^{-1}$, and, surprisingly, no clear correlations with any meteorological parameter could be identified; only total Hg content in the soil samples seemed to always correlate with the estimated fluxes. The largest fluxes were usually derived from early morning measurements, but sometimes increased fluxes were obtained at nighttime or in the afternoon. Such results may indicate that some of the measured gradients were artifacts of advection of contaminated air from local sources, and cannot be used to derive fluxes. Scaling of the results to the whole Carson River Superfund site, Nevada, USA, yielded emissions on the order of 150 to 400 kg $year^{-1}$. These studies also showed elevated atmospheric Hg concentrations over a naturally mercury-enriched geothermal area, but these were much smaller than over the mine tailings and no fluxes were derived (Gustin et al. 1996). More extensive measurements over a nearby geothermal site confirmed air concentrations as high as 50 $ng\ m^{-3}$ while fluxes measured with portable soil chambers ranged from approximately 10 to 1000 $ng\ m^{-2}\ h^{-1}$, much lower than seen over the mine tailings (Gustin and Lindberg 1997). More data will be necessary to confirm that fluxes over naturally enriched soils are generally below those over contaminated soils.

Another type of Hg-contaminated soils are wetlands adjacent to Hg-contaminated waterbodies which have stored large amounts of Hg as a result of sedimentation of contaminated particulate matter during periodic floodings. It is a well-established remediation technique to construct wetlands in order to remove contaminants from polluted river systems due to the wetlands' high organic carbon accumulation (Patrick 1994). In the case of Hg, however, there are indications that floodplains are not only initial sinks of deposited Hg, but also long-term diffuse sources of atmospheric Hg since the high water content and high productivity of wetlands seem to generate ideal conditions for the transformation of nonvolatile into volatile Hg species (Wallschläger et al. 1997c). There are two prominent case studies for this kind of diffusely Hg-polluted wetlands, namely East Fork Poplar Creek (EFPC) in Tennessee, USA (Lindberg et al. 1995), and the river Elbe in Germany (Wallschläger et al. 1997a). The former is a small waterbody that was contaminated directly from a single large industrial point source, while the latter is the third largest river in Germany and comprises a much larger area that became contaminated by much larger absolute amounts of Hg discharges from the entire industrial area along its course. However, both study areas are relatively similar in that they are floodplain ecosystems which show similar degrees of Hg contamination. Therefore, it seems appropriate to compare the results of investigations on air-soil exchange of Hg at both sites.

The EFPC floodplains has an area of 250 ha and is estimated to contain a total Hg pool of 80 t. Soil Hg concentrations range from 4 to 60 $\mu g\ g^{-1}$ in less contaminated and 50–200 $\mu g\ g^{-1}$ in more contaminated areas (Lindberg et al. 1995). Solid-phase speciation experiments seem to reveal that most of the deposited Hg is present in the forms of HgS (cinnabar and meta-cinnabar) with minor amounts of elemental Hg(0) (Barnett et al. 1995). Mercury volatilization from these floodplain soils was studied by the first micrometeorological approach to be developed for Hg, the modified Bowen-ratio method (MBR) (Lindberg et al. 1995). Naturally, a net Hg emission from these soils was observed, and fluxes were found to be between 10 and 200 $ng\ m^{-2}\ h^{-1}$ over the study site containing an average of $4.3 \pm 0.3\ \mu g\ g^{-1}$ Hg. Since the applied technique has a fairly large "footprint area" (i.e., is representative of a large area surrounding the actual measurement site), these fluxes are partially influenced by air masses passing over more contaminated areas of the floodplain. Higher fluxes were correlated with wind directions coming from the more contaminated parts of the floodplains, and lower fluxes came from the less contaminated regions. The total annual emission from the EFPC floodplains to the atmosphere was estimated to be between 1 and 10 kg.

The Elbe floodplains have an estimated area of 1100 km^2 and total deposited Hg amount of 1500 t (Wallschläger and Wilken 1997). Average Hg concentrations in the soils range from 1 to 10 $\mu g\ g^{-1}$ and most of the Hg is bound by high molecular weight organic matter (Wallschläger et al. 1997a). In these studies, flux chamber measurements were employed to estimate Hg volatilization, but they were also compared to the MBR method by measuring the concentration gradient of atmospheric Hg in the soil atmosphere boundary layer and adopting the micrometeorological parameters from the EFPC floodplains (Lindberg et al. 1995). Additionally, Hg volatilization was estimated from soil air concentration data assuming laminar diffusion. Results for Hg emissions from soils containing 10 $\mu g\ g^{-1}$ range from 28 to 260 $ng\ m^{-2}\ h^{-1}$ and are thus comparable to the fluxes measured at EFPC. However, it was noted that the individual applied techniques did not agree very well due to their inherent methodological differences, and that there is a need for intercomparison studies (Wallschläger et al. 1997c). The overall annual emission of Hg from the Elbe floodplains to the atmosphere was estimated to be on the order of 1 t $year^{-1}$, thereby representing a significant, but not dominant Hg source on the regional scale.

Micrometeorological studies over background forest soils at Walker Branch Watershed (WBW), Tennessee, USA, led to the conclusion that these soils can act both as sources and as sinks of atmospheric Hg, although emission events were observed more frequently and had a larger magnitude (Kim et al. 1995). Considering the wet deposition inputs, it appears that background soils may not significantly change the atmospheric Hg pool, but small variations in the mass balance could have large impacts on the global scale. Hg emission rates measured averaged 7.5 ± 7.0 ng m^{-2} h^{-1} over soils containing 0.35–0.82 μg g^{-1} total Hg. These concentrations, however, are slightly elevated compared to the background concentrations generally assumed today, indicating that true background soils may indeed be neutral towards surface exchange of atmospheric Hg. Studies over background forest soils in Sweden (Xiao et al. 1991) found fluxes of lower magnitudes, but these flux chamber measurements were somewhat hampered by blank problems (Kim and Lindberg 1995). However, it was noted that net Hg emissions occur during summer at a rate of 0.3 ± 0.4 ng m^{-2} h^{-1}, while net deposition was observed in winter at a rate of 0.9 ± 0.4 ng m^{-2} h^{-1}. Unfortunately, no Hg soil contents are given to correlate with the flux magnitudes. In a related study using flux chamber methods, background forest soils in Tennessee, USA showed Hg emission fluxes between 2 and 7 ng m^{-2} h^{-1}, while open field soils in the same region exhibited greater volatilizion rates of between 12 and 45 ng m^{-2} h^{-1}, although the open field sites had a smaller Hg soil content than the forest sites. This difference was attributed to increased solar irradiation over unshaded soils. These results give an estimate of 1000 T Hg $year^{-1}$ which are emitted globally from all background soils to the atmosphere, with two thirds coming from sunlight-exposed open field soils (Carpi and Lindberg 1998).

It has been noted that vast areas of open-field agricultural soils in the world are treated with municipal sewage sludge as fertilizer. In studies of sewage sludge-amended soils, it was found that sludges containing 7.3 ± 2.5 μg g^{-1} Hg when applied to background soils result in a release on the order of 25 ± 10 ng m^{-2} h^{-1} when kept in the shade. However, this emission increased dramatically when soils were exposed to sunlight and emissions rose to an average of 460 ± 120 ng m^{-2} h^{-1}. In these experiments, solar radiation was shown to induce Hg emissions more directly than indirectly via soil temperature changes (Carpi and Lindberg 1997), and that the Hg^{2+} contained in the sludge must be reduced in situ to Hg(0) when exposed to sunlight, perhaps by photoreduction reactions. It was calculated that in the EU and the USA alone, sludge-amended soils anually release 5 t of Hg to the atmosphere, making them a small source on the global scale, but important on a regional scale in otherwise uncontaminated areas where this practice is common. In addition, during the same studies, it was shown that the sewage sludge-amended soils also release methylated Hg compounds to the atmosphere at a rate of 12 to 24 pg m^{-2} h^{-1}, identifying the first-ever emission of gaseous MeHg to the atmosphere (Carpi et al. 1997). If upscaled, sewage soil treatment in the EU and USA would liberate 1 kg of MeHg annually. Considering that vast areas outside the Western world may be impacted by sludge applications, these soils could have an important regional or global contribution.

All studies show that the Hg emission flux is more complex than simple phase transfer of Hg(0), because the phase transfer enthalpy (derived from temperature dependencies) is always twice as high as for the pure compound, indicating that formation and/or transport processes slow down the volatilization rate. The whole process has been suggested to be a combination of volatilization of existent elemental Hg(o) and reduction of Hg^{2+} in soil solutions or in water layers on soil minerals (Lindberg et al. 1998). Observations of increased Hg volatilization from contaminated floodplains during rain events led to the suggestion of a two-step mechanism comprising the same reactions: a small initial displacement of soil air containing Hg(o) and DMM (Wallschläger et al. 1995) followed by formation of volatile Hg compounds in the liquid phase as a result of reduction of Hg(II) and dismutation of MeHg (Wallschläger et al. 1997b). These studies also calculated that direct reduction of wet-deposited Hg(II) is not likely to be a major source of Hg volatilization. Soil moisture may play a role in mercury emissions, however. Using a roof to eliminate rainfall over soils at Walker Branch Watershed (WBW), a strong decrease in soil Hg emissions was measured after 6 weeks of no rainfall (Advokaat and Lindberg 1996). In fact, these soils became a net sink, exhibiting consistent mercury uptake compared to continued Hg emission from adjacent control soils. Rainfall was not the source of the mercury, however, as the authors once again induced mercury emissions by simply irrigating the treatment plot with distilled water.

There has been increasing interest in understanding the driving forces behind air-surface exchange of Hg, especially over soils. In these numerous investigations, several key factors besides Hg content and speciation were identified as influencing the Hg emission rates from soils. At EFPC, it was demonstrated that Hg fluxes increase exponentially with both soil and air temperature (Lindberg et al. 1995). Annual and diel cycles were observed with highest fluxes in summer and in the afternoon and lowest fluxes in winter and at night (Lindberg et al. 1995; Kim et al. 1995). These observations are probably coupled to temperature cycles, but also indirectly to biological cycles associated with them. Positive correlations were observed between emission rates and wind speed as well as relative humidity and turbulence (Kim et al. 1995). Solar radiation, temperature, and soil moisture were identified as key parameters affecting Hg emissions from soils, and it was shown that solar radiation induced the formation of elemental Hg(o) in surface soils (Carpi and Lindberg 1997). Rain events also increase Hg flux to the atmosphere, supporting that soil moisture is one important parameter in the formation of volatile species, probably in liquid-phase reactions (Wallschläger et al. 1997b).

6
Global Fluxes of Mercury

The combustion of fossil fuels to produce heat, steam, and electricity is the major anthropogenic source of atmospheric emissions of mercury on a global scale (Nriagu and Pacyna 1988). Other important anthropogenic sources include waste incineration, mining, and smelting. Globally, the emission of mercury from natural surfaces (re-emissions plus natural source emissions) is comparable to these anthropogenic emissions (Lindqvist et al. 1991). Whether the emitted Hg(0) originates from parent geological materials or from past and recently deposited atmospheric mercury, the net effect of these processes on the global mercury cycle is the same: to increase the atmospheric mercury pool available for redistribution and redeposition.

The global total atmospheric emission of mercury has been estimated to be in the range 5000 to 15 000 t $year^{-1}$ (Lindqvist et al. 1991; Fitzgerald 1996; Sukhenko and Vasiliev 1996). The first quantitative worldwide estimate of the annual industrial inputs of 16 elements to the environment including mercury has been prepared by Nriagu and Pacyna and is summarized in Table 17 (Nriagu and Pacyna 1988). This table also demonstrates that mercury inputs to the air are comparable with direct inputs of the element to the aquatic environment and are almost a half of the direct releases to soils. These estimates also support that combustion of fossil fuels to produce heat and electricity is the major anthropogenic source of atmospheric emissions of mercury on a global scale. Pirrone et al. (1996) estimated an annual increase in anthropogenic mercury emissions into the atmosphere in the order of 3 to 5% for developing countries. This number is comparable to the anthropogenic emissions in the western industrialized countries before 1989; however, emissions in those countries have remained on a stable level since then (Pirrone et al. 1996).

Table 18 shows a number of recent estimates for anthropogenic mercury emissions into the atmosphere on a global scale and for various regions of the world.

Table 18 suggests that <1000 t $year^{-1}$ or 25% of total emissions come from western industrialized countries. In comparison to that number, Nriagu and Pacyna (1988) have estimated the global natural mercury emission to be 3000 t $year^{-1}$, which corresponds to an emission rate of about 6 μg km^{-2} $year^{-1}$. Fitzgerald (1986) estimated the oceanic mercury reemission alone to be 2300 t $year^{-1}$, emphasizing the degree of uncertainty involved in the estimate of natural Hg (re)emissions. Nevertheless, natural emissions of mercury are a significant source to the atmospheric burden and are of the same magnitude as

Table 17. Global anthropogenic inputs of mercury to the environment. (Nriagu and Pacyna 1988)

Environmental compartment	Source category	Emission (t $year^{-1}$)
Air	Combustion of fuels	700–3800
	Industrial manufacturing	150–200
	Refuse incineration	200–2100
	Subtotal	1000–6100
Water	Direct releases	1000–6700
	Dumping of sewage sludge	1000–3000
	Atmospheric deposition	400–1800
	Subtotal	600–8800
Soil	Direct releases[a]	1000–10 700
	Atmospheric deposition	600–4300
	Subtotal	1600–15 000

[a] Includes sewage sludge applications.

Table 18. Estimated anthropogenic emissions of mercury to the atmosphere

Area	t year $^{-1}$	Reference
Global	3600–4500	Fitzgerald (1996)
Global	4500	Sukhenko and Vasiliev (1996)
Global	910–6200	Nriagu and Pacyna (1988)
Europe	600–1100	Lindqvist et al. (1991)[a]
Europe	630	Pacyna (1996)[b]
USA	650	Lindqvist et al. (1991)[a]
USA	200	US EPA (1994)
USA	150	Porcella et al. (1996)
Canada	30	Lindqvist et al. (1991)[a]
Canada	50	Porcella et al. (1996)
Mexico	80	Porcella et al. (1996)
Brazil	160	Lacerda et al. (1995)
Siberia	90	Sukhenko and Vasiliev (1996)

[a] And references therein.
[b] Updated since the opening of the Eastern block.

the anthropogenic emissions. It is unclear to what extent anthropogenic activities have affected natural emission rates. This is particularly important considering that approximately 200 000 tons of mercury have been deposited to soils by anthropogenic emission sources since 1890 (Expert Panel 1994). It would require only a small fraction of this part to be reemitted to rival other global sources (Lindberg 1996).

Gas exchange processes have recently been studied in detail in terrestrial systems, especially forests where dry deposition is found to be comparable to or greater than wet deposition (Iverfeldt and Lindberg 1996; Rea et al. 1996). However, still newer data suggest that forests could be an important conduit for mercury emissions from soils (Lindberg 1998). Forests cover ~4 $\pm 10^9$ ha of the Earth's surface. Many forests are experiencing important disturbances and the fate of vast quantities of mercury stored in these ecosystems is not well understood. The mechanisms of the exchange of Hg(o) over forests have been studied extensively (Lindberg et al. 1992; Hanson et al. 1995) and indicate that foliar surfaces in forest landscapes are dynamic exchange surfaces that can function as a source or sink for Hg, and that Hg(o) exchange is stomatally controlled to some extent. The net exchange of Hg(o) with foliage depends upon the leaf temperature, leaf surface conditions (wet or dry), soil gas Hg(o) levels, atmospheric oxidants, and biological factors such as tree and leaf age and foliar nutrition. Lindberg (1996) has observed that at background mercury levels wet vegetation is more conducive to airborne Hg uptake than are dry surfaces. Mercury emissions from forest ecosystems and its global estimates are summarized in Table 19. The spatial scaling was based on forest type distributions given by Dixon and Wisniewski (1995).

It is evident from Table 19 that biogenic emissions of Hg(o) from forests may exceed similar emissions from background soils on a global basis. Based on the limited study, Lindberg et al. (1998) suggest that terrestrial emissions and reemissions may be comparable to or exceed those from oceans. When one

Table 19. Mercury emissions from forests and its global estimates. (After Lindberg et al. 1996)

Region	Forest type	Forest area (in 10^{12} m^2)	Estimated global emissions (in t $year^{-1}$)
High latitude	Boreal	1.37	10–100
Mid latitude	Temperate	1.04	300–700
Low latitude	Tropical	1.76	540–1200
All forests[a]		4.17	850–2000
Soils – non forested[a]			500–1100
Soils – forested			80–160
Total net emission of Hg(0)			1400–3200

[a] From Carpi and Lindberg (1998).

considers that these estimates do not yet include emissions from either Hg-contaminated or Hg-enriched soils and surface deposits, it becomes clear that an accurate assessment of air/surface exchange processes on a global scale will be critical to the refining of global mercury cycling models.

The global wet deposition of mercury to the ocean has been estimated from the measured concentrations of Hg in precipitation, coupled with estimates of ocean area and precipitation (Mason et al. 1994a). Unfortunately, there are no comparable regional to global scale estimates of dry deposition, although it is certainly a highly significant component of the global cycle (Iverfeldt and Lindberg 1996) The global wet deposition input of mercury to the oceans at different latitudes has been summarized in Table 20. The mercury fluxes in individual latitudinal regions were found to range from 0.13 μg m^{-2} $year^{-1}$ for Antarctica to 7.7 μg m^{-2} $year^{-1}$ for the northern mid latitude region (30–70°). The fluxes show the influence of anthropogenic activity which is largely confined to the Northern Hemisphere. It is also evident from the table that the fluxes are at least a factor of 2 higher in the Northern Hemisphere compared to the corresponding Southern Hemispheric regions. Because of low particulate

Table 20. Estimated average wet deposition to the oceans and to land for various latitudinal regions. (Data after Mason and Fitzgerald 1996)

Latitude	Precipit. (m/year)	Ocean area × 10^{12} m^2	Global input (t $year^{-1}$)		Flux (μg m^{-2} y^{-1})	
			Ocean	Land	Ocean	Land
70–90° N	0.16	11.7 (47)	8	14	0.68	0.1
30–70° N	0.79	51.4 (50)	400	800	7.7	15.8
10–30° N	0.97	56.6 (69)	540	500	9.5	19.8
10°N–10° S	1.54	67.9 (77)	320	260	4.7	12.7
10–30° S	1.0	63.2 (77)	260	180	4.1	9.6
30–70° S	0.96	105.1(94)	400	60	3.8	8.5
70–90° S	0.07	3.1 (20)	0.4	2	0.13	0.32
Total			1928	1816		

Note: Values in brackets are % of surface covered by ocean.

mercury concentrations over the oceans, dry deposition fluxes are not considered to be significant (Mason et al. 1994). However, recent measurements of RGM in air (Lindberg and Stratton 1998) suggest the need to reassess dry deposition near the continental margins.

It should be noted that a significant decrease in mercury deposition in Scandinavia has been reported during the last few years (Iverfeldt 1995) and reductions in concentrations in the remote marine atmosphere have also been reported (Slemr 1996). However, accurate long-term data to assess actual trends in atmospheric mercury concentrations do not yet exist. A long-term, fixed network of continuous global monitoring sites (oceanic as well as land-based) is essential for spotting global trends (Expert Panel 1994). Recently, results from continuous monitoring of atmospheric mercury have been reported for Alert in the Canadian Arctic (since 1995) (Schroeder and Schneeberger 1996), Mace Head at the west Irish coast (since 1995) (Ebinghaus et al. 1996) the Wank summit in the Bavarian Alps, Germany (since 1990) and Cape Point, South Africa (since 1995) (Slemr et al. 1996).

7
Summary and Conclusions

Compared with present knowledge on environmental and health effects of mercury, much less information is available on the fluxes of the element and its compounds to the air, water, and soils. Globally, it can be concluded that mercury inputs to the air are comparable with direct inputs of the element to the aquatic environment and are almost a half of direct releases to soils. The main objective of this section is to summarize present knowledge on natural and anthropogenic mercury fluxes to the atmosphere from various sources and their relative importance on regional and global scales.

The major anthropogenic source of atmospheric mercury emissions on a global scale is the combustion of fossil fuels to produce heat, steam, and electricity. Other important anthropogenic sources include waste incineration, mining, and smelting. The speciation of anthropogenically emitted mercury depends upon the nature of the source of emission and includes gaseous elemental mercury and gaseous or particulate oxidized forms of the metal. Available information on mercury emissions to the air in North and parts of South America, Western Europe, and parts of Asia allows a crude estimate of regional mercury emissions for vast areas of the globe. However, more information on mercury emissions is needed on a global scale, especially for China and many developing countries.

Anthropogenic emissions of mercury are of the same order of magnitude as the natural emissions. It has been estimated that approxiamtely 200 000 tons of mercury have been emitted by anthropogenic sources since 1890 and the major fraction has been deposited to soils. It is unclear to what extent anthropogenic activities have affected natural emission rates.

The assessment of truely "natural" mercury sources and their relative importance compared to direct anthropogenic emissions and indirect (re)-emissions is a fundamental problem in studying the global balance and cycling of mercury in the environment. This knowledge is critical to our understanding of the fate of mercury since so-called natural emissions cannot be controlled or reduced in most cases.

In general, the distinction between natural and "quasinatural reemissions" of mercury (that which was formerly deposited from the atmosphere to surfaces) is difficult to discern and may be regarded as an unresolved problem. In addition, the so-called natural emissions also include mercury that was previously deposited from natural sources. It can be assumed that mercury that evades from these types of sources is generally in the elemental form. Estimates of global natural fluxes suggest a total of 700 t $year^{-1}$ degassing from soils, with 500 t $year^{-1}$ originating from the mercuriferous belts. However, more recent measurements may suggest larger natural emissions. Degassing of mercury from mineralization zones is a major contribution to the total continental mercury emissions and seems to be comparable in magnitude to emissions from volcanic and geothermal activity.

Soils located close to large atmospheric mercury sources are clearly sinks while these sources are active. In the long run, however, these soils may turn into important sources when the original emissions are discontinued. Globally, hundreds of such sites exist and their contribution to regional mercury emissions could be significant. It has been suggested that fluxes over contaminated soils are generally higher from those over naturally enriched soils. The existence of a "compensation point" for soils has been proposed, meaning that soils below a certain Hg(o) concentration (or probably: soil gas concentration) tend to adsorb mercury, while soils above that concentration emit mercury depending on the Hg(o) concentration in the overlying air. For forest soils, a recent study has demonstrated that fluxes could change from net emission to net deposition depending on the soil moisture status. Solar radiation, temperature, and soil moisture were identified as key parameters affecting mercury emissions from soils. A recent study gives an estimate for mercury emissions from background soils of 1000 t $year^{-1}$ on a global basis, with two thirds coming from sunlight-exposed open-field soils. There is evidence that the treatment of agricultural soils with municipal sewage sludge can be an important factor for mercury emissions into the atmosphere at least on a regional scale in otherwise uncontaminated areas where this practice is common. Emissions of mercury from forests may exceed similar emissions from background soils on a global basis. It has further been suggested that terrestrial emissions and re-emissions may be comparable to or exceed those from oceans.

The emission of mercury from oceans to the atmosphere is certainly occurring; a number of studies have measured volatile mercury species (primarily DGM) in ocean waters well above saturation levels. However, since deposition is also occurring at the same time, a major question is whether or not emissions exceed deposition. More detailed and larger-scale studies are necessary for the characterization of the oceans or the ocean margins as either absolute sinks or sources for atmospheric mercury. It is especially important to finally obtain real flux measurements for Hg emissions from the oceans in order to facilitate comparison with terrestrial exchange processes. In addition, considering the enormous part of the Earth's surface covered by oceans, it is of vital importance to understand if oceans only reemit mercury that is transported by rivers or deposited from the atmosphere, or if they actually contribute actively to the global mercury budget as a net source, since there is evidence that mercury from oceanic sediments is released to the overlying water during diagenetic processes.

Wind seems to be an important driving force behind gas exchange at environmental surfaces. Mercury fluxes to the atmosphere are temporarily increased by windy conditions and waves. Mercury emissions from waters are generally estimated using measured concentrations of atmospheric mercury and DGM in the water phase. However, recent results from field measurements have indicated that these types of model calculations may considerably underestimate mercury volatilization from lakes.

References

Advokaat EM, Lindberg SE (1996) Effect of rainfall exclusion on air/surface exchange rates of mercury over forest soils. In: Ebinghaus, R, Petersen G, Tümpling Uv (eds.) 4th Int Conf on Mercury as a global pollutant, Book of Abstracts, Hamburg, Aug 4–8, 1996, 458 pp (GKSS Forschungszentrum Geesthacht GmbH, Max-Planck-Str, D-21502 Geesthacht, Germany)

Airey D (1982) Contributions from coal and industrial materials to mercury in air, rainwater and snow. Sci Total Environ 25:19–40

Alberts JJ Schindler JE, Miller RW, Nutter, DE (1974) Elemental mercury evolution mediated by humic acid. Science 184:895–897

Amyot M, Mierle G, Lean DRS, McQueen DJ (1994) Sunlight-induced formation of dissolved gaseous mercury in lake waters. Environ Sci Technol 28:2366–2371

Amyot M, Lean DRS, Mierle G, McQueen, DJ (1995) Volatilization of Hg from lakes mediated by solar radiation. Proc 1995 Canadian Mercury network workshop, York University, Toronto, Ontario, Canada

Axenfeld F, Munch J, Pacyna JM (1991) Belastung von Nord und Ostsee durch ökologisch gefährliche Stoffe am Beispiel atmosphärischer Quecksilberkomponenten. Dormier GmbH, Report for the German Environmental Protection Agency. Umweltbundesamt, Berlin

Baeyens W, Leermakers M (1996) Particulate, dissolved and methylmercury budgets for the Scheldt estuary (Belgium and The Netherlands), In: Baeyens W, Ebinghaus R, Vasiliev, O. (eds) Global and regional mercury cycles: sources, fluxes and mass balances. NATO-ASI Series 2. Environment vol 21. Kluwer, Dordrecht, The Netherlands, pp 285–301

Baeyens W, Leermakers M, Dedeurwaerder H, Lansens P (1991) Modelization of the mercury fluxes at the air-sea interface. Water Air Soil Pollut 56:731–744

Baeyens W, Ebinghaus, R, Vasiliev O (eds) (1996) Global and regional mercury cycles: sources, fluxes and mass balances. NATO-ASI Series 2 Environment vol 21. Kluwer Dordrecht, The Netherlands

Bahlmann E (1997) Untersuchungen einer GC/AFS Kopplung zur Bestimmung flüchtiger Quecksilberspezies in der Elbe. Magisterarbeit Universität Lüneburg, Lüneburg

Bargagli R, Barghigiani C, Siegel BZ, Siegel SM (1989) Accumulation of mercury and other metals by the lichen, *Parmelia sulcata*, at an Italian mine site and a volcanic area. Water Air Soil Pollut 45:315–327

Barkay T, Liebert C, Gillman M (1989) Environmental significance of the potential for mer (Tn21)-mediated reduction of Hg(II) to Hg(0) in natural waters. Appl Environ Microbiol 55:1196–1202

Barnett MO, Harris LA, Turner, RR, Henson TJ, Melton RE, Stevenson RJ (1995) Characterization of mercury species in contaminated floodplain soils. Water Air Soil Pollut 80:1105–1108

Bergström J (1983) Mercury retention in electrostatic filters and in flue gas desulfurization. KHM Rep 89, Swedish Hydropower Ltd, S-162 87 Vällingby, Sweden

Bloom NS, Watras, CJ (1988) Observations of methylmercury in precipitation. Sci Total Environ 87/88:199–207

Bloom NS, Prestbo EM, Tokos JS, Von der Geest E, Kuhn ES (1996a) Distribution and origins of mercury species in the Pacific Northwest atmosphere, In: Ebinghaus R, Petersen G, Tümpling Uv (eds) 4th Int Conf on Mercury as a global pollutant, Book of Abstracts, Hamburg, Aug 4–8, 1996, p 90 (GKSS Forschungszentrum Geesthacht GmbH, Max-Planck-Str, D-21502 Geesthacht, Germany)

Bloom NS, Prestbo EM, Von der Geest, E (1996b) Determination of atmospheric gaseous Hg(II) at the pg/m^3 level by collection onto cation exchange membranes, followed by dual amalgamation/cold vapor atomic fluorescence spectrometry. In: Ebinghaus, R, Petersen, G, Tümpling Uv (eds) 4th Int Conf on Mercury as a global pollutant, Book of Abstracts, Hamburg, Aug 4–8, 1996, p 190 (GKSS Forschungszentrum Geesthacht GmbH, Max-Planck-Str, D-21502 Geesthacht, Germany)

Branches FJP, Erickson TB, Aks SE, Hryhorczuk DO (1993) The price of gold: mercury exposure in the Amazon rain forest. Clin Toxicol 31:295–306

Brosset C (1983) Mercury in flue gases from coal combustion. IVL Rep EM 999, 1983-02-15, IVL PO Box 47086, S-402 58 Göteborg, Sweden

Brosset C, Lord E (1991) Mercury in precipitation and ambient air – a new Scenario. Water Air Soil Pollut. 56:493–506

Camargo JA (1993) which source of mercury pollution? Nature 365:302

Carpi A, Lindberg SE (1997) The sun-light mediated emission of elemental mercury from soil amended with municipal sewage sludge. Environ Sci Technol 31:2085–2091

Carpi A, Lindberg SE (1998) Application of a Teflon dynamic flux chamber for quantifying soil mercury flux: tests and results over background soil. Atmos Environ (in press)

Carpi A, Lindberg SE, Prestbo EM, Bloom NS (1997) Methyl mercury contamination and transport in municipal sewage sludge amended soil. J Environ Qual (in press)

Carpi A, Lindberg SE, Prestbo EM, Bloom NS (1998) Global and regional impacts of elemental and methyl mercury emitted by soils to the atmosphere. J Env Qual (in press)

Chu P, Porcella DB (1995) Mercury stack emissions from U.S. electric utility power plants. Water Air Soil Pollut 80:135–144

CIMELCO (1991) Proyecto Binacional Puyango-Tumbes, Consorcio Cimelco Construtores, Tumbes, Peru

Cleary D (1996) Mercury contamination in the developing world: problems and solutions. In: Ebinghaus R, Petersen G, Tümpling Uv (eds) 4th Int Conf on Mercury as a global pollutant, Book of Abstracts, Hamburg Aug 4–8, 1996, p 3 (GKSS Forschungszentrum Geesthacht GmbH, Max-Planck-Str, D-21502 Geesthacht, Germany)

Coquery M (1995) Mercury speciation in surface waters of the North Sea. Neth J Sea Res 34:245–259

Cossa D, Gobeil C (1996) Speciation and mass balance of mercury in the lower St. Lawrence estuary and Saguenay Fjord (Canada). In: Ebinghaus R, Petersen, G, Tümpling Uv (eds) 4th Int Conf on Mercury as a global pollutant, Book of Abstracts, Hamburg, Aug 4–8, 1996, p 458 (GKSS Forschungszentrum Geesthacht GmbH, Max-Planck-Str, D-21502 Geesthacht, Germany)

Cossa D, Coquery M, Gobeil C, Martin JM (1996) Mercury fluxes at the ocean margins. In: Baeyens W, Ebinghaus R, Vasiliev O (eds) Global and regional mercury cycles: sources, fluxes and mass balances. NATO-ASI Series, vol 21. Kluwer, Dordrecht, The Netherlands, pp 229–247

Cramer SW (1990) Problems facing the Philippines. Inter Mining, July 1990, 29–31

Deslauriers M, Schroeder WH (1995) Personal communivation of draft data. Environment Canada, Hull, Quebec, Canada and atmospheric environment service. Downsview, Ontario, Canada

Dixon RK, Wisniewski J (1995) Global forest systems: an uncertain response to atmospheric pollutants and global climatic change. In Grennfelt P, Rodhe H, Thornelof E, Wisniewski J (eds) Proc 5th Int Conf on Acidic deposition. Kluwer, Dordrecht pp 101–110

Dyvik F (1995) Mercury as trace contaminant in non ferrous metal production. Occurrence, distribution, waste and gas treatment and mercury free sulphuric acid production by the use of the Boliden/Norzink process for gas. In: Proc Academy of Certified Hazardous Materials Managers, Conf and Annu Meet, Aug 1–4, Rochester, New York, 15 pp

Ebinghaus R, Krüger O (1996) Emission and local deposition estimates of atmospheric mercury in north western and central Europe. In: Baeyens W, Ebinghaus R, Vasiliev O (eds) Global and

regional mercury cycles: sources, fluxes and mass Balances. NATO-ASI Series 2. Environment vol. 21. Kluwer, Dordrecht, The Netherlands, pp 135–159
Ebinghaus R, Wilken RD (1993) Transformations of mercury species in the presence of Elbe river bacteria. Appl Organomet Chem 7:127–135
Ebinghaus R, Kock HH, Jennings SG, McCartin P, Orren MJ (1995) Measurements of atmospheric mercury concentrations in northwestern and central Europe – comparison of experimental data and model results. Atmos Environ 29:3333–3344
Ebinghaus R, Kock HH, Jennings SG, Spain TG (1996) Continuous measurements of total gaseous mercury at Mace Head, Ireland. In: Ebinghaus R, Petersen G, Tümpling eds. 4th Int Conf on Mercury as a global pollutant, Book of Abstracts, Hamburg, Aug 4–8, 1996, p 464 (GKSS Forschungszentrum Geesthacht GmbH, Max-Planck-Str, D-21502 Geesthacht, Germany)
Ebinghaus R, Jennings SG, Schroeder WH, Berg T, Donaghy T, Ferrara R, Guentzel J, Kenny C, Kock HH, Kvietkus K, Landing W, Mazzolai B, Munthe J, Prestbo EM, Schneeberger D, Slemr F, Sommar J, Urba A, Wallschläger D, Xiao, Z (1998) International field intercomparison measurements of atmospheric mercury species at Mace Head, Ireland. Atmos Environ (in press)
Edner H, Faris GW, Sunesson A et al. (1991) Lidar search for atmospheric atomic mercury in Icelandic geothermal fields. J Geophys Res 96D:2977–2986
Expert Panel on Mercury Atmospheric Processes (1994) Mercury atmospheric processes: a synthesis report. Worksh Proc, EPRI/TR-104214, Tampa, Florida, Sep 1994
Fairbridge RW (1966) The encyclopedia of oceanography. Reinhold Publishing New York
Ferrara R, Maserti B, Petrosino A, Bargagli R (1988) Mercury levels in rain and air and the subsequent wash-out mechanism in a central Italian region. Atmos. Environ 20:125–128
Ferrara R, Maserti BE, Breder R (1991) Mercury in abiotic and biotic compartments of an area affected by a geochemical anomaly (Mt. Amiata, Italy) Water Air Soil Pollut 56:219–233
Ferrara R, Maserti BE, Deliso A et al. (1994) Atmospheric mercury emission at Solfatara volcano (Pozzuoli, Phlegraean Fields – Italy). Chemosphere 29:1421–1428
Ferreira RH, Appel L (1991) Fontes e Usos de Mercurio no Brasil. CETEM/ CNPQ, Rio de Janeiro, 27 pp
Finlayson-Pitts BJ, Pitts JN (1986) Atmospheric chemistry: fundamentals and experimental techniques. John Wiley, New York
Fitzgerald WF, (1986) Cycling of mercury between the atmosphere and oceans. In: The role of air-sea exchange in geochemical cycling. NATO, ASI, Reidel, Dordrecht, The Netherlands, pp 363–408
Fitzgerald WF (1993) Global biogeochemical cycling of mercury. In: Allan RJ, Nriagu JO (eds) Proc 9th Int Conf Heavy Metal Environ 1:320–323, Toronto, Sept 12–17 1993, CEP Consultants, Edinburgh
Fitzgerald WF, (1995) Is mercury increasing in the atmosphere? The need for an atmospheric mercury network (AMNET). Water Air Soil Pollut 80:245–254
Fitzgerald WF (1996) Mercury emissions from volcanoes. In: Ebinghaus R, Petersen G, Tümpling Uv (eds) 4th Int Conf on Mercury as a global pollutant, Book of Abstracts, Hamburg, Aug 4–8, 1996, p 87 (GKSS Forschungszentrum Geesthacht GmbH, Max-Planck-Str, D-21502 Geesthacht, Germany)
Fitzgerald WF, Mason RP (1996) The global mercury cycle: oceanic and anthropogenic aspects, In: Baeyens W, Ebinghaus R, Vasiliev O (eds) global and regional mercury cycles: sources, fluxes and mass balances. NATO-ASI Series 2. Environment vol 21. Kluwer, Dordrecht, The Netherlands, pp 85–108
Fitzgerald WF, Mason RP, Vandal GM (1991) Atmospheric cycling and air-water exchange of mercury over mid-continental lacustrine regions, Water Air Soil Pollut 56: 745–767
Fitzgerald WF, Mason RP, Vandal GM, Dulac F (1994) Air-water cycling of mercury in lakes. In: Watras CJ, Huckabee JW (eds) Mercury pollution: integration and synthesis. Lewis, Boca Raton, pp 203–220
Galbreath KC; Zygarlicke CJ (1996) Mercury speciation in coal combustion and gasification. Environ Sci Technol 30:2421–2426
Gustin MS, Lindberg SE (1997) Natural sources of mercury emissions to air. Proc EPRI Conf on Managing hazardous air pollutants. Washington, DC, Nov 1997 (in press)
Gustin MS, Taylor GE, Leonard TL, Keislar RE (1996) Atmospheric mercury concentrations associated with geologically and anthropogenically enriched sites in Central Western Nevada. Environ Sci Technol 30:2572–2579
Hall B, Schager P, Lindqvist O (1991) Chemical reactions of mercury in combustion flue gases. Water Air Soil Pollut 56:3–14

Hanson PJ, Lindberg SE, Tabberer TA, Owens JG,. Kim KH (1995) Foliar exchange of mercury vapor: evidence for a compensation point. Water Air Soil Pollut 80:373–382

Heisler S, Pai P, Joshi A (1995) Current status of mercury emission rate estimates for assessment of source receptor relationships. May 8, 1995 draft report to EPRI. ENSR Consulting and Engineering, Camarillo, California

Helwig A, Neske P (1990) Zur komplexen Erfassung von Quecksilber in der Luft. I. Aktuelle Aufgaben der Luftreinhaltung. VDI Berichte 838, VDI Verlag Düsseldorf, pp 457–466

Hudson RJM, Gherini SA, Fitzgerald WF, Porcella DB (1995) Atmospheric influences on the global mercury cycle: a model based analysis. Water Air Soil pollut 80:265–272

Hultberg H, Munthe J, Iverfeldt A (1995) Cycling of methylmercury and mercury – responses in the forest roof catchment to three years of decreased atmospheric deposition. Water Air Soil Pollut 80:415–424

Ikingura J (1994) Mercury contamination from gold-mining in Tanzania. In: Evaluation of the role and distribution of mercury on ecosystems with special emphasis on tropical regions. SCOPE, Rio de Janeiro (unpublished)

Ikingura JR, Mutakyahwa (1995) Sources of mercury contamination and exposure in Tanzania. Proc Int Conf on Mining and environment in eastern and southern Africa, 23–27 Oct 1995, Mwanza, Tanzania

INPE (1992) Deforestation in the Amazon. Inst Nac Pesq Espaciais, SJ Campos

Iverfeldt Å (1991) Mercury in forest canopy throughfall water and its relation to atmospheric deposition. Water Air Soil Pollut 56:553–564

Iverfeldt Å (1995) Mercury chemistry in air and its atmospheric residence time. Proc 1995 Canadian Mercury network workshop, York University, Toronto, Ontario, Canada

Iverfeldt Å, Lindberg SL (1996) Working group report on terrestrial mercury cycling. In: Baeyens W, Ebinghaus R, Vasiliev O (eds) Global and regional mercury cycles: sources, fluxes and mass balances. NATO-ASI Series 2. Environment vol 21. Kluwer, Dordrecht, The Netherlands, pp 543–546

Iverfeldt Å, Munthe J, Brosset C, Pacyna J (1995) Long-term changes in concentration and deposition of atmospheric mercury over Scandinavia. Water Air Soil Pollut 80:227–233

Jacobs LA, Klein SM, Henry EA (1995) Mercury cycling in the water column of a seasonally anoxic urba lake (Onondaga Lake, NY). Water Air Soil Pollut 80:553–562

Jainsinki SM (1995) The materials flow of mercury in the United States. Resour Conserv Recycl 15:145–179

Junge CE (1977) Basic considerations about trace constituents in the atmosphere as related to the fate of global pollution. In: Fate of pollutants in the air and water environments, part 1, vol 8. Wiley, New York, pp 7–23

Keeler G, Glinsorn G, Pirrone N (1995) Particulate mercury in the atmosphere: its significance, transport, transformation and sources. Water Air Soil Pollut 80:159–168

Kim JP, Fitzgerald WF (1986) Sea-air partitioning of mercury in the equatorial Pacific ocean. Science 231:1131–1133

Kim KH, Lindberg SE (1995) Design and initial tests of a dynamic enclosure chamber for measurements of vapor-phase mercury fluxes over soils. Water Air Soil Pollut 80:1059–1068

Kim KH, Lindberg SE, Meyers TP (1995) Micrometeorological measurements of mercury vapor fluxes over background forest soils in eastern Tennessee. Atmos Environ 29:267–282

Klusman RW, Matoske CP (1983) Adsorption of mercury by soils from oil shale development areas in the Piceance Basin of northwestern Colorado. Environ Sci Technol 17:251–256

Klusman RW, Webster JD (1981) Meteorological noise in crustal gas emission and relevance to geochemical exploration. J Geochem Explor. 15:63–76

Kovalevsky AL (1986) Mercury-biogeochemical exploration for mineral deposits. Biogeochemistry 2:211–220

Kuivala A (1984) Paper presented at the 32nd Meeting of GDMB Zinkfachausschuss in Kokkola, Finland

Lacerda DB (1997) Evolution of mercury contamination in Brazil. Water Air Soil Pollut 97:249–255

Lacerda LD, Prieto RG, Marins RV, Azevedo SLN, Pereira MC (1995) Anthropogenic mercury emissions to the atmosphere in Brazil. In: Proc Int conf on Heavy metals in the environment Hamburg, vol I, pp 379–382

Lantzy RJ, Mackenzie FT (1979) Atmospheric trace metals: global cycles and assessment of man's impact. Geochim Cosmochim Acta 43:511–525

Leavander T (1987) Emissions of mercury to air in Sweden during the period 1860–1987. Swedish Environmental Protection Agency, S-171 85, Solna, Sweden
Lindberg SE (1996) Forests and the global biogeochemical cycle of mercury: the importance of understanding air/vegetation exchange processes. In: Baeyens W, Ebinghaus R, Vasiliev O (eds) Global and regional mercury cycles: sources, fluxes and mass balances. NATO-ASI Series, vol. 21. Kluwer, Dordrecht, The Netherlands, pp 359–380
Lindberg SE, Stratton WJ (1998) Atmospheric mercury speciation: concentrations and behavior of reactive gaseous mercury in ambient air. Environ Sci Technol 32:49–57
Lindberg SE, Turner RR (1977) Mercury emission from chlorine-production and solid waste deposits. Nature (Lond) 268:133–136
Lindberg SE, Andren AW, Harriss RC (1975) Geochemistry of mercury in the estuarine environment. In: Cronin LE (ed) Estuarine research: chemistry, biology, and the estuarine system, vol. 1. Academic Press, New York, pp 64–107
Lindberg SE, Jackson DR, Huckabee JW, Janzen SA, Levin MJ, Lund JR (1979) Atmospheric emission and plant uptake of mercury from agricultural soils near the Almadén mercury mine. J Environ Qual 8:572–578
Lindberg SE, Meyers TP, Taylor GE, Turner RR, Schroeder WH (1992) Atmospheric/surface exchange of mercury in a forest: results of modeling and gradient approaches. J Geophys Res 97:2519–2528
Lindberg SE, Kim K-H, Meyers TP, Owens JG (1995) A micrometeorological gradient approach for quantifying air/surface exchange of mercury vapor: tests over contaminated soils. Environ Sci Technol. 29:126–135
Lindberg SE, Meyers TP, Munthe J (1996) Evasion of mercury vapor from the surface of a recently limed acid forest lake in Sweden. Water Air Soil Pollut 85:2265–2270
Lindberg SE, Meyers TP, Miles C (1997) Everglades mercury air/surface exchange study (E-MASE) application of field methods and models to quantify mercury emission from wetlands at the everglades nutrient removal project (ENR). Ist Annu Rep to South Florida Water Management District, West Palm Beach
Lindberg SE, Hanson PJ, Meyers TP, Kim K-Y (1998) Micrometeorological studies of air/surface exchange of mercury over forest vegetation and a reassessment of continental biogenic mercury emissions. Atmos Environ (in press)
Lindqvist O (1986) Fluxes of mercury in the Swedish environment: contributions from waste incineration Waste Manage Res 4:35–44
Lindqvist O, Jernelöv A, Johansson K, Rodhe H (1984) Mercury in the Swedish environment, global and local sources. SNV PM 1816, Swedish Environmental Protection Agency, S-171 85 Solna, Sweden
Lindqvist O, Johansson K, Aastrup M, Andersson A, Bringmark L, Hovsenius G, Hakanson L, Iverfeldt A, Meili M, Timm B (1991) Mercury in the Swedish environment – recent research on causes, consequences and corrective methods. Water Air Soil Pollut 55:261
Mackay D, Wania, F, Schroeder WH (1995) Prospects for modeling the behavior and fate of mercury, globally and in aquatic systems. Water, Air, Soil Pollut 80:941–950
Mason RP, Fitzgerald WF (1991) Mercury speciation in open ocean waters. Water Air Soil Pollut 56:779–789
Mason RP, Fitzgerald WF (1993) The distribution and biogeochemical cycling of mercury in the equatorial Pacific Ocean. Deep-Sea Res 40:1897–1924
Mason RP, Fitzgerald WF (1996) Sources, sinks and biogeochemical cycling of mercury in the ocean. In: Baeyens W, Ebinghaus R, Vasiliev O (eds) Global and regional mercury cycles: sources, fluxes and mass balances. NATO-ASI-Series 2. Environment vol 21. Kluwer, Dordrecht, The Netherlands pp 249–272
Mason RP, Fitzgerald WF, Morel FMM (1994a) The biogeochemical cycling of mercury: anthropogenic influences. Geochim Cosmochim Acta 58:3191–3198
Mason RP, O'Donnell J, Fitzgerald WF (1994b) Elemental mercury cycling within the mixed layer of the equatorial Pacific Ocean. In: Watras CJ, Huckabee JW (eds) Mercury pollution: integration and synthesis. Lewis, Boca Raton, pp 83–97
Mason RP, Morel FMM, Hemond HF (1995a) The role of microorganisms in elemental mercury formation in natural waters. Water Air Soil Pollut 80:775–787
Mason R.P, Rolfhus KR, Fitzgerald WF (1995b) Methylated and elemental mercury cycling in surface and deep ocean waters of the North Atlantic. Water Air Soil Pollut 80:665-677

McNerney J, Buseck PR (1973) Geochemical exploration using mercury vapor. Econ Geol 68:1313–1320
Meji R (1991) The fate of mercury in coal-fired power plants and the influence of wet flue-gas desulphurization. Water Air Soil Pollut 56:21–33
Midwest Research Institute (MRI) (1993) Locating and estimating air emissions from sources of mercury and mercury compounds. EPA-454/R-93-023, USEPA, Sept, 1993, Chicago
MME (1992) Base Energetica Brasileria, Ministerio das Minas e Energia, Brasilia
Ming G (1994) In Evaluation on the role and distribution of mercury in ecosystems with special emphasis on tropical regions. SCOPE, Rio de Janeiro (unpublished)
Mukherjee AB (1989) The release of Cd and Hg into the Finnish environment. Ministry of environment. Helsinki, Finland, Rep No 64, 93 pp
Mukherjee AB, Innanen S, Verta M (1995) An update of the mercury inventory and atmospheric mercury fluxes to and from Finland. Water Air Soil Pollut 80:255–264
Munthe J (1992) The aqueous oxidation of elemental mercury by ozone. Atmos. Environ, 26A:1461–1468
Munthe J (1993) Mercury in the atmosphere – emissions, transformations, deposition and effects. Report prepared for the worksh on Emissions and modelling of persistent organic pollutants and heavy metals, Durham, North Carolina, 6–7 May 1993
Munthe J, Hultberg H, Iverfeldt Å (1995) Mechanisms of deposition of methylmercury and mercury to conferous forests. Water Air Soil Pollut 80:363–371
Nico LG, Taphorn DC (1994) Mercury in fish from gold-mining regions in upper Cuyuni River system, Venezuela, Fresenius Environ Bull 3:287–292
Nriagu JO (1989) A global assessment of natural sources of atmospheric trace metals. Nature 338: 47–49
Nriagu JO, Pacyna JM (1988) Quantitative assessment of worldwide contamination of air, water and soils by trace metals. Nature 33:134–139
Nriagu J (1994) Mechanistic steps in the photoreduction of mercury in natural waters. Sci. Total Environ 154:1–8
Obolensky AA (1996) Natural mercury sources in the environment: contribution of Siberia In: Baeyens W, Ebinghaus R, Vasiliev O (eds) Global and regional mercury cycles: sources, fluxes and mass balances. NATO-ASI-Series 2. Environment vol 21. Kluwer, Dordrecht, The Netherlands, pp 453–462
Pacyna JM (1993) Atmospheric deposition of heavy metals to the Baltic Sea. In: Proc Int conf on heavy metals in the environment Toronto, vol I, pp 93–96
Pacyna JM (1994) Atmospheric emissions of anthropogenic mercury in Europe in 1990. NILU technical report prepared for the Dutch TNO, Norwegian Institute for air research, Kjeller, Norway
Pacyna JM (1996) Emission inventories of atmospheric mercury from anthropogenic sources. In: Baeyens W, Ebinghaus R, Vasiliev O (eds) Global and regional mercury cycles: sources, fluxes and mass balances. NATO-ASI Series 2. Environment vol. 21. Kluwer, Dordrecht, The Netherlands, pp 161–178
Pai P, Karamanchandani P, Seigneur Ch (1997) Simulation of the regional atmospheric transport and fate of mercury using a comprehensive Eulerian model. Atmos Environ 31:2717–2732
Patrick WH (1994) From wastelands to wetlands. J Environ Qual 23:892–896
Petersen G, Iverfeldt Å, Munthe J (1995) Atmospheric mercury species over central and northern Europe. Model calculations and comparison with observations from the Nordic air and precipitation network for 1987 and 1988. Atmos Environ 29 (1): 47–67
Petersen G, Munthe J, Bloxam R (1996) Numerical modeling of regional transport, chemical transformations and deposition fluxes of airborne mercury species. In: Baeyens W, Ebinghaus R, and Vasiliev O (eds) Global and regional mercury cycles: sources, fluxes and mass balances. NATO-ASI Series, 2. Environment vol. 21. Kluwer, Dordrecht, The Netherlands, pp 191–218
Pfeiffer WC, Lacerda LD (1988) Mercury inputs into the Amazon Region, Brazil. Environ Technol Lett 9:325–330
Pirrone N, Keeler GJ, Nriagu JO (1996) Regional differences in worldwide emissions of mercury to the atmosphere. Atmos Environ 30 (17):2981–2987
Poissant L, Casimir A (1996) Water-air and soil-air exchange rate of total gaseous mercury measured in southern Québec (Canada). In: Ebinghaus R, Petersen G, Tümpling Uv (eds) 4th Int Conf on Mercury as a global pollutant, Book of Abstracts, Hamburg Aug, 4–8, 1996, p 136 (GKSS Forschungszentrum Geesthacht GmbH, Max-Planck-Str, D-21502 Geesthacht, Germany)

Porcella DB (1994) Mercury in the environment: biogeochemistry. In: Watras CJ, Huckabee JW (eds) Mercury pollution: integration and synthesis. Lewis, Boca Raton, pp 3–19

Porcella DB (1995) The atmospheric mercury pool. Proc 1995 Canadian Mercury network workshop, York University, Toronto, Ontario, Canada

Porcella DB, Chu P, Allan MA (1996) Inventory of North American mercury emissions to the atmosphere: relationship to the global mercury cycle. In: Baeyens W, Ebinghaus R, Vasiliev O (eds) Global and regional mercury cycles: sources, fluxes and mass balances. NATO-ASI Series 2. Environment vol 21. Kluwer, Dordrecht, The Netherlands, pp 179–190

Prestbo EM, Bloom NS (1996) A global view of the sources and sinks for atmospheric organic mercury, In: Ebinghaus R, Petersen G, Tümpling Uv (eds) 4th Int Conf on Mercury as a global pollutant, Book of Abstracts, Hamburg, Aug 4–8, 1996, p 31 (GKSS Forschungszentrum Geesthacht GmbH, Max-Planck-Str, D-21502 Geesthacht, Germany)

Prestbo EM, Bloom NS, Hall B. (1995) Mercury speciation in coal combustion flue gas: methodology, intercomparison, artifacts and atmospheric implications. Water Air Soil Pollut 80:145–158

Priester M (1993) Tecnologia de meio ambiente para mini-mineraçao e sua difusao. A experienca de um projeto GTZ para evitar emissoes de mercurio na mineraçao primária de ouro no sul da Colombia. In: Mathis A, Rehaag, R, (orgs) Consequencias da Garimpagem no Ambito Social e Ambiental da Amazonia. FASE/Buntstift e.V./Katalise, Belém, pp 102–112

Prospero JM, Charlson RJ, Mohnen V, Jaenicke R, Delany AC, Moyers J, Zoller W, Rahn K (1983) The atmospheric aerosol system: an overview. Rev Geophys Space Phys 21:1607–1629

Ramel C (1996) SCOPE project Evaluation of the role and distribution of mercury on ecosystems with special emphasis on tropical regions. In: Baeyens W, Ebinghaus R, Vasiliev O (eds) Global and regional mercury cycles: sources, fluxes and mass balances. NATO-ASI Series, 2. Environment vol. 21. Kluwer, Dordrecht, The Netherlands, pp 505–513

Rantalahti R (1996) Outokumpu Oy, Harjavalta (personal communication.)

Rasmussen PE (1994) Current methods of estimating atmospheric mercury fluxes in remote areas. Environ Sci Technol 28:2233-2241

Rea AW, Keeler GJ, Scherbatskoy T (1996) The deposition of mercury in throughfall and litterfall in the Lake Champlain Watershed: a short-term study. Atmos Environ 30:3257–3263

Reich S (1995) Untersuchungen zur Speziesanalytik von Quecksilber in Meerwasser. Diplomarbeit, Fachhochschule Hamburg, Hamburg

Rogers RD (1979) Volatility of mercury from soils amended with various mercury compunds. Soil Sci Soc Am J 43:289–291

Schmolke SR, Wängberg I, Schager P, Kock HH, Otten S, Ebinghaus R, Iverfeldt Å (1997) Estimates of the air-sea exchange of mercury derived from the Lagrangian experiment. Summer 1997. Ist Annu Meet EU MAST III Project BASYS, Sept 29–Oct 1, 1997, Warnemünde, Germany

Schroeder WH (1996) Estimation of atmospheric input and evasion fluxes of mercury to and from the Great Lakes, using Lake Ontario as an example. In: Baeyens W, Ebinghaus R, Vasiliev O (eds) Global and regional mercury cycles: sources, fluxes and mass balances. NATO-ASI Series 2. Environment vol 21. Kluwer, Dordrecht, The Netherlands, pp 109–122

Schroeder WH, Lane DA (1988) The fate of airborne toxic pollutants. Environ Sci Technol 22:240–246

Schroeder WH, Schneeberger D (1996) High-temporal-resolution measurements of total gaseous mercury in air at Alert, Northwest Territories, Canada. In: Ebinghaus R, Petersen G, Tümpling Uv (eds) 4th Int Conf on Mercury as a global pollutant, Book of Abstracts, Hamburg, Aug 4–8, 1996, p 128 (GKSS Forschungszentrum Geesthacht GmbH, Max-Planck-Str, D-21502 Geesthacht, Germany

Schroeder WH, Munthe J, Lindqvist O (1989) Cycling of mercury between water, air and soil components of the environment. Water Air Soil Pollut 48:337–347

Schroeder WH, Lindqvist O, Munthe J, Xiao ZF (1992) Volatilization of mercury from lake surfaces. Sci Total Environ 125:47–66

Slemr F (1996) Trends in atmospheric mercury concentrations over the Atlantic Ocean and the Wank summi, and the resulting constraints on the budget of atmospheric mercury. In: Baeyens W, Ebinghaus R, Vasiliev O (eds) Global and regional mercury cycles: sources, fluxes and mass balances. NATO-ASI Series 2. Environment vol. 21. Kluwer, Dordrecht, The Netherlands, pp 33–84

Slemr F, Langer E (1992) Increase in global atmospheric concentrations of mercury inferred from measurements over the Atlantic Ocean. Nature 355:434–437

Slemr F, Schuster G, Seiler W (1985) Distribution, speciation and budget of atmospheric mercury. J. Atmos. Chem 3:407–434

Slemr F, Robertson P, Brunke E (1996) Monitoring of atmospheric mercury at Cape Point and Wank. In: Ebinghaus R, Petersen G, Tümpling Uv (eds) 4th Int Conf on Mercury as a global pollutant, Book of Abstracts, Hamburg, August, 4–8, 1996, p 127 (GKSS Forschungszentrum Geesthacht GmbH, Max-Planck-Str, D-21502 Geesthacht, Germany)

Sloss L (1995) Mercury emission and effects – the role of coal. IEA coal research, London, Report No IEAPER/ 19, pp 1–39

Southworth GR, Turner RR, Peterson MJ, Bogle MA (1995) Form of mercury in stream fish exposed to high concentrations of dissolved inorganic mercury. Chemosphere 30(4):779–787

Strachan WMJ, Eisenreich SJ (1988) Mass balancing of toxic chemicals in the Great Lakes: the role of atmospheric deposition. International Joint Commission, Windsor, Ontario, Canada

Stratton WJ, Lindberg SE (1995) Use of a refluxing mist chamber for measurement of gas-phase water-soluble mercury (II) species in the atmosphere. Water Air Soil Pollut 80:1269–1278

Sukhenko SA, Vasiliev OF (1996) A regional mercury budget for Siberia and the role of the region in global cycling of the metal. In: Baeyens W, Ebinghaus R, Vasiliev O (eds) Global and regional mercury cycles: sources, fluxes and mass balances. NATO-ASI Series 2. Environment vol 21. Kluwer, Dordrecht, The Netherlands, pp 123–134

Swain EB, Engstrom DR, Brigham ME, Henning TA, Brezonik PL (1992) Increasing rates of atmospheric mercury deposition in midcontinental North America. Science 257:784–787

Swaine DJ (1990) Trace elements in coal. Butterworth, London 276 pp

Thibodeaux LJ (1979) Chemodynamics: environmental movements of chemicals in air, water and soil. J Wiley, New York, pp 45–68

Torres EB (1992) Environmental and health survelliance of mercury in small scale gold processing industries in the Philippines. In: Proc Int Symp Epidemiol Stud Environ Pollut and health effects of methylmercury. Natl Inst for Minamata Disease, Kumamoto, pp 56–65

Turner RR, Lindberg SE (1978) Behavior and transport of mercury in a river-reservoir system downstream of an inactive chloralkali plant. Environ Sci Technol 12:918–923

US EPA (1994) Mercury study report to Congress (draft: 12-12-94).OAQPS/ORD vol 2–7, Research Triangle Park, North Carolina

Vandal GM, Mason RP, Fitzgerald WF (1991) Cycling of mercury in temperate lakes. Water Air Soil Pollut 56:791–803

Vandal GM, Fitzgerald WF, Rolfhus KR, Lamborg CH (1995) Modeling the elemental mercury cycle in Pallette Lake, Wisconsin, USA. Water Air Soil Pollut 80:529–538

Vandal G, Fitzgerald WF, Rolfhus K, Russ C, Lamborg C (1996) Sources and cycling of mercury and methylmercury in Long Island Sound: preliminary results. In: Ebinghaus R, Petersen G, Tümpling Uv (eds) 4th Int Conf on Mercury as a global pollutant, Book of Abstracts, Hamburg, Aug 4–8, 1996, p 458 (GKSS Forschungszentrum Geesthacht GmbH, Max-Planck-Str, D-21502 Geesthacht, Germany)

Varekamp JC, Buseck PR (1983) Hg anomalies in soils: a geochemical exploration method for geothermal areas. Geothermics 12:29–47

Varekamp JC, Buseck PR (1986) Global mercury flux from volcanic and geothermal sources. Appl Geochem 1:65–73

Vogg H, Braun M, Metzger M, Schneider J (1987) The specific role of cadmium and mercury in municipal solid waste incineration. Waste Manage Res 4:65–74

Wallschläger D (1996) Speziesanalytische Untersuchungen zur Abschätzung des Remobilisierungs-potentials von Quecksilber aus kontaminierten Elbauen. PhD Thesis, University of Bremen and GKSS Research Center, Geesthacht, Germany

Wallschläger D, Wilken RD (1997) The influence of floodplains on mercury availability, In: Dwyer JF, Doane TR, Hinman ML (eds) Environmental toxicology and risk assessment: modeling and risk assessment, vol 6. ASTM STP 1317, American Society for Testing and Materials, West Conshohocken, pp 179–196

Wallschläger D, Hintelmann H, Evans RD, Wilken RD (1995) Volatilization of dimethylmercury and elemental mercury from river Elbe floodplain soils. Water Air Soil Pollut 80:1325–1329

Wallschläger D, Desai MVM, Spengler M, Wilken RD (1997a) Mercury speciation in floodplains and sediments along a contaminated river transect. J Environ. Qual (in press)

Wallschläger D, Kock HH, Schroeder WH, Lindberg SE, Ebinghaus R, Wilken RD, (1997b) Mechanism and significance of mercury volatilization from contaminated floodplains of the German river Elbe. Atmos Environ (submitted)

Wallschläger D, Kock HH, Schroeder WH, Lindberg SE, Ebinghaus R, Wilken RD, (1997c) Intercomparison of different experimental approaches to quantify gaseous mercury emissions from contaminated floodplain soils to the atmosphere. Environ Sci Technol (submitted)

Watras CJ, Bloom NS, Hudson RJM, et al. (1994) Sources and fates of mercury and methylmercury in Wisconsin lakes. In: Watras CJ, Huckabee JW (eds) Mercury pollution: integration and synthesis. Lewis, Boca Raton pp 153–177

Wheatley B, Wyzga R (eds) (1997) Mercury as a global pollutant: human health issues. Kluwer, Dordrecht, The Netherlands

Wilken RD, Hintelmann H (1991) Mercury and methylmercury in sediments and suspended particles from the river Elbe, North Germany. Water Air Soil Pollut 56: 427–437

Xiao Z, Munthe J, Schroeder WH, Lindqvist O (1991) Vertical fluxes of mercury over forest soil and lake surfaces in Sweden. Tellus 43B:267–279

Xiao Z, Sommar J, Wei S, Lindqvist O (1997) Sampling and determination of gase phase divalent in the air using a Kcl-coated denuder. Fresenius J Anal Chem 358:386–391

Yagolnitser MA, Sokolov VM, Rabtsev AD, Obolensky AA, Ozerova SA, Sukhenko S, Dvurechenskaya (1996) Industrial mercury sources in Siberia. In: Baeyens W, Ebinghaus R, Vasiliev O (eds) Global and regional mercury cycles: sources, fluxes and mass balances. NATO-ASI Series 2. Environment vol 21. Kluwer, Dordrecht, The Netherlands, pp 429–440

Yshuan YS (1994) In: Evaluation on the role and distribution of mercury in ecosystems with special emphasis on tropical regions. SCOPE, Rio de Janeiro (unpublished)

Zapata JQ (1994) Environmental impacts study of gold mining in Nueva Esperanza in the Madeira River Bolivian-Brazilian border. In: Environmental mercury pollution and its health effects in Amazon River Basin. Natl Inst Minamata Disease and Inst Biophysics of the Univ Federal do Rio de Janeiro, Rio de Janeiro, pp 23–24

Mercury Mines in Europe: Assessment of Emissions and Environmental Contamination

R. Ferrara

1 Introduction

Most of the mercury-containing mineralization follows a line encompassing the earth: the circum-Pacific belt and the Mediterranean Tethyean belt.

The Mediterranean basin is included in this belt; even if it occupies only 2–3% of the earth's surface, it makes up approximately 65% of the world's mercury resources (Fig. 1).

The three most important cinnabar deposits, which have provided a large percentage of the entire world production (800 000 tons), are situated in the European continent and are, in order of importance, Almadén (Spain – 280 000 tons), Idrija (Slovenia – 150 000 tons) and Mt. Amiata (Italy – 100 000 tons).

The gradual decline in mercury prices (about $600 per flask in 1965; $120 per flask in 1975) caused a crisis with a present very low production in Almadén and a total interruption of the mining activity in the Idrija and Mt. Amiata districts.

Generally, cinnabar rocks were transported from the different mine sites to the smelting plant, where they were crushed, dried, and roasted to 650–700 °C. The released mercury vapours were condensed in cooling towers and then flasked inside the complex. The roasted cinnabar, still containing variable low amounts of mercury, was piled around the refining complex, where it has formed high hills over the years.

In the old furnaces the mercury loss was about 20–40% of the production; following advances in technology, over the past 30–40 years, the losses were reduced to 2–3%.

The long period of mining and processing of the ore has left scarred landscapes even in the areas where the mining operations have been terminated.

The poor recovery of the past has contributed, through the smelter stacks and flues by gaseous and particulate matter, to the dispersion of the metal in the atmosphere, soil and waters.

In this chapter we examine the environmental situation of the three most important mines in Europe, (as appears from the available literature) with the intention of pointing out the differences between the areas where the activity is still present and those in which it ceased years ago.

A great contribution to the knowledge of the mercury problems in the Mediterranean basin has been made by the numerous researches conducted in

Environmental Science
Mercury Contaminated Sites (ed. by R. Ebinghaus et al.)

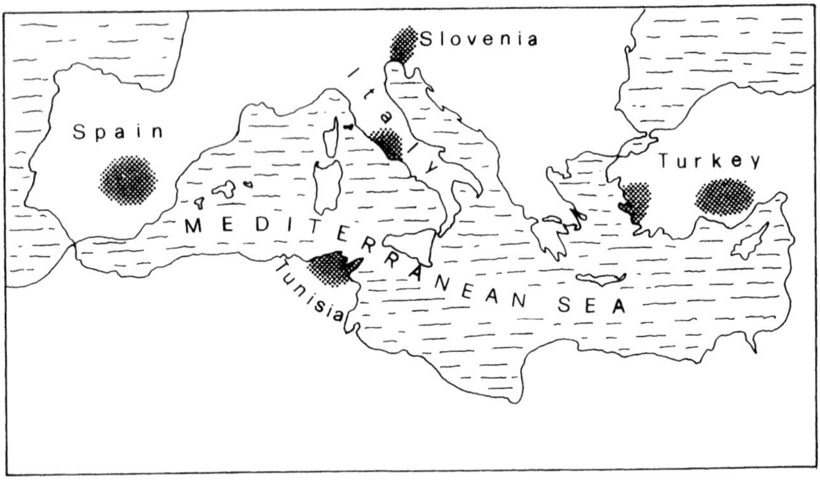

Fig. 1. Cinnabar deposits in the Mediterranean basin

the past 10 years (Bargagli et al. 1987; Siegel et al. 1987; Barghigiani et al. 1989; Ferrara et al. 1991; Palinkaš et al. 1995; Ferrara et al., 1998b in press). The majority of the works are based on studies that deal with the amount and extention of environmental contamination with the use of bioindicators such as lichens (Ferrara et al. 1988; Bargagli 1990), plants (Gaggi et al. 1991; Bacci et al. 1994; Hanson et al. 1995; Lodenius 1994) or animals (Gnamuš et al. 1995). Available data for individuating or characterizing the flux emission from the main mercury sources are very scarce in literature.

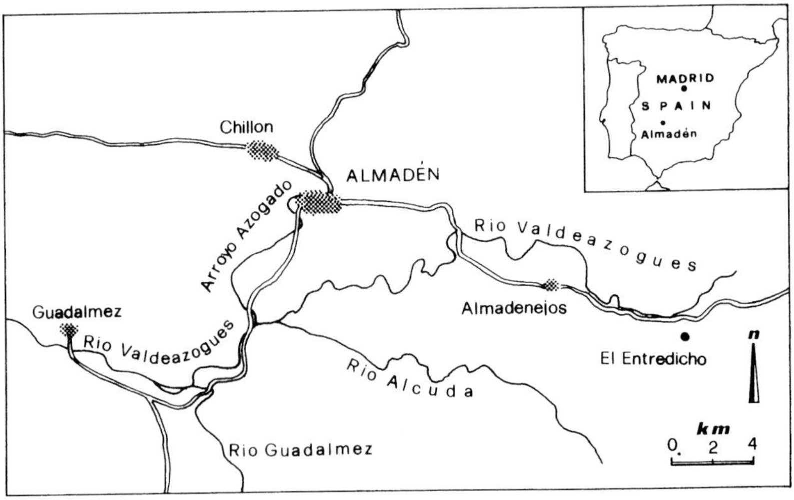

Fig. 2. The Almadén district (Spain)

Only in the past few years has remote sensing LIDAR (light detection and ranging) technique been applied to mercury detection (Edner et al. 1992, 1993, 1994; Ferrara et al. 1992). This has for the first time made possible a reliable evaluation of mercury emissions from both point sources and extended ones.

With this technique of laser spectroscopy it has been possible to determine not only the value of the emissions from the stacks of roasting furnaces but also those due to degassing processes from the roasted cinnabar banks. Of particular interest is the determination of the emission values obtained from mining structures that are still in activity or are in an abandoned state (Ferrara et al. 1998a, in press).

2 The Almadén District

2.1 Location

The Almadén district (Spain) is a world-scale exceptional case of mercury metallogeny. Approximately one third of all the mercury consumed by man during all his history has come from this district.

The mining area is located in the central part of the Iberian Peninsula, 300 km south of Madrid (Fig. 2). The most important village in this area is Almadén, with a current population of 9000.

The climate is semiarid and temperate, with an annual rainfall of 50 cm. Winds blow predominantly from the west and southwest in the shallow valley where the village is located.

The mercury deposits, described in detail by Ortega and Hernandez (1992), have been mined without interruption over the past 2000 years. The three most important mines are: Almadén, El Entredicho (an open pit) and Las Cuevas, with an overall average grade of 3.5% of mercury. The mercury mineralization consists mainly of cinnabar, but contains also native mercury.

Production data indicate that the total production up to 1991 was 7.5 million flasks, 7 million of which were produced solely from the Almadén mine. A maximum annual production of 82 000 flasks was reached in 1942; during the past 20 years, average production has been about 40 000 flasks per year.

The main mine, the ore-roasting furnaces, and the distillation plant are located close to the village of Almadén; the main galleries run beneath it, and the outlets of the ventilation systems are close to the houses. The roasted cinnabar banks (3 700 000 tons of roasted ore, containing a mean percentage of mercury of 0.15%) surround the mine complex and present an overall area of the order of 3×10^5 m^2.

While extensive literature is available on the structure and genesis of mercury deposits in the Almadén zone (Julivert et al. 1972; Saupé 1990), the only prominent studies on mercury cycling and transport in the terrestrial and aquatic systems are those performed by ORNL (Oak Ridge National Laboratory;

Hildebrand et al. 1980) from 1974 to 1977 and that reported by Ferrara et al. (1998a, in press) conducted from 1993 to 1994.

2.2 Main Sources of Atmospheric Mercury

In the remarkable report published by ORNL, the impact of the mining and refining activity on both terrestrial and aquatic environments is discussed. No data are, however reported on mercury concentrations in the air and on atmospheric fluxes from the main sources, probably because of the scarce analytical methodologies available some 20 years ago. Due to this lack of information, the field campaigns performed by the Institute of Biophysics (National Research Council, Italy) and the Lund Institute of Technology (Univ. of Lund Sweden) were mainly devoted to the determination of mercury emissions, using both flux chamber (Xiao et al. 1991) and LIDAR techniques developed at the University of Lund: this laser spectroscopy technique in fact allows a direct and total valuation of emissions from both spatially extended non-homogeneous sources and point sources.

A total mercury flux of about 800 g/h was determined above the village of Almadén (Ferrara et al. 1997) on September 24th 1993 when all the furnaces were in operation. The most important mercury vapour sources and an estimate of the mercury emitted from these individual sources present in the district are:

- stack of the ore-roasting furnaces (~600 g/h),
- area of the refining complex (~100 g/h),
- roasted cinnabar banks (~80 g/h),
- ventilation systems of the mine (~15 g/h),
- contaminated soil around the Almadén village (~5 g/h).

The total mercury flux from the Almadén district represents around 0.1% of the present global anthropogenic emission rate (Hudson et al. 1995).

2.3 Mercury in the Air over Almadén

In Fig. 3 the distribution of atmospheric mercury concentration over the Almadén village is reported together with the position of the sampling stations. As was predictable, the highest mercury concentrations (20 μg/m^3) were measured inside the refining and smelting plant of Almadén (St. 1). Particularly high values (5 μg/m^3) were found also close to the ventilation systems of the mine galleries (Sts. 16, 17, 18). The particular orography of the site and the direction of the prevailing winds during the summer and autumn months favour the occurrence of high levels above the village. In the main square (St. 26) values of the order of 1 μg/m^3 were often observed. The determinations performed at the

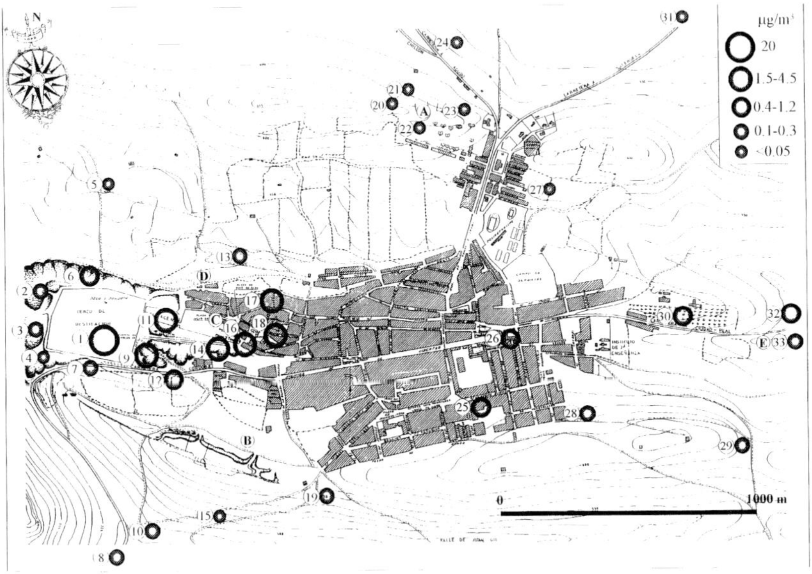

Fig. 3. Atmospheric mercury concentration (μg/m³) over the Almadén village (*St. 1* mine complex; *Sts. 2, 3, 4, 6, 7, 9, 14* roasted cinnabar deposits)

control station Alamillo, located 20 km southeast of Almadén in a nonmineralized area, yielded a value of 6 ng/m³, which is about two to three times the normal background values (Lindqvist and Rodhe 1985).

The high emissions from the roasting furnaces condition the mercury levels present in the air; when there is no emission from the stack (because of ordinary maintenance or inactivity) mercury levels in the Almadén village drop to one third of those observed when the furnaces were in operation.

2.4 Mercury in the Soil

Lindberg et al. (1979) reported that total mercury in the surface soil ranged from 0.5 to 260 μg/g, with highest concentrations occurring near the mine (<1 km). At distances >14 km, higher levels were found in soils along the direction of the prevailing winds from the mine/smelter area. According to these authors, the Almadén soil is more influenced by weathering of residual mercury ore deposits and more highly impacted by atmospheric deposition of mercury than the control soil.

A detailed description of the mercury deposit in the Almadén syncline that shows no sign of exhaustion is reported by Ortega and Hernandez (1992). A particularly interesting case of soil rich in mercury is the open pit El Entredicho, located 15 km east of Almadén, which during the past few years has contributed 75% of the company's total mercury production.

2.5 Mercury in the Sediment

In the years 1974-1975, the ORNL measured mercury concentration in the sediment >1000 μg/g in the rivers near the mine; these values decreased downstream from the rivers. The remarkable levels present in the sediments are a clear index of the intense activity of the mining complex of Almadén. In the study performed by the Institute of Biophysics, the value of mercury in the sediment at the confluence of the rivers Arroyo Azogado and Rio Valdeazogues was of the order of 650 μg/g. The sediment of the Rio Gualdamez, sampled at about 15 km from the mine complex, presented concentration values of 3–4 μg/g.

2.6 Mercury in the River Water

In Table 1 the mercury concentrations of surface water (mean annual value) measured during the years 1988–1993 by the laboratory of Minas de Almadén are reported.

The annual decrease in mercury concentration in the waters is attributable to both the productivity reduction and the installation of an abatement plant. The maximum value reported by the ORNL (1974–1977) is higher than 300 μg/l. The determination performed by the Institute of Biophysics in the years 1993–1994 yielded concentration values in the range 10–20 ng/l, with a maximum value of 80 ng/l at the confluence of the rivers Arroyo Azogado and Rio Valdeazogues.

2.7 Mercury in the Biota

In the ecological survey of the concentration and distribution of mercury in terrestrial and aquatic life performed by the ORNL, the mean concentration in plants ranged from >100 μg/g within 0.5 km of the mine to 1 μg/g 20 km distant from the mine.

Total mercury concentration in fish muscle tissue decreased with distance downstream from the mine. Levels ranged from 2.4 μg/g in *Barbo* to 0.3 μg/g (boga) at the control station. In the recent field campaign performed by the Institute of Biophysics on plants (Maserti et al. 1996) mercury concentrations

Table 1. Mercury concentration (μg/l) in the waters of the rivers Arroyo Azogado and Rio Valdeazogues

	1988	1989	1990	1991	1992	1993
Arroyo Azogado	0.114	0.093	0.008	0.016	0.001	<0.001
Rio Valdeazogues						<0.001

resulted comparable to those found by the ORNL. The highest values (20 μg/g) were measured in *Spartium junceum*, collected inside the mine complex.

3
The Idrija Mine

3.1
Location

The Idrija mine (Slovenia) is located in a shallow valley at 430 m above sea level between the Idrijca river and its tributary Nikova (Fig. 4). Mercury was

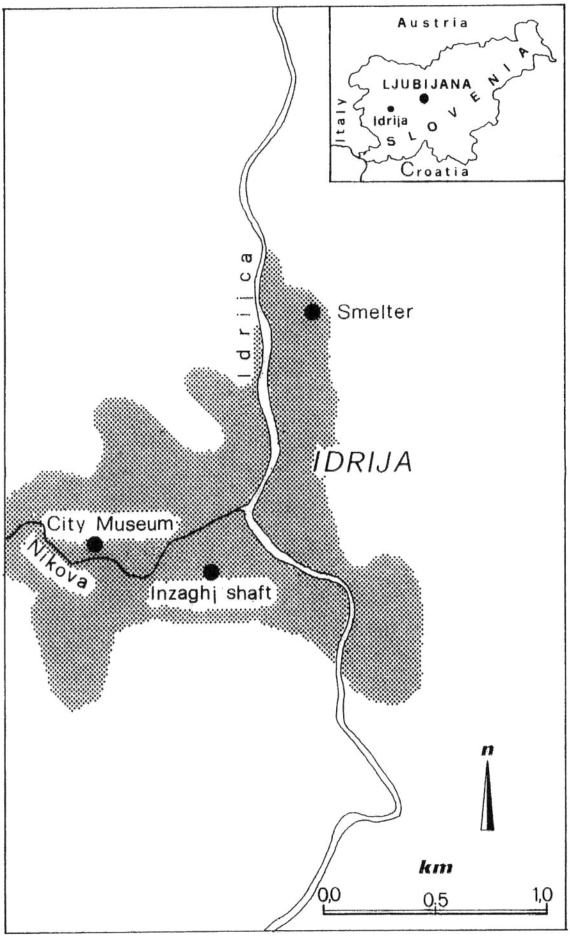

Fig. 4. The Idrija district (Slovenia)

discovered accidentally around the year 1490. In the 17th and 18th centuries, the Almadén mine (Spain) was closed as a consequence of numerous fires, and mercury production was completely substituted by the Idrija mine. Mining activity continued until 1977, with a total production of 150 000 tons of mercury. In 1983 mercury mining started to operate again but with a minimal production, and was definitively terminated in the year 1994.

In the Idrija mine mercury is present both as native mercury (~30%), and in minerals, mostly cinnabar (70%). The percentage of mercury in the rock was around 10% in 1820 and then decreased to 0.6–0.7% in 1914. In about 500 years of activity, 700 km of galleries have been dug, to a depth of 400 m (Kobal 1994).

At present and for about 10 years, 100 miners have worked in the mine and will continue to do so to complete the work necessary for the closure of the mine. Some galleries are being flooded with water, others are being backfilled with gravel, to avoid the collapse of the hill, and to limit damage by environmental contamination. Close control of the health state of these miners is operated by Kobal and Dizdarevic (1996). The old miners' houses have disappeared like many mine facilities (the stack of the furnace plant has been taken down). The Idrija landscape is returning to its original state, thanks to the lack of roasted cinnabar deposits: the roasted ore was used in the past for the building houses and streets, while in recent times it was discharged into the river Idrijca that transported it, through the Isonzo river to the Adriatic Sea, and in particular towards the coastal area of Trieste.

Much interesting research has been carried out on the Idrija mine, but unfortunately the results have often been published in local reviews or were otherwise not available in the current literature (Kosta et al. 1975; Kobal 1994; Gnamuš et al. 1995). Recently, an excellent work was published by Palinkaš et al. (1995) on the ecological impact of mining activity. A complete report of the meeting on the research Idrija as a Natural and Anthropogenic Laboratory (in which are reported numerous studies on geology, on mercury in the abiotic and biotic compartment and on the effect of metal on the workers' health) is in the final phase of preparation.

3.2 Main Sources of Atmospheric Mercury

During the period of greatest activity, the main emission source of mercury was obviously the stack of the distillation plant which emitted 20 kg/day of mercury, associated mainly with the atmospheric particulate matter. A second source was made up by the two systems of ventilation of the galleries, at present still in action, and around which a mercury concentration of the order of 2500 ng/m^3 has been measured.

The emission of mercury from the ground, due to the presence of geogenic cinnabar (and native mercury) has been increased by the enrichment caused by the emission of the roasting plant.

In order to distinguish geogenic and anthropogenic Hg loads in these soils, Biester et al. (1997) investigated Hg phases by a temperature-controlled-pyrolysis technique. The results that mercury is preferentially bound to soil organic matter demonstrate the importance of mercury deposition on the mercury content in the soil. In spite of the numerous studies performed on the Idrija mine, there are no available data on the flux values of mercury from the sources present today in the study area.

3.3 Mercury in the Air over Idrija

During the period of maximum productivity, the concentrations of mercury in the air ranged between 0.1 and 10 $\mu g/m^3$. The highest values were determined near the distillation plant (8.5 $\mu g/m^3$), while in the city centre values of the order of 0.6 $\mu g/m^3$ were measured. In 1991, about 14 years after the end of major producing activity, the levels of atmospheric mercury ranged between 10 and 500 ng/m^3.

Of particular interest are the data published by Gosar et al. (1997), obtained with the Zeeman atomic absorption spectrometer PA-915 of Russian origin, on September 24th 1994. The spectrometer was mounted in a car and the mercury concentration was measured every second. The geochemical map of the mercury distribution is reported in Fig. 5. The highest values of atmospheric mercury concentrations (4078 ng/m^3) were measured near the smelter and close to the ventilation shaft (2346 ng/m^3). Levels above 50 ng/m^3 were found in the town centre.

A drastic drop in mercury in the air was also observed by Kobal (1994) after the production of the mine and the smelter had almost halted. The fact that mercury contents in the air are high is not only due to the influence of mercury vapour from the above-mentioned sources, but also because of dumps and the use of smelter slag in the construction of buildings, and because of the presence in a part of the soil of rocks containing native mercury.

3.4 Mercury in the Soil

Palinkaš et al. (1995) report the total mercury concentration measured on 100 soil samples collected at 56 sampling sites spaced irregularly along the Idrijca valley and the slopes of the surrounding hills. The maximum value (87.6 mg/kg) was determined in the mining area and the vicinity of the smelter. The mean value of mercury concentration in the soil in the studied area was 15.4 mg/kg, with a minimum concentration value of 0.4 mg/kg.

The distribution of mercury along four vertical profiles suggests a predominant influence of the fallout on the accumulation of the pollutant.

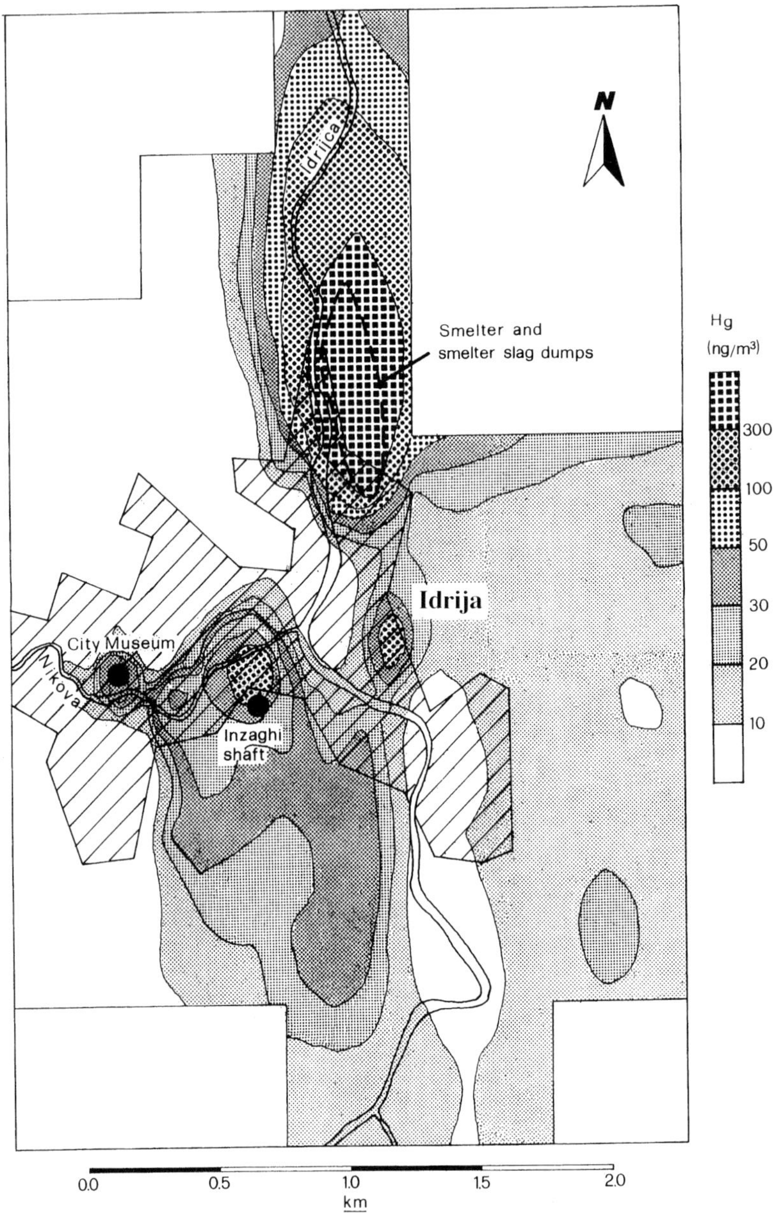

Fig. 5. Atmospheric mercury concentration (ng/m^3) over the Idrija district (Gosar et al. 1997, with permission by Chapman & Hall)

3.5
Mercury in the Sediment

Stream sediment (active sediment from the stream bed in regular contact with stream water) was sampled by Gosar et al. (in press) in July 1991 and 1995 along the Idrijca and Šoca rivers.

Due to the fast flow of the water of the Idrijca river, a large part of polluted sediments is carried downstream towards the river Šoca and finally into the Adriatic Sea (Kosta et al. 1975). The amount of mercury in the region from Idrija to Spodnja Idrija is rather high (300–1000 mg/kg). Downstream, from Spodnja Idrija on, the sediments contain less mercury (5–300 mg/kg).

Overbank (floodplain) sediment is produced when major floods occur in a river system, and can reach a thickness of up to a few metres. A vertical section of this sediment reflects the history of sedimentation back through time. A sample of recent overbank presents a mercury concentration of 120 mg/kg in the upper part of the profile and reaches a value of 240 mg/kg at the depth of 80 cm; then the concentration decreases again to around 120 mg/kg. The increasing concentration of mercury is probably due to the different production levels of the mining and the smelter. The profile was not deep enough to reach the deposits present before the beginning of the mining activity.

3.6
Mercury in the Springs and River Water

Mercury levels (10–31 ng/l) have been determined at five springs (used for the municipal water supply system), of which four are in the town and one is the spring of the Idrijca river.

The data reported for 16 sites (7 km upstream, 18 km downstream, and in the mining area) show extremely low concentration (3–20 ng/l); Palinkaš et al. (1995) believe that these low values are due to the high pH (7.5–8.3) of well-buffered waters, a characteristic of karst hydrology.

3.7
Mercury in the Biota

In 1992 Lupšina et al. published a review on the total mercury and methylmercury in a species of epiphyte lichen (*Hypogymnia physodes* L. Nyl.) sampled in an area which also included Idrija. The highest mercury concentration was found near the smelting house (115–188 µg/g d.w.), in comparison with the areas in the centre of Idrija (17–48 µg/g) and the surrounding area (2–7 µg/g). The lowest concentration was found at the control site in a mountain area (0.4 µg/g). According to the findings of others workers (Bargagli et al. 1987), a good correlation was found by the authors between mercury in the air and lichens. The fraction of methylmercury varied from 0.06 to 3.7%.

An extended text on the results of 5 years' ecological research in the natural environment on the topic concerning wildlife ecotoxicology and the influences of mercury species on biota and terrestrial environment has been prepared by Gnamuš (1998). The same author and coworkers (1995) report data on mercury levels in different organs of wild deer sampled in the surrounding areas of the mine. Generally, the measured concentrations are 100 times higher than that of the controls.

4 The Mt. Amiata Mine

4.1 Location

The mineralized volcanic area of Monte Amiata is located in the Tuscan region (central Italy), about 120 km northwest of Rome (Fig. 6). This region, covering approximately 400 km^2, is drained by several rivers such as the Albegna, the Fiora and the Paglia. The maximum elevation of the mountain is 1738 m. At Bagno Vignoni, Bagni S. Filippo and Saturnia, there are sulphurous thermal springs at a temperature of 50 to 60 °C.

The cinnabar deposits were first worked by the Etruscans and Romans, who used the ore as a red pigment for paint and cosmetics. To protect local workers and the environment, the mines were closed by the Romans, and the mercury was brought from Almadén (Spain). Mt. Amiata mines were rediscovered in 1868. Intense mining activity went on for about a century and reached a maximum production of 3×10^6 flasks of mercury in 1969. The ore, almost exclusively cinnabar, had a mercury content of 0.6 to 2%. In 1948, 42 mines and 4 distillation plants of mercury were active. Since 1970 there has been a drastic reduction in activity and many mines have been closed. An exception is the mine of Abbadia S. Salvatore, where production continued until 1980.

Even though 15 years have passed since mining activity completely ceased, the extensive banks of roasted cinnabar, which are residues of the working still rich in mercury, and the presence on the territory of the abandoned mining structures without any land reclamation action, represent even today a problem for environmental pollution.

The Amiata district is characterized also by the presence of geothermal activity. These steam-dominated or "high enthalpy" geothermal systems are well suited for electric power production because they supply superheated steam (over 200 °C) that can be admitted directly into a turbine. At present five, geothermal power plants (20 MW each) are in operation in the surroundings of Piancastagnaio. These hydrothermal systems, located in the highly mineralized area, make the power plants an important source of atmospheric mercury.

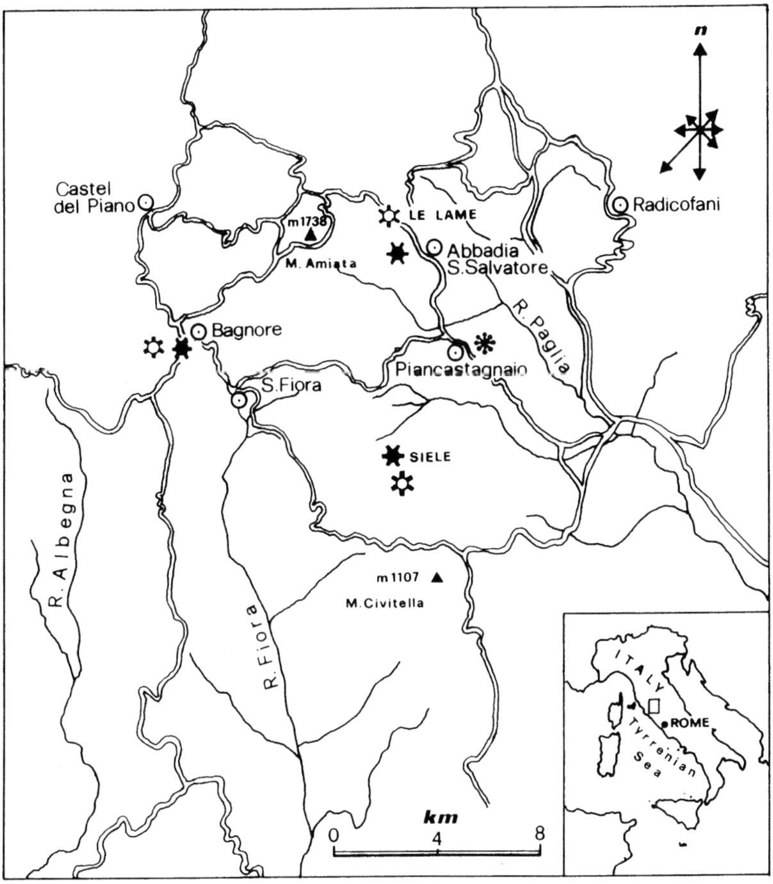

Fig. 6. The Mt. Amiata district (Italy)

4.2
Main Sources of Atmospheric Mercury

In 1990, Bargagli, using epiphytic lichens, stated that contaminated soils around mining and smelting plants constitute the main source of environmental pollution. Emission from geothermal power plants is remarkable, but the diffusion of hot stream probably causes less local contamination.

Only recently, using either the flux chamber technique or the remote sensing technique called LIDAR, Ferrara et al. (1998) have been able to individuate and quantify mercury fluxes. Of particular interest is the determination carried out with the LIDAR system.

On an annual basis, the main atmospheric mercury source is the emission from the five geothermal power plants (~1100 kg/a; Ferrara et al. 1994) not only because of its high value but also because this remains constant during the year,

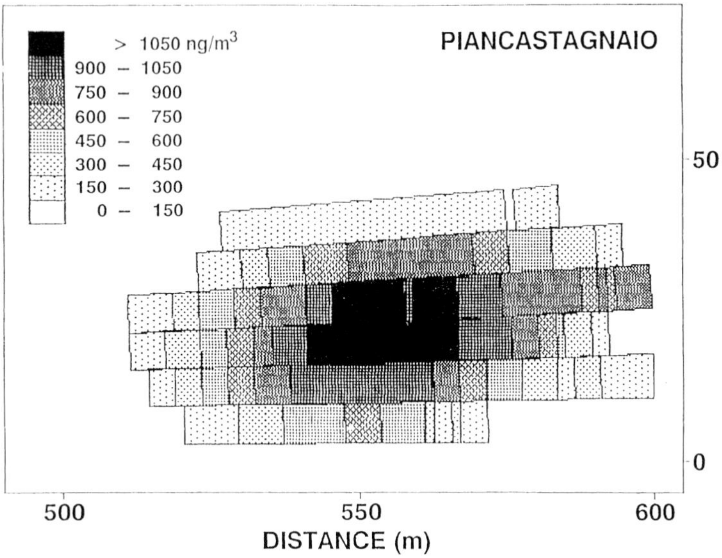

Fig. 7. Mercury levels (ng/m^3) in a vertical section of the plume of a geothermal power plant, measured by the LIDAR technique. Along the x axis is reported the height above the ground; along the y axis the distance from the LIDAR system is reported

not being dependent on the ambient temperature. In Fig. 7 an example is reported of the mercury distribution in the plume of a plant determined by the LIDAR system. In the month of July 1995 the abandoned mine structures of the smelting complex of Abbadia San Salvadore emitted an amount of mercury

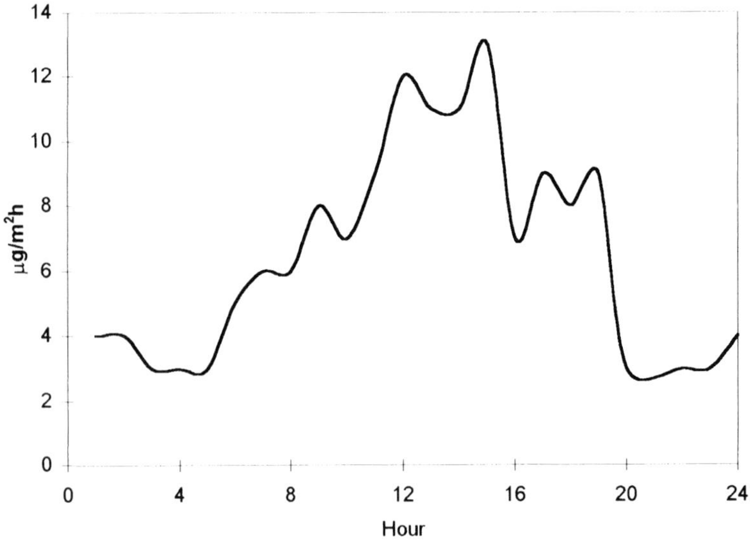

Fig. 8. Daily variation of the mercury degassing rate over a roasted cinnabar bank (July 1995)

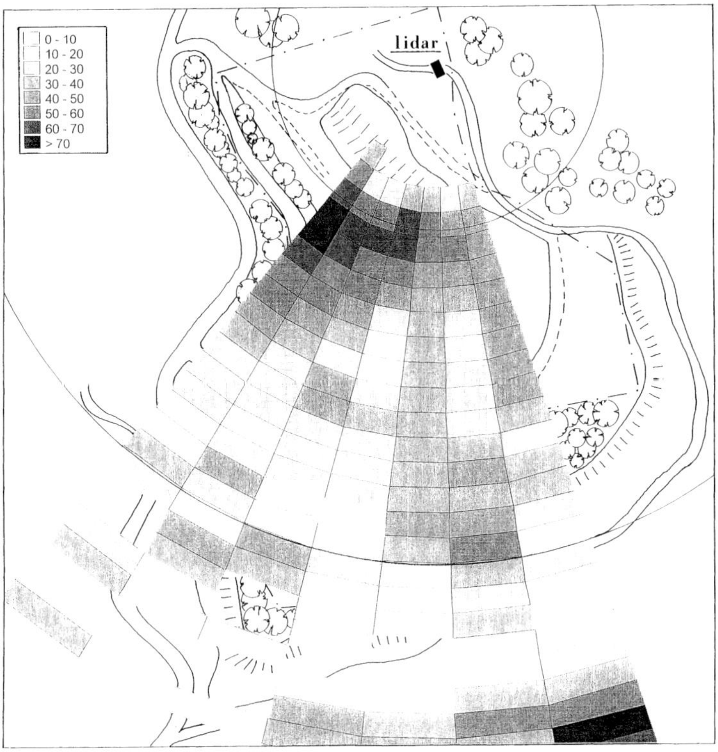

Fig. 9. Atmospheric mercury levels (ng/m^3) in a horizontal section measured 10 m above a roasted cinnabar bank by the LIDAR technique (July 1995). The indicated mercury concentrations within each box refer to the mean concentrations obtained by integrating several hundreds of distinct data points

(100 g/h) comparable to those of the geothermal power plants (Ferrara et al. 1998b). Relatively low emissions come from the extended roasted cinnabar banks (1–2 g/h), which, in any case, during the winter months are quite negligible. Further studies have shown that such fluxes depend not only on the temperature of the ground surface layer, but also on the deeper layer; this explains the lower flux values observed during the winter months compared to the summer ones even when surface soil temperatures reached the same values. In the graph of Fig. 8, an example is reported of the daily variation of the mercury degassing rate measured over the roasted cinnabar bank in July 1995. In Fig. 9 an example is reported of the concentration distribution of the atmospheric mercury measured about 10 m above the ground with the LIDAR technique on the same roasted cinnabar bank (July 1995). Using these horizontal (or vertical) sections and by integrating the value of the mercury concentration on the scanned area, it is possible to obtain the emission value, once the velocity and the direction of the wind are known.

4.3 Atmospheric Mercury Concentration

Several data of atmospheric mercury concentration are available in the literature, performed after the cessation of mining and refining activity (Ferrara et al. 1982; Breder and Flucht 1984; Ferrara et al. 1986).

The range of mercury levels in the air is very wide (10–1000 ng/m^3) and varies with the characteristics of the sampling sites and the presence and type of mercury sources. High mercury concentration values were found near the exhaust pipe of the ventilation system of the mines (~500 ng/m^3) and above the deposits of roasted cinnabar (30–200 ng/m^3). The highest mercury values were measured around and inside the area of abandoned mine structures (1000–15 000 ng/m^3; Ferrara et al. 1998).

Of course, mercury levels are strictly dependent on the ambient temperature and consequently show daily and seasonal behaviour. In Fig. 10 a typical daily trend of atmospheric mercury concentration, measured over a roasted cinnabar deposit, is reported.

The observations made near the geothermal power plants confirm the presence of a distinct mercury emission from hot steam wells (from 900 to 10 ng/m^3). The atmospheric mercury levels decrease sharply with increasing distance from the geothermal power plant (500–600 m) in an asymmetric distribution depending on the prevailing wind direction (Ferrara et al. 1994).

The mean daily values of atmospheric mercury determined in the winter season for the Amiata district proved to be less than 5 ng/m^3, with values of the order of 10 ng/m^3 in the surroundings of the main sources. This distribution is basically the same in the summer; the concentrations are significantly higher due

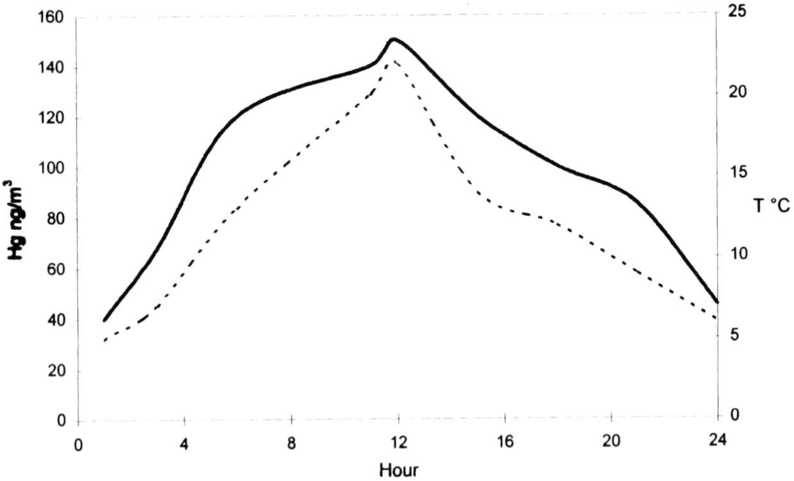

Fig. 10. Daily variation of atmospheric mercury concentration ( ) and soil temperature (- - -) over a roasted cinnabar bank (May 1987)

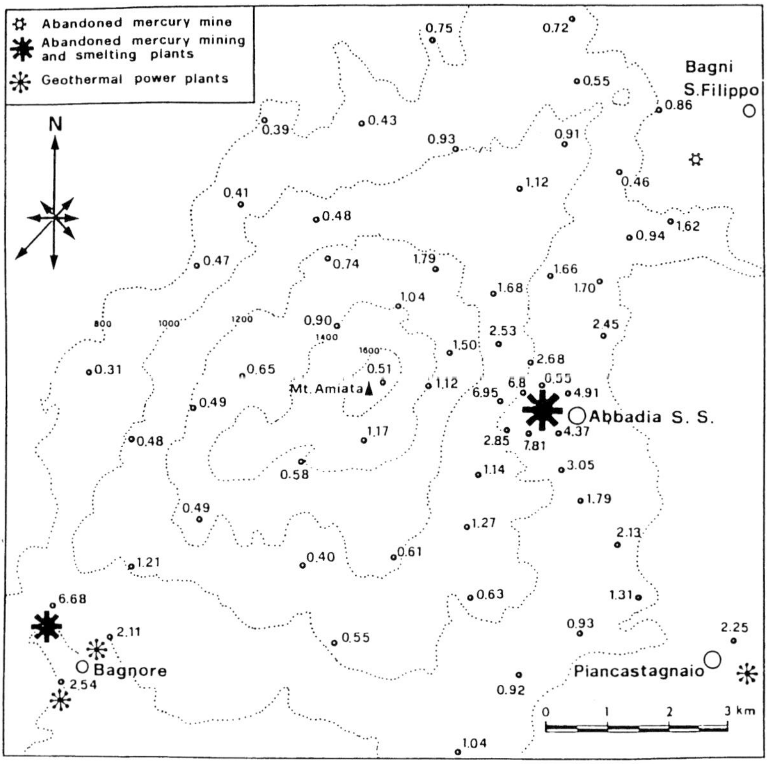

Fig. 11. Total mercury concentrations (μg/g d.w.) in surface soils. (Cheremisinoff 1990)

to the higher ambient temperature. Background values with a range of 4–7 ng/m^3 were observed in wooded areas while values on the order of 20–30 ng/m^3 were measured in the surroundings of the main structures and geothermal power plants.

4.4
Mercury in the Soil

In Fig. 11 the distribution of total mercury concentration in surface soils of the Amiata district is reported (Bargagli 1990). In the soil, several parameters like particle size, pH, organic matter and clay content affect mercury retention, but it is clear that the entire eastern side of Mt. Amiata is very rich in mercury. Concentrations range from 0.24 to 220 μg/g d.w. The lowest values are still about three to four times that of the background areas.

4.5 Mercury in the River Water and Thermal Springs

During mining activity in the years 1960–1970, the mercury concentration in the wastewaters discharged into the rivers was estimated to be approximately 180 ng/l (Bacci et al. 1978). Some data are available from 1967 (Dall'Aglio 1967) on mercury levels in freshwater, but one should keep in mind that reliable techniques for mercury determination in water have been available only in the past 10 years. Since 1980 mercury levels in the courses of the Albegna, Fiora and Paglia rivers, from the source to the estuary have been determined (Ferrara et al. 1983). At the sources of the rivers the mercury concentration are quite low (~10 ng/l) and comparable to those obtained for other water courses of the Italian peninsula. At the estuaries, the waters instead are rich in particulate matter in suspension and can reach concentrations of the order of 10–250 ng/l. The contribution of mercury from thermal springs localized at Bagno Vignoni and Bagno San Filippo to the environmental contamination proved to be negligible (Barghigiani et al. 1989).

4.6 Mercury in the Sediments

The concentration of mercury in surface sediment from the upper course of the Paglia river (the most polluted) displayed a value of 10 μg/g (Ferrara et al. 1991). The mercury content in the Fiora and Albegna sediment, taken upstream from the mixing zone with the Tyrrhenian Sea, ranges from 5 to 7 μg/g. Similar concentrations were measured by Baldi and Bargagli (1982) in the sediments of the coastal areas near the mouths of the two rivers.

4.7 Mercury in the Biota

Foliose lichens belonging to the genus *Parmelia* have been used as biomonitors of air mercury concentration in the Amiata district (Bargagli 1990). Results show that mercury distribution in lichen corresponds closely to that in surface soil, indicating that the ore deposits and mining and smelting of mercury constitute the major source of pollution.

The lowest concentration in *Parmelia sulcata* (0.31 μg/g d.w.) is two times higher than that usually found in samples collected in background areas. Maximum mercury content (7.80 μg/g) was found in the zone most affected by air vented from mine shafts. Of particular interest are the data available on mushrooms that show high concentration with a methylmercury percentage that can be as high as 15% (Bargagli and Baldi 1984). Mercury levels in agricultural products and their relations with soils and air concentrations are reported by Barghigiani and Ristori (1994).

5
General Remarks

Of the three most important mining districts in Europe, only the one in Almadén (Spain) is still active with a discontinuous mercury production and much lower than in the past. In Idrija (Slovenia) and Mt. Amiata (Italy) extraction and smelting activity practically ceased many years ago, but their effects due to mercury environmental contamination are still visible today.

In Fig. 12 the average concentration of mercury on the surface soil as a function of the distance from the mine centre, is reported for the three districts considered (Bargagli 1990). The high production of the biggest mining district in the world (Almadén) is reflected by the highest mercury content in the soil and by its dispersal pattern on the territory. The mining districts of Idrija and Mt. Amiata present a quite similar soil mercury/distance relationship which is a consequence of their comparable mercury production.

The values reported in Fig. 12 are an indication of the mercury enrichment of the soil caused by mining activity, mainly from the stack emissions of the roasting furnaces. This enrichment seems to be at present one of the main causes of mercury environmental pollution.

The recent application of the remote sensing LIDAR technique for determining atmospheric mercury has allowed the study of emissions in two of the mining districts cited (Almadén and Mt. Amiata). As a consequence it has been possible

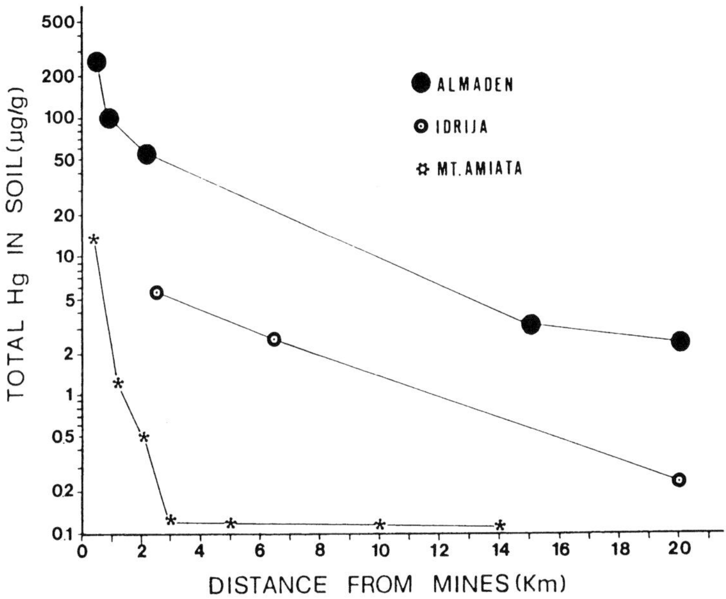

Fig. 12. Average concentrations of mercury in surface soils as a function of the distance from 3 mines (Almadén, Idrija, Mt. Amiata). (Chereminisoff 1990, with permission by Gulf Publishing Co.)

to evaluate the contribution of the single sources present, which is fundamental knowledge required for a rational environmental policy aimed at a pollution containment and recovery.

Just a few years ago it was the general belief that in mining districts in which all activity had ceased, the air ventilated from shafts and natural degassing from the ground and from the large piles of roasted cinnabar were the most significant sources of atmospheric mercury.

Recent results have instead made it obvious that the abandoned mining structures (in particular the smelting and condensation towers) are the main emitting sources, with values 20–30 times higher than those of the cinnabar banks.

Plants grown in these mineralized areas present noticeably higher values of mercury concentration than the same species from unpolluted sites; but these organisms show no clear morphological or physiological alterations. This finding could be due to the presence of a mercury tolerance mechanism allowing plants to overcome the toxic effect of the intracellular metal.

The rivers which drain the catchment of the three districts do not present particularly high mercury concentrations in the dissolved phase (an exception are the waters coming directly from the active plant of Almadén). The suspended particulate matter and the sediments show, on the other hand, high levels, and their effect on the aquatic cycle is quite evident at the mouth of the river and along the relative coast line. In fact, in these areas, the sediments are greatly enriched by metal and through the food chain this has caused the presence of consistent mercury concentrations in organisms belonging to higher rings of the food chain.

References

Bacci E, Leonzio C, Renzoni A (1978) Mercury decontamination in a river of Mount Amiata. Bull Environ Contam Toxicol 20:577–581

Bacci E, Gaggi C, Duccini M, Bargagli R, Renzoni A (1994) Mapping mercury vapours in an abandoned cinnabar mining area by azalea (*Azalea indica*) leaf trapping. Chemosphere 29:641–656

Baldi F, Bargagli R (1982) Chemical leaching and specific surface area measurements of marine sediments in the evaluation of mercury contamination near cinnabar deposits. Mar Environ Res 6:69

Bargagli R (1990) Mercury emission in an abandoned mining area: assessment by epiphytic lichens. Encyclopedia of environmental control technology. Gulf Publishing Company, Houston, Texas, pp 613–640

Bargagli R, Baldi F (1984) Mercury and methylmercury in higher fungi and their relation with the substrata in a cinnabar mining area. Chemosphere 13:1059–1071

Bargagli R, Iosco FP, Barghigiani C (1987) Assessment of mercury dispersal in an abandoned mining area by soil and lichen analysis. Water Air Soil Pollut 36:219–225

Barghigiani C, Ristori T (1994) Mercury levels in agricultural products of Mt. Amiata (Tuscany, Italy). Arch Environ Contam Toxicol 26:329–334

Barghigiani C, Siegel BZ, Bargagli R, Siegel SM (1989) The contribution of mercury from thermal springs to the environmental contamination of Mt Amiata. Water Air Soil Pollut 43:169–175

Biester H, Hess A, Müller G (1997) Mercury phases in soils and sediments in the Idrija mining area. Report of the Meeting Idrija as a Natural and Anthropogenic Laboratory Idrija 24-25 May 1996. Idrija Mercury Mine, Idrija, Slovenia (in press)

Breder R, Flucht R (1984) Mercury levels in the atmosphere of various regions and locations in Italy. Sci Total Environ 40:231–244

Chereminisoff PN (ed) (1990) Encylopedia of environmental control technology, Vol 4. Hazardous waste containment and treatment. Gulf Publ Co, Houston, Texas, 631 pp

Dall'Aglio M (1967) The abundance of mercury in 300 natural water samples from Tuscany and Latium. In: Ahrens LH (ed) Origin and distribution of the elements. Pergamon Press, Oxford, pp 1065–1081

Edner H, Ragnarson P, Svanberg S, Wallinder E, De Liso A, Ferrara R, Maserti BE (1992) Differential absorption lidar mapping of atmospheric atomic mercury in Italian geothermal fields. J Geophys Res 97:3779–3786

Edner H, Ragnarson P, Svanberg S, Wallinder E, Ferrara R, Maserti BE, Bargagli R (1993) Atmospheric mercury mapping in the cinnabar mine area. Sci Total Environ 133:1–15

Edner H, Ragnarson P, Svanberg S, Wallinder E, Ferrara R, Cioni R, Raco B, Taddeucci G (1994) Total fluxes of sulfur dioxide from the Italian volcanoes Etna, Stromboli, and Vulcano measured by differential absorption lidar and passive differential optical absorption spectroscopy. J Geophys Res 99:18827–18838

Ferrara R, Petrosino A, Maserti BE, Seritti A, Barghigiani C (1982) The biogeochemical cycle of mercury in the Mediterranean. Part 2: Mercury in the atmosphere, aerosol and in rainwater of a northern Tyrrhenian area. Environ Technol Lett 3:449–456

Ferrara R, Seritti A, Barghigiani C, Torti M, Petrosino A, Breder R (1983) Distribution and environmental significance of mercury near a Mediterranean geochemical anomaly. Proc 2nd Int Symp Environmental Pollution and its Impact on life in the Mediterranean Region Heraklion, Crete (Greece) Sept 6–9, 1983, pp 203–211

Ferrara R, Maserti BE, Petrosino A, Bargagli R (1986) Mercury levels in rain and in air and the subsequent washout mechanism in a central Italian region. Atmos Environ 20. 125–128

Ferrara R, Maserti BE, Bargagli R (1988) Mercury in the atmosphere and in lichens in a region affected by a geochemical anomaly. Environ Technol Lett 9:689–694

Ferrara R, Maserti BE, Edner H, Ragnarson P, Svanberg S, Wallinder E (1991) Mercury in abiotic and biotic compartments of an area affected by a geological anomaly (Mt Amiata, Italy). Water Air Soil Pollut 56:219–233

Ferrara R, Maserti BE, Edner H, Ragnarson P, Svanberg S, Wallinder E (1992) Mercury emissions into the atmosphere from a chlor-alkali complex measured with the lidar technique. Atmos Environ 26A:1253–1258

Ferrara R, Maserti BE, De Liso A, Edner H, Ragnarson P, Svanberg S, Wallinder E (1994) Could the geothermal power plant at Mt. Amiata (Italy) be a source of mercury contamination? In: Watras CJ, Huckabee JW (eds) Mercury pollution integration and synthesis. Lewis, Boca Raton, pp 601–607

Ferrara R, Maserti BE, Andersson M, Edner H, Ragnarson P, Svanberg S (1997) Mercury degassing rate in some mediterranean areas. Water Air Soil Pollut 93:59–66

Ferrara R, Maserti BE, Andersson M, Edner H, Ragnarson P, Svanberg S, Hernandez A (1998a) Atmospheric mercury concentrations and fluxes in the Almadén district. Atmos Environ (in press)

Ferrara R, Mazzolai B, Edner H, Svanberg S, Wallinder E (1998b) Atmospheric mercury sources in the Mt. Amiata area, Italy. Sci TOTAL ENVIRON 213:13–23

Gaggi C, Chemello G, Bacci E (1991) Mercury vapour accumulation in azalea leaves. Chemosphere 22:869–872

Gnamuš A (1998) Biomonitoring of mercury contamination in the Idrija mining area: is wildlife particularly vulnerable? Environ Toxicol and Chem – Special Annual Review Issue on Wildlife Ecotoxicology (submitted)

Gnamuš A, Horvat M, Stegnar P (1995) Der Quecksilbergehalt von Rehwild und geästem Laub für die Bewertung der Umweltbelastung im Bergbaugebiet von Idrija – Eine Fallstudie aus Slowenien. Z Jagdwiss 41:206–216

Gosar M, Pirc S, Šajn R, Bidovec M, Mashyanov NR, Sholupov SE (1997) Distribution of mercury in the atmosphere over Idrija, Slovenia. Environ Geochem Health 19:101–110

Gosar M, Pirc S, Bidovec M Mercury in the Idrijca river sediments. Report (Meet.) Idrija as a Natural and Anthropogenic Laboratory, 24–25 May 1996, Idrija Mercury Mine, Idrija, Slovenia (in press)

Hanson PJ, Lindberg SE, Tabberer TA, Owens JG, Kim K-H (1995) Foliar exchange of mercury vapor: evidence for a compensation point. Water Air Soil Pollut 80:373–382

Hildebrand SG, Huckabee JW, Sanz Diaz F, Jansen SA, Solomon JA, Kumar KD (1980) Distribution of mercury in the environment at Almadén, Spain. Report ORNL/TM-7446 Oak Ridge National Laboratory Oak Ridge, Tennessee

Hudson RJM, Gherini SA, Fitzgerald WF, Porcella DB (1995) Anthropogenic influences on the global mercury cycle: a model-based analysis. Water Air Soil Pollut 80:265–272

Julivert M, Fontboté JM, Ribeiro A, Conde LEN (1972) Mapa tectónico de la peninsula Ibérica y Baleares a escala 1: 1.000.000, IGME Madrid

Kobal A (1994) Das Quecksilberbergwerk und die damit verbundene Umweltbelastung in der Stadt Idrija. Zentralbl Arbeitsmed 44:211–216

Kobal A, Dizdarevic T (1996) The health safety program for workers exposed to elemental mercury at the mercury mine in Idrija. 4th Int Conf on Mercury as a global pollutant, Hamburg, Germany, 4–8 Aug, 1996, p 16

Kosta L, Byrne A, Zelenko V, Stegnar P, Dermelj M, Ravnik V (1975) Studies on the uptake, distribution and transformations of mercury in living organisms in the Idrija region and comparative areas. Vestn Slov Kem Drus 21:49–76

Lindberg SE, Jackson DR, Huckabee JW, Janzen SA, Levin MJ, Lund JR (1979) Atmospheric emission and plant uptake of mercury from agricultural soils near the Almadén mercury mine. J Environ Qual 8:572-578

Lindqvist O, Rodhe H (1985) Atmospheric mercury – a review. Tellus 37B:136–159

Lodenius M (1994) Mercury in terrestrial ecosystems: a review. In: Watras CJ, Huckabee JW (eds) Mercury pollution integration and synthesis. Lewis, Boca Raton, pp 343–354

Lupšina V, Horvat M, Jeran Z, Stegnar P (1992) Investigation of mercury speciation in lichens. Analyst 117:673–675

Maserti BE, Ferrara R, Panichi MA, Storni M (1996) Mercury concentrations in plants and soil of the cinnabar mineralized area of Almadén (Spain). 4th Int Conf on Mercury as a global pollutant, Hamburg, Germany, 4–8 August, 1966, p 140

Ortega Gironés E, Hernandez Sobrino A (1992) The mercury deposits of the Almadén syncline, Spain. Chron Rech Min 506:3–24

Palinkaš LA, Pirc S, Miko SF, Durn G, Namjesnik K, Kapelj S (1995) The Idrija mercury mine, Slovenia, a semi-millennium of continuous operation: an ecological impact. Environ Toxicol Assess, Taylor & Francis, London, pp 317–341

Saupé F (1990) Geology of the Almadén mercury deposit, Province of C. Real, Spain. Econ Geol USA 85:482-510

Siegel SM, Siegel BZ, Barghigiani C, Aratani K, Penny P, Penny D (1987) A contribution to the environmental biology of mercury accumulation in plants. Water Air Soil Pollut 33:65–72

Xiao ZF, Munthe J, Schroeder WH, Lindqvist O (1991) Vertical fluxes of volatile mercury over forest soil and lake surfaces in Sweden. Tellus 43B:267–279

Mercury Contamination from New World Gold and Silver Mine Tailings

L.D. Lacerda and W. Salomons

1 Introduction

The input of mercury by industrial point sources into the environment has decreased drastically in both developed and developing countries. The remaining issues, in particular in temperate climates, are the accumulated levels of mercury in hot spots like industrial sites and contaminated soils and sediments. The appropriate technology for cleanup of hot spots is, in principle, available (see the various chapters in this Vol.). Furthermore, new technologies are available that completely exclude or limit the use of mercury.

The issue in the developing countries is quite different. Although emissions for point sources have decreased here also, the "diffuse" use of mercury in gold mining has increased over the past years and is likely to increase in the future. Its attractiveness lies in being a reliable and cheap technology easy to operate on the individual scale. Furthermore, mercury is used by large groups of individuals in many widely dispersed and often remote areas. Controlling this type of use is more difficult compared with controlling mercury discharges from (few) industrial point sources. This type of pollution abatement has a strong socioeconomic dimension. Therefore, notwithstanding the recent regulations making the acquisition and use of Hg in many countries more difficult, this type of technology will only be limited by the exhaustion of appropriate gold and silver deposits or by a drastic decrease in gold prices. Both alternatives are unlikely to occur in the next 50 years, and new prospecting areas are expected to be opened in particular in northern South America and Southeast Asia. Thus the issues with mercury-contaminated tailings sites will remain, and their area will increase in the foreseeable future. This chapter gives an overview of the available knowledge on the behavior and impact of mercury in tailings from the "first" and "second" gold rushes.

2 Past and Present Use of Mercury in Gold and Silver Mining

Mercury amalgamation was the major gold and silver mining technique until the beginning of this century. This technique is believed to have consumed over

Environmental Science
Mercury Contaminated Sites (ed. by R. Ebinghaus et al.)

300 000 tons of Hg from 1540 to 1900, mostly in the silver production in Spanish colonial America and in the North American gold rush in the second half of the last century (Nriagu 1994). Also, significant mining occurred in Southeast Asia and Australia (Bycroft et al. 1982; James 1994). In recent years, this technique has been resuscitated in the tropics, where it represents a simple, cheap, and reliable technique for gold production, ready to be used by a growing population lacking proper jobs in most developing countries in the tropics (Lacerda et al. 1995). Present global annual emission of Hg to the environment from this activity may reach up to 500 tons (Lacerda and Marins 1997).

A typical characteristic of these small, artisanal mining operations is the low efficiency of the process due to the intuitive use of Hg. For example, Hg emission factors (the amount of Hg released to the environment to produce 1 kg of gold) in Amazon gold mines may range from 1.1 to 5 (Mallas and Benedicto 1986; Pfeiffer and Lacerda 1988), although the average emission factor is approximately 1.3 (Lacerda et al. 1995). In colonial Spanish America and in last-century North American gold mines, these emission factors typically ranged from 1 to 2 (Fisher 1977; Nriagu 1994). Most of the Hg is emitted to the atmosphere during the burning of the Hg-Au amalgam. However, from 30 to 45% of the total Hg emitted ends up in river banks, soils, and drainages. As a result, Hg-rich tailings are a common feature left by ancient and recent gold and silver mining operations using Hg amalgamation.

A major problem of such tailings is their long time span as a significant source of Hg to the environment. Tailings deposited over 100 years ago and containing up to 500 μg Hg g^{-1} are still present in central Canada (Lane et al. 1988). In the Carson River Valley, Nevada, over 100 000 tons of tailings containing nearly 7000 tons of Hg and with concentrations of up to 4900 μg g^{-1} are continuously reworked along the valley (Lechler and Miller 1993). Other important old tailings from gold and silver mining particularly enriched in Hg are the Falson-Natomas region, California (Prokopovich 1984), the Lerderberg River gold fields, Australia (Bycroft et al. 1982), and Nova Scotia, Canada (Lane et al. 1988). Small tailing deposits, however, are scattered world-wide. These old mining tailings, together with those produced by the "New Gold Rush", represent a significant long-term environmental threat. However, it should be realised that the impact of mercury stored in tailings deposits is only part of the whole issue of mercury release from gold and silver mining operations (Lacerda and Salomons 1998).

3
Mercury Distribution in Gold Mining Tailings

In most gold and silver mining areas, Hg is deposited and accumulated in mine tailings. An important question is how Hg is eventually transported to, and spread through pristine adjacent environments (Allan 1995). Although only few studies have dealt with this subject, a general pattern of Hg distribution in tailings can be stated, based on data gathered from old and recently deposited tailings.

- In the majority of tailings Hg is present as metallic Hg. Water-soluble Hg is generally less than 0.1% of the total Hg content in tailings (Andrade et al. 1988; CETEM 1989; Lacerda et al. 1990; Ramos and Costa 1990; Lechler and Miller 1993). A larger fraction of Hg forms is only extractable with aqua regia, indicating that these forms are associated with residual minerals or as elemental Hg° (Ching and Hongxiao 1985). A fraction of Hg lost to tailings may also be amalgamated with gold grains (Lane et al. 1988; Fuge et al. 1992).
- Mercury distribution in tailings is a function of the distribution of mining activities in the area rather than any geochemical process. In areas where old preconcentration ponds existed, Hg concentration can reach a few micrograms per gram of tailings material, while the mean Hg content in the whole tailings body is similar to local background values (Prokopovich 1984; CETEM 1989).
- Different mining techniques may result in different Hg concentration and distribution in tailings. Typically, these concentrations range from less than 0.5 $\mu g\ g^{-1}$ to 4900 $\mu g\ g^{-1}$ (dry weight) (Andrade et al. 1988; Ramos and Costa 1990; Lechler and Miller 1993). For example, when Hg amalgamation is performed in specific sites in the mining field, these sites will show the highest Hg concentrations in the tailings (CETEM 1989). When the amalgamation step is performed during the gravimetric concentration, as in riffles, Hg tends to be spread through tailings, showing a more even distribution (CETEM 1992). More uniform distribution of mercury concentrations in tailings can also occur when sodium chlorine is used in the extraction process (Lechler 1993).
- Mercury present in gold mining tailings can represent an important source of contamination even centuries after the cessation of the mining operations (Lane et al. 1988; Fuge et al. 1992; Bycroft et al. 1992; Lechler and Miller 1993).

The four characteristics listed result in a very low reactivity and bioavailability of Hg in tailings, but turn these deposits into a long-lasting contamination source.

A detailed study has been carried out on the Hg concentrations in the Tanque dos Padres tailings, central Brazil, obtained from the analysis of a 200 sample points grid (CETEM 1989). The Hg distribution was extremely inhomogeneous, showing background values in over 95% of the samples ($<0.20\ \mu g\ g^{-1}$). The few samples with anomalous high Hg content (4 to 25 $\mu g\ g^{-1}$) showed no distribution pattern and were not correlated with any geochemical or geological aspect of the tailings, but correspond to former amalgamation ponds at the mining site (CETEM 1989). Prokopovich (1984) measured Hg concentrations ranging from 0.036 to 0.2 $\mu g\ g^{-1}$ in over 60 samples of last-century dredge tailings in the Folsom-Natomas gold field, California. In three samples randomly taken, however, Hg concentrations reached 37.5; 21.5, and 22 $\mu g\ g^{-1}$.

The distribution of mercury in gold and silver mine tailings hampers any generalization. Tailing deposits should be considered as unique, at least in relation to the spatial distribution of Hg concentrations. However, the mechanisms of Hg release from tailings are probably very similar, regardless of site, due to the similarity of mining operations and the fact that only a single mercury species enters the process (Hg^o).

4 Mercury Emissions from Contaminated Tailings

4.1 Mercury Degassing from Contaminated Land and Tailings

Emission from contaminated soils, waters, and tailings involves the degassing of Hg vapor to the atmosphere. The evasional fluxes of Hg^o have an important geochemical significance, since they can reach up to 15% of the atmospheric deposition over natural areas (Vandal et al. 1993). Tailings are by far the most important site of Hg degassing, since they can reach Hg concentrations of over 5000 $\mu g\ g^{-1}$. In certain areas such as the Folsom Canal, California, degassing from tailings is a potential source of Hg to humans living in urban developments built on last-century dredge tailings from gold mining (Prokopovich 1984).

Rates of Hg degassing from natural, noncontaminated areas are in general less than 10–20 ng m^{-2} day^{-1} or 1–3 $\mu g\ m^{-2}$ $year^{-1}$ (Fitzgerald et al. 1991; Table 1). Wind speed is a key factor controlling Hg degassing from natural soils and water surfaces, the absence of plant cover on tailings may therefore increase Hg degassing.

Degassing from silver and gold mine tailings typically ranges from 100 to 1500 $\mu g\ m^{-2}$ $year^{-1}$, much higher than over mineralized areas containing cinnabar deposits, where degassing rates can reach up to 60 μg Hg m^2 $year^{-1}$ (Rasmussen 1994). These rates, however, are similar to those measured over Hg mining tailings, such as Almadén, Spain, where degassing may reach extremely high values of up to 2600 μg Hg $m^{-2}year^{-1}$ (Ferrara and Maserti 1994).

Apart from gaseous Hg, tailings may also emit to the atmosphere heavily contaminated dust particles. Extremely high Hg concentrations have been reported in this compartment. For example, reported values from the Amazon gold mining, reach 36 $\mu g\ g^{-1}$ and 250 $\mu g\ g^{-1}$ of dust (Malm et al. 1991; Thornton et al. 1992, respectively). This compartment shows a very short residence time in the atmosphere. However, it has been reported as a significant source of human Hg exposure, through inhalation, in gold mining centers in central Brazil (Hacon 1966).

Contaminated soils and tailings tend to show degassing rates linearly proportional to Hg concentrations in the first 15 cm of soil (Lechler 1993), similar to soils over geological anomalies (Rasmussen 1994). Surface temperature also exhibits a very strong influence on Hg^o degassing from contaminated soils. At the Carson River tailings, Nevada, USA, Lechler and Miller (1993) estimated Hg efflux to the atmosphere in excess of 500 ng m^{-2} day^{-1} (c.a. 180 $\mu g\ m^{-2}$ $year^{-1}$) (Table 1). These authors found a proportion between degassing rates and Hg concentration in tailings close to linearity from concentrations ranging from 0.02 to 350 μg Hg g^{-1} dw of tailings. Similarly, high degassing rates were measured over contaminated soils with Hg content typical of gold mine tailings (ca. 3000 μg g^{-1}) in Oak Ridge, Tennessee (Kim et al. 1993).

As a result of the continuous degassing of Hg from tailings, Hg concentrations in the air over them tend to be abnormally high. In general, Hg concentrations in air over tailings reach orders of magnitude of the expected concentrations in natural air (Table 2), and can be similar to concentrations in indoor atmospheres in chlor-alkali work places or over cinnabar deposits. In general, these concentrations can reach more than 10 μg g m^{-3}, at least four orders of magnitude higher than natural background concentrations (Table 2).

Interesting are the concentrations found over contaminated tailings in the Carson River Valley (Miller et al. 1993), which, although deposited in the 1840s, still generate atmospheric Hg concentrations in the same order of magnitude as present-day mining sites in the Amazon, and at least one or two orders of magnitude higher than background concentrations (Table 2).

Mercury speciation in the degassing flux from tailings is significantly different from emissions from natural noncontaminated sites. Whereas natural degassing from soils typically presents an organomercurial component, which may represent up to 10 to 20% of the total (Mora et al. 1993), over tailings this

Table 1. Mercury degassing rates from diverse Earth surfaces, compared to contaminated areas from gold mining sites

Surface type	Degassing (μg m^{-2} a^{-1})	Reference
Noncontaminated soils	3.2–9.5	Rasmussen (1994)
	1.0–3.0	Fitzgerald et al. (1991)
Soils over mineralized sites	9.5–56.0	Fitzgerald et al. (1991)
Over temperate lakes	26–87	Schroeder et al. (1992)
Over the open ocean	0.3–6.5	Baeyens et al. (1991)
Over tailings-contaminated soils, Oak Ridge, USA	130–1400	Kim et al. (1993)
Over cinnabar ore, Almadén	2600	Ferrara and Maserti (1994)
Over last-century gold mining tailings, Nevada, USA	180	Lechler (1993)

Table 2. Atmospheric Hg concentrations measured in urban and rural areas under the influence of gold mining compared with other gold-producing and industrialized areas

Location	Hg (μg m^{-3})	Reference
Over tailings, Central Brazil	0.08	Tümpling et al. (1993a,b)
Over old tailings, Nevada, USA	0.01–0.04	Lechler and Miller (1993)
Over tailings, Dixing, China	14.9	Yshuam (1994)
Over Comstock tailings, USA	0.23	Gustin and Leonard (1994)
Teles River mining site, S. Amazon	0.01–3.05	Hacon et al. (1995)
Over cinnabar ores	0.01–0.09	Ferrara et al. (1982)
Background, rural values	0.001–0.015	Kothny (1974)

fraction is negligible, due to the low biological activity typical of tailings and the high concentration of residual metallic Hg^{0}.

In a study of gold mining tailings in central Brazil, Tümpling et al. (1993, 1995) showed total Hg concentrations in air just over tailings to range from 76 to 85 ng m^{-3}, with only 3 to 6 ng m^{-3} (less than 0.1%) of organomercurials. In the same region and over the same geological formation, Hg concentrations in air over noncontaminated soils average 2.8 ng m^{-3}, with 0.4 ng m^{-3} (ca. 15%) of organomercurials (Tümpling et al. 1993). However, these results are only preliminary, and detailed studies on Hg speciation in the degassing flux from tailing should still be performed.

4.2
Mercury Export from Tailings to Adjacent Aquatic Systems

Strong gradients have been frequently found in both sediments and water Hg concentrations from point sources of Hg, such as tailings. In general, these gradients are very steep, indicating the low dispersability of Hg from the releasing points. For example, in small creeks draining mine tailings in central Brazil, Andrade et al. (1988), found that Hg contamination in the Crixás deposit, Goiás State, decreases from a maximum value of ca 16 μg g^{-1} at the tailings to the local background of ca 0.7 μg g^{-1} in the first 200 m along the drainage. The same behavior was found by Lacerda et al. (1991) at the Tanque dos Padres tailings, Mato Grosso State. Mercury concentrations decrease from values up to 3.0 μg g^{-1} in the tailings to local background levels (~0.04 μg g^{-1}) in the first 100 m along the drainage.

Figure 1 shows Hg concentrations in sediments of rivers draining tailings deposits in Poconé, central Brazil (Lacerda et al. 1991a). It is clear that Hg is very refractory to transport since its concentrations drop to nearly local background values only a few kilometers from tailings.

Similar results have been found in other tailings-affected drainages in the Crixás mining site, Goiás State, central Brazil and in the Cumaru mining site, northern Amazon (Andrade et al. 1988; Ramos and Costa 1990).

Although Hg leaching from tailings seems a slow process, through the years the migration of tailings themselves may be the most important Hg dispersal mechanism. In the Carlson City mining district over the past century Hg-rich tailings were transported over 100 km from the Six Mile Canyon and tributaries to the Carlson River Valley, forming an Hg-rich 1.5 km^{2} fan area (Lechler and Miller 1993; Miller et al. 1993). In the Falsom-Natomas region, California, Prokopovich (1984) suggested that groundwater migration could export Hg from 1868 dredge tailings, based on anomalous concentrations found in groundwater samples from that area. In Nova Scotia, Canada, however, in groundwater samples collected in last-century tailings, with Hg content ranging from 1.1 to 19.9 μg g^{-1}, no contamination of groundwater could be detected (Lane et al. 1988). Creek waters draining tailings with up to 40 μg g^{-1} of Hg in the Waverly area showed Hg concentrations always below 50 μg l^{-1} (Mudroch and Clair 1986). Surface and groundwater samples collected from over 21 stations in creeks and tailings in North Carolina also showed concentrations below 20 $\mu g\ l^{-1}$ (Callaham et al. 1994).

Tümpling et al. (1995), working in creek waters draining tailings in central Brazil, found that volatile Hg^{0} is typically less than 5% of the total dissolved Hg content in water. Easily reducible Hg corresponds to 20 to 50%, and the other fractions in dissolved form are refractory compounds.

Lacerda et al. (1991a) studied Hg dynamics in a creek draining a contaminated tailing deposit during a storm in the Poconé mining site, central Brazil. Significant changes in Hg concentrations in water and suspended particles, as well as in major physicochemical parameters in creek water draining the tailings, occurred during the storm event. The water became progressively more acidic, with pH dropping from 6.8 to 5.1. Redox potential also increased due to the entry of rainwater supersaturated in oxygen, average Eh became more oxidizing, increasing from 280 to 430 mV 10 min after the start of the storm. The content of suspended particles also showed a dramatic increase, reaching values up to ten times higher than during dry conditions, increasing from 6 mg l^{-1} to 66 mg l^{-1}. These changes occurred just a few minutes after the onset of rain. Mercury concentrations in suspended particles increased from <0.02 $\mu g\ g^{-1}$ to 0.61 $\mu g\ g^{-1}$.

The results obtained show the capacity of rain to erode fine particles, enriched with Hg, from the tailings, suggesting that erosion, followed by Hg transport associated with suspended particles, is the main pathway of Hg contamination of the tailings drainage. Prokopovich (1984) found a significant relationship between Hg content and the clay fraction in old California tailings. The fact that dissolved Hg concentrations remained constant during the storm suggests that only small amounts of Hg, if any, are remobilized into solution and that Hg concentration in water is independent of mercury content in sediments (Andrade

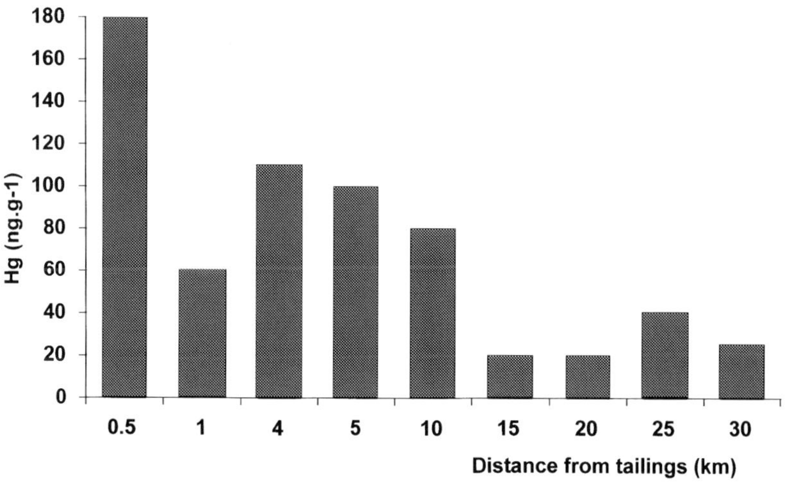

Fig. 1. Mercury concentrations in sediments of a creek draining tailings deposits in Pocone, central Brazil

et al. 1988). Nriagu et al. (1992), studying Hg partitioning between dissolved and particulate forms in the Madeira River waters, also showed that dissolved mercury concentrations are independent of its concentrations in suspended sediments. Mudroch and Clair (1986) found the same Hg distribution pattern in waters draining tailings in Canada. They have also suggested that contamination of sediments was due to transport and deposition of suspended particles brought into the drainage by erosion of mining wastes. Similar behavior was reported by Lechler and Miller (1993) for tailings in the Carson River Valley, USA, and by Nelson et al. (1977) for mineralized areas in Alaska.

Most studies have found no significant changes in dissolved mercury concentrations, even during strong storm events. Apparently, only small amounts of Hg are remobilized into solution and Hg concentrations in water appear to be independent of Hg content in sediments. Thus, controlling waste erosion and suspended particle transport by means of dams and impermeabilization and other protecting devices to tailings should be quite efficient in preventing Hg contamination in areas surrounding tailing. Among the many techniques available to stabilize tailings deposits is revegetation (Archer et al. 1988; Ernst 1988; von Michaelis 1988). However, few data exist which are applicable to Hg-rich gold and silver mine tailings.

5 Impact on Biota

5.1 Terrestrial Vegetation

Terrestrial vegetation interacts with both soil and the atmosphere in the acquisition of Hg (Siegel et al. 1987). We have shown (Lacerda and Salomons 1998) the importance of the atmospheric Hg in the geochemical cycles in gold mining areas. Terrestrial plants can takeup Hg from the atmosphere directly through gas exchange at the stomatal level or through membrane exchange with wet and dry precipitation (Browne and Fang 1978). These processes are likely to be enhanced in tropical vegetation, due to the efficient mechanisms of chemical elements acquisition from the atmosphere mediated by epiphyllic organisms, in particular algae and lichens (Jordan et al. 1980) at the crown leaf level, and those mediated by the high diversity and biomass of ephiphytes typical of most tropical ecosystems (Benzing 1981; Nadkarni 1984). Therefore, tropical terrestrial vegetation may provide rapid plant colonization with a self-sustaining plant community able to efficiently decrease Hg mobilization from tailings to the atmosphere. On the other hand, since Hg export from tailings to adjacent drainages is particularly enhanced by erosion, terrestrial vegetation is a key factor in protecting the tailings deposit integrity.

Some studies have been done on the vegetation cover of old mining tailings. At the Oldham tailings, Nova Scotia, vegetation coverage exceeds 100%, with species richness ranging from 12 to 15 species m^{-2}, which is higher than that found in abandoned pasture in the same region (Lane et al. 1988). Many of the plant species recorded in the area are metal-tolerant species or metal-tolerant genotypes derived from local populations. All species analyzed in these tailings presented abnormally high Hg concentrations ranging from 0.2 to 3.5 $\mu g\ g^{-1}$ (Lane et al. 1988). All terrestrial plant species studied presented higher Hg concentrations in roots.

Siegel et al. (1985), studying the vegetation on British Columbia tailings, found that different concentration factors (enrichment factors) among species occurred only at low Hg concentrations in soils. When concentrations increased over 0.05 $\mu g\ g^{-1}$, enrichment factors were relatively constant among taxa, suggesting their being metal-tolerant genotypes. Although only a few studies were done in Hg-contaminated tailings, it seems that the behavior of the vegetation is similar to any other vegetation growing on metal-rich tailings.

Many more data, however are needed for a better comprehension of the fate of Hg in terrestrial vegetation affected by gold mining.

5.2 Aquatic Vegetation

Aquatic macrophytes are more likely to accumulate Hg than terrestrial plants. The availability of Hg to these plants will be a function of the dissolved Hg in waters. Under the more acid freshwater conditions typical of tropical systems, Hg availability should be high, and certainly much higher than in the soil or sediment solution.

Different habitat of aquatic macrophytes may result in different Hg concentrations. Since the roots may effectively avoid Hg uptake from soils and sediments, floating and submerged species, with roots in the water column, may absorb Hg also directly from water through leaves and shoots. This has been shown both between and within plant genera (Kelly 1988).

Aquatic macrophytes growing on tailings-influenced areas in Canada presented elevated Hg shoot concentrations ranging from 0.18 to 0.55 $\mu g\ g^{-1}$ dw (Lane et al. 1988), with root concentrations much higher than in shoots. Lacerda et al. (1991b), studying Hg concentrations in an emergent macrophyte *Potenderia lanceolata* along gold mining tailings drainage in central Brazil, found no significant correlation between Hg levels in the plant and in bottom sediments. However, the highest values measured in the plant were generally found associated with high sediment values. They suggested that the plants preferably incorporate Hg^{2+}, and since most Hg exported from these tailings was strongly bound to suspended particles, its availability for plant uptake would vary according to the environmental conditions of the drainage which would control Hg release from suspended particles rather than the bulk Hg concentration in sediments. In a Siberian gold district, little discharged metallic Hg was assimilated by aquatic plants of local drainages (Tayrayev 1991).

Plants growing on Nova Scotia tailings drainages presented root:shoot ratios ranging from 1.04 to 11.13 (Lane et al. 1988), and reached 6.17 in the floating

Table 3. Comparison between root and shoot mercury concentrations ($\mu g\ g^{-1}$) in aquatic macrophytes from gold mine tailings

Species	Shoot	Root	Root:shoot	Reference
Equisetum fluviatile, Nova Scotia, Canada	0.45	0.47	1.04	Lane et al. (1988)
Juncus articulatus, Nova Scotia, Canada	0.55	2.69	4.89	Lane et al. (1988)
Juncus pelocarpus, Nova Scotia, Canada	0.54	6.11	11.13	Lane et al. (1988)
Dulichium arundinaceum, Nova Scotia, Canada	0.19	0.67	3.53	Lane et al. (1988)
Leerzia oryzoides, Nova Scotia, Canada	0.18	1.56	8.67	Lane et al. (1988)
Eichornia azurea, Teles Pires mines, South Amazon	0.06	0.37	6.17	CETEM (1992)
Potenderia Lanceolata, central Brazil	0.14	–	–	Lacerda et al. (1991a)

macrophyte *Eichornia crassipes* from gold mining sites in South Amazon (CETEM 1992).

The higher content of Hg in roots relative to shoots in freshwater macrophytes is a well-known phenomenon, and also occurs for other trace metals and terrestrial plants. (Kelly 1988). Roots act as an efficient barrier to Hg uptake. Mercury can complex to sulfydril groups at the root surface (Siegel 1973) and adsorb to deposited iron and manganese plates (Kozlowski 1984; Crowder and StCyr 1991; Gambrell 1994). Also, release of Hg^0 vapor from shoots has been reported for many plant species (Siegel et al. 1974; 1985; 1987) and results in less Hg in the leaves (Table 3).

The understanding of the role played by vegetation of gold mine tailings is hampered by lack of information on almost all significant mining sites. Since most of these operations are taking place in the tropics, and tailings are left frequently still surrounded by pristine ecosystems, it is of extreme urgency to gather more information on the role played by vegetation on the migration of mercury contamination from gold and silver mine tailings.

5.3 Animals

Some few animal species other than fish have been analyzed in Hg-contaminated sites from gold mining areas, in order to find biological monitors of Hg contamination. Among them, freshwater snails have been studied more intensively, since these organisms proved to be very good monitors in many trace metal contamination situations.

Freshwater snails are omnivorous animals, feeding on bottom sediments and plant surfaces. Therefore Hg concentrations are generally higher in the guts of these animals than in muscle tissues. Figure 2 shows the distribution of Hg in the

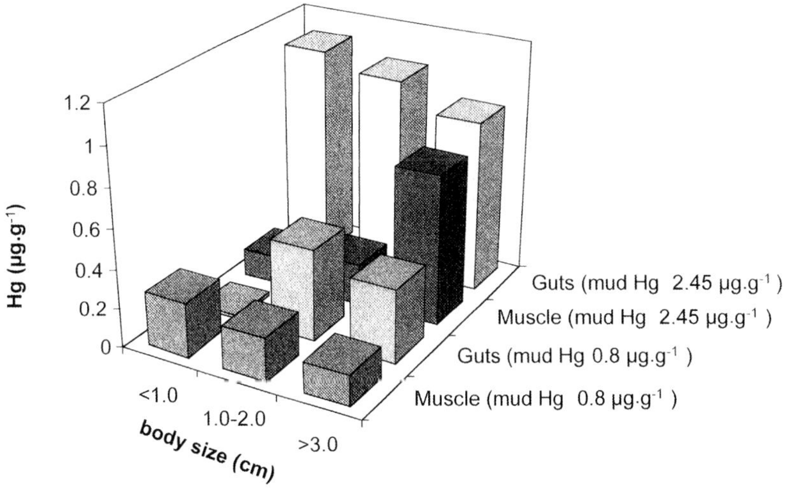

Fig. 2. Mercury distribution in different organs of the freshwater snail *Hemisinus tuberculatus* Spix of different size, and collected in two differently contaminated sites along the Teles Pires River, Southern Amazon, after Farid 1992

guts and muscle tissues of *Hemisinus tuberculatus* Spix, a typical and abundant snail from the neotropics, from different sites along the Teles Pires River, southern Amazon (Farid 1992). Guts have higher Hg, concentrations than muscle tissues, and reflect the general Hg content of fine bottom sediments from each site. Mercury content in guts did not correlate significantly with animal size. A positive relationship, however, was found between body size and Hg content in muscle tissue at the more contaminated site, but not at the less contaminated site.

Hg distribution in *Ampullarius* sp. collected in a contaminated site draining gold mining tailings in central Brazil ranged from 0.43 $\mu g\ g^{-1}$ in the smallest individuals to 0.95 $\mu g\ g^{-1}$ in the larger. There is a significant positive correlation between Hg concentration and shell size ($r = 0.938$, $p < 0.05$), indicating that size is an important factor determining Hg concentrations in this species. Positive size-Hg concentration relationships have been previously reported for other mollusk species contaminated by Hg (Huckabee et al. 1979; Mohlenberg and Riisgard 1988), and have been attributed to the accumulation of the long biological residence time of methyl-Hg.

In a study of freshwater snails distributed in the Tucuruí Reservoir, SE Amazon, Aula et al. (1994) found no significant correlations between Hg content and body size or mass. Mercury concentrations were low and quite variable (11 to 170 $\mu g\ kg^{-1}$) and seemed independent of Hg concentrations in bottom sediments. The results from these two snail species from different mining sites suggest that Hg methylation processes are currently occurring in both areas, at least in sites of elevated Hg content, since Hg concentrations in both water and sediments are, in general, very low, even in the most contaminated sites of these drainages (Lacerda et al. 1990; Farid 1992).

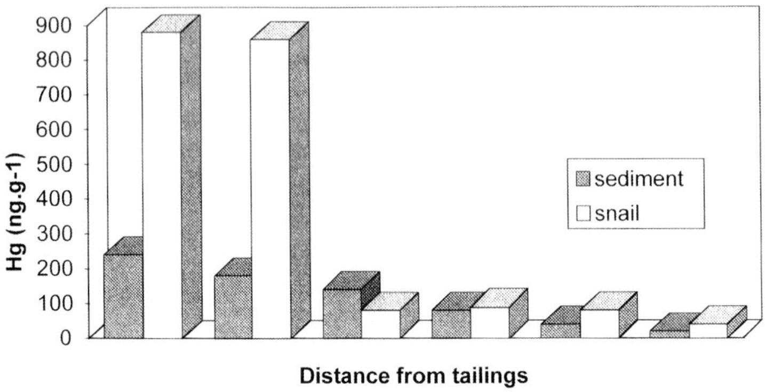

Fig. 3. Mercury concentrations in freshwater snails (*Ampularius* sp.) along a contaminated creek draining tailings deposits in Pocone mining site, central Brazil

Mercury concentrations in the snail *Ampullarius* sp. were determined along a contaminated drainage of the Tanque dos Padres tailing deposit in relation to sediment content and distance from tailings. The results are summarized in Figure 3. The results showed that samples collected close to the tailings presented much higher concentrations, decreasing sharply as the distance from the tailings increases, similarly to Hg distribution in bottom sediments. Proportional response to environmental Hg levels has been reported for many mollusk and macrophyte species and is one of the main reasons for using these organisms as biological monitors for heavy metal contamination. The rapid decrease in Hg concentrations in these organisms, however, reflects the low mobility of Hg in the area and also suggests that exported Hg is probably under a low bioavailable form in the drainage.

Much information is available on the mercury concentrations in fish. However, available evidence shows that tailings deposits are only indirectly responsible for accumulation in fish (Lacerda and Salomons 1998).

6 Summary

Gold and silver mining using the amalgamation technique for extraction of gold results in tailings deposits that contain mercury as a pollutant. Mercury in the deposits is heterogeneously distributed and reflects the distribution of mining activities. Tailings deposits act as a significant source of mercury for the atmosphere. Concentrations over tailings deposits are as high as or higher than over cinnabar deposits. The emission of mercury takes place not only from the more recent deposits in tropical areas, but also from long-abandoned mining sites in the US and Canada. Tailings deposits apparently act as a long-term source for atmospheric mercury. The main process governing the dispersal of mercury from the tailings is the erosion of mercury-contaminated particles by gravity, rain, storm events, and subsequent reworking of the deposits in river or lake systems.

Terrestrial and aquatic vegetation on tailings deposits shows elevated mercury levels. Indicator organisms like snails and mollusks reflect the mercury concentrations in the tailings deposits.

It is expected that the use of mercury for gold and silver mining will continue in the future in tropical areas. Unfortunately, although current information shows an impact on the environment, little information is available on long-term risk assessment in these areas, let alone on available measures to remediate or contain the tailings.

References

Allan R (1995) Impact of mining activities on the terrestrial and aquatic environment with emphasis on mitigation and remedial measures. In: Salomons W, Förstner U (eds) Heavy metals: problems and solutions. Springer, Berlin Heidelberg New York, pp 119–140

Andrade JC, Bueno MIMS, Soares PV, Choudhuri A (1988) The fate of mercury released from prospecting areas (Garimpos) near Guarinus and Pilar, Goiás (Brazil). An Acad Bras Cinc 60:293–303

Archer I, Marshman NA, Salomons W (1988) Development of a revegetation program for copper and sulphide-bearing mine wastes in humid tropics. In: Salomons W, Förstner U (eds) Chemistry and biology of solid wastes. Springer, Berlin Heidelberg New York, pp 166–184

Aula I, Braunsweiller H, Leino T, Malin I, Porvari P, Hatanaka T, Lodenius M, Juras A (1994) Levels of mercury in the Tucuruí reservoir and its surrounding area in Pará, Brazil. In: Watras CJ, Huckabee JW (eds) Mercury pollution: integration and synthesis. Lewis Publ, Boca Raton, pp 21–40

Baeyens W, Leermakers M, Dedeuwaerder H, Larsens P (1991) Modelization of the mercury fluxes at the air sea interface. Water Air Soil Pollut 56:731–744

Benzing DH (1981) Mineral nutrition of epiphytes: an appraisal of adaptive features. Selbyana 5:219–223

Browne CL, Fang SC (1978) Uptake of mercury vapour by wheat: an assimilation model. Plant Physiol 61:430–433

Bycroft BM, Coller BAW, Deacon GB, Coleman DJ, Lake PS (1982) Mercury contamination of the Lerderderg River, Victoria, Australia, from an abandoned gold field. Environ Pollut Ser A 28:135–147

Callaham JE, Miller JW, Craig JR (1994) Mercury pollution as a result of gold extraction in North Carolina, USA. Appl Geochem 9:235-241

CETEM (1989) Relatório Anual do Projeto Poconé. Centr Tecnol Mineral, Rio de Janeiro, 287 pp

CETEM (1992) Diagnóstico preliminar dos impactos ambientais gerados por garimpos de ouro em Alta Floresta/MT: Estudo de caso Centro de Tecnologia Mineral, Rio de Janeiro, p 190

Ching L, Hongxiao T (1985) Chemical studies of aquatic pollution by heavy metals in China. In: Irgolic KJ, Martell AE (eds) Proc Workshop Environm Inorganic Chem, San Miniato, Florida, pp 359–371

Crowder A, St-Cyr L (1991) Iron oxide plaques on wetland roots. Trends Soil Sci 1:315–329

Ernst WHO (1988) Response of plants and vegetation to mine tailings and dredge materials. In: Salomons W, Forstner U (eds) Chemistry and biology of solid wastes. Springer, Berlin Heidelberg New York, 305 pp

Farid LH (org) (1992) Diagnóstico preliminar dos impactos ambientais gerados por garimpo de ouro em Alta Floresta, MT. Ser Tecnol Ambiental CETEM 2:1–190

Ferrara F, Petrodino A, Maserti E, Seritti A, Barghigiani C (1982) The biogeochemical cycle of mercury in the Mediterranean Part II. Mercury in the atmosphere, aerosol and in rain of a northern Tyrrhenian area. Environ Technol Lett 3:449–456

Ferrara R, Maserti BE (1994) Mercury degassing rate in some Mediterranean areas. In: Proc Int Conf Mercury as a Global Pollutant. Whistler BC, Abstract Section 8B

Fisher JR (1977) Silver mines and silver miners in colonial Peru, 1776–1824. Centre for Latin american Studies, Univ Liverpool, Liverpool

Fitzgerald WF, Mason RP, Vandal GM (1991) Atmospheric cycling and air-water exchange of mercury over mid-continental lacustrine regions. Water Air Soil Pollut 56:745–767

Fuge R, Pearce NJG, Perkins WT (1992) Mercury and gold mining. Nature 357:369

Gambrell RP (1994) Trace and toxic metals in wetlands – a review. J Environ Qual 23:883–891

Gill GA, Bruland KW (1990) Mercury speciation in surface freshwater systems in California and other areas. Environ Sci Technol 24:1392–1400

Gustin MS, Leonard GETT (1994) Mercury evasion from contaminated landscape elements, Carson River Superfund Site, Nevada. In: Proc Int Conf Mercury as a Global Pollutant. Whistler BC, Section III, Abstract

Hacon S (1966) Exposicao ao mercúrio da populacao humana da Alta Floresta, Mato Grosso. PhD Thesis, Universidade Federal Fluminense, Niteroi, p 134

Hacon S, Artaxo P, Gerab F, Yamasoe MA, Calixto RC, Conti LF (1995) Atmospheric mercury and trace elements in the region of Alta Floresta in the Amazon Basin. Water Air Soil Pollut 80:273–283

Huckabee JW, Elwood JW, Hildebrand SG (1979) Accumulation of mercury by freshwater biota, In: Nriagu JO (ed) The biogeochemistry of mercury in the environment. Elsevier/North Holland, Amsterdam, pp 277–307

James LP (1994) The mercury tromol mill: an innovative gold recovery technique and a possible environmental concern. J Geochem Explor 50:493–500

Jordan CF, Golley FB, Hall J, Hall J (1980) Nutrient scavenging of rainfall by the canopy of an Amazonian rain forest. Biotropica 12:61–66

Kelly M (1988) Mining and the freshwater environment. Elsevier London, 231 pp

Kim K-H, Lindberg S, Hanson PJ, Meyers TP, Owens J (1993) Application of micrometeorological methods to measurements of mercury emissions over contaminated soils. Proc Int Conf Heavy Metals in the Environment, Toronto 1:328–331

Kothny E (1974) The three-phase equilibrium of mercury in nature. In: Gould RT (ed) Trace elements in the environment. Adv Chem Ser 123:48–91

Kozlowiski TT (1984) Plant responses to flooding of soil. BioScience 34:162–167

Lacerda LD, Marins RV (1997) Antropogenic mercury emissions to the atmosphere in Brazil: the impact of gold mining. J Geochem Explor 58:223–229

Lacerda LD, Salomons W (1998) Mercury from gold and silver mining. A chemical time bomb. Springer, Berlin Heidelberg New York, 146 pp

Lacerda LD, DePaula FC, Ovalle ARC, Pfeiffer WC, Malm O (1990) Trace metals in fluvial sediments of the Madeira River watershed, Amazon, Brazil. Sci Total Environ 97/98:525–530

Lacerda LD, Marins RV, Souza CMM, Rodrigues S, Pfeiffer WC, Bastos WR (1991a) Mercury dispersal in water, sediments and aquatic biota of a gold mining tailings drainage in Pocone, Brazil. Water Air Soil Pollut 55:283–294

Lacerda LD, Salomons W, Pfeiffer WC, Bastos WR (1991b) Mercury distribution in sediment profiles of remote high Pantanal lakes, central Brasil. Biogeochemistry 14:71–77

Lacerda LD, Malm O, Guimarães JRD, Salomons W (1995) Mercury and the new gold rush in the south. In: Salomons W, Stigliani W (eds) Biogeodynamics of pollutants. Springer Berlin Heidelberg New York, pp 213–245

Lane PA, Crowell MJ, Graves MC (1988) Heavy metal removal from gold mining and tailings effluents using indigenous aquatic macrophytes (Phase I). CNMET Spec Publ SP88-23, pp 3–37

Lechler PJ (1993) Mercury vapor sampling at the Carson river superfund site. Int Conf Heavy Metal in the Environment, Toronto 1:377–380

Lechler PJ, Miller, JR (1993) The dispersion of mercury, gold, and silver from contaminated mill tailings at the Carson River mercury superfund site, west-central Nevada, USA. In: Abrão JJ, Wasserman JC, Silva-Filho EV (eds) Proc Int Symp Perspectives for Environmental Geochemistry in Tropical Countries, Rio de Janeiro, 1993, pp 433–436

Mallas J, Benedicto N (1986) Mercury and gold mining in the Brazilian Amazon. Ambio 15:248–249

Malm O, Pfeiffer WC, Souza CMM (1991) Main pathways of mercury in the Madeira River area, Rondonia, Brazil. Proc Int Conf Heavy Metals Environment, Edinburgh 1:515–518

Miller JR, Lechler PJ, Desileta M, Rowland J, Hsu LC, Price JG (1993) Quantity and distribution of mining related trace metals in Lohantan Reservoir, west central Nevada. Proc Int Conf Heavy Metal in the Environment, Toronto 2:251–254

Mohlenberg F, Riisgard HU (1988) Partitioning of inorganic and organic mercury in cockles *Cardium edule* (L) and *C. glaucum* (Bruguière) from a chronically polluted area: influence of size and age. Environ Pollut 55:137–148

Mora SJ, Patterson JE, Bibby DM (1993) Baseline atmospheric mercury studies at Ross Island, Antartica. Antarct Sci 5:323–326

Mudroch A, Clair TA (1986) Transport of arsenic and mercury from gold mining activities through an aquatic system. Sci Total Environ 57:205–216

Nadkarni NH (1984) Epiphyte biomass and nutrient capita of a neotropical elfin forest. Biotropica 16:249–256

Nelson H, Larsen BR, Jenne EA, Sorg DH (1977) Mercury dispersal from lode sources in the Kuskokwim River drainage, Alaska. Science 198:820–824

Nriagu JO (1994) Mercury pollution from past mining of silver and gold in the Americas. Sci Total Environ 149:167–181

Nriagu JO, Pfeiffer WC, Malm O, Souza CMM, Mierle G (1992) Mercury pollution in Brazil. Nature 356:389

Pfeiffer WC, Lacerda LD (1988) Mercury inputs into the Amazon Region, Brazil. Environ Technol Lett 9:325–330

Prokopovich NP (1984) Occurrence of mercury in dredge tailings near Falosm South Canal, California. Bull Assoc Eng Geol 21:531–543

Ramos JEF, Costa MQ (1990) Distribuição de mercúrio em dois garimpos do Estado do Pará. In: Hacon S, Lacerda LD, Pfeiffer WC, Carvalho D (eds) Riscos e consequncias do uso do mercúrio. UFRJ, Rio de Janeiro, pp 70–79

Rasmussen PE (1994) Current methods of estimating atmospheric mercury fluxes in remote areas. Environ Sci Technol 28:2233–2241

Schroeder WH, Lindqvist O, Munthe J, Xiao Z (1992) Volatilization of mercury from lake surfaces. Sci Total Environ 125:47–66

Siegel SM (1973) Metal ions in biological systems. Marcel Dekker, New York

Siegel SM, Puerer N, Speitel T (1974) Release of volatile mercury from vascular plants. Physiol Plant. 34:174–182

Siegel SM, Siegel B, Lipp P, Kruckberg A, Towers GHN, Warren H (1985) Indicator plant-soil mercury patterns in a mercury-rich mining area of British Columbia. Water Air Soil Pollut 25:73–79

Siegel SM, Siegel BZ, Barghigiani C, Aratani K, Penny P, Penny D (1987) A contribution of the environmental biology of mercury accumulation in plants. Water Air Soil Pollut 33:65–72

Tayrayev TT (1991) Man-made dispersion train of gold and mercury in Golet-Taiga terrain. Dokl Akad Nauk SSR 317:719–722

Thornton I, Cleary D, Worthington S (1992) Mercury contamination in the Brazilian Amazon. A cooperative research study conducted by GEDEBAM. Brazil e Sol 3 (Switzerland and Luxembourg), 45p

Tümpling WV, Wilken RD, Einax J (1993) Pathways and contaminations of mercury in the Pantanal area, Brazil. In: Abrão JJ, Wasserman JC, Silva-Filho EV (eds) Proc Int Symp Perspectives for Environmental Geochemistry in Tropical Countries, pp 403–406

Tümpling WV, Wilken RD, Einax J (1995) Mercury contamination in the northern Pantanal region, Mato Grosso, Brazil. J Geochem Explor 52:127–134

Vandal GM, Fitzgerald WF, Boutron CF, Candelone JP (1993) Variations in mercury deposition to Antartica over the past 34,000 years. Nature 362:621–623

Von Michaelis H (1988) Integrated biological systems for effluent treatment from mine and mill tailings. In: Salomons W; Förstner U (eds) Chemistry and biology of solid wastes. Springer, Berlin Heidelberg New York, pp 99–113

Yshuan YS (1994) Mercury contamination in China. In: Evaluation of the role and distribution of mercury on ecosystems with special emphasis on tropical regions. SCOPE, Rio de Janeiro (unpublished)

Mercury-Contaminated Industrial and Mining Sites in North America: an Overview with Selected Case Studies[1]

R.R. Turner and G.R. Southworth

Mercury (Hg) has been used in highly diverse industrial and mining applications in North America, often with both intentional and unintentional releases to environmental media. While industries producing chlorine and caustic soda using the Hg-cell process, and the earlier extensive use of mercury in the mining of noble elements, have historically been among the most significant sources of such releases, other industries, including the US government, have also contributed substantially. Significant curtailment of intentional releases to aquatic systems in North America began about 1970, but it has only been recently that the legacies of past, and in some cases unrecognized continuing releases have begun to be addressed at many sites. This chapter briefly summarizes typical contamination scenarios at a variety of industrial sites and provides more detailed reviews of selected case studies which illustrate some remarkable consistencies among such sites.

1 Introduction

Mercury has historically found hundreds of industrial uses in North America (Jasinski 1995). However, hardly an industry which used mercury in one way or another has escaped experiencing releases into the environment. Some of the same unique physical (e.g., density, liquid state at ambient temperature, volatility) and chemical properties (e.g., ease of reduction) of mercury which make it so useful in industrial applications have also made it difficult to contain and to recover from the environment. Opportunities for environmental releases of mercury are much greater than for other metals. Even when mercury is used initially in one physical or chemical form, it can be readily transformed to other forms which can more easily escape from industrial processes. For example, in the production of chlorine and caustic soda by electrolysis of brine, mercury in the liquid metal form is used both as an electrode and as a "solvent", or amalgamator, for sodium. Theoretically, at least, there should be little consumption of mercury in the process other than small losses which leave the facility in the products. In fact, prior to about 1970, such facilities typically had to replace 5 to 10% of their mercury stocks annually to sustain production rates and the "average" facility could not account for 50% of the annual losses (EPA 1975). In addition to the obvious volatile losses of the metal from the process, mercury was

[1]This work was partially sponsored by the Oak Ridge Y-12 Plant Health, Safety, and Environment Division. The Y-12 Plant is managed by Lockheed Martin Energy Systems, Inc. under contract DE-AC05-84OR21400 with the US Dept. of Energy

Environmental Science
Mercury Contaminated Sites (ed. by R. Ebinghaus et al.)

lost in wastewater far in excess of that expected from the solubility of the pure metal (about 60 $\mu g\ l^{-1}$). Oxidation to more soluble compounds (e.g., mercuric oxide 53 $mg\ l^{-1}$) plus water-borne transport of microscopic metallic particles greatly enhanced historic losses via wastewater.

Those industries which incorporated mercury or one of its numerous compounds into products represent one potential type of enviromental release. Spills of mercury feedstocks and disposals of off-specification products at fungicide, paint, electrical, and explosives manufacturing facilities are some historical examples. These types of industries have mostly discontinued use of mercury (fungicides, paint) or greatly decreased the quantities (batteries) incorporated. Use of mercury as a preservative in forest and agricultural products has also virtually stopped, but legacies of such former use are still present. One also finds examples where mercury was not used directly in a consumable product but was used to test materials (e.g., porosimetry), to measure processes (e.g., natural gas flow using manometers), or as a functional component of apparatus (e.g., pumps and seals). In some cases, facilities which were built to reclaim mercury and other resources from discarded products (e.g., battery recyclers) have led to severe environmental contamination (e.g., NYSDEC 1992). Lastly, there are numerous cases where mercury was used to manufacture a product but the product(s) did not intentionally contain mercury. Production of caustic soda, chlorine, acetaldehyde, and certain plastics are the main examples.

Environmental regulation of releases of mercury began in the US and Canada about 1970 partly as a general consequence of the environmental activism which began in the late 1960s but also as a very specific consequence of scientific revelations which showed that inorganic mercury could be methylated in the environment and lead to levels in fish and other high-level predators which posed a human health risk. By some accounts (D'Itri 1972), the first significant focus on mercury in North America began in the St Clair River and associated waterways (Lake St Clair and Detroit River) which connect Lake Huron and Lake Erie. In late 1969 and early 1970, fish from these waters were found to be contaminated with mercury from two chloralkali plants, one in the US and one in Canada, which discharged wastewater into this aquatic system. This discovery led very quickly to polling of all industries in the US and Canada (e.g., DOI 1970) which might be using mercury, monitoring of their discharges, and monitoring of fish in the receiving waters. The results showed high discharges of mercury from some industries and associated contamination of fish. Most offending industries very quickly reduced or eliminated discharges, and a few soon ceased operations entirely.

Regulation of industrial discharges in the 1970s and early 1980s continued to focus mainly on aquatic releases until the passage of two important environmental laws [Resource Conservation and Recovery Act (RCRA) in 1978 (amended 1986–1987) and Comprehensive Environmental Response, Compensation and Liability Act (CERCLA) in 1980 (amended in 1986)] in the United States effectively expanded coverage to include disposal of industrial mercury-bearing wastes (RCRA) and restoration of sites contaminated prior to most environmental regulations (CERCLA). These, and similar laws in Canada and Mexico, began a process of characterization and classification of currently generated wastes destined for disposal, and of historically contaminated sites, which continues today.

The purpose of this chapter is to summarize typical contamination scenarios, their environmental consequences, and their remediation or natural recovery, using a series of case studies. Although hundreds of mercury-contaminated sites are known to exist in North America (e.g., Fig. 1), comprehensive information is available in accessible form for only a few dozen sites. Often the most abundant and accessible information pertains to aquatic systems (lakes, streams, marine bays) which received discharges from facilities in their respective watersheds. Facility-specific information is often typically absent in the published literature, or is contained only in the files of site owners, their consultants and governmental agencies. This review chapter includes in most cases only the information that was easily accessible to the authors, including some kindly provided by site owners and governmental agencies. Omission of one or more significant sites is likely but the review is believed to be representative of the diversity of sites present on this continent. By design, the emphasis is on "industrial" sites and not

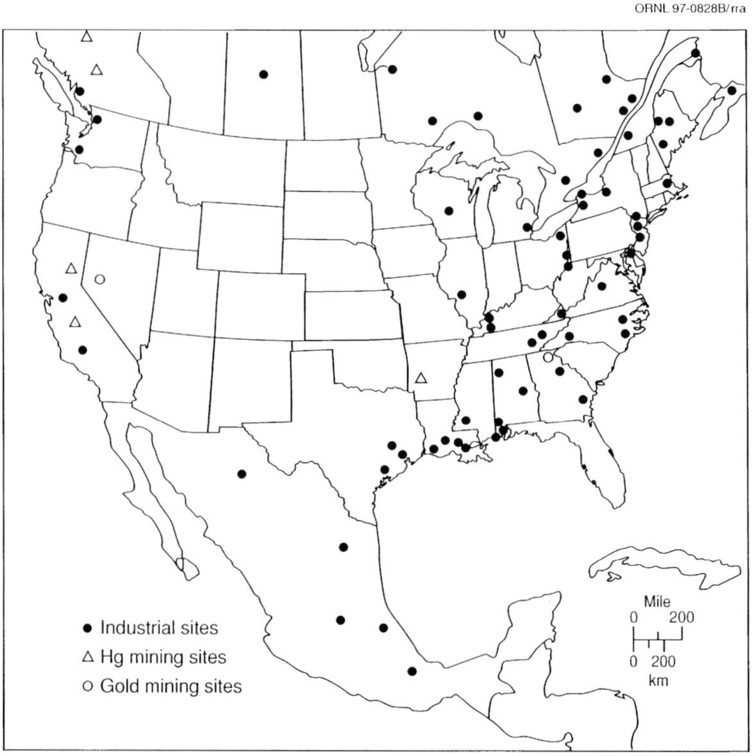

Fig. 1. Locations of some mercury-contaminated sites in North America. Additional known sites in Alaska and Puerto Rico and numerous small sites, such as associated with mercury manometer spills, are not shown

on mercury-mining sites (only two are described), or mining sites where mercury was employed to extract precious metals. Other chapters in this Volume cover examples of the latter sites (e.g., Lechler et al., this vol.). Nor will we discuss the numerous small sites contaminated by spills from mercury manometers associated with natural gas production and transmission systems (see review by Henke et al. 1993). The authors have an extended period of direct experience with two sites (Oak Ridge, Tennessee, and Saltville, Virginia) and thus more detail is presented for these sites.

2 Example Sites and Case Studies

2.1 Lake St. Clair

The St. Clair River forms the uppermost portion of the aquatic system between Lake Huron and Lake Erie. This river, and Lake St. Clair and the Detroit River to the south, are part of an important international waterway. Findings here related to mercury in fish in 1970 led ultimately to the first comprehensive assessment of the impact of mercury releases to aquatic systems and to the first regulation of such discharges in North America. Two chloralkali plants using mercury cells operated along this waterway in 1970 and collectively were discharging kilogram quantities of mercury on a daily basis. Both facilities immediately began aggressively reducing their discharges, with the result that concentrations in fish in the receiving aquatic system decreased rapidly over the next few years (Fig. 2). Eventually, both facilities discontinued the use of mercury cell technology, but no published information could be found describing how on-site contamination at the plant sites was addressed. Mercury in 45-cm walleye in the lake continued to decline through the mid-1980s (Fig. 2), but still exceeds the Canadian limit for fish sold commercially (0.5 $\mu g\ g^{-1}$) and engenders advisories to sport fishermen to limit consumption of larger fish. Walleye of the same size collected upstream (southern Lake Huron) ranged from 0.41 to 0.25 $\mu g\ g^{-1}$ between 1981 and 1992 (Scheider et al., in press) and showed no temporal trend. In addition, concerns persist about the inventory of mercury in sediment downstream of these sites.

2.2 Abbotts Creek/High Rock Lake, North Carolina

Elevated concentrations of mercury were found in 1981 in the sediments (up to 9.5 $\mu g\ g^{-1}$) and fish (up to 3.0 $\mu g\ g^{-1}$, 30% exceeding 1.0 $\mu g\ g^{-1}$ limit) in the Abbotts Creek embayment of High Rock Lake, a 39-km^2 hydroelectric reservoir in central North Carolina (Miller 1983; NCDEHNR 1993). The contamination was traced to a battery manufacturer in the watershed and resulted in issuance of a fish consumption advisory in August 1981. Intensive investigations and monitoring were undertaken jointly by the manufacturer and regulatory agencies. Corrective actions undertaken by the manufacturer over the next decade included (1) capture and treatment of Hg-contaminated groundwater (containing up to 7 mg l^{-1}), (2) upgrade of industrial wastewater pretreatment system for mercury process wastewaters, and (3) removal of Hg-contaminated soil (> 5500 t) at the plant site. Production of mercuric oxide batteries was also discontinued in 1992. Sediment coring in the more lacustrine portions of the embayment demonstrated evidence of burial of previously high surface concentrations under substantially cleaner sediments following actions at the

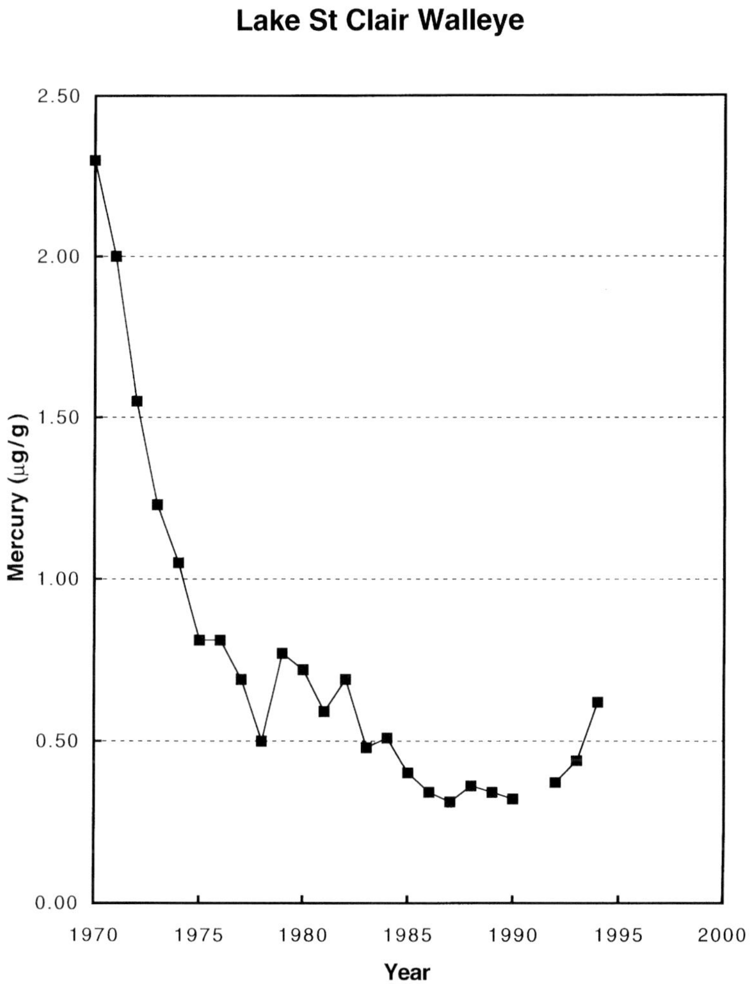

Fig. 2. Temporal trend in mercury in standardized length (45 cm) walleye (wet weight basis) from Lake St Clair, 1970–1994. (Data kindly provided by Ontario Ministry of the Environment)

plant site. Concentrations of mercury in fish in the receiving aquatic system also decreased and the fish consumption advisory was lifted in March 1992. This site is relatively unique in experiencing such a positive response to deliberate corrective actions focused on reducing biological contamination, and is one of few known sites of industrial Hg contamination in North America where a fish consumption advisory was actually lifted. Additional investigative and remedial actions are still underway.

2.3 Oak Ridge, Tennessee

Approximately 11 400 metric tons (Mg) of elemental mercury were shipped from the US government's strategic stockpile to a nuclear weapons plant in Oak Ridge in the early 1950s to support the enrichment of lithium-6, a stable isotope of lithium, which was used in the production of hydrogen bombs. The enrichment process entailed counterflowing liquid mercury and lithium hydroxide. The lighter isotope of lithium, ^{6}Li, partitioned preferentially to the mercury and was thus concentrated. The process otherwise had many similarities to electrolytic production of chlorine and caustic soda (chloralkali) using mercury cell technology. Spills and leaks of elemental mercury occurred during process development as well as during the production phase, which ended in 1963. One step in the process, involving nitric acid cleaning of the mercury, was found to be a significant source of environmental releases, and this step was altered in 1957 to conserve the stock of elemental mercury, but not before nearly 100 Mg were released into the headwaters of a local creek.

This site offers one of the few known situations where a mass balance for a major historical industrial use of mercury is available. As shown in Table 1, approximately 8% of the mercury inventory charged to Oak Ridge was either lost to the environment or could not be accounted for. Some of the unaccounted mercury was most likely never received by the facility, as incoming flasks were only counted, not weighed, with the presumption that each flask contained the typical 35 kg. Some confirmation of the total measured losses to the local creek have been provided by estimates of inventories of mercury in downstream floodplain soils (70 Mg) and reservoir sediments (80 Mg). Similarly, the use of dated sediment cores from the reservoir have independently confirmed the chronology of aquatic Hg releases from the site (Turner et al. 1984), while the annual growth rings of eastern red cedar (*J. virginiana*) trees growing near the plant site and floodplain recorded the atmospheric Hg releases (Turner and Bloom 1994).

Ambient air concentrations of mercury (2 to 1200 ng m^{-3}) have been monitored continuously at the plant site since 1986 (Turner and Bogle 1993), while ambient air monitoring on the floodplain (2 to 120 ng m^{-3}; Barnett et al. 1996) was limited to the periods just before and during soil excavation associated with remediation. In 1993, Lindberg et al. (1995) estimated emissions of mercury from the 100-ha floodplain to the atmosphere using a micrometeorological gradient approach at about 86 ng m^{-2} h^{-1}. They concluded that while this

Table 1. Mass balance of mercury at Oak Ridge Y-12 plant (Data from UCC 1983)

Environmental media	Mercury (Mg)	Percent of inventory[a]
Lost to air (1950–1963)	35	0.32
Lost to creek (1950–1993)	109	1.0
Lost to soil/rock under facility	194	1.8
Lost to headwater pond sediment	6.8	0.06
Not accounted for[b]	590	5.4
Total	935	8.6

[a]Total inventory charged to facility = 10 886 Mg.
[b]Charged inventory minus accounted for inventory plus amounts lost or spilled.

emission was well above background, it was small compared to other atmospheric Hg sources in the region, including coal-fired power plants.

The Oak Ridge site has been intensively investigated over the past 15 years with the result that a vast amount of environmental information is now available (e.g. Table 2). Corrective actions at this site have focused on identifying and eliminating point sources of mercury, as well as on eliminating human health concerns associated with high mercury level (up to 3000 $\mu g\ g^{-1}$) in floodplain soils downstream from the site. Actions included filling and capping an onsite pond (1988), renovation and replacement of subsurface storm and process drains (1986–1993), treatment of basement sump waters (largely composed of groundwater) using activated carbon systems (1993–1995), and bypassing site drainage waters around a small (1-ha) impoundment (1996). As a result of these and other actions, mercury in drainage waters leaving the site has been substantially reduced (from more than 100 g day^{-1} in 1985 to about 15 g day^{-1} in 1996) over the past decade, but concentrations (Fig. 3) still exceed the interim target regulatory limit (200 ng l^{-1}), which was predicted to yield downstream fish concentrations less than 0.4 $\mu g\ g^{-1}$ (limit selected for protection of human health). The state and federal agencies mandate a final instream limit of 12 ng l^{-1}. Whether this limit can be met is the subject of current discussion and the focus of treatment technology evaluations. Over $30 million (US) has been expended over the past decade in characterizing and treating continuing releases of waterborne mercury. While concentrations of mercury in water at the facility boundary have declined, mercury in fish (redbreast sunfish) collected from this location has also declined significantly, but in fish collected further downstream it has either remained constant or shown increases in concentration over the past decade (Fig. 4). It is suspected that as point sources of mercury at the facility have been reduced or eliminated the relative importance of instream sources (sediments, floodplain deposits) has increased and prevented recovery at downstream reaches. The temporal increase 17 km downstream from the facility (Fig. 4) is more enigmatic, but may be related to changes in water quality which enhance the production and/or bioaccumulation of methylmercury. Additional corrective actions are in progress or planned.

Floodplain soils containing approximately 80 t of mercury, distributed over 25 km of creek and 200 ha, were subjected to intensive characterization between 1982 and 1995 with a final decision to excavate all soils over 400 $\mu g\ g^{-1}$ for

Table 2. Summary of recent mercury concentrations in environmental media at the Oak Ridge site

Environmental media	Mean, max, or range of concentrations	Reference
Surface water at facility boundary ($\mu g\ l^{-1}$)	Dry weather = 0.4 to 1.0 Storm flow = 1.0 to 25	LMES (1997) Southworth et al. (1995)
Groundwater ($\mu g\ l^{-1}$)	All samples = <0.01 to 145 "Clear" samples = <0.01 to 1.3	Rothschild et al. (1984); Carmichael (1989)
Sump water ($\mu g\ l^{-1}$)	<0.2 to 1000	LMES (1997)
Recent sediments ($\mu g\ g^{-1}$)	20 to 200	DOE (1994); TVA (1985)
Suspended matter ($\mu g\ g^{-1}$dry wt.)	20 to 1000	LMES (1997)
Surface soils at facility ($\mu g\ g^{-1}$)	Mean = 13, max = 7000	Foley et al. (1989)
Floodplain soils ($\mu g\ g^{-1}$)	0 to 10 cm = 50 to 300 Subsurface max = 3000	Revis et al. (1989) TVA (1985); DOE (1994)
Ambient air ($ng\ m^{-3}$)	On site = 2 to 1200	Turner and Bogle (1993)
Floodplain emissions ($ng\ m^{-2}\ h^{-1}$)	53 to 86	Lindberg et al. (1995)
Reservoir sediments ($\mu g\ g^{-1}$)	0 to 2 cm = 0.5 to 2.0 Subsurface peak = 37	Turner et al. (1984) Olsen et al. (1992)
Sunfish ($\mu g\ g^{-1}$ wet wt.)	Fillets = 0.4 to 1.8	LMES (1997) Peterson et al. (1996)
Prey fish ($\mu g\ g^{-1}$ wet wt.)	Whole body = 0.5 to 4	Hill et al. (1996) LMES (1997)
Invertebrates: ($\mu g\ g^{-1}$ wet wt.)		
Dragonflies and damselflies	0.31 to 1.0	
Crayfish	0.4 to 3.4	G. R. Southworth (unpubl. data)
Kingfishers ($\mu g\ g^{-1}$ wet wt.)	0.6 to 0.8 (muscle)	Baron et al. (1997)
Mink ($\mu g\ g^{-1}$)	104 (hair)[a]	Stevens et al. (1997)

[a] Hg in mink hair from reference site = 5 $\mu g\ g^{-1}$.

disposal in a landfill with leachate collection. This decision and action were driven almost entirely by human health risk assessment focused on exposure of a hypothetical child via soil ingestion, and the need to relieve affected private property owners from future liability. Support for this unusually high cleanup level was provided by studies of mercury speciation and bioavailability in the soil (Revis et al. 1989; Dodd et al. 1994; Barnett and Turner 1995; Barnett et al. 1995, 1997). The soil excavation project was completed in October 1997 at a combined investigation and cleanup cost of about $30 million (US). The floodplain cleanup is not expected to materially affect concentrations of mercury in the stream or aquatic biota and it may be necessary to address this lingering issue in the future if subsequent monitoring continues to demonstrate unacceptable contamination of aquatic biota.

2.4 Pinchi Lake, British Columbia

Mercury mining adjacent to this 55-km^2 lake in north-central British Columbia, Canada, between 1940 and 1944 resulted in disposals of spoil and tailings (calcine) along the shoreline of and in the lake (Fimreite 1970). This mine

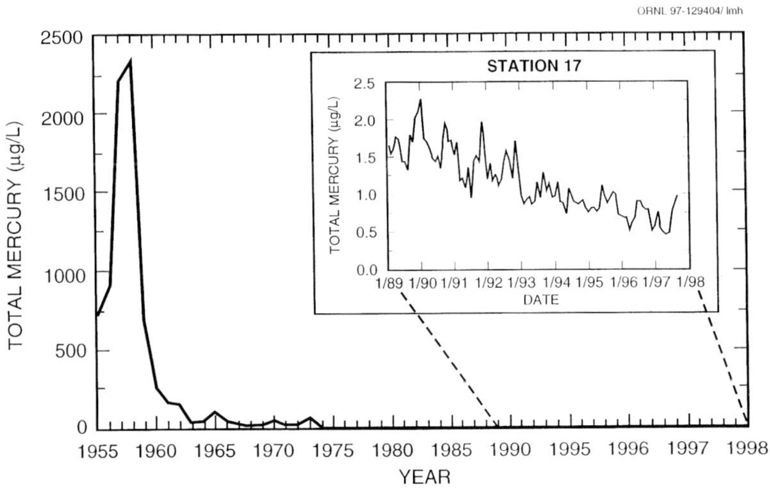

Fig. 3. Temporal trend of total mercury in streamwater from East Fork Poplar Creek, Oak Ridge, Tennessee, near boundary of nuclear weapons plant (stream km 23.4), 1955–1997. Data plotted represent annual or monthly (inset 20–40 measurements per month) averages

produced more than 2000 Mg of mercury from a high-grade cinnabar deposit (Ploufee 1995) during this first period (Siegel et al. 1985). A later phase of mining activity, which began in 1968 and ended in 1975, was conducted under more rigid environmental controls. Nonetheless, high levels of mercury were found (Fimreite et al. 1971) in lake trout (up to 10 $\mu g\ g^{-1}$) and pisciverous

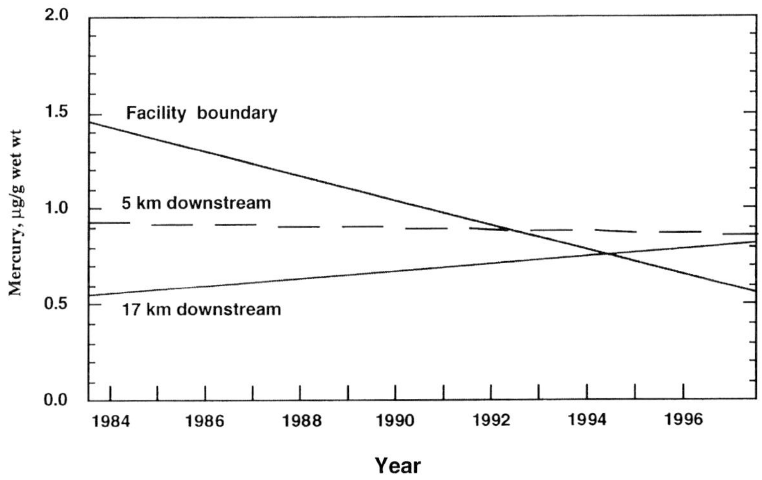

Fig. 4. Regression of mean mercury concentrations in redbreast sunfish fillets versus time, East Fork Poplar Creek at the boundary of the weapons plant, 5 and 17 km downstream. Eight fish were collected two times per year at each location

birds (up to 17 μg g^{-1} in liver of red-necked grebe) collected during the second mining period. Since mining activity ended in 1975, mercury in fish has declined substantially but the lake is still under a fishing advisory (Martin et al. 1995). Siegel et al. (1985) compared mercury in soil and plants near this mine with regional and larger patterns to try to distinguish Hg background values from mining-enhanced values. They concluded that effects of mining seemed to be intense but localized to the vicinity of the mine. Abelson and Gustavson (1979) also compared water quality and sediments in Pinchi Lake with nearby lakes not impacted by mining. No corrective actions beyond those associated with the last period of mining are known to have been undertaken at the site.

2.5
Clear Lake, California

Although somewhat larger (177 km^2) than Pinchi Lake, Clear Lake in northern California also received direct inputs of mercury mining wastes in one relatively small area (Oaks Arm) due to episodic cinnabar mining spanning nearly a century. The last activity ceased in 1957 after 4700 Mg of mercury had been produced from the site with an estimated 100 Mg being deposited in the lake (Suchanek et al. 1993, 1995). The site has been under investigation as a CERCLA site and has already experienced some minor corrective action (completed in 1992) directed at stabilization of mine wastes along the shoreline. Gill and Bruland (1990) measured mercury at several locations in this lake in 1987 and reported finding the highest concentrations (about 100 ng l^{-1}), with most in the particulate fraction, close to the mine site. Total mercury ranged from about 4 to 20 ng l^{-1} elsewhere in the lake. Suchanek et al. (1995) reported that total mercury in sediment (0.27 to 183 μg g^{-1}) and in benthic invertebrates (< 0.5 to 42 μg g^{-1}) decreases systematically (linearly or exponentially) with distance from the mine. Although contamination of fish in the lake has not been exceptionally high in recent years (most predacious fish are below 1.0 μg g^{-1}), the lake remains under a fish consumption advisory. Patterns of mercury in water and biota (plankton and fish) as a function of distance from the mine site have been somewhat enigmatic in showing either no trend or increased concentrations with increasing distance from the mine. Pisciverous wildlife has been monitored for evidence of both contamination with mercury (using feathers, fur, tissues, blood) and nesting success (Suchanek et al. 1997; Elbert and Anderson 1998; Wolfe and Norman 1998). The most recent information does not suggest that these receptors are currently affected by mercury, although older data suggest that some may have been prior to stabilization of the shoreline in 1992. The focus of ongoing investigative work at the lake has recently shifted from the inventory of mercury contained in lake sediments to acidic groundwater inputs of mercury from the mine site which appear to be more important in sustaining surface sediment contamination (Suchanek et al. 1997).

2.6 Onondaga Lake, Syracuse, New York

This alkaline, eutrophic lake received waste effluents from a variety of chemical plants and the city of Syracuse, NY, for more than a century. A chloralkali plant operated on the shores of the lake between 1947 and 1988, releasing an estimated 75 Mg of mercury to the lake over this period. The biogeochemistry of mercury in this lake has been investigated extensively and many publications are available (e.g., Bloom and Effler 1990; Wang and Driscoll 1995; Jacobs et al. 1995; Becker and Bigham 1995; Henry et al. 1995a, b; Bigham and Vandal 1996; Effler 1996). The temporal record (Effler 1996; Bigham and Vandal 1996) of mercury in fish in this lake illustrates an enigmatic pattern sometimes observed elsewhere (see also Oak Ridge and Saltville) in aquatic systems subjected to high mercury discharges which are subsequently greatly reduced: initial rapid decreases followed by leveling off at still elevated concentrations and even increases in concentrations while mercury inputs are still decreasing. This site also nicely illustrates a case where contaminated sediments were long suspected to be largely, or entirely, responsible for slow or incomplete restoration of fish concentrations to safe levels. Detailed studies and modeling eventually demonstrated that "although the sediments of Onondaga Lake contain an enormous mass of total Hg at depth, they are a net sink rather than a source for both total Hg and CH_3Hg to the lake" (Henry et al. 1995a).

2.7 Wabigoon/English/Winnipeg River System, Ontario

This site is perhaps the best-studied example in the world of an aquatic system highly contaminated with mercury from a chloralkali plant and associated wood pulp mill (e.g., Fimreite and Reynolds 1973; Armstrong and Hamilton 1973; Annett et al. 1975; Armstrong and Scott 1979; Jackson and Woychuk 1980; Jackson et al. 1982; Allan et al. 1984; Rudd et al. 1983; Parks et al. 1986, 1989, 1991; Parks and Hamilton 1987). This plant operated from 1963 until 1975 with an estimated loss of mercury to the river of about 10 t. Some of the highest concentrations of mercury ever reported for freshwater fish (northern pike: 27.8 $\mu g\ g^{-1}$, walleye: 19.6 $\mu g\ g^{-1}$, burbot: 24.8 g^{-1}) were measured at this site (summarized in Annett et al. 1975). The history of releases of mercury from this plant are nicely recorded in sediments in Clay Lake, where successive corings have shown the peak being gradually covered by cleaner sediment (W.L. Lockhart, pers. comm.). In spite of this evidence of burial of mercury from historic releases, mercury in fish has not returned to levels allowing lifting of consumption advisories. For example, in 1990, walleye from Clay Lake still averaged over 2 $\mu g\ g^{-1}$, four times higher than the Canadian human health limit. Detailed studies of geographical patterns in mercury in this system revealed some apparent anomalies. For example, Parks et al. (1991) noted that while the overall trend for the entire system shows that mercury concentrations in biota decrease with distance downstream from the former mercury source, this was not true for the Wabigoon River alone. Mercury concentrations in biota actually increased with distance downstream. This pattern was ascribed to increasing bioavailability of mercury with distance. Thus, although water and sediment mercury concentrations decreased in the downstream direction, the fraction that was methylated and thus most bioavailable increased in downstream direction. Data from this site support the hypothesis that general water-quality conditions are far more important in regulating mercury in fish than total mercury in water and sediment.

2.8 Saltville, Virginia

Chemical plants have operated along the banks of the North Fork of the Holston River (NFHR) in Virginia since the late 1800s. Construction and operation of a chloralkali plant from 1952 to 1972 resulted in Hg contamination of the river downstream of the plant and of a large reservoir 100 km downstream. An estimated 100 Mg of mercury was released into the river (NUS 1983). Disposal of mercury waste (approx. 45 Mg; NUS 1983) into large settling ponds containing solid Solvay wastes (from soda ash production) has resulted in extensive seepage of mercury into the river long after the facility was closed and the chloralkali plant building razed. Mercury contamination of fish from the North Fork Holston River in Virginia was first discovered in the late 1960s (EPA 1995; Derryberry 1972; Bailey 1974; Toole and Ruane 1976; Hildebrand et al. 1980a), soon after mercury bioaccumulation became recognized as a concern associated with chloralkali plant discharges. In 1970, in-plant controls at the facility reduced mercury losses to approximately 100 g day^{-1}. All discharges from ongoing operations ceased with closure of the plant in 1972. Four years after the plant was closed, mercury releases to the NFHR, mainly from pond seepage and overflow, were estimated (Lindberg and Turner 1977; Turner and Lindberg 1978) to be 50 g day^{-1}, with comparable air emissions from the surface of the waste ponds.

In 1982–1983 river sediments along approximately 600 m of stream were dredged and the rock bottom grouted to seal in mercury which could not be removed (TVA 1983) The mercury cell building was razed and the site capped with clean fill, including placement of riprap along the river bank adjacent to the former plant site. Other actions were taken to reduce percolation of runoff and direct rainfall through the large waste ponds which had been used to dispose of aqueous wastes from the mercury cell operation. However, considerable (800 LPM) seepage water from this pond, with mercury concentrations around 50 μg l^{-1} continued to be discharged into the adjacent river. This flow was finally directed through an activated carbon-based treatment system in 1995.

Summertime baseflow measurements of total inorganic and methylmercury (unfiltered samples) made by us on samples from NFHR near Saltville in 1996 and 1997 found water-borne inorganic mercury to be relatively low for a contaminated site. Aqueous total mercury averaged 17 ± 6.5 ng l^{-1} on three sampling dates. Methylmercury was 0.19 and 0.49 ng l^{-1} when sampled in 1996 and 1997. Typical values for uncontaminated streams in geologically similar watersheds are 1–5 ng l^{-1} for total mercury and 0.02–0.06 ng l^{-1} for methylmercury. Dry weather total mercury concentrations in NFHR at Saltville in the late 1970s were typically in the range from 100–200 ng l^{-1} (Turner and Lindberg 1978).

Mercury concentrations in fish in NFHR appeared to increase approximately twofold in the period from 1970 to 1975 (Fig. 5). Over this period, the facility owner, now aware of concerns over mercury releases, was taking steps to reduce discharges of mercury. An explanation for this phenomenon was that an increase in bioavailability of mercury in the NFHR system occurred at the same time that mercury releases were being restricted as a consequence of greatly reduced discharges of waste brine (high in NaCl) to the river (Bailey 1974) at the facility. Average dissolved solids in NFHR below the site decreased from about 3500 mg l^{-1} in 1970 to about 400 mg l^{-1} in 1972 (Bailey 1974). Complexation of inorganic mercury with chloride at high chloride concentrations may make mercury less available for methylation by microorganisms (cf. Blum and Bartha 1980; Barkay et al. 1997), but improved water quality could also affect mercury bioaccumulation by promoting development of a more diverse streambed invertebrate fauna and hence longer food chain. Mercury concentrations in fish in NFHR reached a peak in 1976–1979, and have gradually decreased since that time (Fig. 5). The exponential rate of approach to an assumed background concentration (0.20 μg g^{-1} wet wt.) at the two monitoring sites closest to the facility was about 6.5% per year over the period from 1976–1995, suggesting that average mercury concentrations in fish at monitoring sites within several kilometers of the Saltville site would approach (within 10%) background

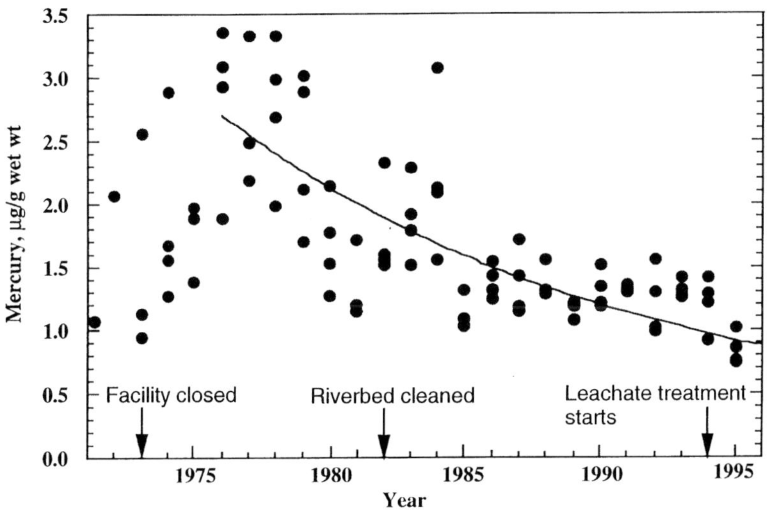

Fig. 5. Temporal trend in mean concentrations of mercury in rockbass and hogsuckers at North Fork Holston River km 126 and 115. Each *point* is the mean mercury concentration in one species at one site. *Curve* is least-squares fit exponential decay [$(Hg)_{fish} = 2.53e^{0.065t} + 20$, $R^2 = 0.61$], 1976–1995

concentrations in 75 years. The completion in 1994 of the leachate collection and treatment system for Pond 5 at the site was followed by a decrease in mean mercury concentrations in fish at the uppermost monitoring sites in 1995. The treatment system eliminates a major remaining source of dissolved mercury inputs to NFHR, and may be effective at accelerating the rate at which mercury contamination in fish decreases in the river.

The downstream pattern of mercury accumulation in fish in the NFHR below the site was approximated by the downstream pattern of dilution of the river flow at Saltville (Fig. 6) in the lower two thirds of the river below the site, but the upper 35 km below the site are an unexplained exception to this pattern. A similar downstream pattern of mercury bioaccumulation in fish was observed in East Fork Poplar Creek in Oak Ridge, Tennessee, in the mid-1980s (LMES 1997); however, that pattern was replaced in the 1990s by one that showed no relationship to downstream dilution of the headwater source. Assuming that the downstream pattern of aqueous total mercury in the NFHR remains similar to downstream dilution of the Saltville input with additional flow as it was in the late 1970s (Turner and Lindberg 1978), it would appear that a five-fold reduction of mercury concentrations at Saltville would be needed for mercury concentrations in fish to return to background (0.20 $\mu g\ g^{-1}$) levels. This would essentially require baseflow mercury concentrations to fall to about 4 $ng\ l^{-1}$, or within the background range for limestone-fed streams in the Appalachian ridge and valley region.

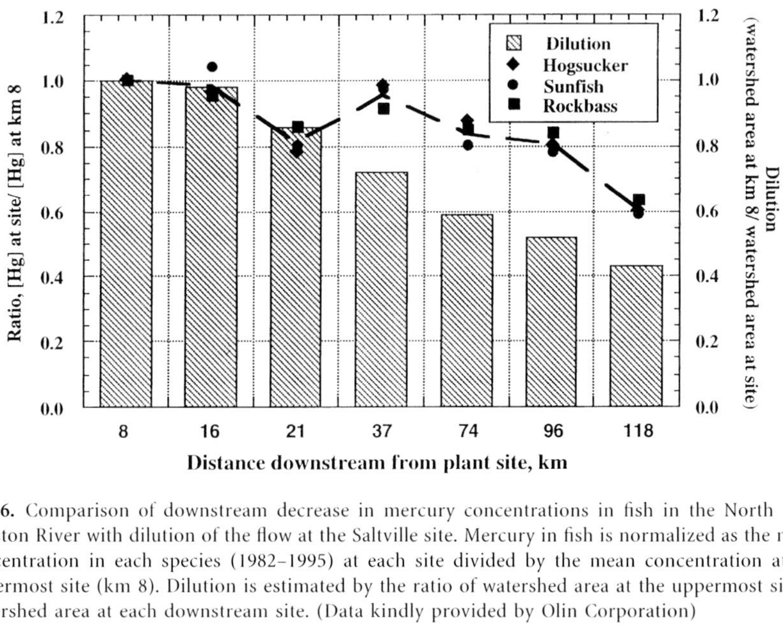

Fig. 6. Comparison of downstream decrease in mercury concentrations in fish in the North Fork Holston River with dilution of the flow at the Saltville site. Mercury in fish is normalized as the mean concentration in each species (1982–1995) at each site divided by the mean concentration at the uppermost site (km 8). Dilution is estimated by the ratio of watershed area at the uppermost site to watershed area at each downstream site. (Data kindly provided by Olin Corporation)

2.9 South River, Waynesboro, Virginia

Mercury contamination in the South River, downstream from Waynesboro, Virginia, has been evident since 1977, but the contamination originated during 1929–1950 when an industrial facility in Waynesboro used mercuric sulfate as a catalyst in producing acetate fiber (Carter 1977; VWRRC 1979). Metallic mercury spills contaminated storm drains at the facility, with an estimated 500–1000 kg released to the South River in the 1930s and 1940s (Carter 1977). Mercury contamination in sediments was evident in the South River, the South Fork of the Shenandoah River, and the Shenandoah River, a distance of 160 miles from the source of contamination. In 1983, mercury concentrations in filets of adult redbreast sunfish (*Lepomis auritus*) from the South River averaged 0.7 $\mu g\ g^{-1}$ wet weight 17 km downstream from Waynesboro (Crimora), and about 0.4 $\mu g\ g^{-1}$ in the South Fork Shenandoah River, 125 km downstream (Nicoletto and Hendricks 1987). Higher mercury concentrations were found in rock bass (*Ambloplites rupestris*), which averaged 0.84 $\mu g\ g^{-1}$ at the Crimora site in 1986 and 1.37 $\mu g\ g^{-1}$ in 1987 (Bidwell and Heath 1993). Fillets of adult redbreast sunfish collected and analyzed by us in 1996 and 1997 at Crimora and Waynesboro averaged 1.1 and 1.4 $\mu g\ g^{-1}$ mercury, respectively.

Aqueous total mercury and methylmercury (unfiltered samples) concentrations were measured by us in the South River in 1996 and 1997 under dry weather summertime conditions. Total mercury at Crimora averaged (± SD) $97 \pm 58\ ng\ l^{-1}$ on three sampling dates, and methylmercury was 2.7 and 0.4 $ng\ l^{-1}$ on two sampling dates.

Continuing contamination of fish and water with mercury decades after discontinuing the use of mercury at industrial facilities on the South River and NFHR highlights the potential for mercury contamination of pathways leading to surface waters to cause long-term contamination of those streams. Remedial efforts at many sites are unlikely to reduce mercury inputs to levels at which contamination of fish and invertebrates approaches background concentrations.

2.10 Marine and Estuarine Sites

Several marine or estuarine sites in North America are known to have experienced significant historic mercury inputs from industrial sources of mercury. These include Chaleur Bay, New Brunswick, Berry Creek near Woodridge New Jersey, Cape Fear River near Wilmington, North Carolina, Turtle Creek near Brunswick, Georgia, Mobile Bay in Alabama, Houston Ship Channel and Lavaca Bay in Texas, Bellingham Bay in Washington, and Howe Sound in British Columbia. Little or no easily accessible information was available for most of these sites.

A "mercury-processing" plant operated adjacent to Berry Creek between 1930 and 1974, apparently releasing as much as 300 Mg of mercury into soils (on-site concentration > 10 000 $\mu g\ g^{-1}$, off-site > 1000 $\mu g\ g^{-1}$) surrounding the site and the adjacent tidal marsh (Lipsky et al. 1981). Results of analysis of mercury in fish and shellfish (Santoro and Koepp 1986) collected from the area in 1979–80 did not suggest an immediate human health hazard (values were elevated but not above levels requiring consumption advisories). These authors and others who conducted investigations at the site (Berman and Bartha 1986) speculated that the sulfide-rich marsh sediments may have reduced the bioavailability of mercury. In contrast to the early findings of relatively low biotic contamination, analysis of biota conducted in the mid 1980s (HMDC 1987, 1989) revealed high average concentrations in a number of fish (white perch, killifish, eel, catfish, pumpkin-seed) and crab. Previously, Berman and Bartha (1986) had speculated that improved water quality due to a planned sewage diversion could potentially increase the bioavailability of mercury. Whether such improved water quality actually accounted for the differences in biotic contamination between the two sampling periods is undetermined. This site is particularly interesting in the context of the hypothesis that general improvements in water quality at such sites may lead to paradoxical trends in biotic contamination. Hopefully, more information about this site will appear in the accessible literature in the future.

Lavaca Bay, a secondary embayment of Matagorda Bay in Texas, was contaminated by mercury releases from a chloralkali plant which operated on the shore of this bay between 1966 and 1979. The plant was razed in 1985. Today, the bay is one of the few marine systems with a partial closure to fishing due to mercury contamination. Releases to the bay were greatly reduced in 1970 after high concentrations of mercury were found in oysters and crabs. Mercury in these organisms decreased rapidly after releases were reduced. Regular monitoring of mercury in fish (red and black drum) began in 1977 and has shown moderately elevated concentrations which have changed little since monitoring began. Concentrations of mercury in sediment, water, and biota (including fish, crabs, oysters, and benthic invertebrates) have been highest rather consistently among many studies at sampling locations closest to the site of the former chloralkali plant (Locarnini and Presley 1996; C. Dobbs, pers. comm.). King et al. (1991) found somewhat higher mercury residues in eggs from terns and black skimmers collected from Lavaca Bay in 1984 compared with a reference area, but residues were not correlated with nesting success. Recent intensive studies at Lavaca Bay have redefined the focus of potential corrective actions. Results of these intensive studies are scheduled for publication in the open peer-reviewed literature (C. Dobbs, pers. comm.). As with some other sites reviewed here, the possibility of continuing releases from the former plant site has been examined more closely and found to be a credible explanation for slow further recovery of the site.

Turtle Creek near Brunswick, Georgia, received effluent from a chloralkali plant between 1957 and 1989. Winger et al. (1993) reported that sediments (with Hg up to 27 $\mu g\ g^{-1}$) and sediment pore water (Hg not measured) immediately below the site were toxic, but they could not unambiguously attribute the toxicity to mercury. The sediments also contained elevated concentrations of polychlorinated biphenyls and other metals. This site is currently under intense investigation and some corrective actions (e.g., soil removal and groundwater treatment) are in progress or have already been completed. The site is one of several chloralkali plants where leaks of caustic soda radically altered groundwater chemistry (highly alkaline pH caused extraction of soil organic matter) and led to unusual partitioning of mercury almost entirely to the dissolved phase.

Bellingham Bay in Washington has received effluent from a chloralkali plant since 1965. No recent information was available for this site. Earlier, Bothner et al. (1975) and Bothner (1973) investigated the spatial and temporal distribution of mercury in bay sediments after releases from the facility were sharply reduced in 1970. Bothner (1973) estimated total releases from the facility to be about 18 Mg prior to 1970. These investigators attempted to measure and model decreases in mercury in the estuarine sediments after point source releases were greatly reduced. Their results suggested that decreases in sediment concentrations (with a half-time of 1.3 years) were mainly due to particle-associated mercury being diluted by sediment-mixing processes and by bottom-current transport of resuspended sediment. They also found that fluxes of dissolved mercury from anoxic sediments were significantly greater than from oxic sediments.

Howe Sound near Squamish, British Columbia, received effluents from a chloralkali plant beginning in 1965 and ending in 1994. Approximately 40 Mg of mercury are estimated (Harbo and Birdwell 1983) to have been discharged into the sound prior to 1970 when the releases were greatly curtailed. Upper Howe Sound was closed to all fishing in 1970 but reopened again to sport fishing in 1978. A long record (1970 to present) of annual monitoring of Dungeness crabs shows rapid decreases from values close to 2 $\mu g\ g^{-1}$ in 1970 to values around 0.2 $\mu g\ g^{-1}$ in 1975, followed by no subsequent changes. The site is currently implementing a number of corrective actions, including containment (slurry wall technology), capture and treatment of groundwater, and demolition of buildings.

3
Nature and Effectiveness of Remediation Strategies

Few mercury-contaminated sites in North America have actually undergone deliberate comprehensive remediation, nor have the consequences of specific corrective actions always been determined, and thus information on the effectiveness of remediation strategies is rather limited. This is not to imply that no corrective actions have been undertaken. In addition to the corrective actions described earlier and summarized in Table 3, many sites have at least experienced default remediation by virtue of having processes which employed mercury discontinued or altered to reduce or eliminate losses to the environment. At a few sites (e.g., Saltville, Oak Ridge), process buildings have been razed or decontaminated, and contaminated soil covered with clean soil or excavated for disposal/treatment. Excavation or capping of Hg-contaminated soils has been a common-corrective action. Excavated soils have been commonly treated to reduce or stabilize mercury by a variety of technologies, but the floodplain soils (approx. 30 000 m^3, >400 $\mu g\ g^{-1}$) from the Oak Ridge site were disposed of by reburial in a landfill with a leachate collection and monitoring system. There are a few known examples where river sediments (e.g., Saltville) were dredged to remove mercury from aquatic environments.

A limited number of sites are currently pumping and treating Hg-contaminated groundwater (GW). In addition to the groundwater treatment at Oak Ridge and Lavaca Bay already described above, treatment is underway, or soon will be at a number of other sites. For example, two spatially well-defined Hg-contaminated GW plumes (up to 40 $\mu g\ l^{-1}$) were found in Centreville, Mississippi, beneath a landfarm for refinery wastewater sludges. The nature of the Hg sources was never identified and could not be unequivocably related to the landfarming activity. Nonetheless, GW recovery systems were installed to capture the plumes. A combination of capture wells and trenching was used to intercept GW greater than the Maximum Concentration Limit (MCL) of 2 $\mu g\ l^{-1}$ for mercury. To date, more than 70 million gallons of GW have been treated successfully with activated carbon at this site. In another case Hg-contaminated groundwater beneath a former chloralkali plant site in McIntosh, Alabama, is being extracted using several wells with a combined pumping rate of 2000 to 2500 LPM. Treatment entails adsorption on activated carbon columns which have not needed to be changed in 3 years [Influent (Hg) = 20 to 50 $\mu g\ l^{-1}$; effluent (Hg) = < 0.2 $\mu g\ l^{-1}$]. This equates to an effective carbon adsorption capacity of approximately 0.18 kg of mercury per 45 kg of carbon (4000 $\mu g\ g^{-1}$). This experience and the one described above suggest that activated carbon is an effective sorbent for mercury, at least for water containing $\mu g\ l^{-1}$ levels of mercury. Although activated carbon seems to be the most common sorbent used to remove mercury, at least one former chloralkali plant site (Squamish, British Columbia) is performing batch treatment of groundwater with precipitants followed by filtration. Influent concentrations at this site are typically 400 $\mu g\ l^{-1}$ and greater than 99% mercury removal is achieved.

As noted elsewhere in this Volume (Hempel and Thoeming), many remediation technologies have been proposed and evaluated at the laboratory or pilot scale but, thus far, few have been demonstrated in North America.

Table 3. Examples of corrective actions implemented at some Hg-contaminated sites

Corrective action	Example site(s) where implemented
Soil excavation, with reburial or treatment	Oak Ridge, Turtle Creek, many others
Groundwater capture and treatment	Oak Ridge, Centreville, Lavaca Bay, McIntosh, Howe Sound
Sediment dredging/removal	Saltville
Soil capping/stabilization	Oak Ridge, Clear Lake
Surface water rerouting	Oak Ridge, Saltville
Rock grouting	Saltville
Storm sewer rehabilitation (cleaning, relining, replacement)	Oak Ridge, many other sites
Slurry wall containment	Howe Sound
Bank armoring/ripapping	Clear Lake, Saltville

4 Summary and Conclusions

Investigation and remediation of Hg-contaminated sites in North America has been a continuing and iterative process in spite of the fact that many sites have been known for decades. Among numerous reasons for these delays in restoration, two figure prominently: (1) the unique properties of Hg, which have complicated identification of appropriate restorative efforts, and (2) the slow evolution of environmental regulations, which has led to only recent reconsideration of relative risks. Presently, many sites in North America are under renewed examination to determine the need for and nature of any corrective action. Information available to the authors from these sites suggests the following general conclusions.

1. The occurrence and environmental significance of small continuing releases of mercury from facilities no longer using mercury, or long closed, has been either not considered or underestimated. Major sites in North America which illustrate this issue include Onondaga Lake, Lavaca Bay, and Clear Lake. Contaminated sediments in all three of these systems were initially considered to be the root cause of slow recoveries. While sediments have not been eliminated as significant sources at these sites, they now at least share the spotlight with continuing releases from the facilities of mercury origin.
2. Partitioning of mercury between dissolved and particulate phases (e.g., in groundwater) can be radically altered in industrial and mining settings resulting in enhanced mobility. Chlorinated effluents (e.g., cooling water) and highly alkaline leachates (spills of caustic, mining calcine) can effectively reverse the strong tendency of mercury to affiliate with solids (e.g., clay minerals, organic matter) and lead to greater relative solubility and unexpected environmental mobility.
3. Mercury concentrations in biota (especially fish, crustacea, and shellfish) have been observed in some cases to decrease rapidly following reductions in waterborne releases of mercury from point sources. Relationships between concentrations of waterborne inorganic mercury and bioaccumulation are poorly understood at contaminated sites, but appear to be highly site-specific in nature. Further reductions may be limited by continuing inputs from facilities as discussed above, by residual contamination which persists in sediments, or by water-quality conditions which are especially favorable to production of methylmercury even when sediment concentrations are approaching background.
4. Spatial and temporal patterns in mercury in biota and environmental media often reveal unexpected or enigmatic trends: apparent temporal increases in mercury in fish as general water quality improves (Oak Ridge, Onondaga Lake, Saltville), lower methylmercury concentrations and lower fractions of total Hg to methylmercury near point sources but increasing downstream (Oak Ridge, Wabigoon River).

5. Natural burial of mercury-contaminated sediments has effectively removed a considerable inventory of mercury from contact with the biosphere at many sites, but especially marine bays, lakes, and reservoirs where bulk sedimentation rates have been relatively high. Uncertainty persists about the permanence of this burial in some cases where extreme natural events, such as hurricanes, or anthrogenic activities, such as dredging, may disinter long-buried inventories.
6. Natural recovery, even augmented by remedial efforts, at many sites where Hg contamination has been widely dispersed is unlikely to reduce mercury inputs to levels at which contamination of fish and invertebrates approaches background concentrations. This is not to imply that full recovery is not eventually possible. Given sufficient money and/or time, even the most difficult sites could be restored (e.g., Minamata Bay, see Kudo and Turner, this Vol.). However, one has to question seriously whether the large commitment of resources to achieve background contamination levels is mandatory to protect human and ecological health.

References

Abelson DHG, Gustavson E (1979) Water quality and sediment analysis for heavy metals in selected lakes in the Pinchi fault area of north central BC Omineca Peace Region, British Columbia Ministry of the Environment, Vancouver

Allan RJ, Brydges T, Dodge D, Hamilton RD, Jeffs DG, Shiaze (eds) (1984) Mercury pollution in the Wabigoon-English river system of Northwestern Ontario and possible remedial measures. Ontario Ministry of the Environment, Toronto, Canada

Annett CS, D'Itri FM, Ford JR, Prince HH (1975) Mercury in fish and waterfowl from Ball Lake, Ontario. J Environ Qual 4(2):219–222

Armstrong FAJ, Hamilton AL (1973) Pathways of mercury in a polluted northwestern Ontario lake, In:Singer PC (ed) Trace metals and metal-organic interaction in natural waters. Ann Arbor Science Publishers, Ann Arbor, pp 131–155

Armstrong FAJ, Scott DP (1979) Decrease in mercury content of fishes in Ball Lake, Ontario, since imposition of controls on mercury discharges. J Fish Res Board Can 36:670–672

Bailey DS, (1974) The occurrence of mercury in the fish and sediment of the North Fork of the Holston River 1970–1972. Basic Data Bulletin 41, Virginia State Water Control Board. Richmond, Virginia. 62 pp

Barkay T, Gillman M, Turner RR (1997) Effects of dissolved organic carbon and salinity on bioavailability of mercury. Appl Environ Microbiol 63(11):4267–4271

Barnett MO, Turner RR (1995) Bioavailability of mercury in East Fork Poplar Creek Soils. Y/ER-215, US Department of Energy Oak Ridge Y-12 Plant, Oak Ridge, Tennessee

Barnett MO, Harris LA, Turner RR, Henson TJ, Melton RE, Stevenson RJ (1995) Characterization of mercury species in contaminated floodplain soils. Water Air Soil Pollut 80:1105–1108

Barnett MO, Owens JG, Lindberg SE, Turner RR (1996) Mercury concentrations in air during the phase I remediation of Lower East Fork Poplar Creek floodplain at the Oak Ridge Y-12 Plant, Oak Ridge, Tennessee. Rep No Y/ER-281, US Department of Energy Oak Ridge Y-12 Plant, Oak Ridge, Tennessee

Barnett MO, Harris LA, Turner RR, Stevenson RJ, Henson TJ, Melton RC, Hoffman DP (1997) Formation of mercuric sulfide in soil. Environ Sci Technol 31(11):3037–3043

Baron LA, Ashwood TL, Sample BE, Welch C (1997) Monitoring bioaccumulation of contaminants in the belted kingfisher (*Ceryle alcyon*). Environ Monit Assess 47:153–165

Becker DS, Bigham GN (1995) Distribution of mercury in the aquatic food web of Onondaga Lake, New York. Water Air Soil Pollut 80:563–571

Berman M, Bartha R (1986) Control of the methylation process in a mercury-polluted aquatic sediment. Environ Pollut (Ser B) 11:41–53

Bidwell JR, Heath AG (1993) An in situ study of rock bass (*Ambloplplites rupestris*) physiology: effect of season and mercury contamination. Hydrobiologia 264:137–152

Bigham GN, Vandal G (1996) A drainage basin perspective of mercury transport and bioaccumulation: Onondaga Lake, New York. Neurotoxicology 17:279–290

Bloom NS, Effler S (1990) Seasonal variability in the mercury speciation of Onondaga Lake, New York. Water Air Soil Pollut 53:251–265

Blum JE, Bartha R (1980) Effect of salinity on methylation of mercury. Bull Environ Contam Toxicol 25:404–408

Bothner MH (1973) Mercury: some aspects of its marine geochemistry in Puget Sound, Washington. PhD Diss, University of Washington, Seattle, Washington, 126 pp

Bother MH, Jahnke RA, Peterson ML, Carpenter R (1980) Rate of mercury loss from contaminated estuarine sediments. Geochim Cosmochim Acta 44:273–285

Carmichael JK (1989) An investigation of shallow groundwater quality near East Fork Poplar Creek, Oak Ridge, Tennessee. Water Resources Investigation Rep 88–4219, US Geol Survey, Nashville, Tennessee

Carter LJ (1977) Chemical plants leave unexpected legacy for two Virginia rivers. Science 198:1015–1020

Derryberry OM (1972) Investigation of mercury contamination in the Tennessee Valley Region. In:Hartung R, Dinman BD (eds) Environmental mercury contamination. Ann Arbor Science Publishers, Ann Arbor, pp 76–79

D'Itri FM (1972) The environmental mercury problem. CRC Press, Cleveland, Ohio, 124 pp

Dodd D, Miller E, Cardenas D, Brown K (1994) Determination of mercury, with speciation, in Poplar Creek soil samples. Task 22410124, US Environmental Protection Agency, Environmental Monitoring Systems Laboratory, Las Vegas, Nevada

DOE (1994) East Fork Poplar Creek/Sewer Line Beltway remedial investigation report. DOE/OR/02-1119&D2&V2, US Department of Energy (DOE), Oak Ridge, Tennessee

DOI (1970) News release, September 16, 1970. US Department of the Interior (DOI), 6 pp

Effler SW (ed) (1996) Limnological and engineering analysis of a polluted urban lake. Springer, Berlin Heidelberg, New York, 832 pp

Elbert RA, Anderson DW (1998) Mercury levels, reproduction, and hematology in western grebes from three California lakes, USA. Environ Toxicol Chem 17(2):210–213

EPA (1975) Materials balance and technology assessment of mercury and its compounds on national and regional bases. US Environmental Protection Agency, EPA 560/3-75-007, Washington, DC

EPA (1995) Saltville waste disposal ponds superfund site, Saltville, Virginia. Record of decision operable unit 2 (pond 5 and pond 6). US Environmental Protection Agency, Region III, Philadelphia, Pennsylvania

Fimreite N (1970) Mercury uses in Canada and their possible hazards as sources of mercury contamination. Environ Pollut 1:119–131

Fimreite N, Reynolds D (1973) Mercury contamination in fish in northwestern Ontario. J Wildlife Manage 37(1):62–68

Fimreite N, Holsworth WN, Keith JA, Pearce PA, Gruchy IM (1971) Mercury in fish and fish eating birds near sites of industrial contamination in Canada. Can Field Nat 85(30):211–220

Foley RD, Carrier RF, Zeighami EA (1989) Results of outdoor radiological and chemical survey of the Y-12 site. Rep Y/TS-600, US Department of Energy, Y-12 Plant, Oak Ridge, Tennessee

Gill GA, Bruland KW (1990). Mercury speciation in surface freshwater systems in California and other areas. Environ Sci Technol 24(9):1392–1400

Harbo RM, Birtwell IK (1983) Trace metal content data for crustaceans and fishes from Howe Sound, British Columbia. Data Rep No 379, Can Dept Fish Aquat Sci, 44 pp

Henke KR, Kuhnel V, Stepan DJ, Fraley RH, Robinson CM, Charlton DS, Gust HM (1993) Critical review of mercury contamination issues relevant to manometers at natural gas industry sites. Topical Rep GRI-93/0117, Gas Research Institute, Chicago, Illinois, 92 pp

Henry EA, Dodge-Murphy LJ, Bigham GN, Klein SM, Gilmour CC (1995a) Total and methylmercury mass balance in an alkaline, hypereutrophic urban lake (Onondaga Lake, New York). Water Air Soil Pollut 80:509–518

Henry EA, Dodge-Murphy LJ, Bigham GN, Klein SM (1995b) Modeling the transport and fate of mercury in an urban lake (Onondaga Lake, NY). Water Air Soil Pollut 80:489–498

Hildebrand SG, Strand RH, Huckabee JW (1980a) Mercury accumulation in fish and invertebrates of the North Fork Holston River, Virginia and Tennessee. J Environ Qual 9:393–400

Hildebrand SG, Lindberg SE, Turner RR, Huckabee JW, Strand RH, Lund JR, Andren AW (1980b) Biogeochemistry of mercury in a river-reservoir system:impact of an inactive chloralkali plant on the Holston River-Cherokee Reservoir, Virginia and Tennessee. ORNL/TM-6141. Oak Ridge National Laboratory, Oak Ridge, Tennessee

Hill WR, Stewart AJ, Napolitano GE (1996) Mercury speciation and bioaccumulation in lotic primary producers and primary consumers. Can J Fish Aquat Sci 53:812–819

HMDC (1987) Mercury in biota – data summary. Environmental Operations Research Laboratory Hackensack Meadowlands Development Commission (HMDC), Lyndhurst, New Jersey

HMDC (1989) Berry's Creek mercury in aquatic biota – data summary. Environmental Operations Research Laboratory Hackensack Meadowlands Development Commission (HMDC), Lyndhurst, New Jersey

Jacobs LA, Klein SM, Henry EA (1995) Mercury cycling in the water column of a seasonally anoxic urban lake (Onondaga Lake, NY). Water Air Soil Pollut 80:553–562

Jackson TA, Woychuk RN (1980) Mercury speciation and distribution in a polluted river-lake system as related to the problem of lake restoration. In:Restoration of lakes and inland waters. Proc Int Symp on Inland waters and lake restoration, Portland, Maine, Sept 1980, Rep EPA 440/5-81-010, US Environmental Protection Agency, Washington, DC, pp 93–101

Jackson TA, Parks JW, Jones PD, Woychuk RN, Sutton JA, Hollinger JD (1982) Dissolved and suspended mercury species in the Wabigoon River (Ontario, Canada):seasonal and regional variations. Hydrobiologia 92:958–960

Jasinski SM (1995) The materials flow of mercury in the United States. Resour Conserv Recyl 15:145–179.

King KA, Custer TW, Quinn JS (1991) Effects of mercury, selenium, and organochlorine contaminants on reproduction of Forster's terns and black skimmers nesting in a contaminated Texas bay. Arch Environ Contam Toxicol 20:32–40

Lindberg SE, Turner RR (1977) Mercury emissions from chlorine-production solid waste deposits. Nature 5616:133–136

Lindberg SE, Kim K-H, Meyers TP, Owens JG (1995) Micrometeorological gradient approach for quantifying air/surface exchange of mercury vapor: tests over contaminated soils. Environ Sci Technol 28(1):126–135

Lipsky D, Reed RJ, Harkov R (1981) Mercury levels in Berrys Creek. Report for New Jersey Department of Environmental Protection, Office of Cancer and Toxic Substances, Trenton, New Jersey

Locarnini SJP, Presley BJ (1996) Mercury concentrations in benthic organisms from a contaminated estuary. Mar Environ Res 41(3):225–239

LMES (1997) Mercury abatement report for the US Department of Energy Y-12 Plant for Fiscal Year 1997, Oak Ridge, Tennessee. Y/ER-297, Lockheed Martin Energy Systems Inc., Oak Ridge, Tennessee

Martin JW, Duncan WFA, Masselink MNM (1995) Mercury concentrations in fish tissue and sediment from Pinchi Lake, British Columbia. Draft Report, Environmental Protection Branch, BC Environment, Prince George, British Columbia

Miller EM (1983) The mercury pollution of High Rock Lake: a case study analysis of policy decisions with implications for environmental mediation. MS Thesis, Duke University, Durham, North Carolina

NCDEHNR (1993) Phase I diagnostic-feasibility study of Abbotts Creek embayment of High Rock Lake, Davidson County, North Carolina. Report by Division of Enviromental Management, North Carolina Department of Environment, Health, and Natural Resources (NCDEHNR), Raleigh, North Carolina, 66 pp

Nicoletto PF, Hendricks AC (1987) Sexual differences in accumulation of mercury in four species of centrarchid fishes. Can J Zool 66:944–949

NUS (1983) Remedial action master plan (RAMP) – Saltville Waste Disposal Site, Smyth County, Virginia. USEPA Work Assignment 01-3V24, Contract 68-01-6699, NUS Corporation, Gaithersburg, Maryland

NYSDEC (1992) Evaluation of off-site contamination associated with a mercury recycling facility: Mercury Refining Company (Colonie, NY), New York State Department of Environmental Conservation, Albany, New York

Olsen CR, Larsen IL, Lowry PD, Moriones CR, Ford CJ, Dearstone KC, Turner RR, Kimmel BL, Brandt CC (1992) Transport and accumulation of cesium-137 and mercury in the Clinch River and Watts Bar Reservoir System. Environmental Restoration Program Rep ORNL/ER-7. Martin Marietta Energy Systems, Oak Ridge, Tennessee, 50 pp

Parks JW, Hamilton AL (1987) Accelerating recovery of mercury-contaminated Wabigoon/English River system. Hydrobiologia 149:159–188

Parks JW, Sutton JA, Lutz A (1986) Effect of point and diffuse source loadings on mercury concentrations in the Wabigoon River:evidence of a seasonally varying sediment-water partition. Can J Fish Aquat Sci 43:1426–1444

Parks JW, Lutz A, Sutton JA (1989) Water column methylmercury in the Wabigoon/English River-Lake system. Factors controlling concentrations, speciation, and net production. Can J Fish Aquat Sci 46:2184–2202

Parks JW, Curry C, Romani D, Russell DD (1991) Young northern pike, yellow perch and crayfish as bioindicators in a mercury-contaminated watercourse. Environ Monit Assess 16:39–73

Peterson MJ, Southworth GR, Crumby WD (1996) Monitoring mercury in fish in a stream system receiving multiple industrial inputs. Environ Monit Assess 40:91–105

Ploufee A (1995) Glacial dispersal of mercury from bedrock mineralization along Pinchi Fault, north central British Columbia. Water Air Soil Pollut 80:1109–1112

Revis NW, Osborne TR, Holdsworth G, Hadden C (1989) Distribution of mercury species in soil from a mercury-contaminated site. Water Air Soil Pollut 45:105–113

Rothschild ER, Turner RR, Stow SH, Bogle MA, Hyder LK, Sealand OM, Wyrick HJ (1984) Investigation of subsurface mercury at the Oak Ridge Y-12 Plant. ORNL/TM-9092, Oak Ridge National Laboratory, Oak Ridge, Tennessee, 258 pp

Rudd JWM, Turner MA, Furutani A, Swick AL, Townsend BE (1983) The English-Wabigoon River system. I. A synthesis of recent research with a view towards mercury amelioration. Can J Fish Aquat Sci 40:2206–2217

Rule JH, Iwashchenko MS (1998) Mercury concentrations in soils adjacent to a former chlor-alkali plant. J Environ Qual 27:31–37

Santoro ED, Koepp SJ (1986) Mercury levels in organisms in proximity to an old chemical site (Berrys Creek, Hackensack Meadowlands, New Jersey, USA). Mar Pollut Bull 17(5):219–224

Scheider WA, Cox C, Hayton A, Hitchin G, Vaillancourt A (in press) Current status and temporal trends in concentrations of persistent toxic substances in sport fish and juvenile forage fish in the Canadian waters of the Great Lakes. J Environ Monit Assess

Siegel SM, Siegel BZ, Lipp C, Kruckeberg A, Towers GHN, Warren H (1985) Indicator plant-soil mercury patterns in a mercury-rich mining area of British Columbia. Water Air Soil Pollut 25:73–85

Southworth GR, Turner RR, Peterson MJ, Bogle MA (1995) Form of mercury in stream fish exposed to high concentrations of dissolved inorganic mercury. Chemosphere 30:779–787.

Stevens RT, Ashwood TL, Sleeman JM (1997) Mercury and PCBs in muskrats (*Ondatra zibethicus*) and mink (*Mustela vision*) from the US Department of Energy Oak Ridge Reservation. Bull Environ Toxicol Chem 58:720–725

Suchanek TH, Richerson PJ, Woodward LA, Slotton DG, Holts LJ, Woodmansee CEE (1993) A survey and evaluation of mercury in sediment, water, plankton, periphyton, benthic invertebrates and fishes with the aquatic ecosystem of Clear Lake, California. Preliminary Lake Study Report, University of California-Davis

Suchanek TH, Richerson PJ, Holts LJ, Lamphere BA, Woodmansee CE, Slotton DG, Harner EJ, Woodward LA (1995) Impacts of mercury on benthic invertebrate populations and communities within the aquatic ecosystem of Clear Lake, California. Water Air Soil Pollut 80:951–960

Suchanek TH, Richerson PJ, Mullen LH, Brister LL, Becker JC, Maxson AE, Slotton DG (1997) The role of the Sulphur Bank Mercury Mine Site (and associated hydrogeological processes) in the dynamics of mercury transport and bioaccumulation within the Clear Lake aquatic ecosystem. Interim Final Report, University of California-Davis

Toole TW, Ruane RJ (1976) Evaluation of the mercury monitoring program in the north Fork Holston River. E-WQ-76-2, Tennessee Valley Authority, Chattanooga, Tennessee

Turner RR, Bloom NS (1994) Reconstruction of historical atmospheric Hg releases using analysis of tree rings in red cedar (*Juniperus virginiana*). Book of Abstracts, Int Conf Mercury as a Global pollutant, 10–14 July 1994, Whistler, British Columbia

Turner RR, Bogle MA (1993) Ambient air monitoring for mercury around an industrial complex. In: Chow W, Connor KK (eds) Managing hazardous air pollutants – state of the art. Electric Power Research Institute, Lewis Publishers, Boca Raton, pp 162–172

Turner RR, Lindberg SE (1978) Behavior and transport of mercury in river-reservoir system downstream of inactive chloralkali plant. Environ Sci Technol 12(8):918–923

Turner RR, Olsen CR, Wilcox WJ Jr (1984) Environmental fate of mercury and 137-cesium discharged from Oak Ridge facilities. In: Hemphill DD (ed) Trace substances in environmental health XVIII. University of Missouri, Columbia, pp 329–338

TVA (1983) Quicksilver cleanup. Tennessee Valley Authority (TVA), Chattanooga, Tennessee. TVA Impact 6:5–7

TVA (1985) Sediment characterization, Task 2, Instream contaminant study. Office of Natural Resources and Economic Development, Tennessee Valley Authority (TVA), Chattanooga, Tennessee, 82 pp

UCC (1983) Mercury at the Y-12 plant – a summary of the 1983 UCC-ND task force study. Union Carbide Corp (UCC) Rep Y/EX-23, US Department of Energy, Oak Ridge, Tennessee

VWRRC (1979) Mercury contamination in Virginia waters: history, issues, and options. Special Rep No 8. Virginia Water Resources Research Center (VWRRC), Virginia Polytechnic Institute and Stae University. Blacksburg, Virginia. 7 pp

Wang W, Driscoll C T (1995) Patterns of total mercury concentrations in Onondaga Lake, New York. Environ Sci Technol 29:2261–2266

Winger PV, Lasier PJ, Geitner H (1993) Toxicity of sediments and pore water from Brunswick Estuary, Georgia. Arch Environ Contam Toxicol 25:371–376

Wolfe M, Norman D (1998) Effects of waterborne mercury on terrestrial wildlife at Clear Lake: evaluation and testing of a predictive model. Environ Toxicol Chem 17(2):214–227

Remediation Techniques for Hg-Contaminated Sites

M. Hempel and J. Thoeming

1 Introduction

Mercury-polluted soil is a very serious problem, which we have to face not only in the highly industrialized regions, but also in developing countries. Although the production and use of mercury has decreased in the Western World since the late 1970s (in the USA from 1632 t Hg in 1972 to ca. 411 t Hg in 1990), in some developing countries the import and use is increasing, e.g. in 1972 Brazil imported 115 t and in 1990 240 t Hg (Hasse 1994; Jasinski 1995).

The sources and uses of mercury with respect to its chemical form are listed below (see Table 1). At places of mercury production or use, we often face problems of environmental contamination. Mercury is generally considered one of the most toxic elements found in the environment. Especially the organic compounds are characterized by high toxicity and genotoxicity. Although mercury is often emitted in the elemental or inorganic form, highly toxic organic compounds can be formed by biological activity or chemical reaction (Hintelmann et al. 1993). In contrast to most of the heavy metal compounds, mercury in the elemental form and also in some of the organic forms is volatile at room temperature. These chemophysical properties must be taken into account during risk assessment.

Managing sites contaminated with mercury requires the assessment of the remediation alternatives which may be applied. This chapter shows remediation techniques either already applied, emerging on the laboratory scale, or just invented and published. It reviews techniques aimed to immobilize or isolate the pollutant and focus on cleanup methods which attempt to clean the soil or at least part of the soil.

2 Remediation Principles

Once investigations of the history of a site, the extent of the contamination, the concentration of mercury and its organic compounds at the source and on its pathways to mankind and environment are finished and risk is assessed, remediation alternatives may need to be considered.

Environmental Science
Mercury Contaminated Sites (ed. by R. Ebinghaus et al.)

Table 1. Sources and uses of mercury with respect to its chemical form (Jasinski 1995)

Chemical form	Source or use
Elemental mercury	Chlorine-alkali-manufacturing Dental fillings Gold mining Electrical equipment (batteries, switches) Instruments (thermometers, barometers)
Inorganic mercury (Hg^{2+})	Electrical equipment (batteries) Skin care products Medicinal products Wood impregnation
Inorganic mercury (Hg^{2+2})	Electrical equipment (batteries) Medicinal products
Organic mercury e.g. phenylmercury, methoxyethylmercury	Fungicides Pesticides Pigments (paints) Medicinal products

Generally, we differentiate between protection techniques, which aim to immobilize or isolate the Hg, and cleanup techniques, which aim to separate the Hg from soil. These cleanup methods could be divided into on site methods, where the contaminated soil is excavated and subsequently treated, and in situ methods, where the soil is treated, e.g. by leaching, in the ground.

3 Cleanup Methods Applied in Practice

3.1 Physical Treatment

Physical treatment of excavated soil is a method, by which the Hg is removed from the soil by wet classification. A simplified process scheme of a wet classification process is shown in Fig. 1. The process basically consists of the following steps: a sorting step where the soil is separated from other material, for example rubble; a dry classification where coarse material like uncontaminated gravel is removed by sieving; an intensive mixing step where the medium and small fractions (e.g. <50 mm) are mixed with water and a separation step where the washed clean medium fraction (e.g. <50 mm and >0.1 mm) is removed from the sludge or fine fraction which is enriched with Hg. The Hg is often concentrated in the fine fraction, which is usually dewatered before leaving the wet classification plant.

In practice, no chemical additives are usually used during the above procedure. Several techniques are available for mixing the medium and small fraction with water. In the mixing step, techniques are used by which the energy

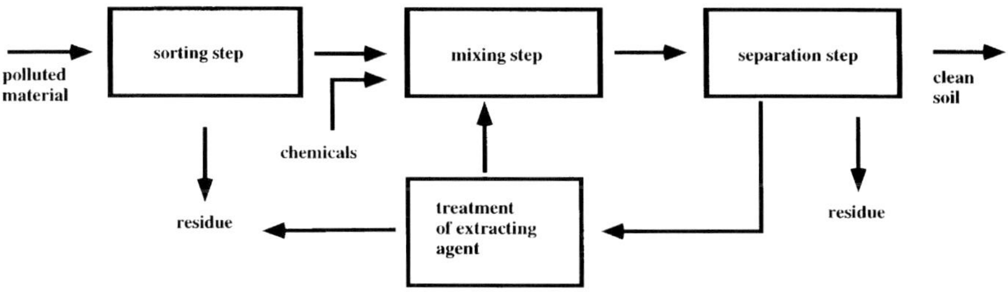

Fig. 1. Simplified process scheme of wet classification

dispenses the particles and clusters of medium and small particles, e.g. high pressure water jet (LRS 1995) or attrition cells or washing helix (Hennig 1993). Several methods, including hydrocyclones, spiral classifiers and fluidized bed separation systems, are available for separating the clean wet soil particles from the contaminated sludge. Generally, these techniques are based on differences in particle size, shape and density as well as settling velocity and surface properties.

In practice, the wet classification of excavated Hg-contaminated soil is applied on a large scale, e.g. 20–40 t/h. Due to the affinity of Hg and its compounds for the organic and clay fraction of the soil, wet classification often works for various ionic and organic Hg compounds and elemental Hg (Bachmann 1993; Hafemeister 1993; Heimhard 1993; Hennig 1993; LRS 1995).

Due to the fact that nearly all silt and clay particles as well as organic substances are concentrated in the sludge, these techniques are limited to soils containing high quantities of sand and gravel and limited quantities of silt, clay and organic substances. A general problem associated with cleaning Hg-contaminated soil is dealing with possible emissions, especially prior to the main processes. In many cases, the unloading station should be enclosed due to personal and occupational health measures. Due to the fact that most of the applied separation techniques are based on differences in particle size, density or settling velocity, light particles, which contain high mercury levels, will be excluded with the so-called clean fraction. These materials are usually relatively light demolition materials like wood, rubber or porous bricks.

3.2
Thermal Treatment of Excavated Soil

Thermal methods are based on the fact that Hg and its compounds may be volatilized. Theoretically, municipal waste incineration plants could be used to treat Hg-contaminated soil; but in contrast to municipal waste, soil has a low heating value. Fuel oil, gas or coal is required to heat up the soil. Furthermore, waste gas treatment of municipal waste burning has different requirements. Generally, municipal waste incineration plants show considerable differences from thermal soil treatment plants. For technical and economic reasons, soil is not treated in municipal waste incineration.

Table 2. Chemical and physical properties of some mercury compounds (Hempel 1993)

Mercury compound	Formula	Density (g/cm^3)	Melting point (°C)	Boiling point (°C)	Vapour-saturated (mg/m^3)	Vapor pressure at (mm Hg)	(°C)	Solubility in H_2O (g/l)
Mercury (element)	Hg^{0}	13.6	−39	356.58	14.36	1.22×10^{3}	20	6×10^{5}
Hg(I)-chloride	Hg_2Cl_2	7.15	400					0.002
Hg(II)-chloride	$HgCl_2$	4.44	276	302		8.34×10^{5}	25	69
Hg(III)-sulfide	HgS		584					1×10^{5}
Methyl-Hg-chloride	CH_3HgCl	4.06	170		66	4.7×10^{3}	20	5
Phenyl-Hg-chloride	C_6H_5HgCl		251		0.06	3.6	35	0.03
Dimethyl-Hg	$(CH_3)_2Hg$	3.07		93		58.6	23.7	

For remediation purposes it is an advantage that various Hg compounds, as well as elemental Hg, are volatile at relatively low temperatures (see Table 2). To remove Hg from excavated soil by thermal treatment, Hg needs to be transformed from a solid phase of the soil to the gas phase by heating. The temperature required for removing Hg and its compounds is theoretically below 600 °C; if HgO is absent, 250 °C should be sufficient. Most thermal soil plants are, however, designed to remove all kinds of pollutants from the soil, e.g. volatile hydrocarbons (petrol, oil, PAH, etc.) and halogenated organic compounds (HCH, PCB, dioxine, furane). Treatment plants work at much higher temperatures (up to 1300 °C) than needed to remove Hg and its compounds from the soil. Most of these all-purpose thermal soil plants use a direct heating system with after-burner, but pyrolysis techniques are also used. A typical soil treatment process of an all-purpose soil treatment plant is shown schematically in Fig. 2. After entering the first oven, the soil is warmed to a temperature of 100 °C in a drying drum; subsequently, the soil enters a heating drum at a temperature of approximately 600 °C and the pollutants are released. Before leaving the plant as a clean matrix, the soil passes through a cooling drum where it is cooled and moistened. The waste gas from the drying and the heating drum enters an after-burner (ca. 800–900 °C). A heat exchanger cools the waste gas to temperature of

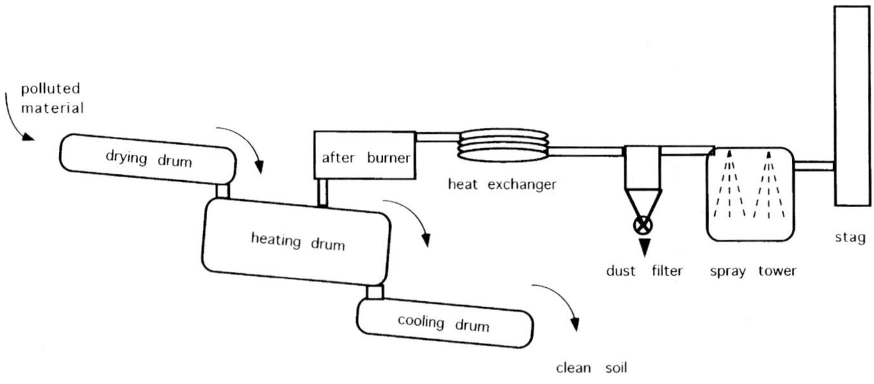

Fig. 2. Simplified process scheme of an all-purpose soil treatment plant

about 150 °C and heats the gas for the heating drum. Subsequently, the gas passes through a dust filter and a spray tower to reduce dust and SO_2 emissions.

A thermal process, optimized for the remediation of Hg-contaminated soil, was developed and constructed by Fortmann and Jahns (1993). The semimobile plant is able to clean approximately 6 t/h. Soil and demolition material is crushed in a special grinding device, which is under air control. The crushed and sieved material enters a rotary kiln at a temperature of 1000 °C, where it rests for a maximum of 3 h. Heating is performed by natural gas. After leaving the kiln, the soil is cooled to a temperature of 80 °C. The waste gas enters an after-burner and subsequently a heat exchanger where the gas is cooled down to 200 °C. After passing through a dust filter, the gas still contains most of the Hg. This is removed by a gas-washing system, which consists of a venturi washer and a two-step spray tower in which the gas is cooled to 40 °C. Hg is removed by chemical oxidation at low pH and subsequent chemical sorption. The gas contains less than 0.05 mg Hg/Nm^3 at the end of the pipe.

Since mechanical treatment is cheaper, most of the remediation companies combine wet classification with subsequent thermal treatment. It is advantageous to enlarge the fraction which can be cleaned by simple wet classification and therefore to minimize the fraction which needs to be treated by thermal techniques.

Renner (1995) separated a fraction with a grain size <1 mm, which is dewatered and subsequently introduced into a steam distillation chamber. The water steam is produced by an external steam generator and is introduced into the chamber by nozzles, which are located above and below a sieve. The soil fraction is moved above the sieve and the steam streams through the contaminated material. A duration time of 0.5 to 1 h at temperatures of 250 to 540 °C are used for different kinds of material, prior to entering a cooling chamber. The steam, which is enriched with pollutant, is pumped out of the chamber and passes through a hot-gas dust filter which prevents a recontamination of the dust. Subsequently, the steam is condensed and the resulting water is treated in a wastewater facility. Before the gas is exhausted to the atmosphere, it passes through a charcoal filter. The technique has been implemented in a pilot plant which has a capacity of 0.5 t/h. Various materials (soil, rubble material, charcoal, wood and oil sludge) with contamination levels between 71 and 13 000 mg Hg/kg were tested. Final concentrations of Hg in the cleaned soil of 0.2 to 11 mg Hg/kg were obtained.

In 1992, Harbauer GmbH installed a large soil-washing and vacuum distillation plant for cleaning mercury-contaminated soil and demolition rubble from the former Marktredwitz chemical works (Hennig 1993). The wet classification plant itself has a capacity of 25 t/h, while the thermal part of the plant has a capacity of 6 t/h. After a crushing and sorting step, which is in-house, the soil is separated by wet classification. An Hg-enriched soil fraction is fed into a vacuum distillation plant and cleaned. In a second stage the fine-grained suspension (<100 μm) is prethickened with the help of flocculating agents to form compressible sludge. After dewatering in a chamber filter press, the sludge is predried. Subsequently, the material enters the vacuum distillation module where it is treated at a pressure of 100 hPa and an average temperature of 310 °C. The waste gases are condensed so that the mercury can be recovered for reuse. The condensed water is treated in a wastewater treatment facility and all air involved in the process is passed through a waste gas-cleaning device. The plant is able to clean soil with an average contamination of 3000 mg Hg/kg to an average concentration of 2 mg Hg/kg.

Another thermal technique combined with vacuum is described in a patent (Bernecker and Ratzel 1988) where mercury-bearing materials are cleaned by distillation. The material is introduced as a suspension or solution by a thin layer combuster under vacuum/low pressure.

Chemical Waste Management Inc. (1992) desorbs Hg from soil at temperatures between 93 and 482 °C in an externally fired rotary dryer. The processed solids are cooled with treated condensed water to eliminate dusting. A nitrogen gas transports the volatilized pollutants out of the dryer to the gas treatment system. The gas first passes a high-energy scrubber to remove particles and part of the organic contaminants. Afterwards, it passes through two condensers in series, where it is cooled to less than 10 °C. The plant can process 120 to 180 t/day. Mercury levels have been reduced from 5100 to 1.3 mg Hg/kg.

A batch system which combines thermal and vacuum techniques is described by Gronholz, (Gronholz 1992; 1995). The main steps of the technique are a thermal pretreatment at approx. 200 °C and 200 hPa with a duration time of 5 to 30 min, followed by the main thermal process, taking place in an externally heated reactor under a vacuum of 0.5 to 1 hPa at temperatures of 250 to 400 °C. The technique was applied to Hg-contaminated soil, drilling sludge and wood, and showed final concentrations of 0.23 to 1.8 mg Hg/kg d.w. from an input concentration of 0.41 to 150 000 mg Hg/kg d.w.

A thermal technique with a chemical pretreatment is described by Weyand et al. (1994). The soil is mixed with chemical additives that facilitate the decomposition of mercury compounds and the formation of elemental mercury. The pretreated soil is heated in two stages, firstly to vaporize the low-temperature volatiles such as water and thereafter at a higher temperature to vaporize/decompose the elemental mercury and mercury compounds present in the soil mixture. The technique has been applied on a pilot plant with an approximate capacity of 113 kg.

Although different principles, techniques and designs are implemented in the above thermal soil treatment plants, some problems and limitations often occur:

1. Rotating parts in the plant, connected with heating devices or in the areas with temperatures differences, require special material, maintenance, stoppages and costs.
2. Material which is entering the heated zone of the plant may calcinate or agglomerate to burned gravels. Evaporation of the Hg captured in these gravels requires a much longer duration time in the oven.
3. In plants with high volumes of gas, Hg separation out of the waste gas could be a problem, or at least expensive. Besides, the separation of the fine dust from gas is a technical challenge. In some plants a recondensation of the treated dust could occur.
4. Contamination of the plant devices itself should be noted. High concentrations of Hg in the surfaces of the devices, for example the oven, could limit the final concentrations of the processed material.

3.3 Immobilization of Contaminated Soil

Due to technical, environmental, geographical, social or financial reasons, a cleanup of contaminated soil is often not feasible. In these cases, methods which are not intended to remove mercury are applied. These techniques of immobilization, isolation, solidification and stabilization aim to reduce the risk to mankind and the ecosystem without a cleanup (Bölsing 1993; Mull 1993; Müller-Kirchenbauer et al. 1993). A large variety of techniques are available and may be classified into the following groups.

3.3.1 *Physical Barriers*

The installation of physical/mechanical barriers helps to reduce the lateral and vertical mobilization of the pollutant. Lateral barriers consist of sealing walls connected to a natural or an artificial aquiclude or impermeable bed as the bottom barrier. A top sealing layer is installed to reduce the ingress of rainwater and the emission of gas and dust. For vertical and horizontal sealing, there are a great variety of techniques and commercial suppliers, showing differences in material, ageing, chemical resistance and costs. Meschede and Vogelsberger (1995) used an impermeable base and cover comprised of two layers of asphalt concrete and a layer of sulphur to act as a mercury immobilization barrier.

3.3.2 *Geohydrological Isolation*

If a natural impermeable bed is not present and a man-made bottom barrier (injection system) is impossible to realize, geohydrological isolation is another possibility to isolate the polluted site. This is usually performed by a series of pumping wells, often combined with a water infiltration system and lateral barriers.

3.3.3 *Solidification and Stabilization*

These on-site techniques involve mixing the contaminated soil with, for example, cement, gypsum, calcium carbonate or lime, along with other binders such as fly ash or furnace slag silicate reagents. Specialized suppliers add reagents to bind with the metal. Some suppliers use silicate compounds to solidify and stabilize contaminated soils and sludges. After pre-treatment where coarse material is separated out, a silicate reagent and water are added, mixed and then poured into confining pits or cast into moulds (Hazardous Waste Control 1992; Silicate Technology Corporation 1992).

Another typical treatment process begins by adding water and the chemical additive to the contaminated soil (Funderburk and de Percin 1992). Subsequently, cement or fly ash is mixed with the suspension. After hardening reaction takes place, and after some hours, a concrete-like mass with low permeability (10^{-9} cm/s) can be deposited at suitable sites. Leaching tests are used for quality control of solidified mass.

3.3.4
Chemical Immobilization

In contrast to solidification and stabilization techniques, there are some methods which aim to immobilize mercury by transforming the existing chemical species into a form of lower toxicity, lower solubility or lower available form. This can be achieved by adding a sulphur-containing chemical additive to the soil (Aqua Control 1992). Hg reacts with the sulphur compounds to form HgS or HgS-containing compounds. The chemical reagent can be added either by spraying onto the excavated soil or by injecting the sulphur-containing solution into the undisturbed soil using high pressure.

Immobilization techniques in general have the disadvantage that the pollutants remain in the environment. Chemical or physical immobilization could be a sufficient treatment for a period of time. The physical barriers may break down or have limited barrier function. Immobilized Hg compounds could be remobilized by changes of pH, redox or by biological activity. There are doubts that sulphur compounds added to Hg in the soil form pure HgS.

4
Emerging and Potential Alternative Cleanup Methods

4.1
Alternative Thermal Techniques

A large variety of modified thermal techniques are emerging, and have been demonstrated on a pilot-scale plant.

A steam distillation method was designed by Luther et al. (1995) for highly contaminated sludges. Mercury is released from the sludge by steam distillation at temperatures below 300 °C and so predrying is not necessary. The mercury-bearing steam is separated from the soil by cyclones. The gas is cleaned by a washing technique, followed by condensation and adsorption on charcoal. The condensed water is cleaned by ultrafiltration and reverse osmosis before reuse.

Koshinski et al. (1996) describe a pilot-scale facility where the contaminated material is blended with additives prior to processing. The additives need to decompose stable mercury compounds, convert mercury-compounds to elemental form, retain the chloride and sulphide components and prevent the release of chloride and sulphur in gaseous form. The blended material is heated in two stages up to 540–650 °C. Gaseous mercury is condensed in a heat exchanger.

Good results were obtained in a 6000-t project: the feed contained between 3000 and 7000 mg Hg/kg, the cleaned soil between 0.44 and 1.9 mg Hg/kg.

A combination of high vacuum (50 mm mercury) and "low" temperature (600 °C) infrared radiation is described by Dagidian and Stolz (1996). The aim of the vacuum-low temperature combination is to depress the boiling points of the Hg compounds. The compounds are desorbed by energy-efficient infrared radiation. In bench scale tests, a feed with a mercury contamination of 0.5 to 615 mg Hg/kg was reduced to <0.002 to 0.03 mg Hg/kg.

Navetta (1996) describes a continuous processing technique where the material is transported in an internally fired, hollow screw desorber, which operates under vacuum levels of 5×10^4 N/m^2 at temperatures between 175 and 650 °C.

4.2
Hydrometallurgical Extraction (Cleanup) Processes

Because of their similarity, all extraction processes that make use of aqueous solutions can be grouped under the general term hydrometallurgical extraction processes. The term hydrometallurgy originates from the field of extractive metallurgy, which involves the treatment of metallic ores. It covers all wet metal extraction and separation processes, operating at low temperatures such as leaching, electrochemical reduction and ion exchange. Wet extraction involves the recovery of the metals from soil using solvent or lixivant (see Fig. 3).

4.2.1
Thermodynamic and Kinetic Considerations

Thermodynamics is the basis of all extraction processes and provides the possibility to ascertain process limitations. It allows equilibrium calculations of (1) species distribution, (2) the solubility of simple or complex salts, (3) the loading limits of leach liquors, and (4) determination of the extent of reaction

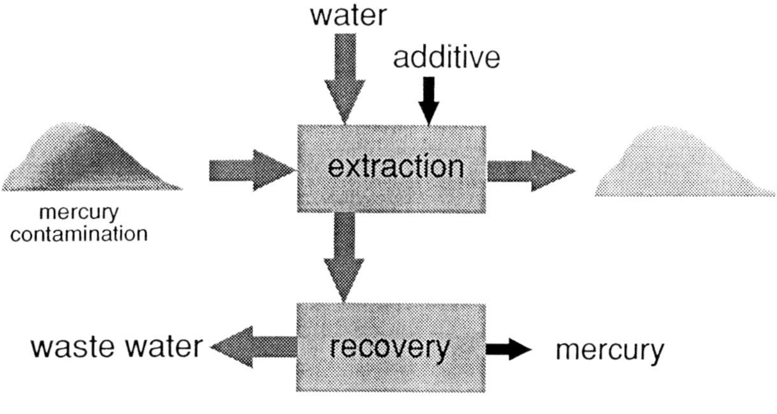

Fig. 3. Principle of hydrometallurgical cleanup processes

and its dependence on concentrations and temperature. While in high-temperature processes the rates of reaction are generally fast enough to assume an equilibrium state for equilibrium processes, in hydrometallurgy the reaction rate is the limitation of thermodynamical or kinetic predictions.

The three fundamental parameters, pH, the redox potential of the solution and the concentration of complexing agents, determine the speciation of mercury and the solubility of mercury. Moreover, these three parameters are responsible for the equilibrium state of all reactions of interest, the reaction rates and also for the equilibrium between adsorption /desorption. The parameters can be controlled by adding appropriate chemicals like acids, oxidizing agents or complexing agents.

4.2.2
Extraction Mechanisms

In soils, mercury is bound or attached to different solid phases. Thus, different extraction mechanisms have to take place to achieve a selective cleanup. The chemical mechanisms involved can be classified as:

1. Desorption of adsorbed species.
2. Oxidation of metallic mercury.
3. Complexing extraction of organically complexed mercuric ions by competing strong complexing agents.
4. Dissolution of precipitated mercury such as mercury oxide (Fig. 4).

It should be noted that the equilibria controlling these mechanisms depend highly on each other. In particular, readsorption or recomplexation of the dissolved mercury species on surfaces or organic matter may cause a shift in the mercury distribution within the different soil phases and lead to a low extraction efficiency. Besides chemical mechanisms, transport mechanisms also control the extraction. Since mercury species can be bound to soil phases which are covered by another phase, a coating of iron oxide for example, diffusion limits the speed of the extraction process to a great extent. Since soil is a porous system, micropores account for most of the surface area of the soil particles. Moreover, it is the inside of these micropores where the adsorbed as well as complexed mercuric ions are located. Thus, in an agitated leaching module, mass transport can be enhanced by increasing the turbulence or by diminishing the particle size by grinding. Another possibility to overcome the problem of mass transport is to apply an electric field. This results in electromigration of the mercuric ions, the dissociation of complexes and the electroosmotic flow of the whole pore water.

4.2.3
Technical Approaches

Leaching Techniques. The term leaching is often applied to the process of extracting a metal from soil. Due to the use of extracting agents, the cleanup of soils by leaching is also called chemical extraction. Approaches for mercury

form Hg(o). More recently, Rugh et al. (1996) described the potential of transgenic plants to remediate Hg-contaminated soil.

5 Conclusions

Due to financial and technological constraints, there are only a few possibilities for treating mercury-contaminated sites. Wet classification is applied in many cases. In the 1980s, diverse types of techniques were tested and used. At present, the authors do not expect significant developments, which suggests that wet classification will be used for a larger number of applications. The technique is realized in small mobile plants as well as in large >100 t/h plants. It is a relatively cheap cleanup method. Regulatory authority permission is often less time-consuming than for thermal techniques.

Thermal methods are now in the stadium of diversification. Different types of techniques, often only on the laboratory scale, have been described. Only a few techniques are applied full scale. Thermal removal has the advantage that it reduces the Hg-containing material to a minimum. There are still problems, e.g. with porous fine material such as bricks and sludge. It is quite time-consuming to obtain regulatory permission for a thermal plant. Besides, the acceptance in the neighbourhood of a thermal plant is typically poor. The measures which have to be taken for occupational health are more expensive than in any other applied technique.

Compared with thermal treatment, hydrometallurgical extraction processes would have the following advantages:

- low costs at small operation scales,
- higher public acceptance,
- less emissions: wastewater can be avoided due to recirculation of the spent liquid,
- potentiality of in-situ treatment in the case of electrokinetic techniques,
- high flexibility for treating multi-metal contaminations,
- larger thermodynamic separation factors, which leads to a lower energy consumption.

In contrast, the disadvantages are the following:

- care has to be taken to avoid corrosion of the plant facilities,
- in the case of chlorination processes, the formation of (1) chlorinated compounds is likely if the soil contains organic matter and also (2) chlorine emissions,
- leaching agents could cause the dissolution of a significant fraction of the soil, e.g. acid dissolution of carbonates could lead to a mass defect of more than 75% in the case of limestone

With respect to different geological and environmental situations, there are a large number of techniques to immobilize mercury in soils and rocks. However, these are only time-limited solutions. There is still a lack of methods to prove long-term stability of immobilized materials. So we have to take into account that environmental changes, e.g. of the pH, redox or salt concentration, may remobilize the mercury. A future cleanup after immobilization could be more expensive, if not impossible, due to the added immobilizing chemicals.

References

Anonymous (1991) Verfahren zum Reinigen kontaminierter Erdmassen und Vorrichtung zur Durchführung des Verfahrens. Ger Offen DE 39 31 017

Aqua Control (1992) Sanierung quecksilberbelasteter Standorte. Information by aqua control Wiehl-Oberbantenberg, Germany

Bachmann A (1993) Soil remediation at Schweizerhalle – a case study. In: Eijsackers HJP Hamers T (ed) Integrated soil and sediment research: a basis for proper protection. Kluwer, Dordrecht, pp 695–712

Ball DL, Boateng DAD (1991) Method for the recovery of mercury from mercury-containing material. US Patent 5,013,358

Ballester A, Otero E, Gonzalez F (1988) Mercury extraction from cinnabar ores using hydrobromic acid. Hydrometallurgy 21:127–143

Barkay T, Saouter E, Turner RR (1995) Microbial activities for the bioremediation of mercury contamination. Proc of the 2nd SETAC World Congr, 5–9 Nov 1995, Vancouver, Canada, 81 pp

Behrends W, Lorenz I (1989) Verfahren zur Dekontaminierung elementares Quecksilber enthaltener Böden. Ger Offen DE 3814684

Bernecker G, Ratzel G (1988) Verfahren zur destillativen Reinigung quecksilberhaltiger Stoffe und/oder Stoffgemische und/oder zur Gewinnung von Quecksilber und Vorrichtung zur Durchführung des Verfahrens. Deutsches Patentamt DE 38 16 282 C1

Beyer K, Krauss R, Fatrzek F (1990) Dekontaminationsverfahren für schwermetallverunreinigte Böden, speziell für Quecksilber. Ger Patent DE 3312986 C2

Bölsing F (1993) Verfestigen. In: Weber HH, Neumaier H (eds) Altlasten. Springer, Berlin Heidelberg New York, pp 279–299

Brown D, Carlson AR (1973) Process for preparing elemental mercury. US Patent 3,732,094

Brunke M, Deckwer WD, Frischmuth A, Horn JM, Lünsdorf H, Rohde M, Röhricht M, Timmis KN, Weppen P (1993) Microbial retention of mercury from waste streams in a laboratory column containing merA-gene bacteria. FEMS Microbiol Rev 11:145–152

Chemical Waste Management Inc (1992) X TRAX thermal desorption. The Superfund Innovative Technology Evaluation Program: Technology Profile, 5th edn. EPA/540/R-92/077 Nov 1992, pp 66–67

Cox CD, Shoesmith MA, Ghosh MM (1996) Electrokinetic remediation of mercury-contaminated soils using iodine/iodide lixiviant. Environ Sci Technol 30(6):1933–1938

Dagidian JV, Stolz CA (1996) High vacuum temperature thermal desorption-remediation of mercury-impacted soil. Proc of the Int Conf on Incineration and thermal treatment technologies; 6–10 May 1996, Savannah, Georgia, pp 207–212

Doering F, Decristofero P (1990) Verfahren zum Dekontaminieren eines schadstoffbelasteten Erdbodenreiches. Ger Patent DE 39 20 342

Fortmann J, Jahns P (1993) Thermische Behandlung. In: Weber HH, Neumaier H (eds) Altlasten. Springer, Berlin Heidelberg New York, pp 209–239

Foust DF (1994) Extraction of mercury compounds from contaminated material and solutions. Int Appl WO 94/009167, PCT

Funderburk R, de Percin P (1992) Dechlorination and immobilization. The Superfund Innovative Technology Evaluation Program: Technology Profile 5th edn. EPA/540/R-92/077 Nov 1992, pp 92–93

Gronholz C (1992) Neu entwickelte Technologie zur Quecksilberentfernung aus Boden und Schlamm. In: Franzius V (ed) Sanierung kontaminierter Standorte 1992 Abfallwirtsch Forsch Praxis 52 (Berl) 213–220

Gronholz C (1995) Die erfolgreiche Behandlung von Hg-kontaminiertem Boden, Bohrschlamm und Holz mit der UTN-Schadstoffrückgewinnungstechnologie. GKSS-Workshop Reinigung Hg-kontaminierter Schluffe, 7.9.1995, GKSS Forschungszentrum Geesthacht GmbH, Geesthacht, Germany, pp 95–100

Hafemeister D (1993) Verfahren zum Reinigen von mit elementarem Quecksilber verunreinigten Haufwerk. Deutsches Patentamt DE 41 35 079 A1

Hansen HK, Ottosen LM, Kliem BK, Villumsen A (1996) Electrodialytic soil remediation. J Chem Tech Biotechnol 70: 67–73

Hasse RF (1994) Der Quecksilberhandel in Brasilien. In: Mathis A, Rehaag R (eds) Gold und die Folgen. Volksblatt Verlag, köln pp 94–113

Hazardous Waste Control (1992) NOMIX technology. The Superfund Innovative Technology Evaluation Program: Technology Profile 5th edn. EPA/540/R-92/077 Nov 1992 pp 100–101

Heimhard HJ (1993) Waschen. In: Weber HH, Neumaier H (eds) Altlasten.Springer, Berlin Heidelberg New York, pp 239–270

Hemmings RL, Bradbury D, Elder GR, Dunn MJ (1992) Removal of hazardous and radioactive metals from soil using electrochemical migration. Proc of Symp on Waste Management 92, vol 2, Tucson, Arizona, 1-5 March 1992

Hempel M (1993) Entwicklung und Anwendung von Methoden zur Gefährdungsabschätzung quecksilberkontaminierter Standorte. PhD, Technical University of Hamburg-Harburg GKSS-Forschungszentrum, Geesthacht, Germany, p 131

Hennig R (1993) Reinigung von quecksilberkontaminierten Böden mit Hilfe eines kombinierten Wasch- und Destillationsverfahrens. In: Arendt F, Annokkée GJ, Bosman R, van den Brink WJ (eds): Altlastensanierung '93, Kluwer, Dordrecht pp1335–1344

Hintelmann H, Hempel M, Wilken RD (1993) Observation of organic mercury species in soils sediments of industrially contaminated sites. Environ Sci Technol 29: 1845–1850

Jasinski SM (1995) The materials flow of mercury in the United States. Resources Conserva Recycl 15: 145–179

Koshinski CJ, McLaughlin JB, Weyand TE (1996) Removal and recovery of mercury from mixed wastes. Proc of the Int Conf on Incineration and thermal treatment technologies, 6–10 May 1996, Savannah, Georgia, pp 181–189

Lagemann R, Pool W, Seffinga G (1989) Electro-reclemation: theory and practise. Chem Ind: 18:585–590

Luther G, Wietstock P, Hempel M (1995) Remediation of mercury contaminated soils and sludges by steam distillation. In: van den Brink WJ, Arendt F (eds) Contaminated soil '95. Kluwer, Dordrecht, pp 1287–1288

LRS (1995) Sanierung einer quecksilberkontaminierten Altlast in Regensburg.Information by Lurgi Renner Sanierungsverfahren GmbH, Regensburg, Germany, pp 10

Meschede T, Vogelsberger R (1995) Remediation of a mercury-contaminated site in Egypt. In: van den Brink WJ, Arendt F (eds) Contaminated soil '95. Kluwer, Dordrecht, pp 237–238

Mull R (1993) Hydraulische Massahmen. In: Weber HH, Neumaier, H. (eds) Altlasten.Springer, Berlin Heidelberg New York, pp 198–208

Müller-Kirchenbauer H, Friedrich W, Günther K, Nußbaumer M, Stroh D (1993) Einkapselung. In: Weber HH, Neumaier H (eds) Altlasten. Springer, Berlin Heidelberg New York pp 168–197

Navetta MS (1996) RTG's Thermal screw processor: recovering mercury from contaminated soils. Proc Int Conf on Incineration and thermal treatment technologies, 6–10 May 1996, Savannah, Georgia pp 219–221

Parks G, Fittinghoff NA (1970) Mercury extraction now possible via hypochlorite leaching. Engin Mining J 6:107–109

Renner J (1995) Erfahrungen bei der Reinigung kontaminierter Böden durch Dampfdesorption mit dem Terrasteam-Verfahren. Information by Lurgi Renner Sanierungsverfahren GmbH, Regensburg, Germany, pp 10

Rugh CL, Wilde HD, Stack NM, Marin-Thompson D, Summers AO, Meagher RB (1996) Mercuric ion reduction and resistance in transgenic *Arabidopsis rhalina* plants. Proc Natal Acad Sci USA 93: 3182–3187

Rulkens WH, Tichy R, Grotenhuis JTC (1995) Sites polluted with heavy metals: current techniques for clean-up and desirable future developments. In: Wilken RD, Förster U, Knöchel A (eds) Proc of the Int Conf on Heavy metals in the environment CEP Consultants Ltd 1995, Edinburgh, pp 10–19.

Sandstede G, Köhling A, Schönbucher A (1993a) Verfahren zur Entfernung von Quecksilber und anderen Schwermetallen aus Erdreich und Schlämmen. Ger Offen DE 42 12 322

Sandstede G, Köhling A, Schönbucher A (1993b) Elektrochemisches in-situ-Verfahren zur Entfernung von Quecksilber und anderen Schwermetallen aus Erdreich, Schlämmen und Wässern. Ger Offen DE 42 10 950

Scheiner BJ, Lindstrom RE, Henrie TA (1972) Extraction of mercury from mercury-bearing materials. US Patent 3,639,222

Silicate Technology Corporation (1992) Solidification and stabilization treatment technology. The Superfund Innovative Technology Evaluation Program: Technology Profile 5th edn. EPA/540/R-92/077 Nov 1992, pp 162–163

Tittel R, Hess W(1992) Verfahren und Anlageprinzip zur Trennung von Quecksilber aus kontaminierten Böden. Deutsches Patentamt DE 42 18 775 A1

Vogelbserger R, Meschede T (1994) Mercury decontamination of a chloralkali plant in Alexandria, Egyp. Proc 87th Annu Meet and Exhibition on Air and waste management, Cincinnati, Ohio, 19–24 June 1994, 94-TP64.01, pp1–10

Wasay SA, Arnfalk P, Tokunaga S (1995) Remediation of a soil polluted by mercury with acidic potassium iodide. J Haz Mat 44:93–102

Weber HH, Neumaier H (1993) Altlasten. Springer, Berlin Heidelberg New York, 395 pp

Weyand TE, Koshinski CJ, Baum W (1994) Removal and recovery of mercury from contaminated soils and debris. Proc 87th Annu Meet and Exhibition of Air and Waste Management, Cincinnati, Ohio, 19–24 June 1994, 94-TP62.01; pp 1–16

Advanced Technology Available for the Abatement of Mercury Pollution in the Metallurgical Industry

A.B. MUKHERJEE

This report deals with the technology developed in the Nordic countries and in other parts of the globe to abate mercury (Hg) pollution from the metallurgical industry. Hg, along with several other chemicals, has received special attention due to its toxicity and bioaccumulation in the environment. Hg is present as a trace constituent in sulphide minerals, coal and other raw materials. At higher temperatures, it volatilizes and accompanies SO_2, thus combining with H_2SO_4 and passing into the environment. Much of this acid is used for the production of fertilizers, causing the entry of Hg into the food web. There are 11 commercially known Hg removal processes from gases in the metallurgical industry, of which 9 are examined here.

1 Introduction

Mercury has been known to society and used since neolithic times. This metal attracts attention and concern due to its extreme toxicity, mobility, bioaccumulation, methylation processes and long-range transport in the atmosphere (Fitzgerald and Clarkson 1991). In recent decades, high concentrations of Hg in lake sediments and fish distant from point sources in the Nordic countries and North America have been documented (Swain and Helwig 1989; Verta 1990; Rusk and Verta 1995). Studies in Canada indicate that aboriginal people suffer from methylmercury (MeHg) poisoning, which bioaccumulates in fish found downstream from historical Hg sources, in natural lakes and hydroelectric reservoirs (Wheatley and Paradis 1995).

The cycling of Hg in the environment is extremely intricate and there are still uncertainties regarding the quantity, chemical species and physical form of Hg from both natural and industrial sources. Due to the toxicity and mobility of Hg, production and its uses in society have been restricted in many countries. However, its concentration in the atmosphere is still increasing by 0.6 to 1.5% a^{-1} (Expert Panel on Mercury Atmospheric Processes 1994; Sloss 1995). Recent estimation indicates that the average concentration (natural + anthropogenic) of Hg in the atmosphere is about 1.5 ng m^{-3}, but global emissions have been decreasing at the rate of 1.3% a^{-1} since the beginning of the 1990s (Pirrone et al. 1996). As per the UN and the OECD action plans, this chemical should be phased out from society within a short period of time; many countries have adopted the agenda to reduce Hg emissions including the Nordic countries, Canada, the United States and Japan (Minnesota Pollution Control Agency 1994; Dyvik 1995).

Environmental Science
Mercury Contaminated Sites (ed. by R. Ebinghaus et al.)

Raw materials used for energy production or in the metallurgical industry contain trace amounts of Hg and other heavy metals. At higher temperatures, Hg vaporizes and passes into the atmosphere in the gas phase. In addition, SO_2 released during the production of Cu, Ni, Zn and Pb, contains Hg which passes into the H_2SO_4 production line. The major part of this H_2SO_4 is used for production of fertilizer and titanium dioxide (TiO_2). Uses of H_2SO_4 are also noted in water-treatment plants and leaching of ores, thus Hg passes into soils, water, and into the food chain. Due to the development of new technology, it is possible to restrict and recover Hg from the metallurgical industry, as discussed below.

2 Mercury in Raw Materials

Hg in coal is mostly associated with minerals in the inorganic fraction. Concentration of the elements in coal are on average equal to the average concentrations in the earth's crust (Meij 1994). In nine European countries, Swaine (1990) observed that the Hg content in coal burnt in utility boilers was 0.3 $\mu g\ g^{-1}$. Occasionally, high concentrations of Hg (1.0 $\mu g\ g^{-1}$) are reported. Hg concentrations in coal vary from mine to mine. On the other hand, sulphide concentrates contain on average 100–300 μg Hg g^{-1} (Habashi 1992b). The concentration of Hg in raw materials used in various facilities differs substantially, as can be seen in Table 1, and it should be borne in mind that huge quantities of raw materials containing trace amounts of Hg are used by different facilities today.

Since the end of the 1960s, problems due to the presence of Hg in sulphide ores such as chalcopyrite, zinc sulphide, galena, iron pyrites, gold and manganese ores have been discussed (Habashi 1978; Dyvik 1995). When these raw materials are treated at high temperatures, most of the Hg volatilizes and then combines with $SO_{2(g)}$. About 30% of the SO_2 is consumed in the production of H_2SO_4 and

Table 1. Mercury content in different types of raw materials

Source category	Origin	Hg content ($\mu g\ g^{-1}$)	Reference[a]
Coal		0.02–1.0	1
Coal	Australia	0.01–0.25	2
Oil	Former USSR	0.005	3
Peat	Finland	0.01	3
Wood	Finland	0.01	3
Zn concentrate	Finland	100–500	4
Cu ores	USA	0.5 (average)	5
Cu concentrate	Finland (domestic)	13	6
	Finland (imported)	110	6
	Sweden	200	7
Pb concentrate	Sweden	50	7

[a] 1, Swaine (1990); 2, Sloss (1995); 3, Mukherjee et al. (1995); 4, Kuivala (1984); 5, Jasinski (1995); 6, R. Rantalahti (1996, pers. comm.); 7, Dyvik (1995).

the rest is released into the atmosphere (Habashi 1992a). Occasionally, the Hg, As, Se and Te present in SO_2 are vented directly into the atmosphere or these elements will pass into the sludge in many countries including the former Soviet Union, Canada, USA, Australia and South America (Dyvik 1995). The global atmospheric discharge of Hg from mining and metallurgical industries was estimated to be between 134 (median value) to 170 t a^{-1} in 1983 (Nriagu and Pacyna 1988; Pirrone et al. 1996), whereas in 1992 it was 267 t a^{-1} (Pirrone et al. 1996). Global consumption of zinc and copper concentrates were 14×10^6 and 35×10^6 t a^{-1} respectively, while the production of H_2SO_4 was 143×10^6 t in 1995, of which 60% was high quality H_2SO_4 according to the market standards (Boliden Chemtrade, Zug, Switzerland, 1996, pers. comm.). If there were no developments in technology, then 20 000 t a^{-1} Hg would enter into the ecosystem from the non-ferrous metallurgical industry alone (Habashi 1992b).

In the past, problems of Hg emissions from ferro-alloys production, scrap remelting, iron and steel facilities and also from the traditional gold-winning process in the Amazon territory have been highlighted (Malm et al. 1990; Habashi 1992a; Dyvik 1995; Cleary 1996). It is necessary to control the Hg emissions from these sources by available technology.

Due to the seriousness of the Hg problem in the metallurgical industry and stringent legal regulations, continuous research since the 1960s in the Nordic countries, Europe, North America and Japan has promoted the development of a number of processes which are commercially adopted to recover or remove Hg from gases in the metallurgical industry. Nine processes have been examined in this study.

3 Removal of Mercury from the Gas Phase

The form of Hg in a process guides the selection of the Hg removal technology. In the metallurgical industry, most of the Hg is released in the elemental form (Hg^0).

The processes for Hg removal from smelter gases are based on precipitation-filtration and adsorption-filtration methods (Habashi 1978). In the former, Hg^0 reacts with concentrated H_2SO_4, $HgCl_2$ or amorphous solid elemental Se to form insoluble Hg slurry compounds in scrubbers which may be filtered to recover Hg and other metals such as Se. In the latter method, gases are filtered through activated carbon, zeolite or selenium. The Hg removal processes are described below.

3.1 The Outokumpu Process

In the early 1970s, Outokumpu Oy, a multimetal production company, developed an Hg recovery process from roaster gases based on precipitation-filtration methods at the Kokkola Zinc Plant in Finland. In this process, the roaster gases at a temperature of 350 °C are washed by 85–90% H_2SO_4 in a brick-lined tower. The temperature of the gases drops from 350 to 200 °C and they are then scrubbed with the same acid to convert the Hg^0 into the sulfate as per the following equation:

$$2\ Hg^0 + H_2SO_4 + 1/2\ O_2 \rightarrow Hg_2SO_4 + H_2O. \quad (1)$$

The Hg-containing acid slurry in the sulphatizing tower is settled in a thickener and is then washed and filtered for the production of metallic Hg and Se at the Kokkola Zinc Plant.

The roaster gases contain moisture which may dilute the acid, causing it to be ineffective as a sulphating agent. The vapour pressure of water in the acid must correspond to the water partial pressure of the gas to be treated (Rastas et al. 1971). To maintain the vapour pressure, the H_2SO_4 concentration is about 90% and its inlet and outlet temperatures inside the Hg-sulphatizer are 40 and 180 °C, respectively. Gas leaving the Hg-sulphatizer contains about 20 μg Nm^{-3} of Hg^0. In the scrubber unit, the gas is further washed with weak H_2SO_4 to bring down the gas temperature from 180 to 70 °C. Table 2 indicates typical data for Hg removal from roaster gases at the Kokkola plant and Fig. 1 is the schematic diagram of the process. This process can be commercially applied for the removal of Hg from smelter and sinter-plant gases and two facilities have been working in Japan since 1970.

In Finland, this process contributed to the recovery of Hg from 70 to 150 t a^{-1} during the period 1984–1994. The high quality H_2SO_4 production according to the market standards is about 300 000 t a^{-1} at the Kokkola Zinc Plant.

3.2 The Boliden and the Boliden-Norzink Processes

In the 1960s, Boliden AB in Sweden and Norzink AS in Norway realized the importance of removal of Hg from SO_2 as Hg vapor finds its way to the H_2SO_4. As

Table 2. Typical data for mercury removal from the Outokumpu and the Boliden-Norzink processes (Rastas et al. 1971; Svens 1985; Dyvik 1990; Habashi 1992b)

Source categories Hg in Zn-concentrate, μg g^{-1}	Outokumpu process 380	Boliden-Norzink process 300
Hg in calcine, μg g^{-1}	5	2–10
Gas volume, $m^3\ h^{-1}$	112 000	80 000
Hg in filter cake after wet purification of gas %	n.a	50–60
Hg in gases after dry ESP, mg m^{-3}	100	100
Hg in gases after wet ESP, mg m^{-3}	n.a.	40
Hg in gases after treatment, mg m^{-3}	<0.2	0.05–0.1
Hg content in acid, ppm	0.2–0.5	0.1–0.2
Total acid production originated from roaster gases, t a^{-1}	300 000	180 000
Hg removal efficiency %	99.5	99.95–99.97

n.a. = not applicable

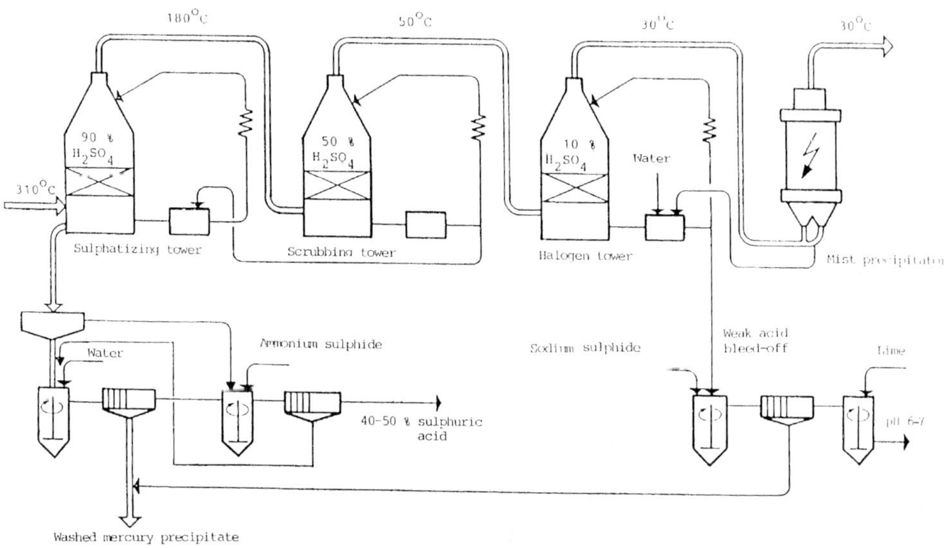

Fig. 1. Outokumpu mercury removal process at Kokkola zinc plant (Svens 1985)

a result of joint research activities, the companies were successful in developing several Hg-removal processes used in the metallurgical industry. These processes are based on systems which are capable of collecting and binding the Hg in the unit where the vapour pressure of Hg must be as low as possible. The developed processes are as follows:

1. Carbon filter
2. Selenium filter
3. Selenium scrubber
4. Acid purification method
5. Mercuric chloride method (Boliden-Norzink process)

These processes will be briefly discussed here, but details can be read elsewhere (Sundström 1975; Habashi 1978; Dyvik 1995).

3.3 Carbon Filter

The process is based on adsorption of $Hg_{(g)}$ on activated carbon. The temperature of towers connected in parallel is maintained at a maximum 50 °C. Carbon in the towers is activated by passing 100% SO_2 before allowing the SO_2 from which the Hg is to be removed. The process claims Hg removal efficiency of 90% or more.

Recently, a study has been made (Schager et al. 1995) on adsorption of Hg by activated carbon and fly ash at 150 °C, respectively. It was observed that retention of Hg by activated carbon in presence of HCl is 99%, whereas it decreases considerably when SO_2 is added to the system. The authors believe that SO_2 reduces HgO to elemental $(Hg^0)_g$, which has a stronger tendency to escape through activated carbon. However, it has been suggested that the reaction of SO_2 with $HgO_{(s,g)}$ will not have any effect on the adsorption of Hg^0 by activated carbon (Zacharwski et al. 1987). The overall reaction is given below:

$$2HgO_{(s,g)} + SO_{2(g,\,ads)} \rightarrow Hg^0 + HgSO_{4(s)}. \quad (2)$$

There is a chance that H_2SO_4 will form due to reaction of SO_3 with the moisture in the gases, causing corrosion problems inside the reactor (Expert Panel on Mercury Atmospheric Processes 1994; O. Lindqvist 1996, pers. comm.).

In this process, the charcoal bed contains many tons of Hg in the form of mercury (I) sulphate. Generally, after every 5 years, the filters are inspected at Boliden mineral in Sweden and the charcoal in the filter is replaced. The used charcoal is stored under controlled conditions for future disposal together with other hazardous waste materials (A. Svedberg 1997, pers. comm. F. Habashi 1997).

It is feasible to remove Hg from the Hg-loaded charcoal by leaching with acid, NaOCl solution or cyanide solution and the Hg can be recovered from the solution by cementation-precipitation process in the presence of iron scrap (F. Habashi 1997).

3.4 Selenium Filter

This process is based on the reaction between Hg^0, Hg_g and Se forming mercury selenite (HgSe) which is a stable compound. The operating temperature is kept to about 110 °C. At higher temperatures, Se will be lost from the filter due to its high vapour pressure (Y. Lundgren 1996). The filter material is porous, inert, and soaked in a solution of selenium dioxide (SeO_2) which is then dried. This compound is converted into activated Se when the SO_2 comes into contact with the filter. The filter element absorbs Hg until its content reaches 10–15%. The Se filter will be inactive in the presence of moist gas. It can be reactivated, but Se loss is inevitable. For this reason, moist gases must pass through mist eliminators to remove liquid droplets and the relative humidity should be lowered before the gas enters into filter units. The Hg collection efficiency varies between 95 and 99.5%.

3.5 Selenium Scrubber

In this process, amorphous solid elemental Se is maintained in the wash liquid containing H_2SO_4 which accelerates the removal of Hg^0 from the SO_2. The acid concentration in the unit must be between 10–60%. At lower concentrations, soluble selenium sulphur compounds are formed whereas at higher concentrations, the elementary selenium is quickly oxidized to selenium dioxide or selenite, thus reducing the Hg removal efficiency. The Hg collection efficiency of the process is about 90%.

3.6 Acid Purification Method

This process has been described in detail by several authors (Sundström 1975; Habashi 1978). It is based on the addition of thiosulphate to sulphuric acid of concentration of 80–85%. Colloidal sulphur is formed according to the following reaction:

$$2H^{+} + S_2O_3^{2-} \rightarrow S + SO_2 + H_2O. \qquad (3)$$

Here, sulphur reacts with Hg to form mercuric sulphide (HgS), which is then filtered off. The acid product is suitable for the production of fertilizers. It is stated that the dosing of sodium thiosulphate should be about 0.5 kg m^{-3} of acid. This process was recently adopted by the Outokumpu Harjavalta Metals Oy to remove Hg from the acid at the copper-smelting plant at Harjavalta, Finland. The acid contains about 0.1 ppm Hg; the Hg removal efficiency is 99.3% (A. Kalliokoski 1996).

3.7 The Boliden-Norzink Process

The Boliden-Norzink process has been in operation at the Zn-concentrate treatment plant in Odda, Norway since the early 1970s. At present, 34 facilities (e.g., Zn plants and several Cu, Pb, and pyrite roasters) worldwide have adopted this process and 8.6×10^6 t a^{-1} of high quality H_2SO_4 according to the market standards is produced by the application of this process (Dyvik 1995). In the process, mercuric chloride complexes are brought into contact in a scrubber where the following reaction occurs:

$$Hg^0_{(g)} + HgCl_2 \rightarrow Hg_2Cl_2. \qquad (4)$$

The reaction (4) occurs within a few seconds and calomel precipitates as slurry.

In the next step, calomel slurry is periodically withdrawn and oxidized by chlorine gas to form $HgCl_2$ according to the following reaction:

$$1/2\ Hg_2Cl_2 + 1/2\ Cl_2 \rightarrow HgCl_2. \qquad (5)$$

Hence, net gas treatment reaction = (4) + (5) i.e.,

$$Hg^0_{(g)} + 1/2\ Cl_2 \rightarrow 1/2\ Hg_2Cl_2. \qquad (6)$$

The scrubber sludge is then electrolyzed to form Hg at the cathode and chlorine gas at the anode.

$$HgCl_2 + 2e = Hg_{(liq)} + Cl_2. \qquad (7)$$

Half of the chlorine gas generated is used for the reaction (5) and the rest is used in the recovery step. Table 2 shows the typical analyses of solids, gases and acids and Fig. 2 is the schematic diagram of the process. The process claims 99.95% recovery of Hg from Zn-roaster gases.

3.8 Medisorbon Process

This process was developed in the early 1990s by Lurgi Bamang GmbH in Germany. The process is suitable for the removal of Hg, dioxins and furans from waste incinerators, hazardous waste treatment plants and metallurgical processes. Three units are in operation in The Netherlands and very successful results have been reported.

The process is based on modified synthetic hydrophobic Y-type zeolite, impregnated with sulphur as an adsorbent for Hg^0 from SO_2-laden gases. Here the metallic Hg is removed by chemisorption process. The hydrophobic Y-type zeolite was developed by Degussa AG. It is reported that this type of zeolite is highly resistant to abrasion, corrosion and the poisoning effects of alkaline and heavy metal particulates in the flue gas stream. The process also claims that exhausted sorbent can be regenerated. During regeneration, Hg is distilled for reuse and the organic pollutants can be destroyed. The operating data for the process is given in Table 3.

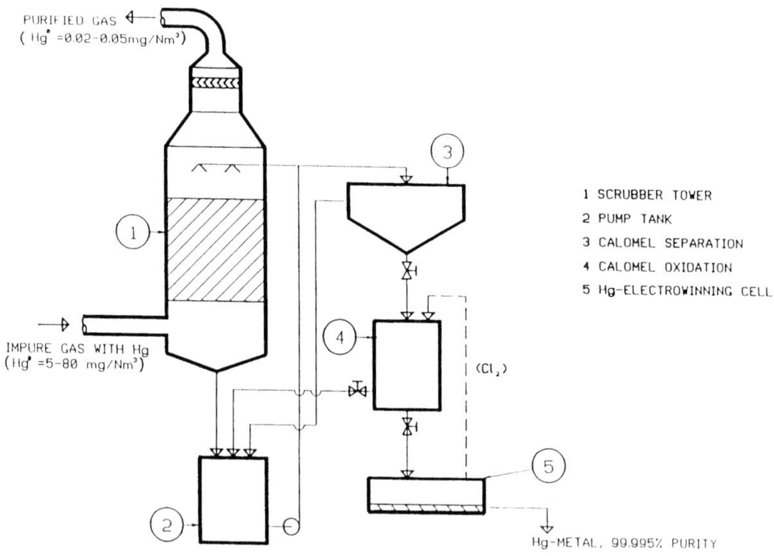

Fig. 2. Boliden-Norzink process for mercury removal from roaster gases. (Dyvik 1995)

Table 3. The key operating data for the Medisorbon process (Lurgi Bamag GmbH 1994, with permission)

Sources	Data
Gas flow rate $m^3\ h^{-1}$	1000–100 000
Dew point °C	85
SO_2, HCl %	20
HF mg m^{-3}	5
Dust load mg m^{-3}	10
Mercury, Hg(O) mg m^{-3}	10
Dioxins (total) µg m^{-3}	1
Espressed as I-TE[a] ng m^{-3}	10
Clean gas data	
Mercury (total) µg m^{-3}	40 (Std design)
	10 (Special design)
Dioxins, I-TE ng m^{-3}	0.05
Removal efficiency %	90–99

[a]Toxicity equivalent as per International System; all data based on dry, 11% O_2.

3.9 Other Processes

There is a thiocyanate-sulphide process in which SO_2-laden gases are washed with a solution of sodium thiocyanate. Here, the major part of Hg dissolves as a complex and the rest precipitates as HgS as per the following reaction:

$$3Hg + 8SCN^- + 4H^+ + SO_2 = 2[Hg(SCN)_4]^{2-} + HgS + 2H_2O. \qquad (8)$$

In addition to the above-mentioned processes, there are also a number of processes such as CENIM, St. Joe and Toho processes to recover Hg. The latter process is developed by the Toho Zinc Company in Japan. In this process, potassium iodide is added to remove Hg as mercuric iodide (HgI_2) from H_2SO_4. The acid concentration must be at least 93% and at 0 °C for a better yield. The main reaction is as follows:

$$Hg^{2+} + 2I^- \rightarrow HgI_2. \qquad (9)$$

4 Mercury Abatement Costs in the Metallurgical Industry

Estimates for the cost of Hg removal by the processes mentioned in this report are not readily available. The National Swedish Environmental Protection Board (SNV) has indicated that the investment cost for removal of Hg by the adsorption-filtration method such as Se filter is in the range of US$7.6–15 million (Innanen 1996). Recently, Petzoldt et al. (1995) also studied economical considerations for offgas treatment by Medisorbon process and their findings are given in Fig. 4. It is quite clear from their study that investment cost increases

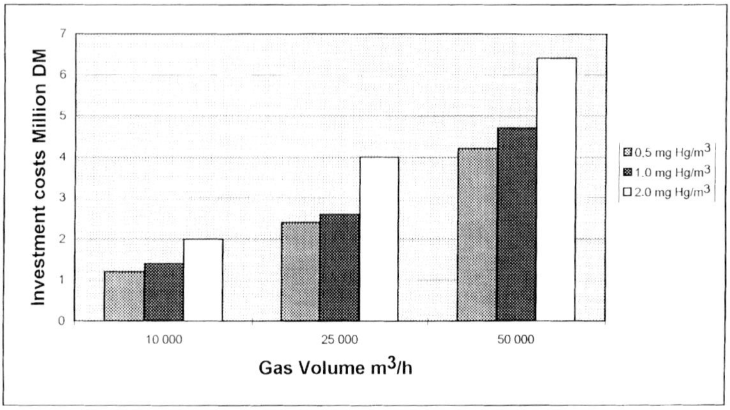

Fig. 3. Investment cost for the Medisorbon process (Petzoldt et al. 1995)

with the increase of gas volume and with the concentration of Hg in the gas to be treated by a process; but it is noted with interest that in the long run the accumulated operating cost of the Medisorbon process using Y-type zeolite is less than the coke injection in their experiment.

However, the investment cost for removal of Hg depends upon the nature of the process, gas volume to be handled, Hg concentration in gas, knowhow transfer, planning, equipment, and the location of the plant. To present a tentative idea for the investment on Hg abatement from sulphide-roasting gases, it is estimated that about US$110 is needed per Nm^3 of gas, based on a gas volume of 110 000 $Nm^3\ h^{-1}$, removal efficiency of 99.5% and the gases after treatment containing < 0.2 mg Hg Nm^{-3}. The investment costs for a facility discharging roaster gases of between 60 000 and 200 000 $Nm^3\ h^{-1}$ can be estimated with an exponent of 0.6 according to the following equation:

$$C_{invest} = 12.1(V/110,000)^{0.6}, \qquad (10)$$

where

C_{invest} = investment costs
V = volume of gases.

Figure 3 indicates how the investment costs vary with the gas volume. It is obvious that investment costs will vary from one process to another. Hence, the value presented in this report is tentative and should be used cautiously.

5 Conclusions

The mining and metallurgical industry produces huge amounts of Hg-containing gases. A part of these gases are used for the production of H_2SO_4 which is then

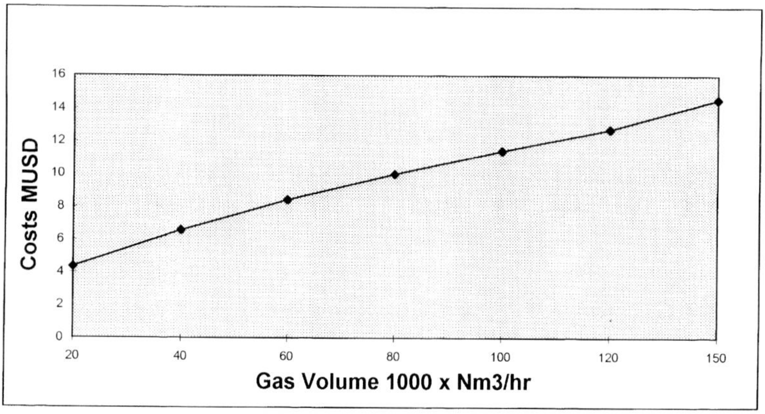

Fig. 4. Tentative investment cost for removal of mercury from sulphide roaster gases

mostly used for the production of plant nutrients. Considering a Hg concentration in sulphide concentrate between 100 and 300 $\mu g\ g^{-1}$, it has been estimated that about 20 000 $t\ a^{-1}$ Hg circulate in the non-ferrous metallurgical processes (Habashi 1992a). The Nordic countries first realized the problems with Hg in the environment and, due to continuous research, it is now possible to produce high quality H_2SO_4 according to the market standards for the production of fertilizers. Recently, a new adsorption process using amorphous Y-type zeolite for capturing Hg from gases has been reported in Germany which may be a significant contribution to the field of Hg removal technology. In addition, investment costs for abatement of Hg pollution from a process depend upon many factors including, gas volume, process, know-how transfer, equipment and location of the plant.

Acknowledgements. I would like to thank Professor Fathi Habashi (Laval University, Québec City, Canada), Outokumpu Oy, Boliden AB, Norzink Technology AS, and Lurgi Bamag GmbH for supplying valuable current information on mercury removal processes. Many thanks are due to Mr. Froystein Dyvik (Norzink Technology AS, Norway) for reviews of the initial manuscript and also to Mr. Kurt Svens (Outokumpu Oy, Finland) for comments on the Outokumpu Process. The assistance received in preparation of this manuscript from the staff members of the Department of Limnology and Environmental Protection, University of Helsinki, is gratefully acknowledged.

References

Cleary D (1996) Mercury contamination in the developing world: problems and solutions. In: Proc 4th Int Conf on Mercury as a global pollutant, Hamburg, 4–8 Aug 1996. p 3

Dyvik F (1990) Rear metals '90. In: Proc Int Symp the Processing of rare metals, 14–16 Nov 1990, Kitakyushi, pp 263–266

Dyvik F (1995) Mercury as a trace contaminant in non-ferrous metal production. Occurrence, distribution, waste and gas treatment, and mercury free sulfuric acid production by the use of the Boliden/Norzink Process for gas. In: Proc of Academy of Certified Hazardous Materials Managers, 1995 Conf and Annu Meet, Rochester, New York, 1–4 Aug 1995, pp 15

Expert Panel on Mercury Atmospheric Processes (1994) Mercury atmospheric processes: a synthesis report. Worksh Proc, Electrical Power Research Institute, Rep No TR-104214, p 2

Fitzgerald WF, Clarkson TW (1991) Mercury and monomethylmercury. Present and future concerns. Environ Health Perspect 99:156–166

Habashi F (1978) Metallurgical plants: how mercury pollution is abated. Environ Sci Technol 12 (13):1372–1376

Habashi F (1992a) Environmental issues in the metallurgical industry – progress and problems. In: Singhal RK, Mehrotra AK (eds) Environmental issues and waste management in energy and minerals production. AA Balkema, Roherdam, pp 1143–1153

Habashi F (1992b) Mercury removal from smelter gases: an update. Int J Environ Issues Mineral Energy prod 1 (1):17–20

Innanen S (1996) Cited in: Issues in environmental mercury contamination: Ontario (Canada), Finland and Sweden. Master's Thesis, Faculty of Environmental Studies, York University, 4700 Keele St, North York, Ontario, Canada, p 300

Jasinski SM (1995) The materials flow of mercury in the United States. Resour Conserv Recycl 15:145–179

Kuivala A (1984) Verfahren zur Wiedergewinnung von Quecksilver bei Outokumpu Oy. Paper presented at the 32nd Meeting of GDMB Zinkfachausschuss, Kokkola, Finland

Lurgi Bamag GmbH (1994) Medisorbon process to remove mercury and dioxins from flue gases by adsorption on Wessalith day. Butzbach, Germany, pp 1–6

Malm O, Pfeiffer W, Souza C, Reuther R (1990) Mercury pollution due to gold mining in the Madeira River basin, Brazil. Ambio 19:11

Meij R (1994) Trace element behavior in coal fired power plants. Fuel Process Technol 39 (1-3):199–217

Minnesota Pollution Control Agency (1994) Strategies for reducing mercury in Minnesota. 520 Lafayette Rd, St Paul, Minnesota 55155, pp 55

Mukherjee AB, Innanen S, Verta M (1995) An update of the mercury inventory and atmospheric mercury fluxes to and from Finland. Water Air Soil Pollut 80:255–264

Nriagu JO, Pacyna JM (1988) Quantitative assessment of worldwide contamination of air, water and soils by trace metals. Nature 333:134–139

Petzoldt O, Actun S, Fell HJ (1995) Adsorptive Abgasreinigung – Wirtschaftliche Betrachtungen. Abfallwirtschafts J 7 (5):310–312

Pirrone N, Gerald JK, Nriagu JO (1996) Regional differences in worldwide emissions of mercury to the atmosphere. Atmos Environ 30 (17):2981–2987

Rastas J, Nyholm E, Kangas J (1971) Mercury recovery from SO_2-rich smelter gases. Eng Min J 172 (4):123–124

Rusk M, Verta M (1995) Concentrations and amounts of methylmercury in water and fish in the limed and acid basins of a small lake. Water Air Soil Pollut 80:577–580

Schager P, Hall B, Lindqvist O (1995) The retention of gaseous mercury on fly ashes. In: Watras CA, Huckabee JW (eds) Mercury pollution: integration and synthesis. Lewis, Boca Raton, pp 621–628

Sloss L (1995) Mercury emission and effects – the role of coal. IEA Coal Research, London, Rep No IEAPER/19:1–39

Sundström O (1975) Mercury in sulfuric acid. Sulphur 116:37–43

Svens K (1985) Outokumpu mercury recovery. Metal News 7 (1):8–12

Swain EB, Helwig DD (1989) Mercury in fish from northeastern Minnesota lakes: historical trends, environmental correlates, and potential sources. J Minn Acad Sci 55 (1):103–109

Swaine DJ (1990) Trace elements in coal. Butterworth, London pp 276

Verta M (1990) Mercury in Finnish forest lakes and reservoirs: anthropogenic contribution to the load and accumulation in fish. Publications of the Water and Environment Research Institute, National Board of Water and the Environment, Helsinki, Finland p 6

Wheatley B, Paradis S (1995) Exposure of Canadian aboriginal peoples to methylmercury. Water Air Soil Pollut 80:3–11

Zacharwski TR, Cherniak EA, Schroeder WH (1987) FTIR investigations of the heterogeneous reaction of HgO(s) with SO_2(g) at ambient temperature. Atmos Environ 21:2327

Mercury Contamination of Minamata Bay: Historical Overview and Progress Towards Recovery[1]

A. Kudo and R.R. Turner

The contamination of Minamata Bay with mercury compounds and the subsequent effects of this contamination on humans and animals who ate seafood from the bay have been described repeatedly elsewhere, but there are few publications which have focused on the physicochemical aspects of the situation and the nature and effectiveness of the work undertaken to restore the bay. This chapter briefly reviews the available English language literature on these subjects and is intended to allow for easier comparison with other similar sites which have not progressed as far in restoration work. Minamata Bay was severely contaminated with mercury from acetaldehyde and vinyl chloride plants between 1950 and 1968. While the predominant form of mercury discharged to the bay was inorganic, some may have been methylated and the origin of highly elevated concentrations in seafood. Somewhat lower but still unacceptably elevated levels of mercury in seafood persisted even after use of mercury at the plants was discontinued. Sediment contamination was mapped in detail in the early 1970s and led to a major dredging and containment project, which began in 1980, to restore the bay. All restoration was completed by 1990. Limited available data from monitoring of mercury in commercially important fish species appear to demonstrate a successful restoration.

1 Introduction

Perhaps no other legacy of man's early ignorance of the price of releasing toxic compounds into the environment is better known than that of Minamata Bay, a small bay on the southwestern shore of Kyushu Island in southern Japan. Characterized as a sleepy fishing community which shared the shores of the bay with a chemical company, Minamata first brought global attention to the perils of consumption of fish and shellfish contaminated with mercury, and gave its name to the disease caused by eating foods containing alkyl mercury (D'Itri 1994). Although acetaldehyde production using inorganic mercury as a catalyst occurred outside of Japan (e.g., Germany, cf. Wilken and Wallschlaeger 1996) there are apparently no comparable situations where releases from this type of industry contaminated food which was the main staple for a community. Much of the published literature on Minamata Bay has rightly focused on the human tragedy which occurred there (e.g., D'Itri and D'Itri 1977; Harada 1982; Ellis 1989; Mishima 1992) and less on the physicochemical characteristics of the contam-

[1]This chapter was prepared while the second author held a Senior Research Associateship under the auspices of the National Research Council – US Environmental Protection Agency National Health and Environmental Effects Research Laboratory

Environmental Science
Mercury Contaminated Sites (ed. by R. Ebinghaus et al.)

ination and its reclamation. Information on the latter subjects (e.g., Ishikawa and Ikegaki 1980; Fujiki and Tajima 1992; Hosokawa 1993) has been slow to appear in English language publications. The purpose of this chapter is to summarize some important characteristics of the site, the nature and extent of remedial actions undertaken, and the response of the system to those deliberate actions, and to natural processes which can achieve similar endpoints. Appropriately, Minamata Bay was the first mercury-contaminated ecosystem in the world in which a major restoration has been attempted. Restorations of a similar kind and magnitude are likely to be contemplated elsewhere and thus the experience at Minamata may hold valuable lessons.

It was the authors' resolve that no major book on mercury-contaminated sites should be published without a chapter covering Minamata Bay. As no papers were presented at the 1996 Hamburg conference which might have served, the authors elected to prepare this review chapter. Little is presented here which has not already been published elsewhere in one form or another. The authors were able to obtain recent data (through 1995) for mercury in fish which allowed extension of a previously published trend chart. The authors have attempted to assemble in one document as much information as possible concerning site characterization, risk assessment, and remediation, and thus provide for easier comparisons with other mercury-contaminated sites in the world. The authors offer a few opinions and speculations for which they alone should be held accountable.

1.1
Hydrography of Minamata Bay and Adjacent Yatsushiro Sea

Minamata Bay is a small coastal inlet (about 300 ha) of the Yatsushiro Sea (also called the Shiranui Sea) on the southwestern coast of Kyushu, Japan (Fig. 1). The only freshwater inflow is a small creek with a flow rate of 0.5 to 1 m^3/s, most of which is discharged from a chemical factory. The maximum tidal range is 3.4 m with an average of 2.23 m, and thus tidal flushing of the bay is considerable. The bay holds about 30 million m^3 of water and has a theoretical water retention time of 2.5 days. The relatively steep topography of the lands surrounding the bay is mirrored in the bay itself which reaches a maximum depth of 30 m and has an average depth of 10 m. Sediments deposited in the bay have been described as organic-rich silts and clays. Calculations of natural water velocities in the bay by Kudo (1992) suggested that bed sediment transport should not be expected but that suspended sediment transport within and from the bay is possible. However, Kudo et al. (1980) pointed out that ship traffic in the bay could resuspend bed sediments and thus lead to export of some bed sediments to the Yatsushiro Sea. In addition, sediment within the intertidal zone may be expected to be vulnerable to resuspension and transport due to wave action and ship wakes.

The Yatsushiro Sea is an inland sea facing the East China Sea with a surface area of 63 600 ha, an average depth of 20 m, and a maximum depth of 47 m. The sea holds about 13 km^3 of water, about 10% of which is exchanged daily with the East China Sea by the tide. Kudo (1992) estimated that surface water velocities caused by the tide were insufficient to transport bed sediments over most of the

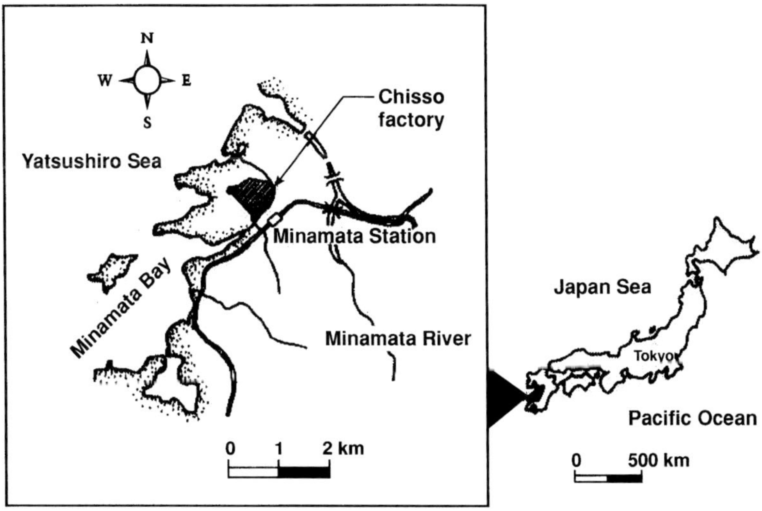

Fig. 1. Location of Minamata Bay with respect to Japan

sea. Kudo et al. (1980) described sediments in the sea as mixtures of fine sand, silt, and clay, with some sea shells.

1.2 Brief Chronology of Site

Prior to about 1930, the only commercial enterprise on Minamata Bay was fishing. At that time, Chisso Co., Ltd began construction of a chemical plant to produce acetaldehyde using mercuric oxide dissolved in sulfuric acid as a catalyst. Vinyl chloride production, which also utilized mercury as a catalyst, began at the site in 1952. Wastewater from the acetaldehyde plant was discharged directly into the Yatsushiro Sea until about 1950, when the discharge was apparently redirected into Minamata Bay. The point of discharge was again changed in September 1958 when it was diverted briefly (1 year) to the Minamata River, which flows into the Yatsushiro Sea just north of Minamata Bay (Fujiki and Tajima 1992). After October 1959, wastewater from both acetaldehyde and vinyl chloride production was routed through a sedimentation pond (Hachiman pool) prior to discharge into Minamata Bay. In June 1966 the wastewater flow system became closed-loop or recirculating. Acetaldehyde production was finally stopped in May 1968. Beginning in January 1969, wastewater from vinyl chloride production was treated to remove mercury and then routed to a pond (Yabusa pool). Vinyl chloride production was finally halted in March 1971. As discussed widely in other publications (e.g., D'Itri and D'Itri 1977; Ellis 1989; Mishima 1992), the first human victims of Minamata disease were identified as early as 1953, but more than a decade elapsed before the epidemic was finally acknowledged officially and serious corrective action undertaken. Portions of

Minamata Bay were dredged several times between 1964 and 1970 to construct and maintain a navigation channel (Fujiki and Tajima 1992).

The mercury levels in bay sediments and fishery were characterized beginning around 1959 and continued until 1974 when the Kumamoto Prefecture, the jurisdictional government, issued an extensive restoration plan for the bay. The plan focused on dredging all sediments from the bay above 25 ppm in mercury content. The most contaminated area of the bay was designated to be surrounded by a barrier (i.e., contained) and used as the disposal site for dredged sediments. Dredging commenced in 1980 and ended in December 1987, but the entire restoration process (including final closure of the contained area) was not completed until March 1990.

1.3 Mercury Releases to the Environment

Total mercury losses in wastewater from the acetaldehyde production process, which was the main source of environmental releases, can only be estimated based on a few reported (see review by Fujiki and Tajima 1992) mercury concentrations in wastewater (e.g., 20 ppm) and other information (Arima 1979 as cited in Nakanishi 1992). The temporal trend in such estimated losses is given in Fig. 2, which shows mercury discharges peaking twice (around 1940 and again in the late 1950s) over the 36-year period of plant operation. Mercury losses from vinyl chloride production, which began around 1952, are also included in Fig. 2. Total losses of mercury to Minamata Bay have been given officially as "150 tons or more", with some estimates (Akima 1979 as cited in Nakanishi 1992) placing the losses to the bay somewhat higher at 190 to 225 tons.

1.4 Form(s) of Mercury Released

Methylmercury, found in fish and shellfish, and the cause of so much suffering in Minamata, was identified fairly early in the epidemic as the toxic agent but some

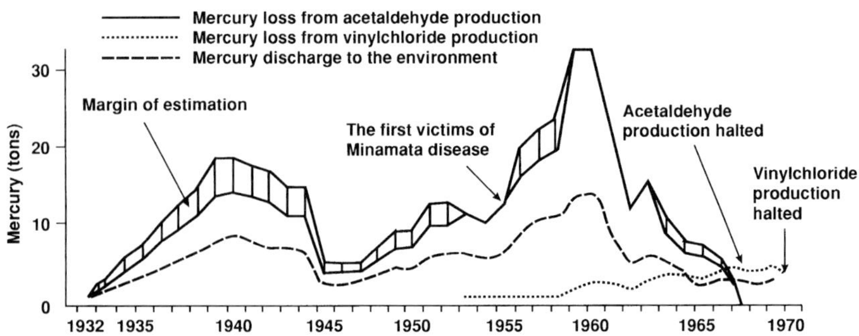

Fig. 2. Estimated mercury loss and discharge to the environment from Chisso Co. Ltd. in Minamata. (After Environment Agency of Japan 1992)

controversy continues even today (Nakanishi 1992) as to what fraction of the mercury in biota in the bay originated from direct discharges of methylmercury from the process and what fraction was produced in the bay via biomethylation. The latter mechanism was unknown and unsuspected prior to 1968 (Jensen and Jernolov 1969). Unquestionably, methylmercury was shown (see review by Fijiki and Tajima 1973) to be an unintentional byproduct of acetaldehyde production, but analytical chemical technology available in the late 1950s and early 1960s precluded identification and quantitation of methylmercury in the wastewater effluent. Irukayama (1966) was the first to positively identify methylmercury in sludge samples from the Chisso plant. In addition, Fijiki and Tajima (1992) note that wastewater from the vinyl chloride production process in May 1968 contained 40 µg/l total mercury and 0.3 µg/l methylmercury (0.75% methyl). Research conducted by these same authors in the 1960s and early 1970s did not support the hypothesis that inorganic mercury was methylated in bay sediments. Instead, their results pointed to direct discharges as the origin of methylmercury in fish and shellfish.

As is so well known today, exceedingly small quantities of methylmercury can contaminate biota to levels unsafe for consumption by humans and wildlife. If one assumes that only 0.1% of the total mercury released from the Chisso plant was methylmercury, this quantity represents an enormous mass of methylmercury with the potential to contaminate easily all the fish and shellfish in the bay to levels far exceeding those observed during the height of the epidemic. As discussed subsequently, the rapid rate at which the mercury content of fish and shellfish in the bay decreased after action was taken to halt further direct discharges suggests that direct discharges of methylmercury may indeed have been the most important aspect of the episode. Biomethylation of inorganic mercury in the bay, while unimportant during the height of the epidemic, most likely increased in importance after direct discharges of mercury were halted.

2 Distribution of Mercury in Sediments

The distribution of mercury in bay sediments began to be studied in detail in 1963 (see Tables IV and V in Fujiki and Tajima 1992), with studies continuing well into the early 1970s. Information summarized from other sources by Kudo (1992) indicates that the first measurements of mercury in sediment took place in 1960. One of these early samples was from a creek near the Chisso plant and contained over 2010 mg/kg total mercury on a "wet weight" basis. Other samples collected at the same time indicated decreasing concentrations with distance from the mouth of the creek out into the bay. For example, total mercury (wet basis) decreased in the sequence 133, 59, 40, 19, and 12 mg/kg at 0.4, 1, 2.5, 3, 3.5 km, respectively, from the mouth of the creek (see Fig. 14.10 in Takizawa 1979). Extensive measurements (342 locations) of mercury in bay sediments were undertaken in 1973 by the Kumamoto Prefecture and resulted in the first detailed map (see Fig. 1 in Yoshida and Ikegaki 1977) of the areal distribution of mercury in the bay and the adjacent Yatsushiro Sea (see Fig. 1 in Nishimura and Kumagai 1983). In 1973, the mercury level in sediments exceeded 15 mg/kg everywhere in the bay, and a level as high as 600 mg/kg was found at the innermost part of the bay (Nishimura and Kumagai 1983). Later (1983 and 1987), analyses of surface sediments from Minamata Bay (3 locations) and the Yatsushiro Sea (15 locations) were reported by Sakamoto et al. (1995). Values for the samples from the bay were 27 to 30 mg/kg, while those from the sea were 0.41 to 5.5 mg/kg. Hosokawa (1993) provided a detailed map (see his Fig. 9) of the areal distribution of mercury in bay sediments after completion of all dredging in 1988. The maximum value was 8.75 mg/kg but most values were less than 5.0 mg/kg.

Published information on the distribution of mercury with depth in Minamata Bay sediments is sparse. Fujiki and Tajima (1992) present some data (see their Table V) which indicate that the thickness of contamination in the 1960s was in excess of 3 m in some locations. Similarly, Ishikawa and Ikegaki (1980) mention that contaminated deposits in the innermost portion of the bay were 4 m thick. This large thickness implies a very rapid rate of bulk sedimentation in some portions of the bay, or perhaps unfortunate selection of coring sites over underwater dredge spoil. In any case, other authors (e.g., Kudo et al. 1980) give the thickness of contamination as "over 50 cm". Unfortunately, the data presented by Fujiki and Tajima (1992) are insufficient (depth intervals are too large) to detect any chronological pattern in mercury deposition in the bay. Nor have any other published data been found in the English language literature which would permit a temporal reconstruction of mercury deposition in the bay. Such a sediment "record" was probably present in undisturbed bay sediments, and would have been valuable to retrieve, but has now been almost certainly obliterated by the bay-wide dredging which began in 1980.

2.1
Forms of Mercury in Sediments

Some of the earliest measurements of the forms of mercury in Minamata Bay sediments appear to have been made in the early 1960s. Fujiki and Tajima (1992) cite references as early as 1964 as sources of information to the effect that "mercury compounds in the muds were inorganic, such as mercury sulfide and oxide, and the highest concentration of methyl mercury in the mud was 0.015 ppm as mercury (wet weight)". Once methylmercury was identified as the main contaminant of concern in fish and shellfish, it became the focus of further measurements in sediment but little was ever found (Fujiki et al. 1977; Yoshida and Ikegaki 1977; Nishimura and Kumagai 1983; Fujiki and Tajima 1992) in bay sediments using the methods available through the mid-1970s. For example, Nishimura and Kumagai (1983) note that alkylmercury was not detected (detection limit 0.001 mg/kg) in any of 263 sediment samples from Minamata Bay and Yatsushiro Sea. In minor contrast to this report, Hosokawa (1993) states that, while most sediments contained less than 0.005 mg/kg methylmercury, the maximum concentration was 0.03 mg/kg. Sediment methylmercury concentrations reported for other mercury-contaminated marine systems (see review in Kudo et al. 1980) tend to be somewhat higher than those reported for Minamata Bay, suggesting possible low recovery of methylmercury from Minamata Bay sediments during analysis, rather than an anomalous content of this form of mercury. Even today, the extraction and analysis of this form of mercury from sediments is not straightforward (Bloom et al. 1996; Liang et al. 1996).

Sakamoto et al. (1995) reported results of mercury speciation analyses, based on selective sequential chemical extraction, applied to Minamata Bay sediments. The forms reported included "organic"mercury, mercury(II)oxide, mercury(II) sulfide, and residual mercury. In this case, "organic" mercury represented sediment mercury extractable with chloroform and analyzed as total mercury, and thus does not necessarily represent alkylated forms. In a similar vein it is prudent to consider all the forms reported from such sequential extraction techniques as "operationally defined". The samples (3 sites, 26.7 to 35.6 mg/kg total Hg) analyzed by Sakamoto et al. were collected from Minamata Bay in September 1983 and from Yatsushiro Sea (15 sites, 0.43 to 5.49 mg/kg total Hg) in January 1987. The predominant form of mercury in all samples was HgS (42–93% of total Hg). Organic Hg in Minamata Bay samples ranged from 0.35 to 0.85 mg/kg, while those from the Yatsushiro Sea ranged from 0.4 to 0.12 mg/kg. Organic Hg generally accounted for more of the total mercury in Yatsushiro Sea sediments (2.6 to 12.8%) than in Minamata Bay sediments (1.1 to 2.4%).

2.2 Redistribution of Sediment Mercury Beyond Minamata Bay

An annual monitoring effort to track possible redistribution of mercury in sediments beyond Minamata Bay into the Yatsushiro Sea began in 1975 (Kudo et al. 1980) and has been continued (Kudo and Miyahara 1988, 1992; Miyahara et al. 1988; Kudo 1992). The most recent sampling campaign (23rd in the series) took place in June 1996. Twenty-four locations extending to more than 50 km from Minamata Bay were established in 1975 and then resampled using identical methods and analytical techniques. Analysis of the results of this monitoring by the involved investigators has shown substantial movement of mercury out of Minamata Bay and redeposition in the Yatsushiro Sea. Kudo (1992) estimated that the average annual movement of mercury out of the bay was 3.7 tons, and thus over several decades since the bay was contaminated a significant fraction (perhaps 10 to 30%) of the 150 to 225 tons of mercury in bay sediments was dispersed and "diluted" in the adjoining sea. Following the initial surveys in 1975 the level of mercury in sediment from the Yatsushiro Sea closest to Minamata Bay increased dramatically. At the station located 8 km from the Hiyakken-mon (where bay and creek meet) sediment mercury levels increased in the sequence, 0.61, 3.5, and 10 mg/kg, in 1975, 1978, and 1984, respectively (Kudo 1992). More recent data (unpubl.) collected annually since 1990 have shown a gradually decreasing trend at all stations.

Kumagai and Nishimura (1978) also estimated transport of mercury-contaminated sediment out of Minamata Bay. They used data for depth profiles of particulate Hg in the water column measured in November 1975 and an assumed water exchange rate of 0.4 to 1.2 million m^3/day to estimate an export of 20 to 60 kg/year. This estimated range is much lower than that calculated by Kudo (1992) based on the increase of sediment mercury burden (average 3.7 tons/year) in the Yatsushiro Sea. Even using the ten fold higher water exchange rate given by Kudo (1992) for the bay does not bring these estimates of mercury export from the bay any closer. Regardless of the reasons for the discrepancy, sediment transport out of the bay has been an important process.

As discussed in several of the publications by Kudo and his coworkers, natural forces have played the dual roles of both dispersing and burying mercury-contaminated sediments from Minamata Bay. A record rainfall event (200-year return interval) occurred in 1982 producing millions of tons of storm water borne soil eroded from the hilly lands surrounding the bay and sea which blanketed the bed sediments of the Yatsushiro Sea. The result was burial of more contaminated surface sediments and lower levels of mercury in surface sediments in the years following the event. This natural event occurred within the period when artificial decontamination (dredging) of bed sediments in Minamata Bay was underway and thus its effects on the rate of recovery of Minamata Bay itself cannot be determined.

3
Mercury in Fish and Shellfish

The mercury content of fish and shellfish from Minamata Bay during the 1960s and early 1970s has been summarized in many publications (e.g., Fujiki and Tajima 1973, 1992; Nishimura and Kumagai 1983; Environment Agency of Japan 1992). Figures 3 and 4 display the general temporal trends observed through the early 1970s. No more recent data on mercury in shellfish from Minamata Bay could be found in the English language literature. Fujiki and Tajima (1992) updated the situation for fish through 1989 by stating that only 16 species of fish, out of a total of 87, continued to contain over 0.4 mg/kg total mercury (the Provisional Regulatory Standard). The intensive fish monitoring program prescribed by the Kumamoto prefectural government (see section below) for the period of dredging and thereafter generated considerable data, little of which has been seen thus far in English language publications . A glimpse of some of these data was provided by Nakanishi (1992). The graphs of the temporal trend in the content of mercury in two important commercial fish species through 1990 given in her report have been updated here through early 1996 (Fig. 5) using data provided to the authors by the Kumamoto Prefecture.

Overall, the available information on biological contamination in Minamata Bay shows remarkably rapid decreases in mercury content of both fish and shellfish following cessation of direct discharges of mercury to the bay. Subsequently, the responses appear to have been slower or absent until restoration work in the bay was commenced in the early 1980s. As shown in

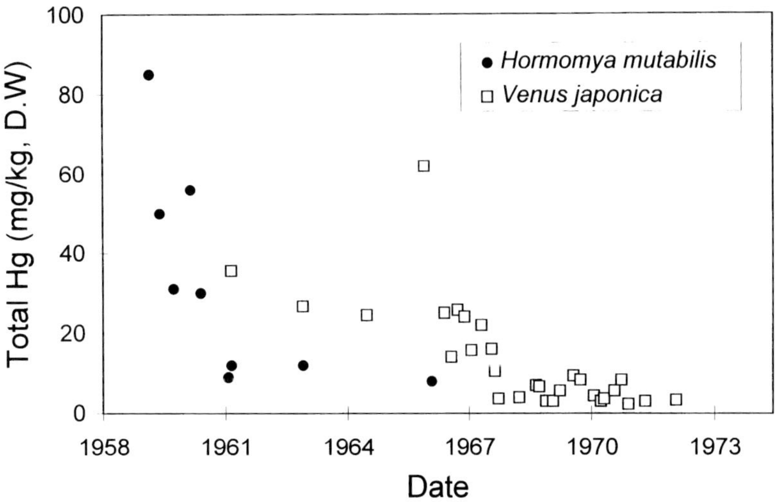

Fig. 3. Temporal pattern in mercury (total) content of two shellfish species from Minamata Bay, 1960–1972. *H. mutabilis* lives in rocks (intertidal zone) while *V. japonica* inhabits sediments. (Data from Fujiki and Tajima 1992)

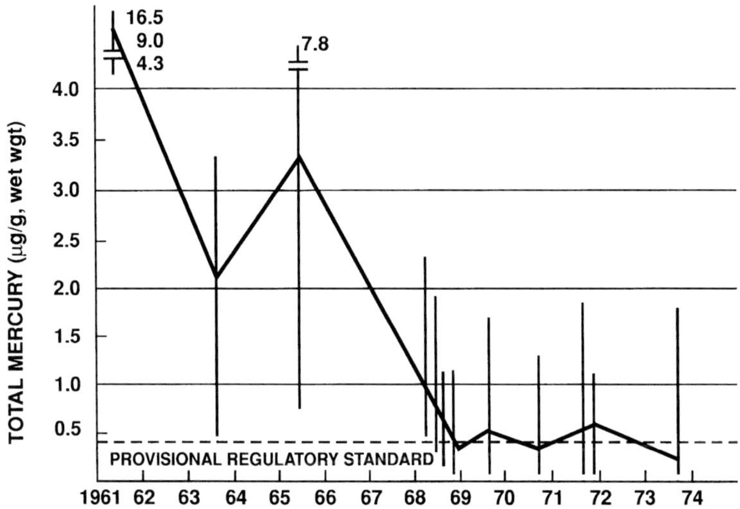

Fig. 4. Temporal pattern in mercury (total) content of fishes (data for all species combined) from Minamata Bay, 1961–1973. *Vertical bars* are ranges among all fish. (After Environment Agency of Japan 1992)

Fig 5. the mercury content of scorpion fish (*Scorpaena*) began to decline rapidly about 1982 when actual dredging work began, slowed or stopped declining in the last half of the 1980s, and then rapidly decreased again beginning about 1990. The

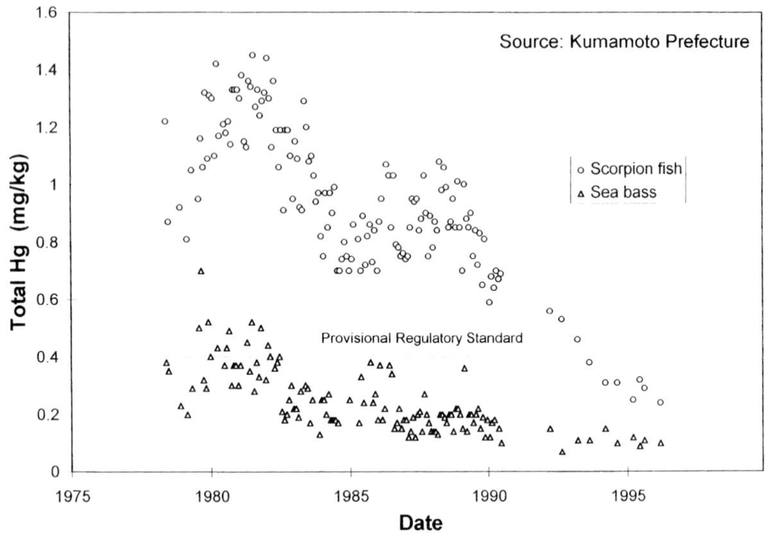

Fig. 5. Temporal pattern in mercury (total) content of two commercially important fish species from Minamata Bay. Scorpion fish (*Scorpaena)* is a territorial benthic feeder while sea bass (*Lateolabrax japonics)* is a migratory pelagic feeder. (Data from Nakanishi 1992 and Kumamoto Prefecture)

mercury content of both sea bass (*Lateolabrax japonicus*) and scorpion fish have been below the Provisional Regulatory Standard since 1993. At least for these commercially important fish species the restoration effort has been highly successful.

4 Sediment Decontamination Project

Although substantial reductions in the degree of mercury contamination of fish and shellfish had been observed in Minamata Bay in the 1960s after direct mercury discharges were halted, the high levels of mercury in bay sediments and persistence of unacceptable levels of contamination in some fish (i.e., >0.4 mg/kg) and shellfish required further corrective action. In 1971 the Kumamoto prefectural government commissioned Kumamoto University to evaluate viable methods for the treatment of the bottom sediments in Minamata Bay (Ishikawa and Ikegaki 1980). An interdisciplinary committee was formed incorporating experts in medical sciences, chemistry, biology, civil engineering, as well as government authorities and other “stakeholders”. Ishikawa and Ikegaki (1980) provide the first English, language summary of the detailed plan to restore Minamata Bay.

4.1 Establishment of Cleanup Level

The Japanese government promulgated a "tentative standard for removal of mercury-contaminated bottom sediments" in 1973. This standard took into consideration dissolution and diffusion of mercury out of bottom sediments, mixing of sediment-derived mercury with surface water and mercury accumulation in food chains. The detailed basis for this standard is difficult to extract from the sparse English language literature on the subject. An equation incorporating tidal range, sediment-to-water transfer rate and a safety factor (100 in fishing zones) is given in Fujiki and Tajima (1992) and thus the standard seems to be somewhat "site-specific", at least for marine environments. For sediments in Minamata Bay the standard value was calculated as 25 mg/kg.

4.2 Nature and Extent of Decontamination

Based on the standard value of 25 mg/kg approximately 1.5 million m^3 of sediment distributed over 2 km^2 of the bay was identified as contaminated above the standard. Because contamination was more serious in terms of both mercury concentrations and sediment thickness in the inner area of the bay, the inner area covering about 0.6 km^2 was enclosed with a line of highly watertight revetments and used as a disposal area for dredge spoil from the rest of the contaminated portions of the bay. Figure 6 illustrates the respective areas dredged and enclosed for disposal. The thickness of sediment to be dredged ranged up to 100 cm with

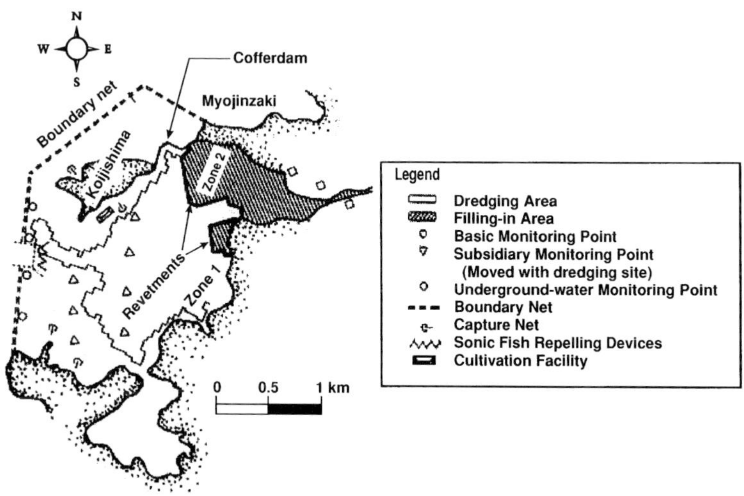

Fig. 6. Map of Minamata Bay showing areas dredged (*Zone 1*) and filled (*Zone 2*), as well as other features (boundary nets, cofferdam, monitoring points) of the remediation completed in 1990. (After Hosokawa 1993)

an average thickness of approximately 50 cm. Hosokawa (1993) provided an excellent summary of the details of the dredging operation as well as an outline of the construction of the revetment cells used to surround the reclamation area. The Japanese pioneered the use of a sand layer up to 1.5 m thick to stabilize and compact sludge (Fujino 1977). The technique was used in Minamata Bay to stabilize the bottom prior to placement of the revetments and to stabilize and compact the dredge spoil in the reclaimed area (Hosokawa 1993).

4.3 Controlling/Monitoring Effects of Restoration

When the plans to restore Minamata Bay were first presented to the public in the mid-1970s, there was great fear (and apparently some objection to the project) that the planned dredging would exacerbate the continuing problem of unacceptable mercury levels in some biota, result in a reappearance of the severe poisonings of the past, and cause spreading of the problem into the adjoining sea. As stated by Hosokawa (1993), "it was essential to get citizen's support and trust for field works... We had to use every possible technique in order to minimize secondary pollution due to the project itself." Ishikawa and Ikegaki (1980) first detailed some of the ameliorative strategies aimed at controlling secondary pollution, but noted that the plan was under a provisional injunction as of October 1979 from some Minamata residents who continued to doubt its safety.

Some of the main elements of the ameliorative strategy eventually followed included: (1) use of boundary nets to exclude nonresident fishes from, and to retain resident fish within, the bay during the dredging operation, (2) construction of a temporary cofferdam across the northern opening into the bay to reduce water circulation in the bay, (3) installation of an underwater sonic fish-repelling device in front of the only opening in the boundary net which permitted vessels to enter and exit the bay, (4) use of suction dredges without cutter heads to minimize resuspension and loss of contaminated sediment at the working face, (5) active treatment of effluent from the dredge spoil disposal area, (6) establishment of rigorous performance standards for all operations which could affect water quality in the bay and beyond. The last element included continuous monitoring of water turbidity both at the suction head, which was also fitted with precision depth sounders and video camera, and at the treated effluent discharge point (Hosokawa 1993). Turbidity had been shown earlier (Yoshida and Ikegaki 1977) to be a good surrogate for total mercury concentration and thus continuous turbidity monitoring was a cost-effective tool to assess, and thus to control, secondary mercury contamination.

An intensive fish-monitoring program was also undertaken in order to track mercury contents of fish caught both at several locations within the bay and at key locations outside the boundary nets where fish might be taken for public consumption (Nakayama et al. 1992). The program included fish (porgy, *Sparus macrocephalux*, and croaker, *Umbrina*) which were cultivated inside the bay and monitored at 10-day intervals over the duration of the dredging work for total mercury content (Hosokawa 1993). Fish monitoring has also been continued after all dredging and reclamation work was completed (e.g., see Fig. 5).

5
Lessons Learned and Implications for Other Sites

It is to be hoped that the most important lesson from Minamata Bay, concerning the great human costs of unregulated releases of a toxicant into the environment, has been learned by the global community. However, the current extraction of gold by garimpos in South America using the primitive amalgamation technology (Lacerda and Salomons 1991), and some recent examples of imprudent discharges of mercury in less environmentally conscious countries as cited by D'Itri (1991), may give some doubt to this hope. There are a number of other lessons, including some still known almost exclusively to the Japanese people who planned, executed and/or observed the restoration of the bay. Some of the latter lessons and other experiences for this site will hopefully be shared in the future via more English and Japanese language publications. The authors of this chapter gleaned the following lessons from the available literature:

1. Cleaning up contaminated aquatic systems can be extremely expensive and take a long time. Hosokawa (1993) placed the total cost for the Minamata Bay project at 48 500 million yen (500 000 000 US dollars). More than a decade was required to plan and execute the restoration.
2. Direct and early involvement of all stakeholders in such restoration projects, including especially the public, is a critical requirement if proposed engineering remedies are to be accepted, and those implementing the remedies trusted. Well before the present "Superfund" process in the United States was formulated to include such stakeholders in selecting remedies, the Japanese government was facing a distrusting and litigative citizenry from the very community they sought to assist.
3. The benefits of natural decontamination should not be discounted or undervalued. Kudo (1992) placed a considerable monetary value on the dilution and burial of Hg-contaminated sediment in the Yatsushiro Sea that attended the occurrence of a major natural hydrologic event (200-year rainfall). However, burial of all sediments with mercury levels greater than 25 mg/kg was unlikely ever to occur in Minamata Bay without intervention because much of the most highly contaminated sediment was exposed at low tide and thus subjected to continuous resuspension and transport into deeper water. It can only be speculated at this point that the use of a criterion for sediment dredging based on water depth might have reduced cost and been as effective as the use of the provisional standard to guide dredging. It seems likely that the Japanese considered such an option, perhaps enhanced by artificial burial under a sand layer, and simply rejected it in favor of a more secure remedy. As proven by a number of studies, the mercury in bay sediments was bound largely in insoluble sulfides and effectively isolated from

the biosphere. This beneficial conversion of soluble mercury from the industrial discharges was another form of natural decontamination or immobilization.

4. Large-scale restoration of aquatic systems can be accomplished with sufficient resources and great attention to minimizing adverse impacts of the restoration activity itself. It seems likely that the Japanese learned not only what worked well in the Minamata Bay project but also what failed to work or worked poorly. Clearly, the project met the goal of removing all sediment in the bay above 25 mg/kg, and seems also to have reduced the fish content of mercury below the provisional regulatory standard of 0.4 mg/kg. Those contemplating similar projects may also wonder how well some of the ameliorative strategies worked, e.g., use of boundary nets and sonic fish-repelling device, use of cutterless dredges and continuous turbidity monitoring. Were these simply cosmetic strategies or were they effective (e.g., Nakayama et al. 1992) in accomplishing their intended purpose? In addition, the authors of this review chapter wondered how the restoration project was affected by the 200-year rainfall that occurred in the midst of the dredging activity.

The accessible information on Minamata Bay prompts many questions which may never be answered, or which can only be answered by scientists who have been intimately involved with the site. Some questions have already been posed above for future authors on this subject to contemplate answering. Key additional questions follow:

1. What were the fate and effects of mercury released into Minamata Bay prior to 1950? Mercury began to be used at the Chisso plant in the early 1930s with an early period of apparently substantial releases centered in the early 1940s (Fig. 2). While these releases were routed directly into the Yatsushiro Sea and not Minamata Bay, it is surprising that no human victims who obtained food from the affected area of the Yatsushiro Sea during that early period were ever identified. Paradoxically, when plant discharges were temporarily rerouted to the Yatsushiro Sea in 1958 as a mitigative strategy, new human victims were identified and the discharge was returned to Minamata Bay.
2. Why was so little methylmercury ever found in bay sediments? Was this entirely an artifact of the inferior analytical protocols available at the time or was some biochemical decontamination process (e.g., Barkay et al. 1992) operating to minimize the net production and accumulation of this form of mercury during the period of highest releases of inorganic mercury from the plant?
3. Were direct discharges of methylmercury from the Chisso plant the chief source of methylmercury in seafood until the processes which used mercury were shut down? In spite of the research focused on this question in the 1960s and later, controversy persists and handicaps attempts to synthesize generic conceptual models of root causes of environmental disasters (e.g., Ellis 1989).

Acknowledgements. The authors are indebted to Dr. Junko Nakanishi of the Institute of Environmental Science and Technology, Yokohama National University, for her assistance in obtaining recent fish data from the Kumamoto Prefecture and for constructive comments on the manuscript. George Southworth and Steve Lindberg of Oak Ridge National Laboratory in Oak Ridge, TN USA also provided constructive comments on the manuscript. The author (A.K.) thanks Dr.

Shojiro Miyahara, Nagasaki University for his cotribution. This research has been partially supported by the Ministry of Education (Monbusho), Japan.

References

Barkay T, Turner RR, Saouter E, Horn J (1992) Mercury biotransformations and their potential for remediation of mercury contamination. Biodegradation 3:147–159

Bloom NS, Coleman JA, Barber L (1996) Artifact formation of methyl mercury during extraction of environmental samples by distillation. Fresen J Anal Chem 358:371–377

D'Itri FM (1991) Mercury contamination – what we have learned since Minamata. Environ Monit Assess 19:165–182

D'Itri FM (1994) Minamata disease. In: Cunningham WP, Ball T, Cooper TH, Gorham E, Hepworth MT, Marcus AA (eds) Environmental encyclopedia. Gale Research, Detroit, Michigan, pp 521–522

D'Itri PA, D'Itri FM (1977) Mercury contamination: a human tragedy. Wiley-Interscience, John Wiley New York, 311 pp

Ellis D (1989) Environments at risk: case histories of impact assessment. Springer, Berlin Heidelberg New York, 329 pp

Environmental Agency of Japan (1992) Our intensive efforts to overcome the tragic history of Minamata disease. Environmental Health Department, Environment Agency of Japan, 18 pp

Fujiki M, Tajima S (1973) The pollution of Minamata Bay and the neighboring sea by factory waste water containing mercury. In: Coulston F, Korte F, Goto M (eds) New methods in environmental chemistry and toxicology. International Academic Printing Co, Tokyo, pp 217–229

Fujiki M, Tajima S (1992) The pollution of Minamata Bay by mercury. Water Sci Technol 25:133–140

Fujiki MR, Hirota R, Yamaguci S (1977) The mechanism of methyl mercury accumulation in fish. In: Management of bottom sediments containing toxic substances: Proc 2nd US-Japan Experts' Meet, Oct 1976, Tokyo. EPA-600/3-77-083, pp 89–95

Fujino S (1977) Using sand to cover dredge spoils containing mercury. In: Management of bottom sediments containing toxic substances. Proc 2nd US-Japan Experts' Meet Oct 1976, Tokyo. EPA-600/3-77-083, pp 144–154

Harada M (1982) Minamata disease. organic mercury poisoning caused by ingestion of contaminated fish. In: Jallife EFP, Jellife DB (eds) Adverse effects of foods. Plenum, New York, pp 134–148

Hosokawa Y (1993) Remediation work for mercury-contaminated bay sediments. Water Sci Technol 28:339–348

Irukayama K (1966) The pollution of Minamata Bay and Minamata disease. In: Proc 3rd IAWPR Conf, vol 3, Munich, Germany, Pergamon Press, Oxford, pp 153–165

Ishikawa T, Ikegaki Y (1980) Control of mercury pollution in Japan and the Minamata Bay cleanup. Water Pollut Control Fed J 52:1013–1018

Jensen S, Jernelov A (1969) Biological methylation of mercury in aquatic organisms. Nature 223: 753–754

Kudo A (1992) Natural and artificial mercury decontamination – Ottawa River and Minamata Bay (Yatsushiro Sea). Water Sci Technol 26:217–226

Kudo A, Miyahara S (1984) Mercury dispersion from Minamata Bay to Yatsushiro Sea during 1975–1980. Ecotox Environ Safety 8:507–510

Kudo A, Miyahara S (1988) Effect of decontamination project at Minamata Bay, Japan – dramatic decrease of mercury dispersed into Yatsushiro Sea. Ecotox Environ Safety 15:339–343

Kudo A, Miyahara S (1991) A case history: Minamata mercury pollution in Japan – from loss of human lives to decontamination. Water Sci Technol 23:283–290

Kudo A, Miyahara S (1992) Predicted restoration of the surrounding marine environment after an artificial mercury decontamination at Minamata Bay, Japan – economic values from natural and artificial processes. Water Sci Technol 25: 141–148

Kudo A, Miyahara S, Miller DR (1980) Movement of mercury from Minamata Bay into Yatsushiro Sea. Prog Water Technol 12:509–524

Kumagai M, Nishimura H (1978) Mercury distribution in seawater in Minamata Bay and the origin of particulate matter. J Oceanogr Soc Jpn 34:50–56

Lacerda LD, Salomons W (1991) Mercury in the Amazonia: a chemical time bomb. Report sponsored by the Dutch Ministry of Housing, Physical Planning and Environment. Chemical Time Bomb Project, The Netherlands, 46 pp

Liang L, Horvat M, Cernichiari E, Gelein B, Balogh S, Clarkson TW (1996) A simple solvent extraction technique for elimination of matrix interferences on determination of methyl mercury in environmental and biological samples by ethylation/GC/CVAFS. Talanta 43:1883–1888

Mishima A (1992) Bitter sea – the human cost of Minamata disease. Kosei Publishing, Tokyo, 231pp.

Miyahara S, Schintu M, Kauri T, Kudo T (1988) Twelve years of observations of mercury concentrations in Yatsushiro Sea (Japan). Water Sci Technol 20:193–197

Nakanishi J (1992) Interview with Dr. Hirokatsu Akagi: the blind spot of mercury contamination survey. Water Rep 2:8–10

Nakayama Y, Kyuma K, Hirota R, Fujiki (1992) Treatment of bottom sediment containing mercury and monitoring method using fishes. In: Management of bottom sediments containing toxic substances. Proc 13th US-Japan Experts' meet, April 1992, Pub ADA 253 003, Water Resources Support Center, Fort Belvoir, Virginia, pp 1–22

Nishimura H, Kumagai M (1983) Mercury pollution of fishes in Minamata Bay and surrounding water: analysis of pathway of mercury. Water Air Soil Pollut 20:401–411

Sakamoto H, Tomiyasu T, Yonehara N (1995) The contents and chemical forms of mercury in sediments from Kagoshima Bay, in comparison with Minamata Bay and Yatsushiro Sea, southwestern Japan. Geochem J 29:97–105

Takizawa Y (1979) Epidemiology of mercury poisoning. In: Nriagu JO (ed) The biogeochemistry of mercury in environment. Elsevier-North Holland Biomedical Press, Amsterdam, pp 325–365

Wilken R-D, Wallschlaeger D (1996) The Elbe River: a special example for a European river contaminated heavily with mercury. In: Baeyens W, Ebinghaus R, Vasilier O (eds) Global and regional mercury cycles: sources, fluxes and mass balances. Kluwer, Dordrecht, pp 317–328

Yoshida T, Ikegaki Y (1977) A study of the behavior of mercury-contaminated sediments in Minamata Bay. In: Management of bottom sediments containing toxic substances. Proc 2nd US-Japan Experts' meet, Oct 1976, Tokyo. EPA-600/3-77-083, pp 127–142 green non-sulphur bacteria, green sulphur bacteria, Spirochetes, *Planctomyces*, *Cytophagas*, Gram-positive bacteria, Purple bacteria, Cyanobacteria, Flavobacteria, *Thermotoga* and *Aquifex*; (2) Domain Archaea: extreme halophiles, *Methanosarcina*, *Methanobacterium*, *Methanococcus*, *Thermococcus*, *Thermoproteus* and *Pyrodictium*; and (3) Domain Eukarya: Entamoebae, slime moulds, animals, fungi, plants, ciliates, flagellates, trichomonads, microsporidia and diplomonads.

The phylogenetic tree, considering these three domains, revealed that since the appearance of the "universal ancestor", two different lines of evolution have, evolved one in the direction of the Domain Bacteria, and another in the direction of the Domain Eukarya. However, this evolutionary tendency changed abruptly, giving rise to another branch from the ongoing line to Domain Eukarya, and the Domain Archaea appeared.

Irrespective of considerations about these evolutionary lines, it is not possible to describe the surface structures for all living organisms, even if we consider the updated phylogeny that considers only three domains. However, based on the available information on the chemistry of some well-studied biological surfaces, some groups of organisms can be selected for study and, based on the analysis of their surfaces, it is possible to associate their biochemistry with the uptake of mercury or any other heavy metal, giving an indication of the reaction surface sites in certain groups of organisms.

5 The Cytoplasmic Membrane and its Potential Reaction Sites

The cytoplasmic membrane completely surrounds the cells, acting as a highly selective barrier. It is a phospholipid bilayer whose hydrophobic groups are directed inward, while the cytoplasm and the hydrophilic groups are arranged outwards. This selective barrier also presents several embedded proteins, being similar in overall structure for both prokaryotes and eukaryotic cells. Nevertheless, even though the cytoplasmic membrane is not the outermost layer in the cells, its basic structure provides some potential binding sites for mercury and other heavy metals. As the cell wall, located outside the cytoplasmic membrane, does not act as a selective barrier, heavy metal ions can be transported into the cytoplasmic membrane. Thus, its overall structure can be discussed indicating potential binding sites.

Basically, the cytoplasmic membrane can be considered as constituted of lipids and proteins, mainly. Both components of this membrane contain potential binding sites for the uptake of mercury. A schematic representation of a phospholipid molecule is represented in Fig. 1.

It can be observed that the non-polar head of the phospholipid is of a hydrophobic nature, not presenting any sort of binding sites for the uptake of mercury or any other heavy metal. However, in the polar head of the molecule there is a phosphate group, negatively charged, acting as a potential mercury ligand. As previously mentioned, the cytoplasmic membrane is a bilipid

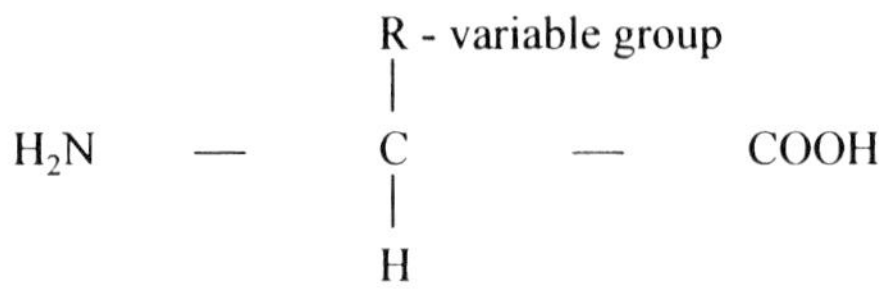

Fig. 3. General representation of an amino acid

of interaction. However, the R-groups of the amino acid residues can actively bind to mercury, for instance, being the strength of this interaction related to the nature of the R-group.

A two-dimensional representation of an amino acid sequence composing a protein (ribonuclease) is given below. This serves as an indication on how the composition of a protein can contribute for the uptake of heavy metals (Fig. 4). Of course, if one considers the tridimensional structure of a protein or enzyme, side reactions between constitutive amino acid residues can take place, thus changing the availability of the interaction sites. Bold areas indicated in the amino acid series of the molecule are those that probably can interact with heavy metals due to the composition of their R-groups.

It is important to emphasize that the proteins associated with the cytoplasmic membrane are not the primary contactors of the metal, due to the presence of the cell wall surrounding the membrane. The cell wall, however, is not usually selective in relation to metallic ions, thus, a specific or non-specific transport of the metal can occur towards the cytoplasmic membrane, and the association with phospholipids and amino acid residues must be considered as well.

6
The Surface of Gram-Positive and Gram-Negative Cells

In order to cope with the concentration of dissolved solutes, most microbial cells developed the cell wall, mainly responsible for their strength and shape of microbes.

There are, between the two major classes of bacteria, marked differences between Gram-positive (G^+) and Gram-negative (G^-) cells. These basic differences are closely related to the chemical nature of their surface structures, and also the way in which these surfaces are arranged.

The external layer of G^- bacterial cells is far more complex than G^+ cells. In G^+ cells, outside the cellular membrane, a rigid layer appears, the peptidoglycan layer. In G^- cells, outside the cellular membrane, there is also a peptidoglycan layer, many times thinner than that of G^+ cells. Beyond, outside this thin layer, there is an outer membrane, basically of a lipopolysaccharide nature.

Some additional structures, such as sheaths and flagellas, will not be discussed in this chapter, due to the fact that they are present under special conditions in some specific groups of cells.

Knowledge of the chemical structures of such layers is a key factor for the understanding of the interaction of metal ions and microbial cells.

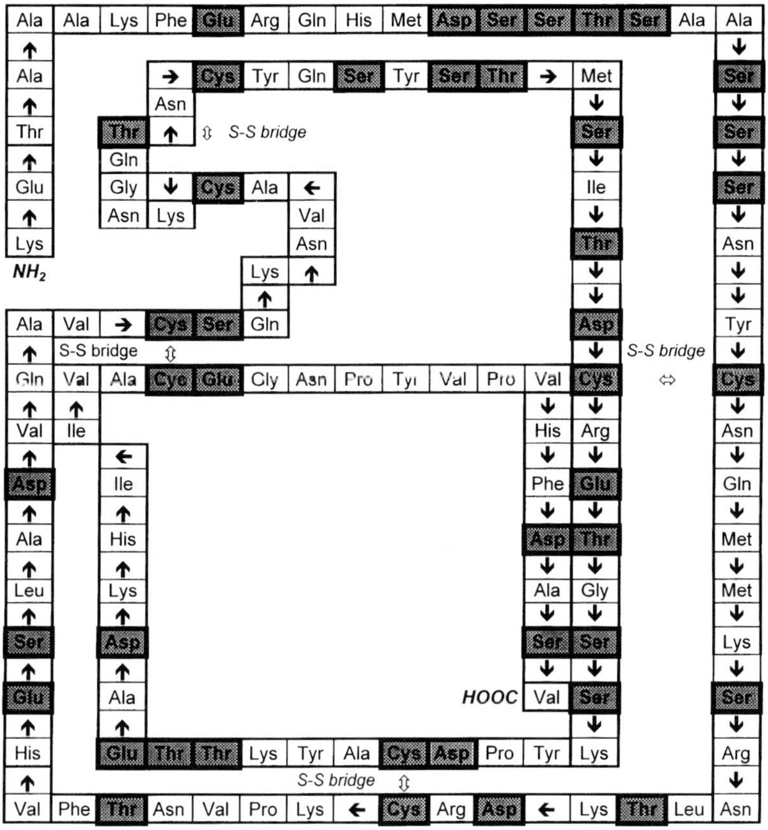

Fig. 4. Amino acid sequence in the ribonuclease molecule

As a common structure for both G^+ and G^- cells, the first structure to be discussed is the peptidoglycan, constituted of sugar derivatives associated to amino acid groups. These amino acid groups are L-alanine, D-glutamic acid, meso-diaminopimelic acid (or lysine, in some cases) and D-alanine. This tetrapeptide structure is stabilized through peptide bonds, connecting the different amino acids through reactions between carboxyl and amine groups from the amino acids. Chemical binding between N-acetylglucosamine and N-acetylmuramic acid themselves is not enough to provide the cells with the rigidity they need; thus, the association with the previously mentioned amino acid groups can give cells the strength they need to keep suitable solute concentrations inside the cells. The three basic building blocks (the two sugar derivatives and the tetrapeptide sequence), of the repeating unit peptidoglycan can be represented as follows in Figs. 5, 6 and 7.

It can be seen that these basic building blocks provide metal binding sites, both through the glucosidic derivatives as well as through the peptide sequence, both components of the peptidoglycan molecule. If the peptide chain is

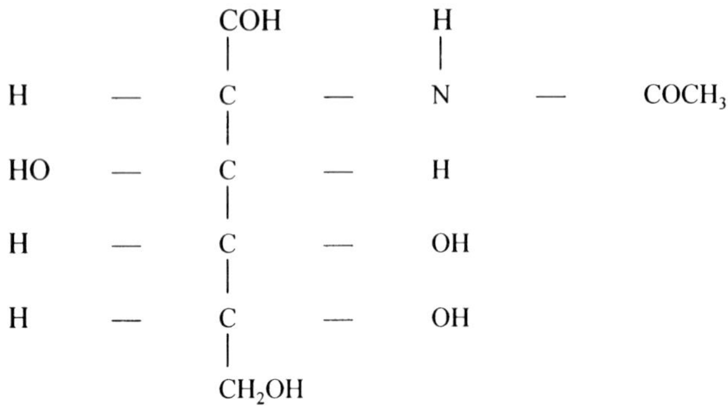

Fig. 5. N-acetylglucosamine molecule

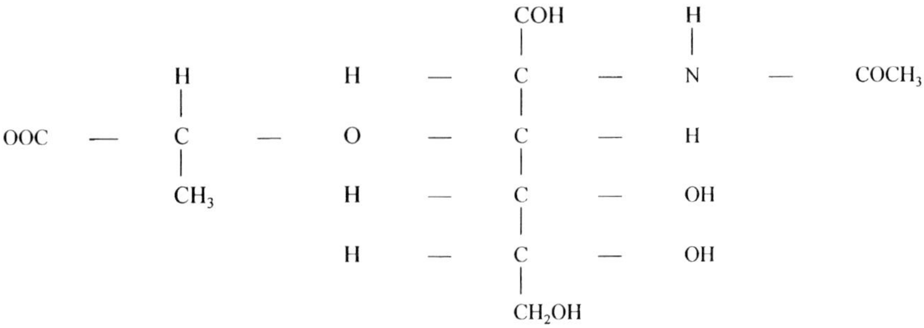

Fig. 6. N-acetylmuramic acid molecule

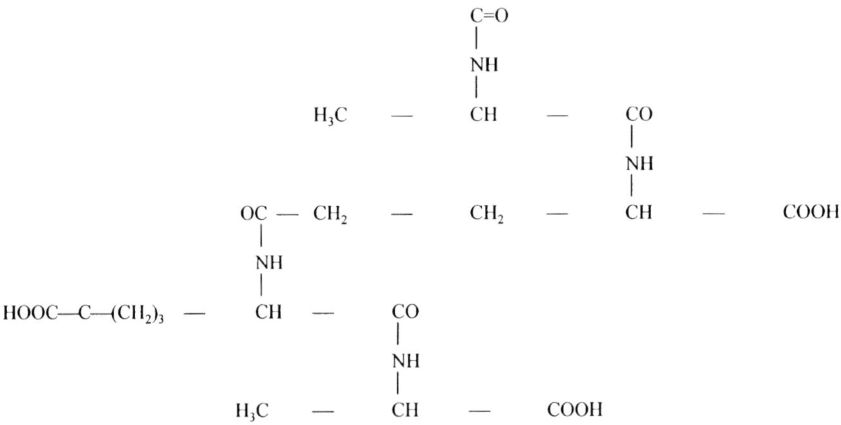

Fig. 7. The tetrapeptide chain

considered, it is observed that carboxyl and amine groups can take part in metal sorption processes. However, the whole structure of the peptidoglycan sheet is based on the connection among all the structures previously presented. The monomeric units are arranged in such a way to form the peptidoglycan layer, including glucosidic bonds, peptide bonds and also specific bridges between peptide chains. Some differences between the peptide chains from G^+ and G^- cells must also to be considered, a fact that can be reflected in their metal binding capacity. For instance, there is a direct bridge between meso-diaminopimelic acid and D-alanine in some G^- cells; however, in G^+ cells the bridge between L-lysine and D-alanine is composed of five glycine residues, providing G^+ cells with a much stronger structure, as expected for G^+ cells.

Figures 8 and 9 represent the basic structure of the peptidoglycan chain, considering the connections between the glucosidic residues and the amino acid sequences. These schematic representations considers the structures of *Escherichia coli* (a typical G^- bacterium), and *Staphylococcus aureus* (a typical G^+ bacterium). For convenience, these specific structures will be considered as typical G^- and G^+ peptidoglycan chains. It must be stressed that peptidoglycan structures are present only in the Domain Bacteria, and a wide diversity of peptidoglycan structures can occur, but, in view of the present overview the basic general structures presented can be considered as universal for G^- and G^+ cells.

Beyond this peptidoglycan layer, G^- bacteria present an outer layer, usually known as outer membrane, whose nature is basically lipopolysaccharidic. The basic difference between this outer membrane and the cytoplasmic membrane is that the former presents polysaccharides and proteins associated to the lipid fraction. The presence of this external structure markedly differs these cells from G^+ cells.

A characteristic component of G^+ cells is teichoic acid, an acid that occurs associated to the cell wall and whose phosphate groups constitute a key component for the uptake of metals.

The literature reports several studies on the interaction of heavy metals with bacterial surfaces, but several Beveridge's results consider these interactions at the molecular level, that means, a detailed investigation on the chemical structures of bacterial cells in relation to the bioaccumulation process. Beveridge

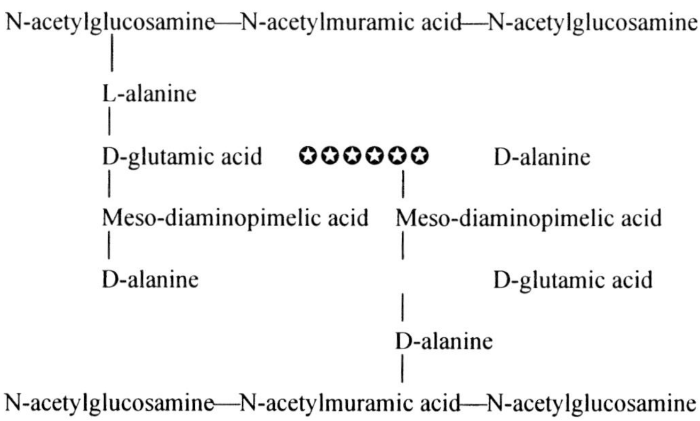

Fig. 8. Basic structure of the peptidoglycan chain, considering the connections between the glucosidic residues and the amino acid sequences (*Escherichia coli*)

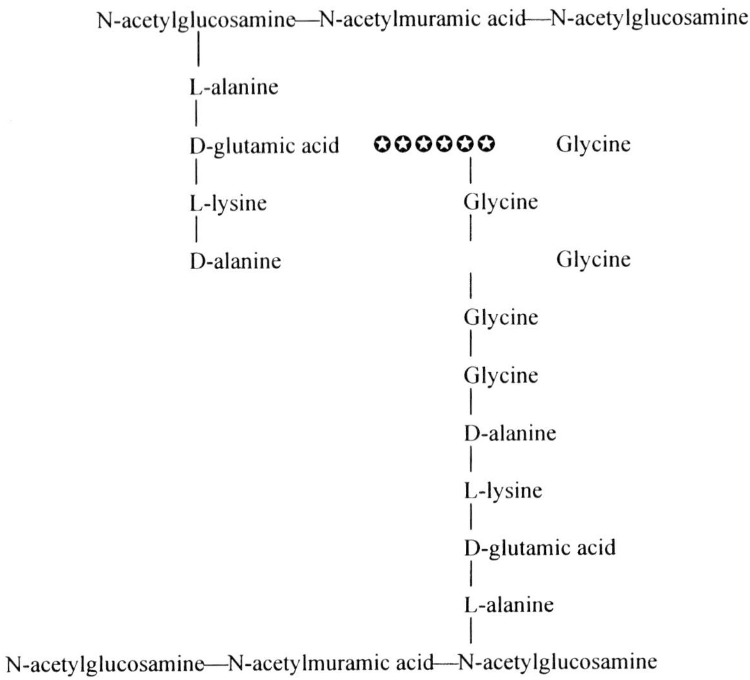

Fig. 9. Basic structure of the peptidoglycan chain, showing the connections between the glucosidic residues and the amino acid sequences (*Staphylococcus aureus*)

and Murray (1976) initially studied the uptake of heavy metal elements by *Bacillus subtilis* cells (G^+) concluding for the presence of selective sites for the deposition of different cations. These studies continued with a view to determining the sites of metal deposition in the cell wall structures of this G^+ bacterium. The same authors concluded for the participation of carboxyl groups as the principal agents in the uptake of heavy metals. The origin of these carboxyl groups is probably in the teichoic acids, associated to the peptidoglycan layers in the cell wall (Beveridge and Murray 1980).

In an excellent review about the ultrastructure of bacterial wall, these surface structures are deeply detailed, providing a better understanding of their possible reaction sites (Beveridge 1981). In this review, Beveridge discussed both the chemical structures of external layers and surface arrays. The same author, again stressing the need for the knowledge of these structures to associate with their natural ability to accumulate heavy metals, has published additional reviews (Beveridge 1988, 1989) emphasizing the role of cellular design.

It is important to mention that bacterial cell walls can be distinct in relation to a specific bacterium and one cannot simply always generalize results from certain bacteria to a whole group encompassing similar genera.

Regarding G^- bacteria, only a few reports are published in the literature, because, due to their lower content of peptidoglycan and the absence of teichoic acids in their structures, their metal uptake capacity is markedly reduced in

considerations, one can envisage that heavy metal elements present a wide range of activity on microbial cells, some biochemical effects, being described in the literature concerning microbial metabolism of any structural change in the cells.

Another methylation mechanism described in the literature involves the synthesis of metal alkyls related to vitamin B_{12}. This relation has been demonstrated through vitamin B_{12}-dependent and non-dependent *Clostridium cochlearium* cells; the dependent cells were able to methylate mercury while the non-dependent cells were not able to do so. In any case, both strains were able to transport Hg^{2+} at the same rate; however, the rate at which the different strains were able to resist its presence was markedly different (Wood and Wang 1983). Hughes and Poole (1989) also considered the involvement of methylcobalamin as the primary route for the non-enzymatic methylation, also mentioning S-adenosylmethionine or 5-methyl-tetrahydrofolate as methyl donors. They also discuss the enzymatic route through inductive systems for the conversion of mercury compounds to volatile species; in this case, the role of mercuric reductase in *Pseudomonas* spp., *E. coli* and *S. aureus*, is mentioned, all plasmid-containing strains and organomercurial lyases, from soil *Pseudomonas* strain K62. The authors discuss the mechanism of resistance involved, indicating that there is a specific plasmid-encoded transport of Hg^{2+} in the direction of the cells, in order to prevent the reaction of the toxic element with the sulphydryl groups present in the membranes. Once again, it is stressed that even though metabolically mediated reactions can take place, the initial step of the incorporation of mercury is not exactly metabolically driven.

These detoxification mechanisms are also discussed in another paper (Summers and Silver 1978). The authors state that there are different detoxification mechanisms, as expected, indicating three different routes for Hg^{2+} detoxification in microbial cultures: (1) synthesis of mercury binding thiols; (2) the existence of permeability barriers that could limit the access of toxic materials against the cells; and (3) elimination of the toxic element for the culture medium. In the detoxification of organomercurials it is proposed a mechanism involving the degradation of the associated alkyl/aryl radical, followed by the reduction of Hg^{2+} to volatile Hg^{0}.

In order not to restrict the discussion on bacterial cells, Overnell (1975) work, attributed to mercury a toxic effect associated with the loss of cell potassium in two marine algae species, *Dunaliella tertiolecta* and *Phaeodactylum tricornutum*. The author explains that, in these algal species, mercury is probably transported through the outer membrane of the cells to the chloroplasts, preventing photosynthetic activity, even though no cell membrane damage has been observed. In this work it is also mentioned that in *Chlorella pyrenoidosa* the primary toxic effect of mercury is to induce leakage of the cells, thus outflowing cell potassium.

For another green microalgae, *Chlorella*, the effect of sublethal concentrations of mercury, added in its culture medium as $HgCl_2$, proved to affect the growth rate. The effect on the growth rate was due to the absorption of mercury, in this case, restricted to 14% in the cell wall (de Filippis and Pallaghy 1976).

Further, de Filippis (1978) observed that cultures of the same microalgae *Chlorella*, with an inducible resistance to mercury compounds, show an enzymatic apparatus to reduce them to volatile mercury; zinc-resistant *Chlorella* cells did not show resistance to the presence of mercury, indicating that a specific enzymatic system should be involved in mercury reduction (de Filippis 1978). Other works also describe the interaction of mercury with microalgal strains (Ben-Bassat and Mayer 1975, 1977; Crist et al. 1981), as previously mentioned.

References

Ben-Bassat D, Mayer AM (1975) Volatilization of mercury by algae. Physiol Plant 33:128–132

Ben-Bassat D, Mayer AM (1977) Reduction of mercury chloride by *Chlorella*: evidence for a reducing factor. Physiol Plant 40:157–162

Bergter EB, Gorin PAJ (1983) Structural chemistry of polysaccharides from fungi and lichens. Adv Carbohyd Chem Biochem 41:67–101

Beveridge TJ (1981) Ultrastructure, chemistry, and function of the bacterial wall. Int Rev Cytol 72:229–317

Beveridge TJ (1986) The immobilization of soluble metals by bacterial walls. Biotechnol Bioeng Symp 16:127–139

Beveridge TJ (1988) The bacterial surface: general considerations towards design and function. Can J Microbiol 34:363–372

Beveridge TJ (1989) Role of cellular design in bacterial metal accumulation and mineralization. Annu Rev Microbiol 43:147–171

Beveridge TJ, Murray RGE (1976) Uptake and retention of metals by cell walls of *Bacillus subtilis*. J Bacteriol 127:1502–1518

Beveridge TJ, Murray RGE (1980) Sites of metal deposition in the cell wall of *Bacillus subtilis*. J Bacteriol 141:876–887

Bharathi PAL, Sathe V, Chandramohan D (1990) Effect of lead, mercury and cadmium on a sulphate-reducing bacterium. Environ Pollut 67:361–374

Bold HC, Wynne MJ (1983) Divisions cyanophyta and prochlorophyta. In: Bold HC, Wynne MJ (eds) Introduction to the algae. Prentice-Hall, Englewood Cliffs, pp 34–59

Brady D, Stoll AD, Starke L, Duncan JR (1994) Chemical and enzymatic extraction of heavy metal binding polymers from isolated cell walls of *Saccharomyces cerevisiae*. Biotechnol Bioeng 44:297–302

Brawley SH, Wetherbee R (1981) Cytology and ultrastructure. The biology of seaweeds. Bot Monogr 17: 248–299

Brock TD, Madigan MT, Martinko JM, Parker J (1994) Molecular systematics and microbial evolution. In: Brock TD, Madigan MT, Martinko JM, Parker J (eds) Biology of microorganisms. Prentice Hall, New York, pp 692–717

Brunker RL, Bott TL (1974) Reduction of mercury to the elemental state by a yeast. Appl Microbiol 27:870–874

Cordery J, Wills AJ, Atkinson K, Wills BA (1994) Extraction and recovery of silver from low-grade liquors using microalgae. Min Eng 7:1003–1015

Crist RH, Oberholser K, Shank N, Nguyen M (1981) Nature of bonding between metallic ions and algal cell walls. Environ Sci Technol 15:1212–1217

da Costa ACA, de França FP (1996a) Cadmium uptake by biosorbent seaweeds: adsorption isotherms and some process conditions. Sep Sci Technol 31:2373–2393

da Costa ACA, de França FP (1996b) The use of seaweeds as immobilization supports for microorganisms used in continuous cadmium biosorption. Biotechnol Tech 10:761–768

da Costa ACA, de França FP (1996c) Cadmium uptake by *Spirulina maxima*. World J Microbiol Biotechnol (In press)

da Costa ACA, de França FP (1997) Biosorption of zinc, cadmium, and copper by a brown seaweed (*Sargassum* sp.) in a continuous fixed-bed laboratory reactor. Bioseparation 6:1–7

da Costa ACA, Leite SGF (1993) The effect of alginic matrix on cadmium uptake by an immobilized green microalgae. Rev Microbiol 24:149–151 da Costa ACA, Teles EMF, Leite SGF (1994) Accumulation of cadmium from moderately concentrated cadmium solutions by *Chlorella* and *Scenedesmus* strains. Rev Microbiol 25:42–45

da Costa ACA, Mesquita LMS, Tornovsky J (1996) Batch and continuous heavy metals biosorption by a brown seaweed from a zinc-producing plant. Min Eng 9:811–824

de Filippis LF (1978) The effect of sub-lethal concentrations of mercury and zinc on *Chlorella*. IV) Characteristics of a general reducing enzyme system for metallic ions. Zentralbl Pflanzenphysiol 86:339–352

de Filippis LF, Pallaghy CK (1976) The effect of sub-lethal concentrations of mercury and zinc on *Chlorella*. I) Growth characteristics and uptake of metals. Zentralbl Pflanzenphysiol 78:197–207

Drews G, Weckesser J (1982) Function, structure and composition of cell walls and external layers. Bot Monogr 19:333–357

Farrel RE, Germida, JJ, Huang PM (1990) Biotoxicity of mercury as influenced by mercury (II) speciation. Appl Environ Microbiol 56:3006–3016

Garnham GW, Codd GA, Gadd GM (1993a) Uptake of cobalt and cesium by microalgal and cyanobacterial-clay mixtures. Microb Ecol 25:71–82

Garnham GW, Codd GA, Gadd GM (1993b) Accumulation of zirconium by microalgae and cyanobacteria. Appl Microbiol Biotechnol 39:666–672

Hu MZC, Norman JM, Faison BD, Reeves ME (1996) Biosorption of uranium by *Pseudomonas aeruginosa* strain CSU: characterization and comparison studies. Biotechnol Bioeng 51:237–247

Hughes MN, Poole RK (1989) Metal toxicity. In: Hughes MN, Poole RK (eds) Metals and micro-organisms. Chapman and Hall, New York, pp 264–270

Macaskie LE, Empson RM, Lin F, Tolley MR (1995) Enzymatically-mediated uranium accumulation and uranium recovery using a *Citrobacter* sp. immobilised as a biofilm within a plug-flow reactor. J Chem Technol Biotechnol 63:1–16

Mishra SP, Chaudhury GR (1996) Kinetics of Zn^{2+} adsorption by *Penicillium* sp. Hydrometallurgy 40:11–23

Nestle NFEI, Kimmich R (1996) NMR imaging of heavy metal absorption in alginate, immobilized cells, and kombu algal biosorbents. Biotechnol Bioengng 51:538–543

Olafson RW, McCubbin WD, Kay CM (1988) Primary and secondary structural analysis of a unique prokaryotic metallothionein from a *Synechococcus* sp. cyanobacterium. Biochem J 251:691–699

Overnell J (1975) The effect of heavy metals on photosynthesis and loss of cell potassium in two species of marine algae, *Dunaliella tertiolecta* and *Phaeodactylum tricornutum*. Mar Biol 29:99–103

Pearson RG (1966) Acids and bases. Science 151:172–177

Pelczar MJ, Chan ECS, Krieg NR (1993) The scope of microbiology. In: Pelczar MJ, Chan ECS, Krieg NR (eds) Microbiology – concepts and applications. McGraw-Hill New York, pp 56–79

Percival E (1979) The polysaccharides of green, red and brown seaweeds: their basic structure, biosynthesis and function. Br Phycol J 14:103–117

Remacle J (1990) The cell wall and metal binding. In: Volesky B (ed) Biosorption of heavy metals CRC Press, Boca Raton, pp 83–92

Sakaguchi T, Tsuji T, Nakajima A, Horikoshi T (1979) Accumulation of cadmium by green microalgae. Eur J Appl Microbiol Biotechnol 8:207–215

Steel KJ (1960) A note on the reaction of mercuric chloride with bacterial SH-groups. J Pharm Pharmacol 12:59–61

Summers AO, Silver S (1978) Microbial transformations of metals. Annu Rev Microbiol 32:637–672

Ting YP, Teo WK, Soh CY (1995) Gold uptake by *Chlorella vulgaris*. J Appl Phycol 7:97–100

Trevors JT, Oddie KM, Belliveau BH (1985) Metal resistance in bacteria. FEMS Microbiol Rev 32:39–54

Valle BL, Ulmer DD (1972) Biochemical effects of mercury, cadmium and lead. Annu Rev Biochem 41:91–127

Volesky B (1994) Advances in biosorption of metals: selection of biomass types. FEMS Microbiol Rev 14:291–302

Volesky B, Holan ZR (1994) Biosorption of heavy metals. Biotechnol Prog 11:235–250

Vymazal J (1987) Toxicity and accumulation of cadmium with respect to algae and cyanobacteria: a review. Toxic Assess: An Int Quart 2:387–415

Wang HK, Wood JM (1984) Bioaccumulation of nickel by algae. Environ Sci Technol 18:106–109

Wilkinson SC, Goulding KH, Robinson PK (1989) Mercury accumulation and volatilization in immobilized algal cell systems. Biotechnol Lett 11:861–864

Winter C, Winter M, Pohl P (1994) Cadmium adsorption by non-living biomass of the semi-macroscopic alga, *Ectocarpus siliculosus*, grown in axenic mass culture and localization of the adsorbed Cd by transmission electron microscopy. J Appl Phycol 6:479–487

Wood JM, Wang HK (1983) Microbial resistance to heavy metals. Environ Sci Technol 17:582A–590A

Xue XB (1989) Why should we be interested in the adsorption of heavy metals by algal surface? Eawag-News 26:3–4

2
Material and Methods

2.1
Field Sampling

2.1.1
Soil

The sampling areas were subdivided according to site and site use (ornamental and vegetable gardens, public and private lawns, playgrounds), soil samples were recovered by means of a cylindrical steel drill down to a maximum depth of 30 cm applying a 1 × 2 or 1 × 1 m grid. Individual samples were filled into brown glass bottles (250 ml) or mixed on site as composite samples in order to obtain a representative of that site (cf. Anonymous 1993a). Pre-treatment of the samples, such as homogenization and separation of partial samples for determination of dry weight, were carried out in the laboratory.

In order to determine the vertical contaminant distribution and to laterally delineate the original Hg release areas, exploratory borings were established to a maximum depth of 6 m, applying a decreasing diameter of 80 to 36 mm to avoid cross-contamination.

2.1.2
Groundwater

In the course of several investigation phases, 6" groundwater monitoring wells were installed which fully penetrate the shallow groundwater down to the base of the aquifer. After purging the wells, the groundwater was sampled by means of a submersible pump via a Teflon hose as rising main.

Sampling was carried out after the conductivity readings of the recovered water stabilized and the water column in the well was exchanged more than three times. The water samples for total Hg determination were stabilized by adding HNO_3 (sample pH < 2) and 0.05% $K_2Cr_2O_7$. In the case of analysis for the various fractions, no reagent was added.

2.2
Analytical Procedures

2.2.1
Total Mercury

Total Hg contents was determined by cold vapour atomic absorption spectroscopy (CVAAS) after digestion of the samples in aqua regia at 160 °C for 3 h (DIN 38414 S7, DIN 38406 E12).

2.2.2
HCl-Extractable Mercury

Following the extraction with 0.07 N hydrochloric acid according to DIN EN 71, Part 3, mercury is determined by means of CVAAS.

2.2.3
Soil Treatment with Digestive Model Juices

Digestive juices are complex mixtures of enzymes, lipids and electrolytes varying in their composition in dependence on nutriments offered.

For the investigation of mercury compounds mobilized from soil samples, an optimized formulation of digestive model juices (Table 1) containing mainly natural compounds as mucin, pepsin, pancreatin and lyophilized bile was used (Rotard et al. 1995).

Table 1. Formulation for digestive model juices

Inorganic components (mg/l)		Organic components (mg/l)	
Saliva			
KCl	895	Urea	200
$KSCN$	200	Uric acid	15
NaH_2PO_4	885	α-Amylase	145
Na_2SO_4	570	Mucin, porcine	50
$NaCl$	290		
Gastric juice			
$NaCl$	2750	Glucose	650
NaH_2PO_4	270	Glucoronic acid	20
KCl	820	Urea	85
$CaCl_2$	200	Pepsin	1000
NH_4Cl	305	Serum albumin, bovine	1000
HCl	1380	Mucin, porcine	18 000
		Glucosamine hydrochloride	330
Duodenum juice			
$NaCl$	7010	Urea	100
$NaHCO_3$	1800	Serum albumin, bovine	1000
Na_2CO_3	1000	Pancreatin, porcine	18 000
KH_2PO_4	80	Lipase, porcine	500
KCl	560		
$MgCl_2$	50		
$CaCl_2$	200		
Bile Juice			
$NaCl$	5250	Bile, porcine, freeze-dried	36 000
$NaHCO_3$	4200	Serum albumin, bovine	1800
Na_2CO_3	1000	Urea	250
KCl	370		
$CaCl_2$	220		

The inorganic and organic component of each model juice was prepared separately in concentrations twice as high as shown in Table 1 and subsequently mixed applying a ratio of 1:1 before treatment of the soil sample.

Five g of the soil sample were shaken at 37 °C according to the flow chart (Fig. 1). When the saliva-stomach stage was terminated, 10 ml of the solution were centrifuged for 1 h at 4600 rpm and 3400 *g*. The supernatant liquid was analyzed for heavy metals by plasma mass spectrometry (ICP-MS) and atomic absorption spectrometry (AAS). Mercury was analyzed by cold vapour AAS and ICP-MS.

To the remaining mixture of saliva, stomach juice and soil sample, duodenum juice and bile juice was added and shaken for 6 h.

After centrifugation for 1 h at 4600 rpm and 3400 *g*, the supernatant liquid was analyzed for heavy metals. For the analysis of mercury by ICP-MS, standards and samples had to be stabilized with sodium bromide.

2.2.4
Solid-Phase Hg Pyrolysis

This method is used to differentiate between Hg compounds that are fixed to soil.

An electric furnace provided with a thermometer in a sample vessel was connected to a detection cell located in an AAS-detector (Perkin Elmer 3030). Sublimating Hg compounds released from the samples were thermally reduced to Hg° by passing a heated glass tube (800 °C) before reaching the detection cell. All samples were heated continuously to 700 °C at a rate of 0.5 °C/s in an N_2 flow of 0.3 l/min. Related to the physical properties of various Hg compounds, Hg is released at different temperatures. Results are presented as Hg-release curves, showing the amount of the released Hg versus the sample temperature (Biester 1997a, b).

2.2.5
Elementary Mercury (Hg°) in Soil Gas

During the exploratory borings, soil gas was gained by means of a membrane pump via a hose in the sampling device at a flow rate of 90 l/h and subsequently measured by CVAAS (mobile Hg monitor, type HG MAK 1200 of Seefelder Meßtechnik). Interference by other gaseous substances also absorbing UV light in the range of 253.7 nm can be compensated for by setting the zero point with the sample gas by means of an absorption cartridge for Hg. Such disturbing volatile substances, however, did not occur in the present study.

2.2.6
Water-Extractable Hg Fractions in the Soil

After extraction with deionized water according to DIN 38414 S4, the water-extractable Hg fraction was divided into Hg°_{aq}, by $SnCl_2$ easy reducible Hg (HgIIa) and total soluble Hg (Hg-tot), determined after oxidation of organic or

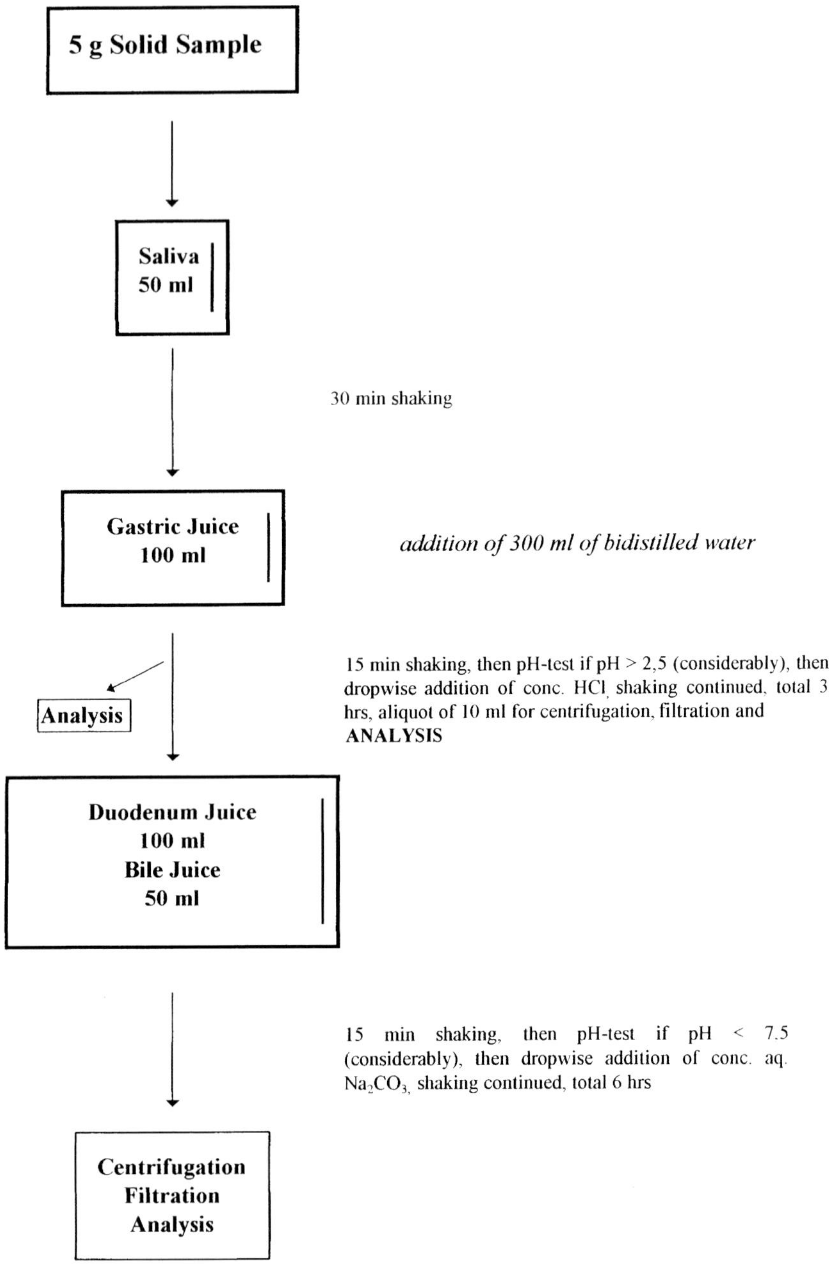

Fig. 1. Flow chart for treatment of solid samples with digestive model juices

colloidal fractions by BrCl or $KMnO_4$ to inorganic Hg followed by reduction (DIN 38406 E12), after the scheme shown in Fig. 2.

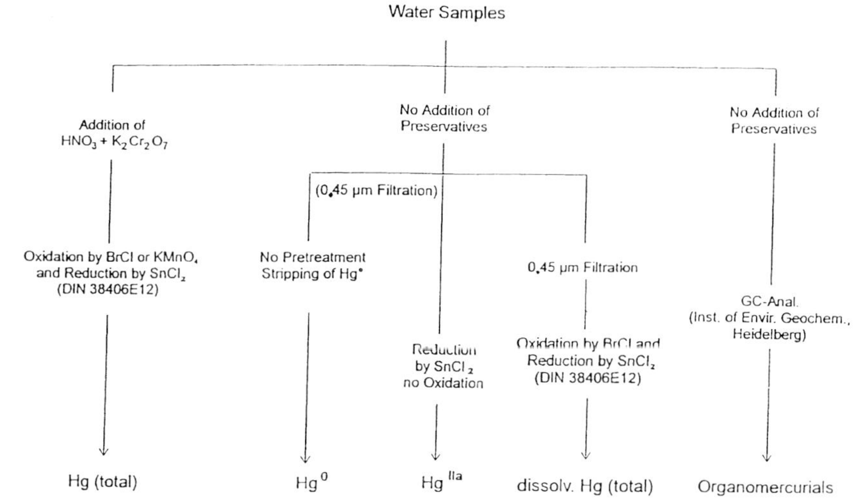

Fig. 2. Determination of operationally defined Hg-fractions

The operationally defined HgIIa-u. HgIIb-fractions are then (Hempel et al. 1992):

HgIIa = sum of dissolved metallic or elemental, easy reducible monovalent and divalent inorganic mercury.
HgIIb = calculated difference between total mercury contents and HgIIa. This fraction is said to include organomercurials, organic Hg complexes and colloidal Hg forms.

2.2.7
Operationally Defined Hg Fractions in Groundwater

As described for the eluate investigations, the Hg fractions HgIIa, HgIIb and Hg° were also determined in the groundwater. Filtration by a 0.45-µm membrane filter was to assist in distinguishing between particulate (per definition) and dissolved/ colloidal Hg, respectively. By determination of Hg° prior to and after filtration, the evaporation effect had to be taken into account.

3
Geology and Hydrogeology

Under a 1–3-m-thick artificial fill consisting of redeposited loess/loess loam and partly building rubble, homogenous loess is encountered down to 5–6 m depth representing the top layer. These slightly clayey, calcic silt sediments [pH(H_2O) = 6.9–8.7] possess a comparatively low content of organic carbon

(0.5–1.2% C_{org}). The underlying fluvial loose gravel deposits forming the upper aquifer contain even less organic substance (mainly below 0.1%). The unconfined groundwater is encountered at a depth of 6 to 10 m below grade. The base of these well-permeable gravels [hydraulic conductivity (k_f) ca. 1×10^{-3} m/s] consists of dense, significantly weathered gravels of low permeability which form the local aquitard.

4
Mercury Distribution in Soil

4.1
Topsoil

The total Hg concencentration is in excess of 10 mg/kg in 90% of the composite samples and still 30% of the results gave values > 100 mg/kg to a maximum of 617 mg/kg. Individual samples even exceeded these values. The evaluation of the individual samples shows concentration variations of up to one order of magnitude, not only between the various sampling areas but also within sampling grids (the smallest sampling grid is 1×1 m, Fig. 3). This might result from the fact that during demolition of the facilities and construction of new buildings, i.e. family homes, highrise buildings and underground garages, the soil (loess) was excavated from the former application area and redeposited. Therefore, today the topsoil is not homogeneously contaminated by Hg and significant pollutions may occur at some distance from former source areas. The thickness of the contaminated soil layer amounts to 0.5–3 m in areas where no direct Hg release occurred and is restricted to the redeposited top soil layer (= anthropogenic fill).

4.2
Deeper Vadose Zone

Unlike the upper redeposited soil layer, the underlying natural soil is contaminated only in the areas of former Hg use. These areas are located in

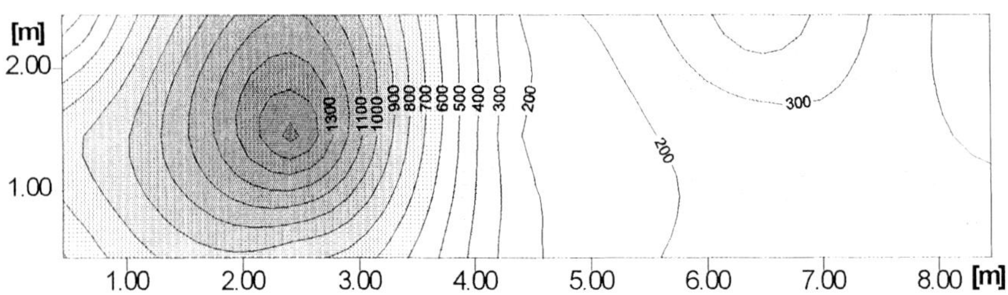

Fig. 3. Hg-concentration distribution in the topsoil (0–30 cm) of a small private lawn. Isolines of Hg(total)-concentrations (mg/kg). 1×1 m sampling grid; mean concentration value: 457 mg/kg Hg(total)

places where sublimate solution could directly penetrate into the ground, i.e. dip basins, storage areas and sublimate production.

Within the centres of contamination, Hg concentrations in the g/kg range were encountered. The contents gradually decrease with increasing depth. The retention capacity of the several-metres-thick loess layer was, however, not sufficient to prevent the gravel aquifer in 6–10 m depth from being contaminated. The insufficient adsorption capacity results from the fact that in alkaline subsoil horizons, which are low in organic C, the adsorption behaviour of Hg is mainly controlled by its direct interaction with an inorganic surface. The low clay mineral contents made translocations within the profile even more probable (Schuster 1991). Beneath the source areas, the unsaturated layer above the aquifer still contained contaminant concentrations of 10 to some 100 mg Hg/kg (Fig. 4).

5 Bioavailability

5.1 Introduction

In addition to the food chain, the direct oral intake of soil particles can play an important role, if dust from polluted soil is whirled up and inhaled or swallowed during games and sports. Experience has shown that the main part of dust particles is moistened in the pharynx and then swallowed. Furthermore, direct oral ingestion of soil particles from hand-to-mouth contact of playing infants can be a crucial exposure path. Hg may be absorbed within the human body after oral ingestion provided that Hg fixed on mineral surfaces is mobilized and dissolved in digestive juices in the gastrointestinal system. These Hg compounds, however, normally show limited dissolution.

The elution of Hg from soil samples with hydrochloric acid (HCl) is used as an economical method to simulate the mobilization of heavy metals by gastric acid. This method does not, however, take into consideration the influence of organic components, such as enzymes, lipids or mucins. The low absorption of dissolved heavy metals in the gastric tract, which is, in comparison, much higher in the intestinal tract, e.g. duodenum and ileum, is also not taken into account. The passage of dissolved heavy metals from the stomach into the duodenum represents a change into a different medium with respect to pH value and electrolyte concentrations. Thus, dissolved heavy metals may be precipitated as hydroxides and subsequently adsorbed to soil particles or indigestible parts of nutriments.

To simulate the mobilization of organic pollutants or heavy metals from soil particles in the gastrointestinal system, several in-vitro models have been developed in recent years (Ruby et al. 1993, 1996; Hack and Selenka 1996). For our investigation the in-vitro model according to Rotard was used (Fig. 1).

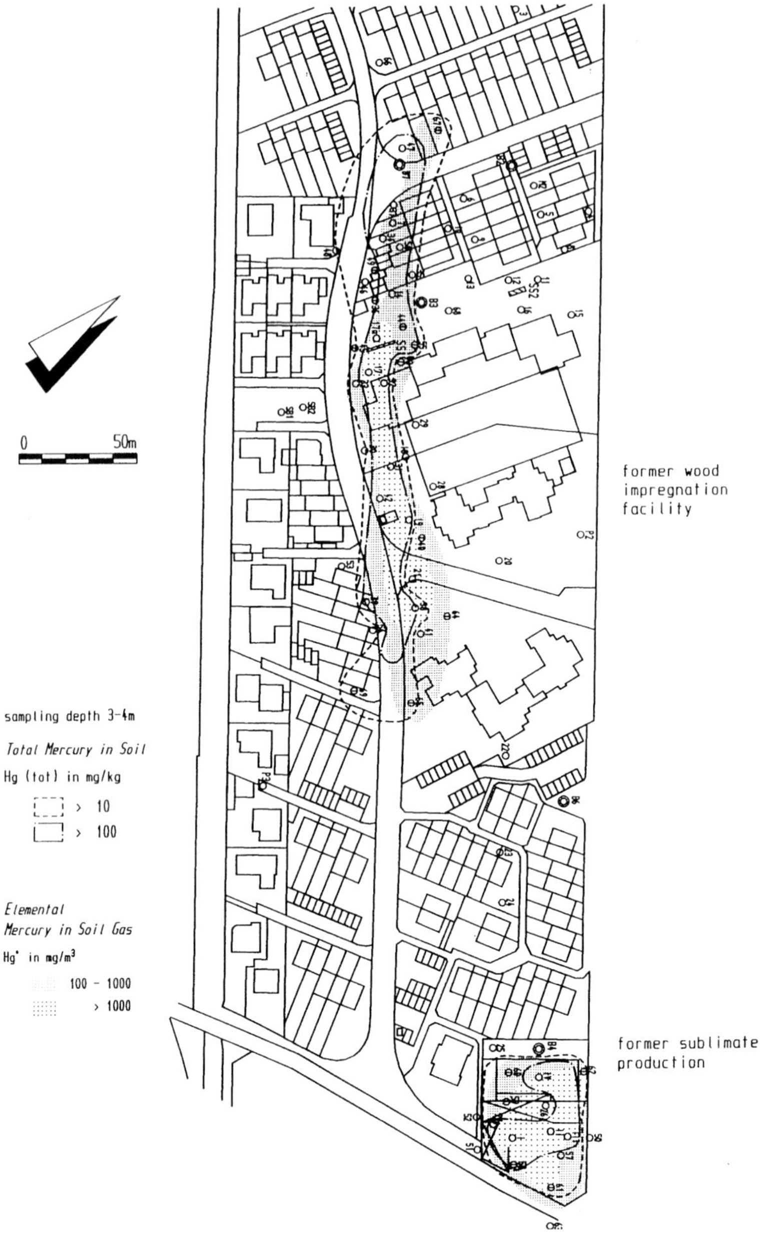

Fig. 4. Distribution of mercury in soil and soil gas of the deeper vadose zone

It is assumed that the heavy metal compounds transferred into the aqueous phase are available for absorption. With a heavy metal content of 100% in the soil obtained from extraction with aqua regia (total Hg), the absorption availability is defined to be the percentage of heavy metals mobilized by digestive model juices. Due to the minor importance of absorption on the gastric stage (Ewers and Schlipköter 1991), the relative mobilization in this stage (in percent) is called the mobilization degree.

5.1.1
Results and Discussion

Topsoil samples recovered from the contaminated area show percentages of HCl-soluble mercury less than 1–2% of aqua regia-extracted Hg (total Hg), if total concentrations are less than 70 mg/kg. At higher concentrations of total Hg the mobilization degree, however, widens dramatically within a range of 5 to almost 20%, whereas there is no strict correlation between HCl-soluble Hg and total Hg (Fig. 5). This might result from the existence of different Hg-binding forms. Especially for higher total Hg concentrations, we can assume that a large portion of the original pure inorganic salt still exists. These mercury chloride residues have not been subjected to any kind of transformation to yield a stronger affinity to mineral and organic material yet.

Table 2 shows the results from extracts with aqua regia, the mobilization (mg/kg) and mobilization degree (%) from the saliva-stomach (gastric) stage as well as the mobilization on the saliva-stomach-duodenum (gastrointestinal) stage and the absorption availability from a total of 17 soil samples. Hg concentrations between 19.8 and 609 mg/kg were determined in the soil samples. The mobilization in the gastric stage ranged between 0.05 and 109 mg/kg with mobilization degrees between 0.25 and 17.9%.

In the gastrointestinal stage, the mobilization of mercury ranged between < 0.2 mg/kg and 43 mg/kg with absorption availabilities between less than 0.1 and 7.1%.

In the gastric stage, the mobilization was found to be relatively high and can be compared with the elution of soil samples with 0.07 N hydrochloric acid (Fig. 5).

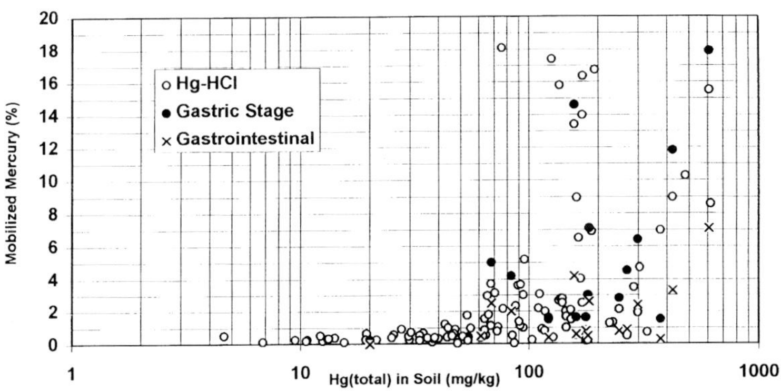

Fig. 5. Percentages of mobilizable mercury by HCl elution and by digestive juices on the gastric stage as well as on the gastrointestinal stage

Table 2. Contents of mercury in the soil samples, mobilizations, mobilization degrees and absorption availabilities on the saliva/stomach (gastric) stage and the saliva/stomach/duodenum (gastrointestinal) stage

Sample no.	Total Hg content aq. reg. extr. [mg/kg]	Gastric stage		Gastrointestinal stage	
		Mobilization [mg/kg]	Mobilization deg [%]	Mobilization [mg/kg]	Absorption avail [%]
1	177	2.8	1.6	≤ 0.2	≤ 0.1
2	297	19	6.4	7.1	2.4
3	158	23	14.6	6.7	4.2
4	609	109	17.9	43	7.1
5	177	2.8	1.6	1.4	0.8
6	246	6.8	2.8	2.0	0.8
7	423	50	11.8	14	3.3
8	161	2.6	1.6	0.9	0.6
9	267	12	4.5	2.5	0.9
10	122	2.0	1.6	0.5	0.4
11	181	5.4	3.0	1.0	0.5
12	374	5.6	1.5	1.2	0.3
13	183	13.0	7.1	4.7	2.6
14	67.6	3.4	5.0	1.7	2.5
15	60.8	0.27	0.4	0.30	0.5
16	19.8	0.05	0.25	≤ 0.2	≤ 0.8
17	82.6	3.5	4.2	1.7	2.0

As expected, the concentrations of dissolved Hg in the digestive juice decrease in the gastrointestinal stage due to the elevated pH value, but the average mobilization is still about 35% with respect to that of the gastric stage.

Although the mobilization of inorganic mercury compounds with 0.07 N hydrochloric acid may be sufficient to assume the worst case of possible absorption of inorganic Hg in the human body, the mercury compounds in the gastric or gastrointestinal system are expected to behave differently. Therefore, it could be more difficult to elute organic mercury compounds with hydrochloric acid, whereas the mobilization by organic components, especially bile juice, might be higher.

In addition, the results show – independently of the used method and its grade of refinement – that the mobilization degree increases overproportionally with increasing total mercury contents.

Today a general factor of 7% Hg absorption following oral ingestion of contaminated soil is often being used to evaluate the risk for human health (ATSDR 1992; Hassauer et al. 1993).

Such practice will tend towards overestimation of actual Hg uptake at lower concentrations and underestimation of higher ones. Even an absorption availability of 20% as a sort of "realistic worst-case assumption" could be too small if Hg is mobilized in the intestinal system in the presence of nutriments. Therefore, more screening tests with digestive model juices are recommended.

6
Mercury Compounds in Soil

6.1
Results and Discussion

The dominating Hg fraction in soil as found by solid-phase pyrolysis is the group of matrix-bound mercury forms which cannot be specified (Fig. 6). These forms may sorb on Fe/Mn-oxyhydroxide coatings covering grain surfaces, they may be fixed to carbonate incrustions or bound to humic matter, i.e. organically bound (e.g. Hodel et al. 1995). Identifiable organic Hg species (organomercurials) such as phenyl-, methyl-, diphenyl-, dimethyl- and diethyl-Hg were determined only at percentages of less than 0.5% of total Hg.

The formation of Hg° in the subsoil could be documented by means of the pyrolysis method and soil gas measurements. Hg droplets could indeed be recognized macroscopically in the soil of the source area in the kyanization building. The soil gas measurements served as a means for rapid delineation on site and for an aimed selection of soil samples for laboratory analysis. Figures 4 and 6 illustrate the practical tracing of the maximum soil contamination by means of direct soil gas measurements on site.

In the former release areas, spatial soil gas contaminations were determined at a depth of 3 m which exceeded 1000 μg of Hg°/m^3 in the centres of contamination. The Hg (total) concentrations of the soil samples from the in-situ loess layer corresponded in general to this pattern of distribution.

The secondary transformation of initially low volatile compounds of cationic Hg(II) may lead to the formation of elementary Hg° ($Hg^{2+} + 2e^{-} \Rightarrow Hg^{\circ}$) either abiotically where reducible reagents are present, such as humic matter or iron and also by microbial mediation (Brosset 1987; Schuster 1991). Due to the high steam pressure of elementary mercury (Hg°), horizontal propagation of mercury may also occur in the vapour phase, as indicated by the results of the soil gas measurements. Considering the results of Fang (1978) on the sorption behaviour of gaseous Hg° on soil, excess of the natural background concentration of Hg in soil (ca. $\leq$0.1 mg/kg) up to concentrations in the lower mg/kg range in loess or fine grains of the gravel sediments may, in this case, also result from the horizontal diffusion of Hg° in the gas phase.

The soil gas measurements, which can easily be carried out, have to be regarded as very useful for several reasons: on the one hand, they may serve as a basis for the risk assessment with regard to the important exposure path via the air, the secondary formation and the propagation behaviour of Hg°, on the other hand, they minimize the investigation efforts during delineation and, finally, they are an important means for interpretation of Hg distribution pattern in the subsoil.

In some of the soil samples, the water-soluble Hg fractions of HgIIa, HgIIb and Hg° were determined. Here, the question was, in which form mercury is preferentially transported to deeper soil layers. Figure 7 illustrates the results of four different sample groups.

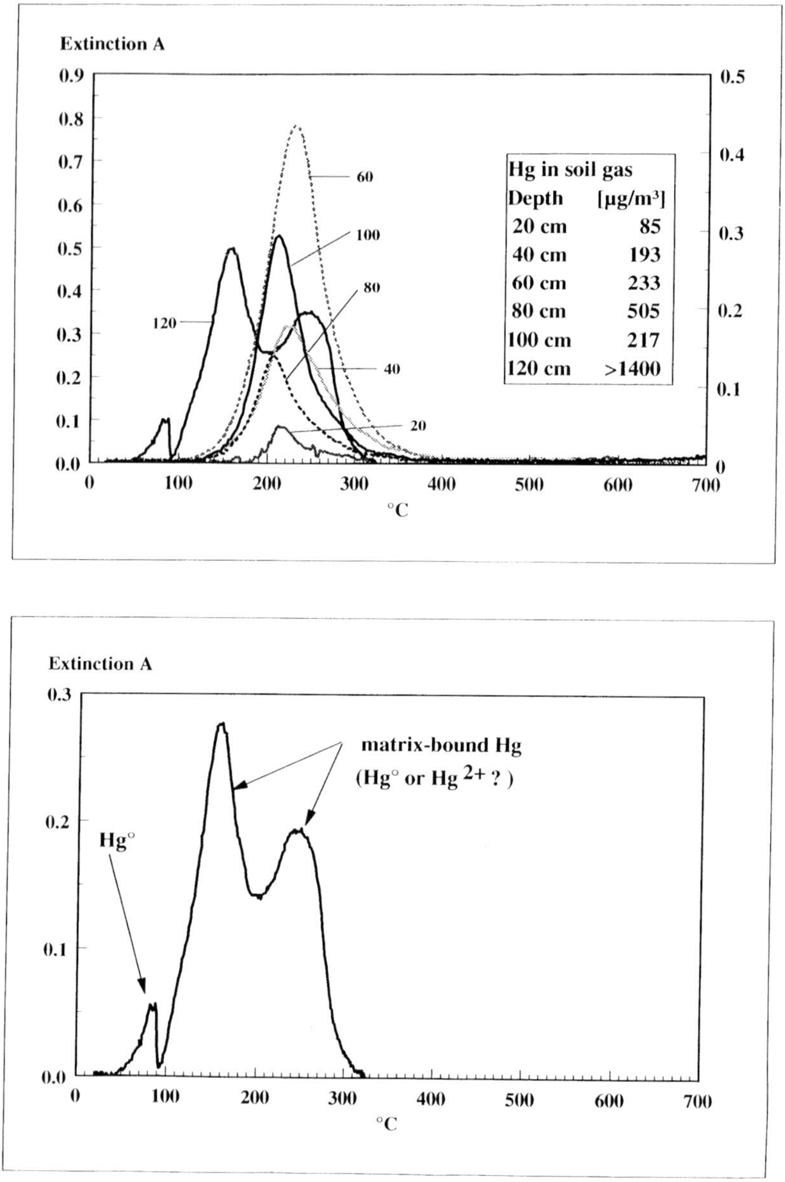

Fig. 6. Determination of Hg-binding forms by solid-phase-pyrolysis

In the storage area the highly contaminated fill material [5390 µg Hg(total)/l in the eluate] contained only 13% mercury in an organic complex form. Remarkable is the high fraction of Hg°. In the underlying loess, the ratio changes in the case of low eluate contents in favour of the water-soluble organic complexed Hg fraction. In the relatively minor contaminated sand/gravel sediments only the

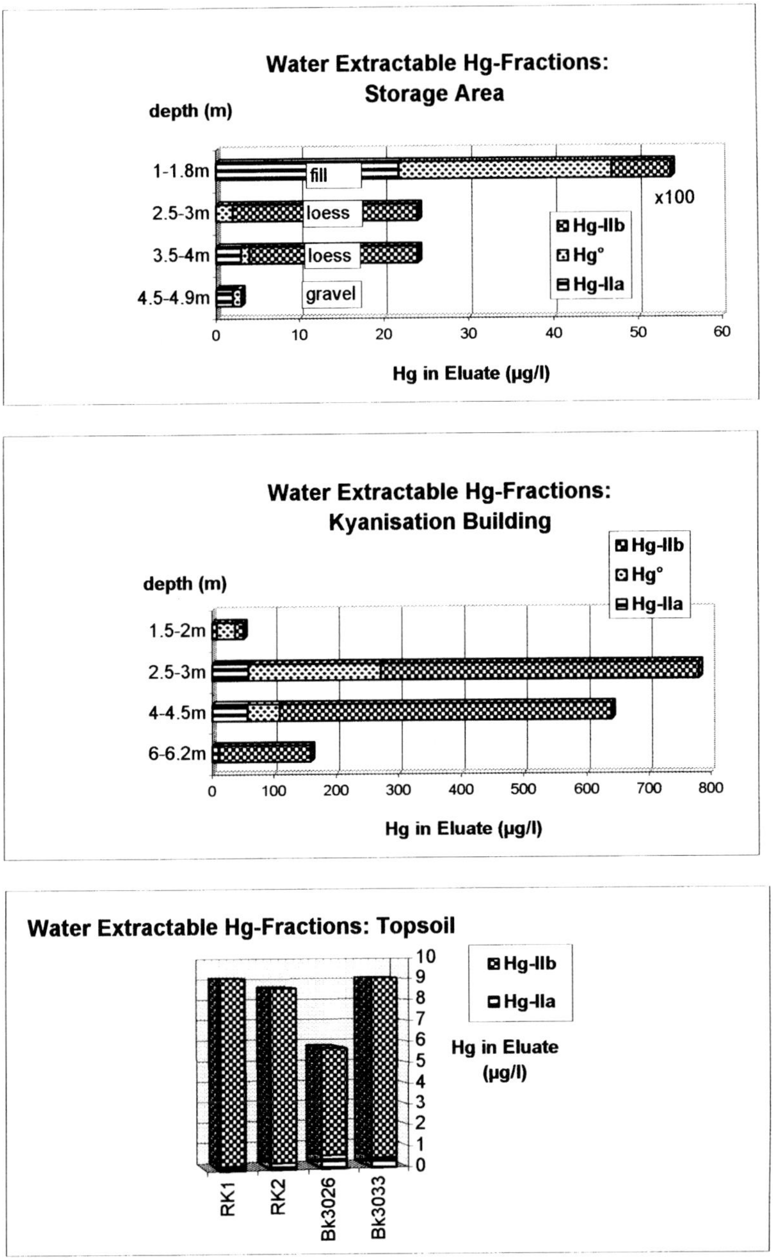

Fig. 7. Water extractable Hg-fractions of different sample depth and areas

easily reducible or inorganic Hg forms were represented at 3 µg Hg/l in the eluate. The soil profile in the area of the former kyanization building demonstrates, in comparison, the effect of an early and long-term Hg release which led to a significant Hg contamination along the soil profile. The Hg concentration in the eluate exceeds by more than one order of magnitude the values of the soil profile from the storage area discussed above. Apart from the sample from the artificial fill, the less reducible, i.e. water-soluble organic Hg complexes are predominant, presumably with humic matter like fulvic acid. Their fraction of 65% in the upper part of the loess layer increased to 83% in deeper soil up to 93% in the sand/gravel sediments. Elementary mercury decreases to background values with increasing depth in the vertical profiles. In the case of low contaminated topsoils, mercury is encountered in an organic complex bound form only. Besides these laboratory results, a perched sample gained from 0.5 m depth contained 28 μg Hg/l, 27.7 μg/l of which in an organic complex form. Hg° was not determined.

In highly contaminated samples, HgIIa is often predominating, presumably mainly in the original form of $HgCl_2$ salt which is dissolved in high concentrations in these samples during the leaching procedure. Inspection samples of loess analyzed for DOC (here as water-soluble organic carbon) determined 10–17 mg DOC/l in the eluate. This confirms that there is sufficient mobilizing organic matter such as fulvic acid to transport most of the soluble Hg fraction in organic complex form, as observed.

7
General Elution Behaviour

The determination of the water-soluble total Hg fraction according to DIN 38414 S4 may be used as a rough indicator for the leachate behaviour of Hg by in-situ percolating water and, thus, for assessment of the risk for groundwater contamination, in case there are no better data available such as leachate samples recovered from lysimeters.

7.1
Results and Discussion

Figure 8, a double logarithmic diagram, illustrates the large range of the Hg concentrations in the eluate in relation to the total contents of mercury in the soil. If the soil samples are subdivided according to origin and geological aspects, a structure can be realized which illustrates the influence of the sample type: compared with the total contents, the highest water soluble Hg fraction is encountered in the gravel/sand sediments recovered from the aquifer underneath the loess sediments. Additionally, this sample type represents the best correlation to total Hg for the following three reasons: homogenous, low contents of organic matter (C_{org} ca. 0.1%), relatively homogenous mineral composition and comparably uniform Hg binding forms. Decreasing elution values were, in general, recorded in loess – artificial fill – topsoil with increasing variation at the same time. This seems to be due to the fact that the sorption capacity of the soil

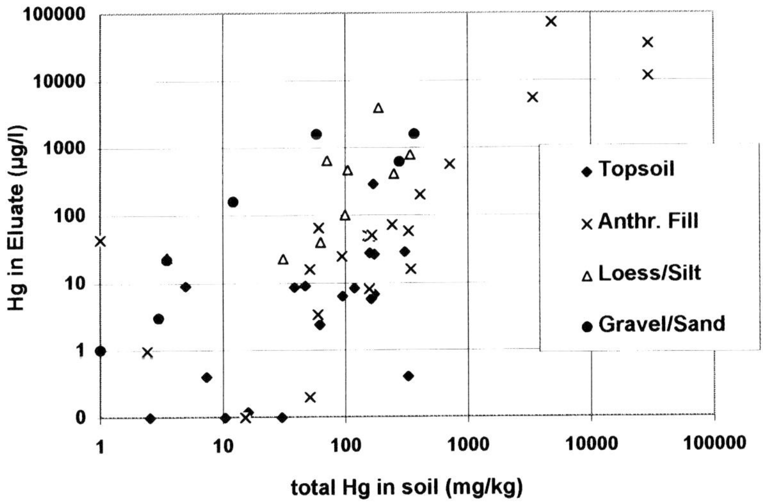

Fig. 8. Water extractable Hg (total) of different soil compartements

increases for Hg with higher contents of organic matter. It can been shown, that the range of variation will reduce, if C_{org} is taken into account. However, it is evident that also other factors such as the Hg-binding forms (Hg°, matrix bound Hg, $HgCl_2$) must have an impact on the eluate behaviour.

By means of eluate values and the groundwater recharge rate, the Hg freights transported by the percolating water to the groundwater were estimated. In this case, a total of 0.5 to a maximum of 3 g of Hg(total) per day has been deduced.

8
Mercury in Groundwater: Contamination and Transport

8.1
Results and Discussion

Narrow contaminant plumes originating from the two separate centers of contamination have developed in the groundwater (Fig. 9). Today, the larger plume is ca. 50 m wide and almost 1 km long, with a maximum concentration of 270 µg total Hg/l (cf. Table 3). Soil samples from the saturated zone indicate Hg concentrations of more than 300 mg adsorbed Hg(total)/kg in the groundwater fluctuation zone and more than 50 mg/kg in deeper levels close to the source area, decreasing to 1 mg/kg at a distance of ca. 1 km downgradient. Within the aquifer, the groundwater has transported an amount of about 4–5 t of Hg from the release area further downgradient.

The balance of the Hg fractions shows that today a higher total Hg freight is transported off the contaminant areas within the groundwater (ca. 10 g of Hg/

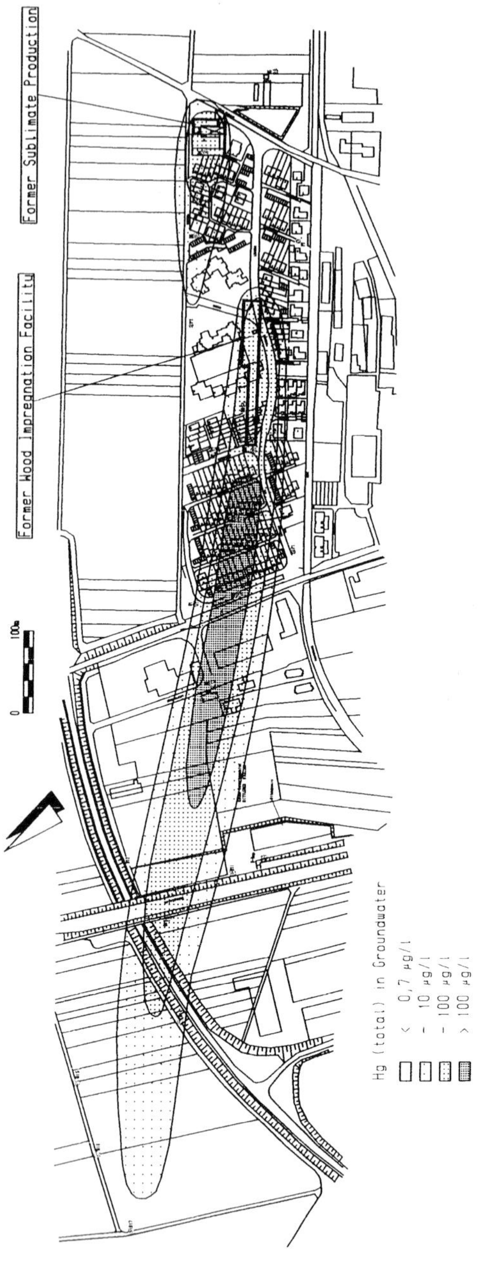

Fig. 9. Plumes of dissolved mercury of groundwater

day) than can be transferred by the groundwater recharge. This is confirmed by investigation results of the aquifer matrix in the immediate downgradient area where a significant Hg abundance was found to exist with still mobilizable fractions in the aquifer and the groundwater fluctuation zone.

Table 3. Concentrations of mercury and its operationally defined fractions in groundwater. Concentration unit: μg/l (ppb)

Date	Hg fraction	B1	B2	B3	B4	B5	B6	B7	B8	B9	P1	P2	P3	B 10	B 11	B 12	B 13
02.03.1995	Hg total	0.6	0.7	40	23	0.2	0.7	69	170	120	0.1	0.3	0.2				
	Hg tot, filtr.	0.6		36	18		1.2	22	130	72							
	Hg IIa	0.3		29	21		0.9	20	100	52							
	Hg°	<0.1		14	13		<0.1	32	34	32							
10.03.1995	Hg total			51				57	250	240							
Innolab	Hg tot, filtr.			51				28	300	280							
	Hg IIa			37				22	90	90							
21.06.1995	Hg total					0.5		10	260	170							
02./03.08.	Hg total	0.9	0.7	100	20	0.1	1.3	60	270	180	<0.1	0.3	0.2	270	<0.1	0.1	22.0
17.06.1995	Hg tot, filtr.	<0.1	0.2	28	18		5.0	27	120	65		<0.1		110			15.0
	Hg IIa	<0.1	0.2	21.2	12.5		4.8	19.3	117	53		<0.1		84	<0.1		10.2
	Hg°	<0.1	0.4	120	30		0.8	70	240	160		0.1		250	<0.1		30.0
	Hg°, filtr.				24									150			13.0
22.-24.05.	Hg total	1.0	0.9	71.1	40	0.1	1.9	60.1	230.7	196.9	<0.1	0.6	0.1	213.4	0.1	0.1	44.4
1996	Hg tot, filtr.	3.2	1.0	82.6	38.7	0.1	2.5	66.2	252.4	203.6	<0.1	0.7	0.1	219.7	0.2	0.1	48.4
	Hg IIa	0.4	0.4	79.9	30.6		0.5	30.0	224.4	187.4		0.1		196.3			33.0
	Hg°	1.5	1.8	4.1	10.9		1.2	13.5	4.8	2.2		1.7		9.5			10.3
25.05.1996	Hg total				36												
	Hg IIa				22												
	Hg°				15												

A remarkable fact is that the concentrations of dissolved mercury initially increase downgradient of the larger source area (kyanization building) before they decrease after some 100 m distance, instead of continuously decreasing right from start of the plume. This can be explained by the time function of various release scenarios such as the termination of quasi-operationally related releases after shutdown of the plant and by former massive releases by accidents, war impacts, etc., as well as by releases still occurring today by leaching of the Hg depot in the groundwater fluctuation zone dependent on fluctuating groundwater levels. With regard to Hg transport behaviour in the groundwater, therefore, retardation factors between 400 to 2000 can be estimated, which characterize the decreased mobility due to sorption to the aquifer material with respect to water flow velocity. However, no constant retardation factor either in time nor in space can be assumed in the case of Hg due to alternating Hg species and concentrations, hydrochemical conditions, mineral surfaces and inner surfaces as well as influences of organic substances.

9 Mercury in Groundwater: Speciation

9.1 Operationally Defined Hg Fractions

9.1.1 *Results and Discussion*

Table 3 summarizes the results of various investigation campaigns. It has to be pointed out that the significantly contaminated groundwater monitoring wells contain mainly easily reducible inorganic mercury (HgIIa), an important fraction of which is present as Hg°. At first sight, the occurrence of dissolved metallic Hg° in groundwater at concentrations of partly over 200 μg/l seems to be improbable because of limiting solubility (cf. Table 3). Richter-Politz et al. (1995), however, have shown that the water solubilitiy of Hg° increases from the already known 60 μg/l (25 °C) in oxygen-free water to the mg/l range in the presence of air/oxygen, as in this case. A balance of these fractions could not be achieved in some of the investigation campaigns due to difficulties during determination of Hg° and particulate Hg. These difficulties, which are still to be investigated, may result from sorption at container walls, from volatile organic matter absorbing in the UV range, from Hg phase transition and species transformation (cf. below) as well as from cross-contamination and evaporation of not stabilized samples. The fraction of the solved organic complex Hg forms gave reliable results, as demonstrated by similar results of various laboratories at various sampling dates. The HgIIb fraction predominates mostly in groundwater monitoring wells with low total concentrations in the range of several μg/l. In the course of a flow path, this fraction has an increasing tendency with decreasing total Hg. This may indicate that the capacity for Hg-complex formation with organic substances dissolved in groundwater (DOC 1–2 mg/l) is exhausted at concentrations of some µg/l of Hg. Individual lab samples did not determine any organomercurials, so that the HgIIb fraction results mainly from these complex compounds and from colloidal Hg compounds.

9.2 Equilibrium Calculations of Inorganic Mercury Species

9.2.1 *Introduction*

Both the described water analysis and direct Hg° measurements in the vapour phase of soil samples from the aquifer zone indicated high contents of dissolved elementary mercury. Plausibility tests were carried out in form of equilibrium calculations to determine whether this can be explained by means of thermodynamic equilibrium between dissolved, inorganic Hg species as well as for determination of species distribution under various conditions. The assumed reactions and constants are from Hem (1970), Lindsay (1979), Andren and Nriagu (1979), Wagman et al. (1982), Behra (1987), Stumm and Morgan (1996). The calculations were performed by MICROQL (Westall 1986; Müller 1993). Initial calculations assumed standard temperatures of 25 °C, i.e. with the reaction constants given in the cited works. As groundwater temperatures, however, deviate with average temperatures of 11 °C, species calculations were also conducted for relevant temperature conditions.

Interpolating known equilibrium constants at standard temperature to another temperature, requires the following equation. If ΔH_r° is not a function of the temperature, what can be assumed in the temperature range of 5 to 35 °C, the temperature dependency of the reaction constants follows in a first approximation the equation (Stumm and Morgan 1996):

$$\ln K_X(T) = \ln K_X(T^{o}) + \frac{H_r^{o}}{R} * \left[\frac{1}{T_o} - \frac{1}{T}\right],$$

where T° is the standard temperature (298.15 K). ΔH_r° is the reaction enthalpy and R is the universal gas constant. Groundwater was treated as a closed system, i.e. the exchange of volatile substances such as Hg° with soil gas has no impact on the species composition.

Because of missing data about the individual components of given DOC and their constants the formation of organic Hg-complexes has not been considered. Mercury does form some very strong organic complexes as, e.g. shown with fulvic acids by Xu and Allard 1991. Some of these are highly soluble in water. Most forms for which data are readily available, however, might be expected to be altered to other more stable and generally less soluble forms in natural water systems.

9.2.2
Results and Discussion

Equilibrium calculations at a standard temperature (25 °C) resulted in the species fractions indicated in Table 4. The system seems to be in a state of transformation between the Hg species of Hg°, $Hg(OH)_2$, $HgCl_2$ and HgOHCl. The significant proportion of Hg° determined is thus plausible even if predominating by 51.2% $Hg(OH)_2$. The result can be compared with stabilization calculations carried out elsewhere under similar hydrochemical conditions (Andersson 1979; Murphy et al. 1994). Under the conditions of the sublimate solution originally released into the subsoil, $HgCl_2$ was predominant due to the high chloride contents. It represents, at the same time, the more mobile Hg form in the unsaturated soil zone. It must be assumed that organic complexes with mercury are of minor importance under such surrounding conditions.

The main features of the aqueous inorganic chemistry of mercury under equilibrium conditions after the transformation of the constants to the measured aquifer temperature of 11 °C is indicated in Fig. 10. Over much of the area Hg° predominates. High pH values and a high redox potential promote the formation of $Hg(OH)_2$. In contrast, at low pH values and a high redox potential, the chlorinated form $HgCl_2$ becomes the most important species. In oxygenated solutions which have high contents of chloride, the solubility of mercury may be greatly increased by the formation of chlorocomplexes. Mercury that enters reduced surrounding conditions can be bound to S to form $Hg(HS)_2^{\circ}$ or HgS_2^{2} , depending on the pH of the solution. A very high electron activity, however, leads to a decomposition of these Hg forms with the result that the neutral form becomes dominant again. Under the hydrochemical conditions measured, this temperature leads to a shift in the species fraction with Hg° predominating now (Table 4)! The analytical results showing a high proportion of Hg° at elevated Hg(tot) concentrations as well as the determination problems (i.e. due to species transformation) can thus be explained.

Table 4. Equilibrium calculations of inorganic Hg species. Calculated proportions in %

	Hg°	$Hg(OH)_2$	$HgCl^{+}$	$HgCl_2$	$HgCl_3$	$HgCl_4^{2}$	HgOHCl	$HgOHCl_2$
Groundwater 25 °C Cl^- 24 mg/l Hg(IIa) 0.1–300 μg/l	20.4	51.2	0.01	13.7	0.07	0	14.6	0
Groundwater 11 °C Hg (IIa) 0.1–300 μg/l	85.9	3.7	0	7.3	0.04	0	3.1	0
Sublimate solution 25 °C Cl^- 1.6 g/l Hg (IIa) 5 g/l	3.1	7.7	0.01	73.5	2.4	0.03	13.2	0.02

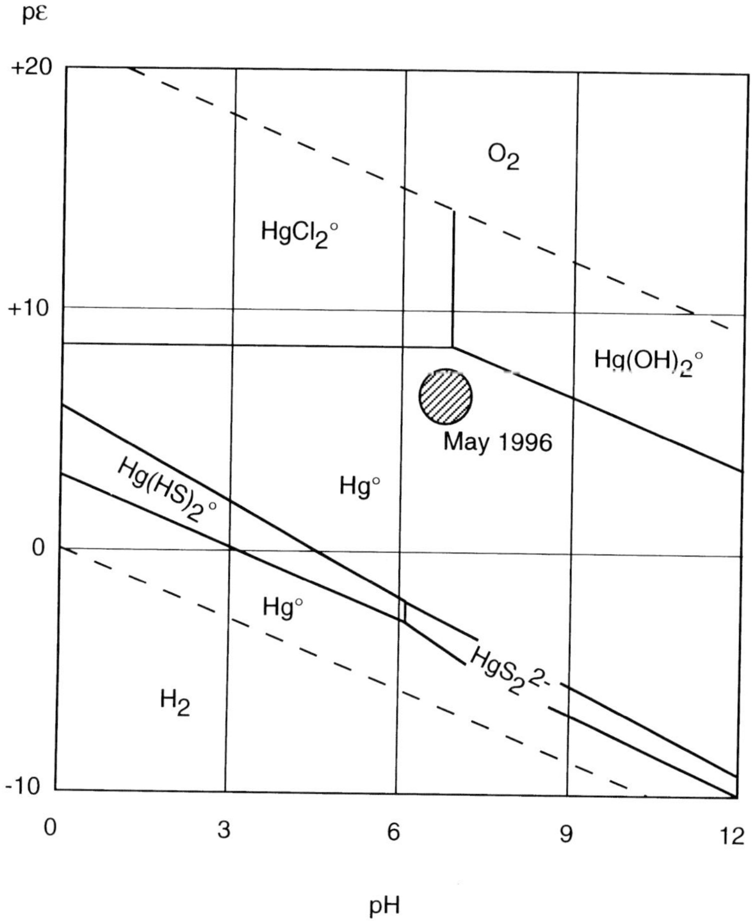

Fig. 10. Fields of stability for aqueous mercury species at 11 °C (with Hg IIa = 1.5×10^{-6} mol dm^{-3}, $Cl_{tot} = 7 \times 10^{-4}$ mol dm^{-3} and $S_{tot} = 2.5 \times 10^{-4}$ mol dm^{-3})

10 Conclusions

In view of risk assessment for humans and the groundwater, as well as in terms of remediation, the question of dominant Hg species is vital as they differentiate significantly in terms of toxicity (cf. resorption availability) and environmental behaviour (transportation paths, mobility).

Together with soil analysis procedures, such as pyrolysis and defined extraction methods, as well as equilibrium calculations of dissolved Hg species, an approximation of the Hg species in soil, leachate and groundwater is made possible. In the soil, Hg seems predominantly to be adsorbed on Mn/Fe-oxihydroxides, humic matter or occluded in carbonate incrustions.

The formation of elementary Hg° in the subsoil could be documented and is hypothesized to be the result of secondary transformation reactions of cationic Hg(II). This process may be due to microbial mediation or to reducible reagents such as humic matter or iron. Therefore, the soil gas measurements for elemental Hg° provide good guidance in the field as well as for sample selection for laboratory analysis and during the interpretation of the gathered data, i.e. the evaluation of horizontal Hg transport in the gaseous phase.

Organically complexed Hg predominates in topsoil leachate, whereas the proportion of easy reducible inorganic forms is highest in the strongly contaminated leachate of the anthropogenic fill. This behaviour is obviously caused by still existing and very soluble $HgCl_2$ salt. Water analysis show that contaminated groundwater mainly contains inorganic Hg species, of which an important part is present as Hg°. This could be confirmed by equilibrium calculations.

Bioavailability tests, using also digestive model juices, give evidence of an overproportional increase of Hg, which is available for absorption in the human body after oral intake of contaminated soil, with increasing total Hg concentrations from less than 1 up to 18% of total Hg. Therefore, a worst-case assumption of 20% bioavailability has been chosen by the public health office to assess the intervention value for human exposure by topsoil contamination. With respect to groundwater contamination level and calculated Hg emissions, our investigations clearly show, furthermore, that the aquifer has to be cleaned up (cf. Anonymous 1993b).

Consequently, a remediation technique has to be developed on the basis of the presented study. This technique has to deal with many difficult handicaps such as the highly heterogeneous distribution of Hg in the topsoil, the contaminated deeper vadose zone in the areas of the former Hg use, the contaminated saturated zone (aquifer) as well as the different Hg behaviour – due to varying species – and the fact that the area has been built over by new family houses, highrise buildings etc.

References

Andersson A (1979) Mercury in soils. In: Nriagu O (ed) The biogeochemistry of mercury in the environment. Elsevier-North-Holland Biomedical Press, Amsterdam, pp 79–112

Andren AW, Nriagu JO (1979) The global cycle of mercury. In: Nriagu O (ed) The biogeochemistry of mercury in the environment. Elsevier, North-Holland Biomedical Press, Amsterdam, pp 1–21

Anonymus (1993a) Zweite Verwaltungsvorschrift des Umweltministeriums zum Bodenschutzgesetz über die Probennahme und -aufbereitung von Bodenproben vom 24.08.1993, veröffentlicht in: Gemeinsames Amtsblatt des Landes Baden-Württemberg (GABL) Nr. 30 vom 29.09.1993

Anonymus (1993b) Orientierungswerte für die Bearbeitung von Altlasten und Schadensfällen, Stand: 12.08.1993, veröffentlicht in: Gemeinsames Amtsblatt des Landes Baden-Württemberg Nr 33 vom 30.11.1993

ATSDR, Agency for Toxic Substances and Disease Registry, US Department of Health and Human Services, Public Health Service (1992) Draft toxicological profile for mercury. Atlanta, Georgia

Behra P (1987) Etude du comportement d'un micropollutant métallique – le mercure – au cours de sa migration à travers un milieux poreux saturé: identification experimentale des mécanismes d'échanges et modélisation des phénomènes. Thèse, Institut de Mécanique des Fluides, Université de Strasbourg, Strasbourg

Biester H, Scholz C (1997a) Determination of mercury binding forms in contaminated soils – Hg-pyrolysis versus sequential extractions. Environ Sci Technol 31:233–239

Biester H, Nehrke G (1997b) Quantification of mercury in soils and sediments – acid digestion versus pyrolysis. Fresenius J Anal Chem 358:446–452

Brosset C (1987) The behavior of mercury in the physical environment. Water Air Soil Pollut 34:145–166

Deutsche Einheitsverfahren zur Wasser-, Abwasser- und Schlammuntersuchung (1984a) Bestimmung der Eluierbarkeit mit Wasser DIN 38414 S4. Beuth, Berlin

Deutsche Einheitsverfahren zur Wasser-, Abwasser- und Schlammuntersuchung (1984b) Königswasseraufschluß zur Wasser-, Abwasser- und Schlammuntersuchung DIN 38414 S7. Beuth, Berlin

Deutsche Einheitsverfahren zur Wasser-, Abwasser- und Schlammuntersuchung (1989) Sicherheit von Spielzeug, Migration bestimmter Elemente (Extraktion mit Salzsäure) DIN EN71. Beuth, Berlin

Deutsche Einheitsverfahren zur Wasser-, Abwasser- und Schlammuntersuchung (1991) Bestimmung von Quecksilber mittels Atomabsorptionsspektrometrie DIN 38406 E12. Beuth, Berlin

Ewers U, Schlipköter W (1991) Intake, distribution, and excretion of metals and metal compounds in humans and animals. In: Merian E (ed) Metals and their compounds in the environment. Verlag Chemie, Weinheim

Fang S C (1978) Sorption of mercury vapor by dry soil. Environ Sci Technol 12/3:285–288

Hack A, Selenka F (1996) Mobilization of PAH and PCB from contaminated soil using a digestive tract model. Toxicol Lett 88:199–210

Hassauer M, Kalberlah R, Oltmanns J, Schneider K (1993) Basisdaten Toxikologie für umweltrelevante Stoffe zur Gefahrenbeurteilung bei Altlasten. Forschungsbericht 10203443/01, Berichte UBA 4/93. Erich Schmidt, Berlin

Hem J D (1970) Chemical behaviour of mercury in aqueous media. US Geol Surv Prof Pap, 713:19–24

Hempel M, Wilken R-D, Miess R, Hertwich J (1992) Untersuchung des Gefährdungspotentials von quecksilberkontaminierten Standorten in Bayern, Schlußbericht F + E von LGA Bayern und GKSS, Nürnberg

Hodel M, Huber L, Lehmann M, Lensing HJ, Herrling B, Frimmel FH, Puchel H (1995) Quantitative Wechselwirkungen zwischen Metallen und Mineralphasen unter Berücksichtigung des Einflusses natürlicher, organischer Wasserinhaltsstoffe und Modellierung des Strofftransports im Grundwasser. Veröffentlicherungen des Lehrstuhls für Wasserchemie und der DVGW-Forschungsstelle am Engler-Bunte-Institut der Universität Karlsruhe 34, Karlsruhe S 227

Lindsay W L (1979) Chemical equilibria in soils. Wiley-Interscience, New York

Müller B (1993) MacμQL, a program to calculate chemical speciation. Limnological Research Center EAWAG, Swiss Federal Institute of Technology, Kastanienbaum

Murphy EA, Dooley J, Windom HL, Smith JRRG (1994) Mercury species in potable ground water in southern New Jersey. Water Air Soil Pollut 8:61–72

Richter-Politz I, Geilhufe Ch, Sela J (1995) Über die Löslichkeit von metallischem Quecksilber in Wasser. TerraTech 2:26–29

Rotard W, Christmann W, Knoth W, Mailahn W (1995) Bestimmung der resorptionsverfügbaren PCDD/PCDF aus Kieselrot-Simulation der Digestion mit Böden. UWSF - Z Umweltchem Ökotox 7 (1): 3-9

Ruby MV, Davis A, Schoof R, Chaney RL, Freeman GB, Bergstrom P (1993) Development of an in vitro screening test to evaluate the in vivo bioaccessibility of ingested mine-waste lead. Environ Sci Technol 27:2870-2877

Ruby MV, Davis A, Kepton JH, Drexler JW, Bergstrom PD (1996) Estimation of lead and arsenic bioavailability using a physiologically based extraction test. Environ Sci Technol 30:422–430

Schuster E (1991) The behavior of mercury in the soil with special emphasis on complexation and adsorption process – a review of the literature. Water Air Soil Pollut 56:667–680

Stumm W, Morgan JJ (1996) Aquatic Chemistry: chemical equilibria and rates in natural waters, 3rd edn. Wiley-Interscience, New York

Wagman DD, Evans WH, Parker VB, Schumm RH, Halow I, Bailey SM, Churney KL, Nuttal RL (1982) The NBS tables of chemical thermodynamic properties: selected values for inorganic and C1 and C2 organic substances in SI units. US National Bureau of Standards. J Phys Chem Ref Data 11, Suppl 2
Westall JC (1986) MICROQL: a chemical equilibrium programm in BASIC. Version 2 for PC's. Report 86-02, Department of Chemistry, Oregon State University, Corvallis
Xu H, Allard B (1991) Effects of a fulvic acid on the speciation and mobility of mercury in aqueous solutions. Water Air Soil Pollut 56:709–717

Relative Importance of Non-Point Sources of Mercury to an Industrialized Coastal System, Sepetiba Bay, SE Brazil

R.V. Marins, L.D. Lacerda, and R.C. Villas Boas

1 Introduction

Estimation of Hg emissions to air, soil and water is the first step in the regional management and control of areas where Hg contamination is potentially significant and no clear point source of this metal exists, such as in diversified industrial parks. Sepetiba Bay basin, SE Brazil, is an example of such a situation. Over 400 industries, from oil-fired power plants to steel and non-ferrous metal manufacturing to plastics and chemicals, have been installed in the region since the late 1960s. Important point sources of metals such as Cd, Zn and Pb have been identified and controlled during the past decade; no significant point source of Hg, however, has been identified in the area. Notwithstanding, Hg is a ubiquitous byproduct of many industries and is also frequently present in urban and domestic wastes (Barcellos and Lacerda 1994; Marins et al. 1996).

With increasing control of known industrial point sources of metals, in particular of Hg, diffuse sources become more important, receiving increasing attention from the environmental authorities (Behrendt 1993; Pacyna 1995). Although diffuse sources of Hg have not been correlated to human health, measured concentrations of Hg in Sepetiba Bay atmosphere (Marins et al. 1996), sediments (Quevauviller et al. 1992) and fish (Kehrig 1995) are doubtless associated with anthropogenic sources of this metal. Diffuse sources are difficult to control and Hg presents very long residence times in coastal waters, thus progressive accumulation of Hg in the bay's biota may occur, with potential health impacts on the local population, still highly dependent on local natural resources.

2 Environmetal Setting of Sepetiba Bay

Sepetiba Bay is a semienclosed waterbody with approximately 420 km^2 of area and average depth of 6 m, located at latitude 23°S and longitude 44°W, at approximately 60 km south of Rio de Janeiro city (Fig. 1). The region is under a hot-humid tropical climate, with mean annual precipitation ranging from 1400 to

Environmental Science
Mercury Contaminated Sites (ed. by R. Ebinghaus et al.)

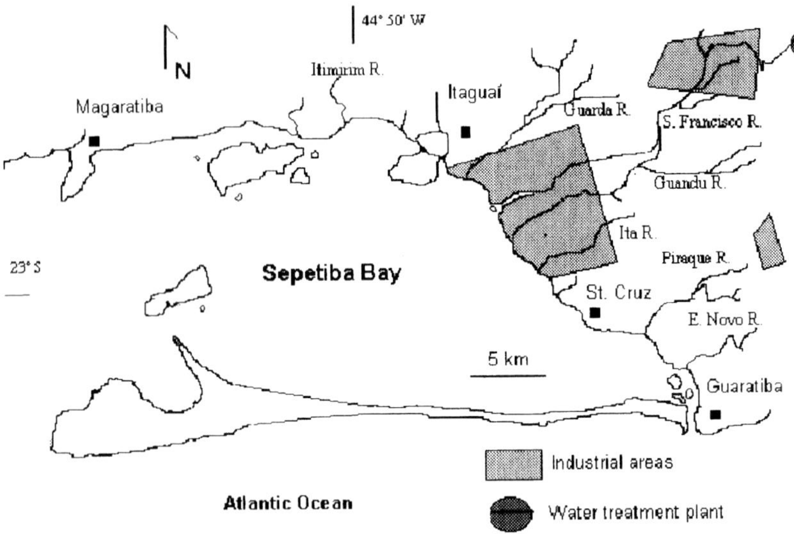

Fig. 1. Map showing location of the study site. Major urban centres (■) and industrial areas

2500 mm, depending on location along the coast. Nine rivers draining the quaternary plain at the northeastern coast of the Bay are responsible for almost the totality of freshwater inputs, reaching an annual flow of 5.7 to 7.6 million m^3. Therefore annual atmospheric precipitation over Sepetiba Bay (0.5 to 10×10^7 $m^3\ a^{-1}$) equals and may even surpass the fluvial inputs, increasing the importance of atmospheric derived pollutants such as Hg (Lacerda et al. 1988; Watts 1990; Barcellos 1995).

The lowlands of the eastern coast of Sepetiba Bay, with good transport facilities, cheap and ample land, good freshwater supply and low population density, became interesting to industrial development, further facilitated by the building in the late 1970's of large harbour facilities. The population of the Sepetiba Bay basin has risen from 600 000 in 1978 to 1.2 million, concentrating mostly along the northeastern portion as a result of industrial growth. This fast and unplanned development resulted in high contamination of the bay, in direct conflict with the necessity of environmental conservation and the sustainable utilization of the bay's natural resources (Barcellos and Lacerda 1994).

Mercury in both inorganic as well as organic chemical forms is known to be widely distributed in the environment. It can reach the coastal environment through many routes such as river inputs, runoff and tides. However, dry and wet atmospheric deposition is frequently the major route of Hg to coastal environments (Miller and Buchanan 1979; Mignon 1993). The atmospheric processes may also significantly affect the global distribution of Hg and provide a means for Hg transfer between the continents and oceans (Mason et al. 1994).

There are few studies dealing with the fate of Hg in Sepetiba Bay. Recent investigations showed increasing Hg concentrations along sediment cores taken from the eastern shore of the Bay, indicating a progressive Hg accumulation in the bay's bottom sediments. Since sediment profiles can be strongly related to Hg loading rates (Rasmussen 1994), it is reasonable to advance increasing Hg inputs to Sepetiba Bay. Therefore, the objective of the present study is to quantify the major sources, inputs and input pathways, and the fate of Hg into Sepetiba Bay.

3 Mercury Emissions to Sepetiba Bay and Basin

Sepetiba Bay basin harbors over 400 industries using a great diversity of technologies and final products which makes the quantification of Hg emissions from individual industrial plants difficult. We studied the most representative of these industries, those which employ over 60 people and are potential sources of Hg (Table 1). This comprises only 20% of the total number of industries but over 90% of the total production (Barcellos and Lacerda 1994). These industries were grouped into five categories: plastics and rubber manufacturing, chemicals, metals manufacturing and metallurgy, iron and steel production, and energy generation. To the industrial sources, three categories of potential urban sources were also added as a sixth group: sewage discharge, urban landfill and fertilizer use, making up six emission groups which were analyzed in detail.

A data base was generated using raw material inputs and production parameters from each industrial sector (Table 1) and emission factors from each major class of production process. The range of values for emission factors is determined by the range of Hg concentrations in the raw material, the technological process used by the industry and the type and efficiency of pollution control processes. Therefore, large variation is expected on emission rates based on these factors (WHO 1982; Nriagu and Pacyna 1988; EPA 1993; Annema et al. 1993). A detailed description of this source assessment is published elsewhere (Marins et al. 1997a). A brief description of that study is given below.

The industrial processes taking place in Sepetiba Bay basin were installed only 30 years ago. They are similar to those from developed countries and follow similar environmental regulations. These facts allow the use of emission factors

Table 1. General characterization of Sepetiba Bay basin potential sources of mercury (ABES 1983; Barcellos and Lacerda 1994). Significance of mercury emissions are from (Straub 1989). Only industries with over 60 employees were included. Details can be found in Marins et al. (1997a)

Source type	No. of plants	Production/ consume ($t\ a^{-1}$)	Emission to air	Emission to waters	Emission to soil
Metal manufacturing	19	175 000	–	×	×
Iron and steel production	3	1 300 000	×	×	×
Oil-fired power plants	1	65 000	×	–	–
Landfills	1	109 500	?	?	×
Chemical industries	16	176 900	–	×	–
Plastic and rubber industry	3	30 900	–	×	–

published in the international literature. However, wherever published emission parameters exist for Brazilian plants and processes, they were preferentially used. The same procedures were used for estimating potential Hg emissions from urban sources.

Estimates of the total annual emissions of Hg from the six different source groups to Sepetiba Bay basin atmosphere, waterways and soils are listed in Table 2. Sewage discharge contributes little to the total Hg load to the basin due to the very low concentrations of Hg in sewage (<0.07 ng l^{-1}) (EPA 1993). In the UK, for example, this source corresponds to less than 0.1% of the total Hg emission (Hutton and Symons 1986). Sewage sludge, a major source of Hg pollution, which may contain up to 5 μg Hg g^{-1} (Nriagu and Pacyna 1988) is not produced in the basin area, due to the absence of such treatment plants. The same is true for refuse incineration which was forbidden in 1977 by a federal law (Barcellos and Lacerda 1994). A direct input of Hg to soils is fertilizer use, but this represents less than 0.1% of the total input to soils.

Municipal solid waste is a significant source of Hg to the environment (EPA 1993; SWA 1993). It is constituted of a heterogeneous mixture of materials found in home, commercial and industrial wastes. Major source of Hg in municipal solid wastes are batteries, discharged electric equipment and wiring, fluorescent lamps, paint residues and plastics (EPA 1993). Mercury concentrations in solid wastes may range from less than 1.0 to 6 μg g^{-1} of wastes (Nriagu and Pacyna 1988; EPA 1993; SWA 1993). Sepetiba Bay basin harbours the third most

Table 2. Mercury emissions to the different compartments of the Sepetiba Bay basin, SE Brazil (kg a^{-1}) (After Marins et al. 1997a)

Source category	Air	Water	Soil
Sewage discharge[a], landfill[b], fertilizer use[c]	–	0–0.002[a]	66–198[b,c]
Plastic and rubber production	–	0–0.003[d]	–
Chemicals production	–	0–0.001[e]	–
Metal manufacturing	–	0.35[f]	18[g]
Iron and steel production	104[h]	?	?
Oil-fired power plants	11–45[i]	–	–
Total	167–201	0.35	84–216

[a]1.2 × 10^6 inhabitants; 200 l inh^{-1} (ABES 1983); 0–0.007 ng l^{-1} (Nriagu and Pacyna 1988).
[b]Itaguaí landfill; 300 t day^{-1} (ABES 1983); 0.6–1.8 g t^{-1} (EPA 1993; SWA 1993). Some authors give higher concentrations of up to 4 g t^{-1} (Hutton and Symons 1986) considering wastes containing a significant fraction of commercial and industrial wastes, rather than mostly household such as in the Sepetiba landfill.
[c]Fertilizer use: 500 t a^{-1}; 0–0.02 g t^{-1} (Nriagu and Pacyna 1988; Barcellos and Lacerda 1994).
[d]Wastewater 12.5 m^3 t^{-1} of product (WHO 1982); 30 900 t a^{-1} (Barcellos and Lacerda 1994); 0–0.08 ng l^{-1} (Nriagu and Pacyna 1988).
[e]Wastewater 2 m^3 t^{-1}, 177 000 t a^{-1} (Barcellos and Lacerda 1994); 0.004–0.3 ng l^{-1} (Nriagu and Pacyna 1988).
[f]2–3% of emissions to soil (Fleister et al. 1984).
[g] 175 000 t a^{-1} of metals produced, 0–0.1 g t^{-1} (WHO 1982; Nriagu and Pacyna 1988; Pacyna 1995; Hutton and Symons 1986).
[h]1 300 000 t of iron and steel (FEEMA 1989), 0.04–0.12 g t^{-1} (OECD 1985; Annema et al. 1993).
[i]550 000 t of oil a^{-1} (Barcellos and Lacerda 1994); 0.02–0.08 g t^{-1} (Chu and Porcella 1995).

important landfill of Rio de Janeiro's metropolitan area. The landfill covers an area of approximately 105 000 m^2 and receives over 10^5 t of urban wastes per year (ABES 1983). Nearly all inputs of Hg go to the soil. However, since the landfill is located along the major tributary of Sepetiba Bay, the Canal de São Francisco, weakly bound forms of Hg can be leached to groundwater or carried out by surface runoff and reach the adjacent rivers (Behrendt 1993). Inputs of Hg to soils from this landfill can range from 66 to 198 kg ga^{-1}, contributing with 75 to 90% of the total Hg inputs to the basin soils, being the most important single source from this urban sources group.

Plastics and rubber manufacturing and the chemical industry, are a major industrial activity in the basin employing over 6000 people (Barcellos and Lacerda 1994). Emissions are mostly to waterways, and account for the third most important source of Hg to waterbodies in the basin. We use general emission factors for this industrial category (EPA 1993), which consider Hg as an accidental component of effluents. However, Hg is frequently used as catalyst in the production of polyurethane and vinyl chlorine and the Hg emission is variable, but has not been reported for Sepetiba Bay industrial park. A detailed survey of the local plastic and chemical industry is advisable since it can represent a significant Hg source.

Metal manufacturing is the major component of the Sepetiba Bay industrial park. Many small and medium-sized metallurgic workshops, as well as three larger plants, employ over 5000 people, producing over 175 000 tons of metals per year, in particular of Al, Zn and Cd. Mercury is a component of liquid and solid wastes from metallurgical plants and these industries are an important diffuse source of Hg in the basin. Annual contribution from this sector to water systems is, however, very small, less than 0.4 kg. Contributions to soils as solid residues are higher than to waters reaching 18 kg a^{-1} and contributing with 10 to 25% of the total Hg inputs to soils. Compared to other regions, however, metal production in Sepetiba Bay basin is a relatively small contributor of Hg. The reason for this is probably the type of ore used by Sepetiba Bay basin plants. Whereas most ores worked as sources of metals are sulphides, particularly enriched in Hg, in Sepetiba Bay basin Al, Zn and Cd are produced from oxides, with very low content of Hg. For example, emission factors from Zn and Cd production from sulphide ores, range from 8 to 45 g t^{-1} (Nriagu and Pacyna 1988), which are orders of magnitude higher than the range used in our estimates, calculated from the use of oxides.

Iron and steel production in Sepetiba Bay basin produces more than 1 million t per year and is the major source of Hg to the atmosphere, mostly due to the technology employed (open furnace-electric arc) and the intense use of scrap metal, particularly enriched in Hg (OECD 1985). Inputs reach 104 kg a^{-1} to the atmosphere. Contribution to soils and waters from large steel and iron production facilities may exist but they are in general small (Hutton and Symons 1986). Apart from the large steel plants, hundreds of small iron metallurgic workshops exist throughout the basin, producing iron houseware and tools and construction material. They certainly contribute with a significant Hg load to the basin, but presently there is no data base available to estimate such a contribution.

Oil-fired plants are another significant source of Hg to the atmosphere over Sepetiba Bay basin, although Hg content in oil is very small (0.02 to 0.08 g t^{-1} (Chu and Porcella 1995). Because the entire fuel supply is exposed to high flame temperatures, essentially all of the Hg contained in the oil is vaporized and exits the furnace with combustion gases to the atmosphere. Also, particle filters are not particularly efficient to retain Hg vapour. The total contribution from this source can range from 11 to 45 kg a^{-1}.

Total estimated contributions of Hg from the industrial park and urban sources to Sepetiba Bay basin range from 251 to 417 kg a^{-1}. The major inputs are to the atmosphere (167–201 kg a^{-1}), and to soils (84–216 kg a^{-1}). Direct inputs to waterways are very small (less than 1.0 kg a^{-1}). However, the large atmospheric and soil load may contaminate fluvial systems through surface runoff. With the recent control of Hg point sources throughout the world, similar situations have been reported in many other coastal areas (Behrendt 1993; Mignon 1993; Pacyna 1995), highlighting the importance of evaluating diffuse sources of Hg to the environment.

4 Mercury Inputs to Sepetiba Bay

4.1 Atmospheric Inputs

To check our estimates based on emission factors, bulk atmospheric deposition of Hg over Sepetiba Bay was measured. Most of the near-surface atmospheric Hg species over both the open ocean and coastal regions are in the vapour phase. Measurements of Hg associated with particles [Hg(p)] in the air have been limited over the past two decades. With the recent advances in instrumental sensitivity and the application of clean techniques, knowledge of Hg(p) concentrations and behaviour has improved. Recent measurements of particulate Hg in several urban / industrial areas documented Hg on large particles and in concentrations similar to those of the vapour phase Hg (ng m^{-3}). Therefore, bulk deposition sampling techniques are the best way of evaluating total atmospheric contribution of Hg to a given system (Marins et al. 1996).

Mercury concentrations in bulk precipitation over Sepetiba Bay ranged from 30 to 164 ng l^{-1}, with an average of 80 ng l^{-1}. This average concentration is from two to three times higher than in pristine areas and suggests a moderate contamination of Sepetiba Bay atmosphere.

Highest concentrations (up to 164 ng l^{-1}) occurred in summer (January and February), whereas lowest concentrations (of 30 ng l^{-1}) occurred in winter (July–September). Higher deposition also occurred in summer when highest Hg concentrations and stronger and more frequent rains occur, leading to more efficient washout of the atmosphere (Iverfeldt 1991). Similar behaviour was reported for other metals in bulk precipitation over Sepetiba Bay (Pedlowiski et al. 1991).

We estimated annual bulk Hg atmospheric deposition over Sepetiba Bay of about 76 μg m^{-2} a^{-1}. This value is in agreement with those reported for moderately industrialized coastal areas (50–90 μg m^{-2} a^{-1}) (Gobeil and Cossa 1993), and up to five times higher than those from continental remote sites of 15 μg m^{-2} a^{-1} (Mason et al. 1994; Table 3).

Considering the basin area of approximately 2500 km^2, this deposition rate gives an annual input from the atmosphere to the basin of ca. 190 kg of Hg. This is in agreement with the estimates we calculated here using emission parameters. This agreement suggests an average residence time of Hg in the local atmosphere of less than 1 year, which is also in agreement with many recent studies on Hg behaviour in the atmosphere (Brosset 1987). Direct atmospheric deposition over the bay is about 30 kg a^{-1}, for a total area of ca. 420 km^2.

4.2 Fluvial Inputs

The final destiny of Hg emitted to the atmosphere and further deposited in Sepetiba Bay basin and of the Hg emitted to soils would be the bay itself. Apart from direct atmospheric deposition, Hg would reach the bay through river discharges. Inputs of Hg from rivers to the bay were assessed by monitoring total Hg concentrations in water of major rivers through 1 year (Table 4). The results showed that total Hg content in major rivers gives a range of concentrations between < 15 to 140 ng l^{-1}, with highest values always occurring in the three major tributaries of the bay, São Francisco, Guandu and Ita rivers (Fig. 1), which cross the most industrialized area of the basin. These tributaries together corresponds to over 90% of the total fluvial inputs to the bay (FEEMA 1989). Total fluvial input to Sepetiba Bay has been monitored by local environmental authorities during the past 30 years and ranges from 5.7 to 7.6×10^9 m^3 a^{-1} (Watts 1990). Over 90% of the total Hg fluvial load enters the bay through the Canal de São Francisco and Guandu rivers. Practically, there is no seasonal variation in fluvial discharge to the bay. This is due to the nearly constant flow of these two rivers, which are artificially controlled to 160 m^3 s^{-1}, by the local water treatment plant located upstream from the major industrial area (Fig. 1; FEEMA

Table 3. Atmospheric Hg deposition over Sepetiba Bay compared to other industrial and pristine areas

Site	Method	Deposition rate (μg m^{-2} a^{-1})	Reference
Sepetiba Bay, SE Brazil, industrialized	Bulk deposition	76	Marins et al. (1996)
Central Brazil gold mines	Bulk deposition	67–151	von Tumpling et al. (1993)
Central Brazil gold mines	Sediment core	90–120	Lacerda et al. (1991)
Southeastern USA, industrialized	Bulk deposition	50	Mason et al. (1994)
Pristine continental areas, between 10° N–10° S	Bulk deposition	13	Mason et al. (1994)

Table 4. Total (dissolved + particulate) mercury concentrations in river waters from Sepetiba Bay basin, SE Brazil (ng l^{-1}).

River system	Total Hg (ng l^{-1}) (range)	Average	n	Water flux ($10^9 l\ a^{-1}$)[a]	Hg flux (kg a^{-1})[b]
Engenho Novo River	<15	<15	3	40–50	< 0.5
Piraque River	19–111	73	4	100–130	7–10
Ita River	69–106	82	4	70–90	5–7
Guandu River	52–103	81	4	170–220	14–18
São Francisco River	83–140	104	3	5000–6500	520–680
Guarda River	57	57	2	330–440	18–25
Itimirim River	<15	<15	3	<2	<0.2
Total	–	–	23	5710–7430	560–740

n = number of sampling campaigns.
[a] After Watts (1990) and FEEMA (1989).
[b] All values are rounded to two significant figures.

1989; Watts 1990). In fact, Sepetiba Bay basin receives a large volume of water from the transposition of an adjacent basin, the Paraíba do Sul river system, a heavily industrialized region north of Sepetiba Basin. Nearly 80% of this river volume is diverted through a hydraulic engineering system to provide potable water to Rio de Janeiro metropolitan area, after crossing two reservoir systems to decrease its suspended particle load (FEEMA 1989; Barcellos 1995). From the water treatment plant, a constant volume of about 5.0×10^9 m^3 a^{-1}, (about 70% of the total fluvial volume reaching the bay) is discharged in the rivers Canal de São Francisco and Guandu (Fig. 1), through a few outlets. This also includes wastewaters from the treatment plant.

Using the total Hg concentration range found in river waters and water fluxes from the sampled rivers, fluvial Hg inputs from major rivers was estimated to range from 560 to 740 kg a^{-1}, with an average of about 650 kg a^{-1} (Table 4). In contrast to the atmospheric inputs, we found no seasonal variation trend for Hg concentrations measured during the four water-sampling campaigns. The import of a large and constant fraction of the total fluvial flow explain the lack of seasonality. This "external" input of fluvial load may result in a significant Hg input from a region "external from the local industrial park" to Sepetiba Bay and basin. However, this potential Hg source to the Bay has not been evaluated in this study.

5
Mercury Distribution in Bottom Sediments

Mercury concentrations and distribution in bottom sediments of major rivers and of the bay's bottom sediments are presented in Fig. 2. The most contaminated rivers are those crossing the industrial area, and the bottom sediment concentrations in these rivers reached 107 and 197 µg kg^{-1} in the Canal

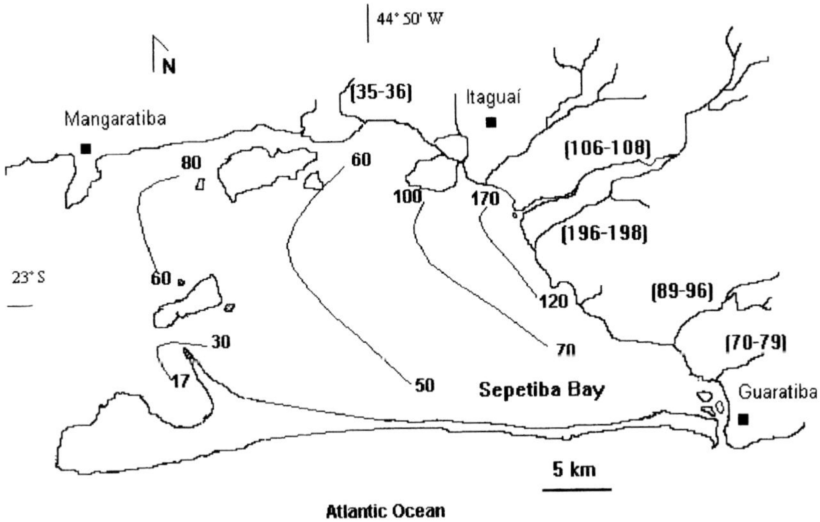

Fig. 2. Mercury concentrations in bottom sediments from major rivers and in Sepetiba Bay bottom sediments (μg kg^{-1} d.w.)

de São Francisco and Guandu rivers, respectively. This result agrees with the total Hg content measured in river waters (Table 4), since those were the rivers which present the highest total Hg concentrations in water. Although Hg contributions are mostly from diffuse sources, their inputs are integrated by rivers, in a way similar to other trace metals with point sources previously studied in this basin (Lacerda et al. 1987; Barcellos 1995), and this explains the results found.

Mercury distribution in the bay's bottom sediments showed a distribution clearly influenced by fluvial inputs, with highest concentrations (120 to 170 μg kg^{-1}) occurring close to the mouth of major rivers. Decrease in Hg concentrations throughout the bay follows surface current pattern, and seems to reflect a dilution of Hg-rich fluvial dominated materials into Hg-poor marine-dominated sediments. This pattern has also been previously described for other trace metals of fluvial origin in Sepetiba Bay (Lacerda et al. 1987; Barcellos 1995). Although Hg concentrations in the most contaminated sediments of the bay are relatively low, when compared to contaminated sediments from other coastal areas in Brazil, which typically reach 1000 μg kg^{-1} (Pfeiffer et al. 1988), the highest concentrations found in Sepetiba Bay sediments are six to ten times higher than the local background measured in the more offshore stations of the bay (Fig. 2).

Mercury distribution in sediment cores from Sepetiba Bay, sampled in shallow mud flats, permitted to estimate Hg deposition rates to bottom sediments. Sedimentation rates were estimated based on excess ^{210}Pb distribution from cores collected at three sites along the northeastern shore of the Bay by Forte (1996) and at two other sites along the eastern coast by Barcellos (1995). Sedimentation rates estimated from these cores ranged from 0.18 to 0.8 cm a^{-1}, with an average of 0.4 cm a^{-1}. Undisturbed sediment profiles taken from nearby sites for this study showed Hg concentrations ranging from 79 to 123 µg g^{-1} at the surface layers of the core to background concentrations, at depths deeper than 30 cm, ranging from 50 to 60 µg g^{-1} (Figure 3).

Mercury deposition rates were calculated according to Eq. (1), after Cundy and Croudance (1995)

$$f(Hg) = \phi \cdot (C_i - C_0) \cdot s \cdot 10, \quad (1)$$

where f(Hg) is the Hg flux in µg m^{-2} a^{-1}, ϕ is the sediment density in g cm^{-3}, C_i is the Hg concentration at depth i, in µg kg^{-1}, C_o is the background Hg concentration in µg kg^{-1}, and s is the sedimentation rate in cm a^{-1}. This equation gives approximate Hg deposition rates ranging from 120 to 385 µg m^{-2} a^{-1}, depending on site and depth of the sample inside of a core. Average deposition rates considering all cores and all depths within cores, corresponding to approximately the last 57 years, was 280 µg m^{-2} a^{-1}. This estimate is considerably higher than those measured in relatively pristine environments. For example, Johnson (1987) estimated Hg deposition rates in 14 Ontario lakes ranging from 5 to 52 µg m^{-2} a^{-1}, whereas Rekolainen et al. (1986) estimated Hg deposition rates in remote Finish lakes ranging from 25 to 50 µg m^{-2} a^{-1}. The estimates for Sepetiba Bay fall in the same range found for other industrialized areas. Contaminated areas in Northern Europe typically shows Hg deposition rates ranging from 120 to 370 µg m^{-2} a^{-1} (Simola and Lodenius 1982; Rekolainen et al. 1986). In gold mining areas in Central Brazil, Lacerda et al. (1991) reported Hg deposition rates in lake sediments reaching 120 µg m^{-2} a^{-1}.

Since sediment deposition patterns and Hg concentrations are spatially variable across the bay, we must consider this rate of Hg deposition for the

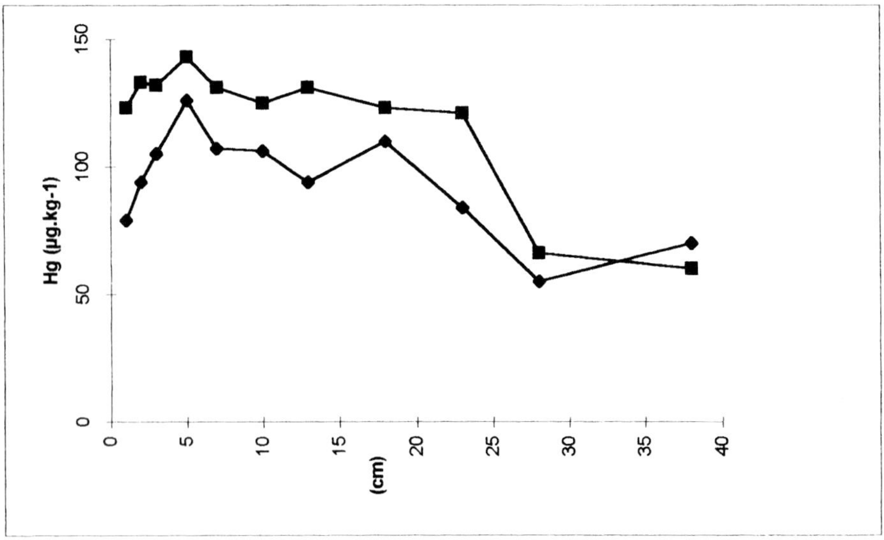

Fig. 3. Mercury distribution in sediment cores from the northeastern coast of Sepetiba Bay

sampled area only. However, shallow mud flats correspond to approximately 70% of the bay's area (ca. 320 km^2; Suguio et al. 1979). Using this area, yearly burial rates of Hg in Sepetiba Bay sediments thus reach 89 kg.

Another assessment of the total Hg amount buried in the bay sediments can be made through estimating the Hg inventory throughout the cores according to Eq. (2), after Krom et al. (1994).

$$C_T = C_X(x_n - x_{n-1})\phi, \tag{2}$$

where C_T is the total amount of Hg in the core in μg cm^{-2}, C_X is the Hg concentration in the depth interval x_n to x_{n-1} in μg g^{-1}, x_n is depth in cm, and ϕ is the sediment density in g cm^{-3}. Using this equation, the total amount of Hg accumulated during the past 57 years in the shallow mud sediments of Sepetiba Bay (roughly corresponding to the top 28 cm of sediment), reaches about 5300 kg, giving an annual burial rate of about 92 kg. This is perfectly in agreement with the annual burial rate estimated using Eq. (1). Therefore, about 90 kg of Hg are annually buried in Sepetiba Bay sediments. The availability of the accumulated Hg in bottom sediments to the biota will depend on the geochemical partitioning of Hg among the different sedimentary phases. Unfortunately, this information is presently unavailable for Sepetiba Bay, although the effect of emergent macrophytes on the releasing of Hg from bottom sediments of the Bay has been recently published by Marins et al. (1997b).

6 Mercury Mass Balance in Sepetiba Bay and Basin

Figure 4 resumes Hg mass balance in Sepetiba Bay and basin based on the average amount of Hg entering each compartment. The use of an average facilitates the understanding of the calculation. It is important to note, however, that these numbers may vary by a factor of 2, if minimum and maximum values are also considered.

Major annual inputs to the basin are from the local industrial and urban sources, reaching 340 kg. Emission to the atmosphere responds for about 53% (180 kg) of the total input, whereas inputs from industrial and urban sources to soils and waters account for 47% (160 kg) of the total. Since the atmospheric deposition measured over the basin (Table 3) is very similar to the estimated emissions to the atmosphere, it probably results from local emissions only.

The output of Hg from the basin to the bay through fluvial inputs reaches an average of 650 kg (Table 4). This input is higher, by a factor of 1.9, than the total industrial and urban inputs of Hg to Sepetiba Bay basin, calculated from emission parameters. Some of this difference may be explained by the large variability in results based on emission databases and emission factors (Nriagu and Pacyna 1988). However, external fluvial inputs from the Paraíba do Sul River basin are artificially transposed to this Sepetiba basin, as described before. Unfortunately, no study has reported on Hg concentrations or Hg concentrations

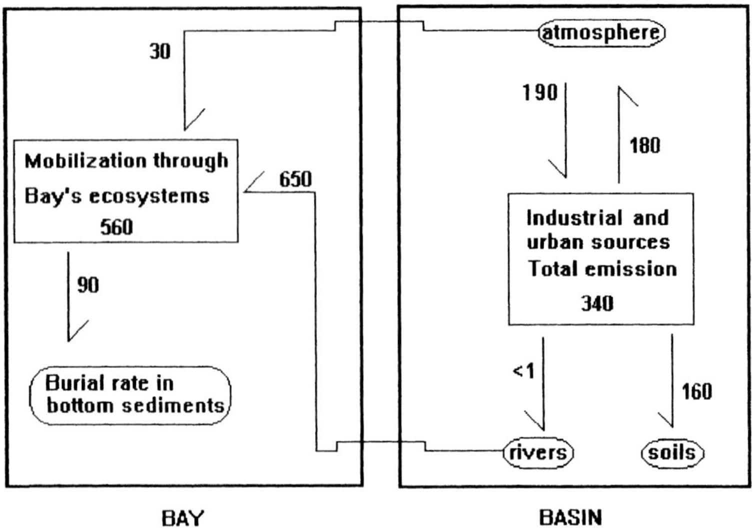

Fig. 4. Mercury mass balance in Sepetiba Bay and basin (kg a^{-1})

load from this route. However, this "external" Hg flux may explain some of the difference between the estimated emission and the calculated fluvial input from the measurements of Hg in the local river waters. This "external" Hg flux may contribute with a significant fraction (ca. 45%) of the total fluvial Hg input to Sepetiba Bay.

The fluvial pathway dominates Hg inputs to Sepetiba Bay, accounting for nearly 95% of the total inputs. Direct atmospheric deposition on the Bay surface contributes with only 5% (30 kg) of the total. Similar balances done for Cd, Zn and Pb also showed the dominance of fluvial inputs over atmospheric deposition (Pedlowiski et al. 1991; Barcellos 1995) in Sepetiba Bay. However, contrary to these metals, Hg contribution from atmospheric deposition over Sepetiba Bay basin is very significant.

Internal burial of Hg in Sepetiba Bay bottom sediments sequester annually 90 kg of Hg. Taking into consideration the total (fluvial + atmospheric) Hg input to Sepetiba Bay, less than 15% of the total Hg load entering Sepetiba Bay is accumulated in bottom sediments. Therefore, approximately 85% (560 kg) of the total Hg input is mobilized through the Bay, being potentially available for entering the Bay's food chains or be eventually exported to the ocean. It is interesting that the only other similar study, to our knowledge, which was carried on in Haifa Bay, Israel (Krom et al. 1994), also concluded that only 10% of the total Hg load to that bay is buried in bottom sediments. Therefore, we can conclude that transport through the water column is the major pathway of Hg in coastal environments, rather than accumulation in coastal bottom sediments.

As shown by the Hg distribution in bottom sediments, a significant fraction of incoming Hg is transported out of the fluvial-influenced area, and an unknown fraction may even be exported to the open ocean, thus keeping a large amount of Hg potentially to food chains. The availability to the biota of this yearly Hg load which is not buried in bottom sediments will depend on methylation processes inside the Bay and on the Hg species dissolved in the Bay waters, which are presently unknown.

Mercury mass balance in Sepetiba Bay and basin is clearly dominated by non-point sources, including exportation from another basin. The identification of the relative importance of such sources however, is only the first step in controlling and monitoring Hg contamination from complexes industrial/urban situations.

Acknowledgements The authors are grateful for the attention and the analytical support by DQI/CETEM/CNPq, and the analysts G. Gonçalves and E. Paiva. Financial support for the present work were from FAPERJ and UFF, and grants from the CNPq and CAPES, Brazil, to the two first authors respectively. Two anonymous reviewers greatly improved earlier versions of this manuscript.

References

ABES (1983) Catálago brasileiro de engenharia sanitária. Associação Brasileira de Engenharia Sanitária e Ambiental, Rio de Janeiro, p 165

Annema JA, Booij H, Ros JPM (1993) Emission and emission factors of heavy metals in The Netherlands. Proc 9th Int Conf Heavy Metals Environment, CEP Consultants,Toronto, pp 267–270

Barcellos C (1995) Biogeodinâmica de Cd e Zn na Baía de Sepetiba. PhD Thesis, Universidade Federal Fluminense, Niteroi, 212 pp

Barcellos C, Lacerda LD (1994) Cadmium and zinc source assessment in the Sepetiba Bay and Basin. Environ Monit Assess 29:183–199

Behrendt H (1993) Separation of point and diffuse loads of pollutants using monitoring data of rivers. Water Sci Technol 28:165–175

Brosset C (1987) The behavior of mercury in the physical environment. Water Air Soil Pollut 34:145–166

Chu P, Porcella D (1995) Mercury stack filter emissions from U.S. electric utility power plants. Water Air Soil Pollut 80:135–144

Cundy AB, Croudance IW (1995) Sedimentary and geological variations in a salt marsh: mud flat environment from mesotidal Hamble estuary, southern England. Mar Chem 51:115–132

EPA (1993) Locating and estimating air emissions from sources of mercury and mercury compounds. United States Environmental Protection Agency, Research Triangle Park, North Carolina, 333 pp

FEEMA (1989) Qualidade de águas do Estado do Rio de Janeiro 1987–1989. Vol IV, Baía de Guanabara e Sepetiba. Fundação Estadual de Engenharia do Meio Ambiente, Rio de Janeiro, 297 pp

Fleister M, Sarofin AF, Fasset DW, Hamman P, Shacklette HT, Nisbet CT, Epstein S (1984) Environmental impact of cadmium: a review. Environ Health Perspect 7:253–323

Forte CMS (1996) Determinação das taxas de sedimentação na porção nordeste da Baía de Sepetiba utilizando datação com o radioisótopo ^{210}Pb em excesso. MSc Diss, Univ Fed Fluminense, Niteroi, p 82

Gobeil C, Cossa D (1993) Mercury inputs to St Laurence estuary. Can J Fish Aquat Sci 50:1794–1797

Hutton MC, Symons C (1986) The quantities of cadmium, lead, mercury and arsenic entering the U.K. environment from human activities. Sci Total Environ 57:129–150

Iverfeldt A (1991) Occurrence and turnover of atmospheric mercury over the Nordic countries. Water Air Soil Pollut 56:261–266

Johnson MG (1987) Trace lement loadings to sediments of fourteen Ontario Lakes and correlations with concentrations in fish. Can J Fish Aquat Sci 44:3–13

Kehrig HA (1995) Estudo da contaminação por mercúrio em corvinas (*Micropogonias furnierii*) em quatro áreas costeiras do Brazil. MSc Thesis, Pontifícia Universidade Católica, Rio de Janeiro, p 132

Krom MD, Kaufman A, Hornung H (1994) Industrial mercury in combination with natural Pb^{210} as time-dependent tracers of sedimentation and mercury removal from Haifa Bay, Israel. Estuarine Coastal Shelf Sci 38:625–642

Lacerda LD, Pfeiffer WC, Fiszman M (1987) Heavy metal distribution, availability and fate in Sepetiba Bay (SE-Brazil) Sci Total Environ 65:163–173

Lacerda LD, Koudstaal R, Blower BT, Pfeiffer WC (1988) The Sepetiba Bay Management Study. IFIAS Research Program in Coastal Management, International Federation of Institutes of Advanced Studies, Rio de Janeiro, 68 pp

Lacerda LD, Salomons W, Pfeiffer WC, Bastos WR (1991) Mercury distribution in sediment profiles of remote high Pantanal lakes, central Brazil. Biogeochemistry 14:71–77

Madsen PP (1981) Peat bog records of atmospheric mercury deposition. Nature 293:127–130

Marins RV, Silva Filho EV, Lacerda LD (1996) Atmospheric mercury deposition over Sepetiba Bay, SE Brazil. J Bras Chem Soc 9:177–181

Marins RV, Lacerda LD, Villas Boas RC (1997a) Mercury emission to Sepetiba Bay and basin, SE Brazil. Ciênc Cult (in press)

Marins RV, Lacerda LD, Gonçalves GO, Paiva EC (1997b) Effect of root metabolism on the post-depositional mobilization of mercury in salt marsh soils. Bull Environ Contam Toxicol 58:733–738

Mason RP, Fitzgerald WF, Morel FMM (1994) The biogeochemistry cycling of elemental mercury: anthropogenic influences. Geochim Cosmochim Acta 58:3191–3198

Mignon C (1993) Riverine and atmospheric inputs of heavy metals to The Ligurian Sea. Sci Total Environ 138:289–299

Miller DR, Buchanan JM (1979) Atmospheric transport of mercury: a technical report. Monitoring and Assessment Research Centre, University of London, London 74 pp

Nriagu JO, Pacyna JM (1988) Quantitative assessment of worldwide contamination of air, water and soils by trace metals. Nature 333:134–139

OECD (1985) Measures to reduce all man-made emissions of mercury to the environment. 1982 information exchange on mercury – summary report. Organization for Economic Cooperation and Development, Paris, 314 pp

Pacyna JM (1995) Global fluxes of metals to the air. Proc 10th Int Conf Heavy Metals in the Environment, CEP Consultants, Hamburg, pp 77–80

Pedlowiski MA, Lacerda LD, Ovalle ARC, Watts PPR, Silva Filho EV (1991) Atmospheric inputs of Fe, Zn and Mn to Sepetiba Bay, Rio de Janeiro. Ciênc Cult 43:380–382

Pfeiffer WC, Fiszman M, Lacerda LD (1988) Heavy metal surveys in Brazilian coastal environments. In: Seeliger U, Lacerda LD, Patchineelam SR (eds) Metals in coastal environments of Latin America. Springer, Berlin, Heidelberg New York, 297 pp

Quevauviller P, Donard OF, Wasserman JC, Martin FM, Schneider J (1992) Occurrence of methylated tin and dimethylmercury in a mangrove core from Sepetiba Bay, Brazil. Appl Organometall Chem 6:221–228

Rasmussen PE (1994) Current methods of estimating atmospheric mercury fluxes in remote areas. Environ Sci Technol 28:2233–2241

Rekolainen S, Verta M, Liehu A (1986) the effect of airborne mercury contents in some Finnish boreal lakes. Publ Water Res Inst Finland 65:11–20

Simola H, Lodenius M (1982) Recent increase in mercury sedimentation in a forest lake attributable to peatland drainage. Bull Environ Contam Toxicol 29:298–305

Straub CP (1989) Practical handbook of environmental control, 2nd edm. CRC Press, Boca Raton. 385 pp

Suguio K, Vieira EM, Barcelos JH, Silva MS (1979) Interpretação ecológica dos foraminíferos de sedimentos modernos da Baía de Sepetiba e adjacências, Rio de Janeiro. Rev Bras Geociênc 9:233–247

SWA (1993) Mercury emissions from municipal solid waste combustors, an assessment of the current situation in the United States and forecast of future emissions. Solid Waste Association of North America, Golden, Colorado, p 227

von Tumpling W, Wilken RD, Einax J (1993) Pathways and contaminations of mercury in the Pantanal area, Brazil. Proc Int Symp Environmental Geochemistry in Tropical Countries. Universidade Federal Fluminense, Niteroi, pp 403–406

Watts PPR (1990) Entradas fluviais de metais pesados para a Baía de Sepetiba e seu comportamento na região estuarina; MSc Thesis, Universidade Federal Fluminense, Niterói, p 148

WHO (1982) Rapid Assessment of Air, Water and Land Pollution. World Health Organization, Geneva, p 181

Monitoring and Assessment of Mercury Pollution in the Vicinity of Electrical Engineering Plants in the CIS

E.P. Yanin and N.N. Moskalenko

1 Introduction

Mercury and mercury compounds are widely used in electrical engineering in the production of different gas-discharge instruments (mercury lamps, mercury valves, indicators), converters, commutators (switches, relay, etc.), photo-resistors, chemical current sources (especially, dry galvanic cells), gas-filled thyratrons and triodes, semiconducting materials, etc. Production of luminescent lamps is of special importance, because in this case mercury is a constituent part of consumer goods. It has been suggested that the main source of the highly toxic mercury pollution in Russia is luminescent lamps (Avrukh 1994).

However, a complete analysis of this problem with respect to the use and production of luminescent lamps is lacking. This has led to a definite exaggeration of the importance of luminescent lamps as the main mercury emission source in the environment, on the one hand, and, on the other, to an underestimation of the lighting industry as a pollution source by other contaminants. The composition of luminescent lamps also includes such chemical elements as cadmium, thallium, lead, zinc, nickel, manganese, aluminium, fluorine, tungsten, etc. Glass manufacturing, being a part of electric lamp plants, is a powerful emission source of lead in the atmosphere. Glass toning technology, where hydrofluoric acid is used, is a source of intense fluorine emission. Furthermore, electrical engineering plants usually produce both luminescent lamps and filament lamps, as well as other lighting sources. Available data show that besides mercury, electrical engineering plants are an emission source of many chemical elements and their compounds into the environment (Burenkov et al. 1993). Intensive pollution of the habitat by different pollutants may be observed in the vicinity of electrical engineering plants. The main aim of the present chapter is to sum up all the data available, including the author's field data, on mercury pollution in the vicinity of electrical engineering plants of the Commonwealth of Independent States. Large lighting plants producing different types of electric lamps, including luminescent lamps, are located in Russia and the CIS.

Environmental Science
Mercury Contaminated Sites (ed. by R. Ebinghaus et al.)

2
Brief Characteristics of Luminescent Lamps

A luminescent lamp is a gas-discharge low-pressure lighting source, where ultra-violet discharge radiation (usually arc discharge) is transformed by luminophors into visible radiation (light). A mercury lamp is the most widespread variety of luminescent lamps (Fedorov 1969, 1992; Rokhlin 1991).

There are two types of mercury lamps – lamps with liquid mercury and amalgam lamps. In the first case, liquid mercury is introduced into a lamp during the process of vacuum lamp processing and evaporates while initiating a discharge. In the second case, mercury amalgam substitutes for liquid mercury. The lamp becomes more acceptable in production and exploitation thanks to the lesser pressure of mercury vapour over the amalgam than over liquid mercury. Nonetheless, the lamp does not become ecologically pure (Fedorov 1992).

Depending on the technology used and lamp type, lamps usually contain 20–300 mg of mercury (Kokinov et al. 1994); in widespread lamp types, 60–120 mg (Fedorov 1992). In general, a lamp contains about 80 mg of mercury (Avrukh 1994). Besides mercury, the lamps' spiral electrodes contain tungsten. The composition of different luminophors includes cadmium, strontium, beryllium, zinc, fluorine, barium, thallium, aluminium, rare earth elements, manganese, tin, lead, silver, etc. (Fedorov 1969).

A large variety of mercury lamps are produced in the world. For example, the list of luminescent lamps produced in former USSR included more than 150 items. Mercury and luminescent lamps are mainly used in industry, agriculture, medicine, scientific research and in the home. They are used for the illumination of streets, flats, public and industrial premises, in medicine and health care, in light-sensitive duplication, in searchlights and explosion-proof luminaires (mine lighting). They are also used for decorative and advertising purposes (colour lamps), at agricultural facilities (special lamps for cattle-breeding farms, for poultry farming, plant selection, etc.), in electrography, etc. Some 50–80% of luminous energy generated by electric power sources in developed industrial countries is provided by luminescent lamps. So far, luminescent lamps are not used on a wide scale in homes. Japan seems to be the only exception. In 1965 up to 80% of apartment houses were illuminated by these lamps, and at present Japan produces three times more luminescent than filament lamps (Fedorov 1992).

3
Global Assessment of Production and Use of Luminescent Lamps: Luminescent Lamps as a Source of Mercury Input in the Environment

The total world production of luminescent lamps reaches 2 billion items a year (Waymaus 1987). If we proceed from the assumption that one lamp contains

100 mg of mercury, then 200 t/a of mercury is used for the production of this quantity. This is the maximum quantity of mercury which may potentially reach the environment (under the condition that all lamps produced in the course of 1 year are disposed of and that the mercury they contain becomes part of the geochemical cycle). It should also be noted that lamp recycling is widespread in some countries. Thus, the actual quantity of mercury involved in the production and use of these lamps that may potentially be disseminated in the environment must be considerably less and not exceed 140–150 t on a global scale.

It should be noted that in the 1980s the annual use of mercury in the former USSR was estimated at 1250 t. The principal part of the mercury was used by ecologically hazardous industries, primarily by chemical plants connected with the production of chlorine, caustic, acetaldehyde, etc. Electrical engineering used a small share of mercury (Table 1). Recently, Avrukh (1994), for example, estimated the share of luminescent lamps in the total use of mercury in Russia at 1%. According to data (Kokinov et al. 1994), the annual mercury use at Russian electric lamp plants is about 20 t.

More than 180 million luminescent lamps are disposed of in the CIS annually. They contain about 18 t of mercury. It is estimated that some 450–500 million luminescent lamps containing about 50 t of mercury are in use in Russia today (Kokinov et al. 1994). According to Avrukh (1994), the annual use of luminescent lamps is about 100 million. Finally they become waste, which corresponds to 10 t of mercury input into the environment (if we proceed from the assumption that all produced lamps are scrapped of during the year). Thus, the mentioned value of 10 t is the probable contribution to environmental pollution in Russia of this toxic element.

It is known that Moscow annually receives nearly 9 million luminescent lamps (Fedorov 1992). Even if we assume that all lamps become waste during the year and are disposed of at refuse tips, then this will amount to 0.9–1 t of mercury (100–120 mg of mercury in one lamp). This value is insignificant for such a metropolitan area as Moscow. This value is comparable, in particular, to the mercury emission of a thermoelectric plant of average capacity using coal or fuel oil. The mercury emission of large thermoelectric plants may be several times higher. For example, the mercury emission of Uglegorskaya thermoelectric plant (Donbass, Ukraine) is over 3 t per year (Dobryansky et al. 1992). According to

Table 1. The distribution of mercury use in the former USSR in 1985 (Yagolnitzer et al. 1993)

Branch of industry	Share (%)	Tons
Electrical engineering	3.1	38.75
Non-ferrous metallurgy	2.0	25.0
Chemical processes	61.0	762.5
Instrument-making	13.8	172.5
Health care	2.2	27.5
Miscellaneous	17.9	223.75
Total	100.0	1250.0

Note: according to recent assessments, primary mercury production in the former USSR totalled 850 t in 1989, 800 t in 1990, 750 t in 1991, 400 t in 1992 (in Russia) (BIKI, 13/1/94).

estimates of one of the present authors, the annual mercury input into the environment from waste dry galvanic elements amounts to 100 tons. Nevertheless, we should not overlook the importance of the issues connected with mercury lamps. No doubt mercury lamps must always be utilized, or at least disposed of at specially prepared waste dumps.

4
Environmental Impact Assessment of Luminescent Lamp Plants

Let us consider the environmental impact of luminescent lamp production, taking several lighting plants of the CIS as an example.

4.1
Saransk, Mordovia, Russia

The largest lighting enterprise in Russia, the joint-stock company Lisma, is located in Saransk, the capital of the Mordovia Republic. It includes the SELZ electric lamp plant producing luminescent lamps and filament lamps, and also SIS-EVS, a plant of special light sources and electrovacuum glass. It should be noted that Saransk (with a population of more than 360 000) is an important industrial centre, the site of many large industrial plants (Dobryansky et al. 1992; Yanin 1996b).

Luminescent lamp production at the Saransk plant was initiated at the beginning of the 1960s. In 1965 annual production totalled 30–35 million items, annual mercury consumption totalled about 5.5 t. To the present, Lisma has produced a billion luminescent lamps, thus potentially providing the environment with 100–120 t of mercury.

Emission of gaseous mercury into the atmosphere totalled 200–400 kg/a during the first 15 years of production. In 1976 coal absorbers were installed that have decreased the emission of mercury vapour to 62 kg/a (average value during the 1980s). Thus, according to official data, total mercury emission during the period of the plant's operation may be estimated at approximately 6 t. In reality, mercury emissions were much higher because of the input of metallic mercury, mercury emission with solid particles and sewage waters. Besides, a large amount of the industrially produced items at Lisma were defective. Nothing is known about the fate of used coal absorbers and defective lamps; probably they were disposed of at city dump sites. A special unit located on the territory of the plant has recently started waste lamp utilization. Mercury emission during the plant's operation totals 10–12 t. This mercury emission at Lisma is minor when compared with data on mercury emission at one of the chemical plants (Temirtau City, Kazakhstan), where emission totals 1200 t during a 30-year period (Yanin 1995; Yanin 1996b).

Nevertheless, the Lisma electrical engineering plants are the main source of mercury and heavy metal pollution in Saransk. Furthermore, the intensity of the city's contamination by chemical elements contained in emissions, sewage and waste of electrical engineering plants is no less than the intensity of mercury contamination. It is clear that the composition of technogenic anomalies in different environmental media depends on the synergistic impact of a whole range of plants. The specific impact of electrical engineering plants may be accurately fixed.

Table 2 represents data on the concentration of chemical elements in the soil at electrical engineering plants. The data adequately show the qualitative composition of technogenic geochemical anomalies. The operation of the glass plant located here has led to high Kc concentrations (Kc is a concentration coefficient of an element over a background concentration level) of lead within the SIS-EVS industrial zone. It is important that the registered concentrations were at extremely high maximum levels for almost all elements. The concentrations of mercury, lead, vanadium and antimony in the soil were much higher than the maximum permissible level allowed in Russia which constitute 2.1 ppm for mercury, 30 ppm for lead, 150 ppm for vanadium and 10 ppm for antimony.

The character and intensity of technogenic impact is visually demonstrated on analyzing the pollutant's distribution in the soil profile. Maximum mercury concentrations in the SELZ industrial zone are found in the upper soil layer, 25–30 cm. Anomalous concentrations can be traced down to a depth of 60–80 cm, with a similar distribution for Zn, Mo, Pb, Cu and Ag. The soil pollution

Table 2. Concentration levels (ppm) of chemical elements in the soil on the territory of the SELZ and SIS-EVS plants

Chemical element	SELZ		Kc	SIS-EVS		Kc
	Mean (ppm)	Range (ppm)		Mean (ppm)	Range (ppm)	
Hg	28.3	0.08–300	472	0.84	0.015–5	14
Sb	60	30–300	60	–	–	–
Ag	1.8	0.05–50	36	0.08	0.05–0.3	1.6
Tl	4	3–5	20	–	–	–
Ba	2530	100–30 000	18	380	100–1500	3
Cd	6	3–50	17	–	–	–
Pb	213	10–1000	14	920	10–10 000	61
Zn	950	80–6000	10	280	50–1000	3.1
Sr	225	30–1000	8	75	30–300	2.8
Mo	8.7	0.5–100	7.3	6.6	1–100	5.5
W	9	5–30	6	11	5–70	7.3
Sn	11.7	1–60	5.6	5.4	3–30	2.6
Cu	198	30–3000	4.1	186	30–1000	3.9
Cr	199	30–2000	2.5	136	30–2000	1.7
Ni	63	20–300	1.8	49	20–150	1.4
V	202	70–600	1.7	345	150–600	2.9

Note: Kc is a concentration coefficient with regard to the background concentration level in the soil, e.g. Kc = Hg concentration in the studied soils/the concentration of Hg in non-contaminated soils (the table lists the mean values of this coefficient).

Table 3. Chemical elements in dust emissions of the SELZ plant

Chemical element	ppm	Kc
Sr	3000.0	86.0
Ba	30 000.0	136.0
Mn	10 000.0	7.7
Cr	500.0	9.0
Ni	200.0	7.7
Co	40.0	2.0
Cu	400.0	7.5
Ag	3.0	60.0
Zn	1800.0	7.7
Cd	More than 1000.0	More than 2800.0
Pb	4000.0	266.0
Sb	More than 10 000.0	More than 10 000.0
Sn	300.0	1429.0
Mo	80.0	100.0
W	300.0	200.0
P	More than 30 000	More than 38.0
As	280.0	43.0
Hg	1000.0	16 666.6

Note: Kc = concentration in dust/soil background concentration.

zone at the SIS-EVS plant is usually limited to a depth of 10–15 cm. At the same time, the presence of a powerful lead emission source has led to deep Pb penetration in the soil profile – up to 90–100 cm. The data on the chemical composition of industrial dust from existing cleaning devices are very impressive (see Table 3).

Data analysis of the structure of mercury pollution in Saransk and the particularities of surface mercury distribution show that the concentration level of this element on most of the urban territory is not critical (Table 4). Background mercury levels in the soil have been registered on 80% of the territory; 10% of the territory is characterized by low-contrast anomalies (Kc = 1.5–3). The most intensive and vast anomalous mercury-polluted zones are

Table 4. The structure of soil pollution by chemical elements in Saransk (% of entire city area)

Chemical element	Kc values					
	< 1.5	1.4–3	3–10	10–30	30–100	>100
Cu	92.8	6.6	0.5	0.1	–	–
Zn	52.5	43.3	3.9	0.2	0.1	–
Ni	81.9	17.1	1.0	–	–	–
Sr	83.0	13.8	2.8	0.4	–	–
Mo	88.6	9.8	1.5	0.3	–	–
Sn	56.2	42.4	1.4	–	–	–
Pb	23.6	29.9	41.9	4.2	0.4	–
Hg	82.0	9.2	6.9	1.5	0.3	0.1

Note: Kc = real concentration in the studied soil/background soil concentration.

found in close proximity to the sites of the SELZ and SIS-EVS plants, as well as to the electric power plant zone (fuel oil and natural gas burning) and the highway (transport of industrial waste to the dump site). Mercury concentrations exceeding the maximum permissible level for soil were registered in the SELZ industrial zone, and to a lesser extent in the SIS-EVS zone, as well as in the residential area close to the SELZ plant.

These particularities of mercury distribution in the soil indicate the presence of local mercury pollution zones in the city. It is significant that substantial mercury concentrations are registered only in the upper soil level (5–10 cm). The most intensive pollution in the city is connected with Pb input from motor transport and glass manufacturing at the SIS-EVS. This resulted in the development of contrasting and intensive Pb soil anomalies. At the same time, quite extensive areas of soil pollution are registered, those contaminated by Zn, Sn, and to a lesser extent by Ni, Sr and Mo. In the SELZ and SIS-EVS industrial zones contrast anomalies of Tl, Cd, Sb and W were revealed. There are local low-contrast technogenic anomalies of Cu, Cr, V, Ba, Bi, Mn, etc. The composition of technogenic anomalies that have developed in the city is correlated with the typical anomalous composition found on the territories of electrical engineering plants, so that we can confirm with certainty that the quality of an urban environment is mainly determined by the activity of electrical engineering plants.

The results of an ambient air-mercury study conducted in Saransk are correlated with the soil pollution data, and corroborate the fact that mercury pollution of the city environment is of local character (Table 5). Concentration levels of mercury vapour were at a background level over practically the entire city territory. Maximum permissible concentration levels were registered in the vicinity of the SELZ plant and the electric power plant, where concentration levels exceeded the background level. Mercury concentration levels were 1.5–3 times higher in the summer than in the autumn in all regions of the city. This can be explained by more active mercury evaporation from the soil during summer than in autumn.

Distribution results of other chemical elements in surface air showed that Pb, Sn, Ni, Cr, V, Cd, Mo, Zn and some other elements are the most hazardous pollutants in the city, i.e. the typical emission elements of electrical engineering (Table 6). Kc of Pb, V, Cd, Mo and Zn constantly exceeded maximum permissible

Table 5. Mercury vapour concentration levels in surface air (ng/m^3)

City district	Summer	Autumn
SIS-EVS	<50–70	<50
SELZ	250–350	70–80
Electric power plant	150–250	100–200
Central	<50–80	<50–60
Residential areas	<50–70	<50
Background	<50	<50
Maximum permissible concentration	300	

Table 6. Chemical elements associations (aerosol form) in surface air

City district	Kc values				
	>100	30–100	10–30	3–10	<3
SIS-EVS	Pb-V-Cd	Cr-Ni-W	Mo	Mn-Zn-Sn	–
SELZ	Sn-Pb-Ni	Zn-Cd	V-Cr	–	Mn

Note: Kc = real concentration/background concentration.

levels. Thus, the impact of electrical engineering plants on the chemical composition of the air lies to a greater extent seen in contamination by the above-mentioned metals, not by mercury. The data shown in Table 6 reflect the environmental impact of other industrial enterprises of the city as well. Nevertheless, the specific impact of electrical engineering plants is clearly visible.

It is harder to evaluate the impact of electrical engineering plants on water systems. The problem is that many different industrial enterprises located in the city discharge sewage waters into the main city river, the Insar, after partial treatment at the city purification plant. Comparative analysis of sludge from the city purification plant, the SELZ and the SIS-EVS plants, on the one hand, and polluted bottom sediments (technogenic sludge) from the Insar River sampled below city limits, on the other, is quite indicative (Table 7). All types of sludge are characterized by lower concentrations of mercury than of many other chemical elements (except for river sludge). Nevertheless, the qualitative composition of all sludge types is relatively close. This indicates the leading role played by electrical engineering in the pollution of water systems.

Mercury concentration levels in the water of the Insar River system did not exceed maximum permissible concentrations for mercury water-soluble forms (Table 8). At the same time, a high level of dissolved mercury was found in a spring polluted by sewage water from the SELZ and some other industrial plants. The major mercury pollution of streams results from the fact that mercury is in a state of solid suspended material, i.e. suspension. Suspended forms of mercury exceed background levels by an order of magnitude.

A water composition study of a small Insar River branch polluted by storm runoff from the SIS-EVS plant showed the presence of extremely high F concentrations (up to 20 mg/l) and high concentrations of Li, Zn, Cu, Ni, Sr and As. This is connected with the use of hydrofluoric acid.

4.2
Poltava, the Ukraine

The country's largest plant producing gas-discharge (mercury) lamps is located in Poltava. During the past several years annual mercury consumption totalled 6 t. The major part of mercury is used for luminescent lamp production (Kozhushko and Koryagin 1994). Presumably, the total mercury emission and annual mercury wastes are comparable to those of the Saransk plants.

Special studies conducted in the vicinity of the plant showed relatively low levels of mercury pollution in the soil and air (Dobryansky et al. 1992).

Table 7. Chemical elements associations in sludges of the city purification plant, Joint-Stock Company Lisma and techr.ogenic sludges

Object	Leading elements and Kc values						
	>300	300–100	100–30	30–10	10–3	<3	Zc
Sludge 1	Bi-Cd	Sn-Ag	Zn-Mo-Hg-W-Cu-Sb	P-Cr-Pb-Ni	As-F-Yb	B-Sr-Be-U	2555
Sludge 2	Zn	Cu-Ni-F-Mo	Sn-Pb	Cr-Cd-W-P	Ag-As-Hg	B-Sn-Sr	1950
Technogenic sludge	Cd	Hg-Mo	Zn-Sn	Cu-W-Ag	Ni-Pb-Cr-Sr-Bi-B	F-V-Fe-Tl-Ga-P-Be-Co	810

Note: Sludge 1 – from the city purification plant, Sludge 2 – from the SELZ, SIS–EVS plants; Kc = real concentration/background concentration in river bottom sediments; Zc – combined pollution coefficient equal to Kc sum.

Table 8. Mercury in surface water, μ/l

Stream	Dissolved Hg	Suspended Hg
The Insar River, downstream from the city	0.05	0.25
The Insar River, mouth (50 km downstream)	0.05	0.14
Spring, sewage waters input from the plants	0.40	0.10
Background	0.07	0.02
Maximum permissible concentration	0.5	None

Nevertheless, almost everywhere concentrations of mercury in the soil exceeded background levels. Mercury concentrations changed from 0.09 to 1.5 ppm, mean concentrations were 0.41 ppm; background concentrations, 0.05 ppm.

Concentration levels of mercury vapour in surface air were 70–125 ng/m^3 near the plant, reducing to background concentrations (10–30 ng/m^3) in the vicinity of the plant, still very high by global standards. Characteristically maximum mercury concentrations in the soil (1.5 ppm) and air (125 ng/m^3) were registered near the main entrance to the plant. This may be the result of active mercury transport by cars, and on clothes and shoes of workers, as well as atmospheric deposition to soil.

4.3
Erevan, Armenia

The EELZ electric lamp plant and the OSTZ experimental lighting plant, incorporated in the Luyis company, are located in the capital of Armenia, Erevan. The input of mercury and other chemical elements into the environment is connected with the operation of these plants. An analysis of the ecological situation is given on the basis of data received by the Moscow Expedition of the Institute of Mineralogy, Geochemistry and Crystal Chemistry (IMGRE) and IMGRE in the mid-1980s under the supervision of Yu.E. Saet (with the participation of L.S. Sokolov, V.A. Soloviev, L.N. Myachina, B.A. Revich, E.S. Kiseleva et al.). The annual mercury consumption of these plants amounted to approximately 4 t in the mid-1980s.

A special research study showed that the most intensive soil mercury pollution in the city is located in the impact zone of the electric lamp plant. Average mercury levels exceed mercury Clarke – average content of an element in the Earth's crust – by more than 16 times. Mercury concentrations reached 0.6–0.9 ppm (Kc = 13–18) on the territory of the lighting plant. Local mercury anomalies were found in residential areas in the vicinity of the plants. Local mercury soil anomalies were also registered along the Tbilisskoye highway. This may be connected with the transportation of wastes and industrial output.

The results of an ambient air-mercury survey of the city showed that the main air pollution source in Erevan is the EELZ plant, especially the department of luminescent lamp production where mercury vapour levels reached 0.06 mg/m^3 (six times higher than the maximum permissible concentration for working areas). Mercury concentrations decreased to 100–3000 ng/m^3 on the territory of

Table 9. Chemical elements associations in the soil on the territories of the EELZ and OSTZ plants

Plant	Kc relative to Clarke			
	>50	30–50	10–30	5–10
EELZ	Pb	–	W	Cu-Ag-Zn-Mo-Sn
OSTZ	–	Pb-Zn-W	Cu-Cr-Ni	Sn-Ag-Mo

the plant, but maximum concentrations were registered near the department of luminescent lamp production. The plant impact zone covered an area of up to 3–4 km to the northeast and southwest of the plant, where mercury vapour levels were 16 times higher than background levels (3–4 ng/m^3). Mercury levels exceeded background levels by 4–16 times at a distance of 5 km, by 2–4 times at a distance of 7–8 km.

The impact of the EELZ and the OSTZ can be seen from the formation of contrasting technogenic anomalies of other chemical elements in the city's soil. The most intensive anomalies were registered on the territories of the plants (Table 9). Pb concentrations in the soil reached 10 000 ppm (625 times higher than Clarke) near the glass manufacturing workshops of the EELZ. On the OSTZ territory Pb concentrations in the soil were 1500–2000 ppm.

Chemical elements mentioned in Table 9 are intensively accumulating in the upper soil level in the vicinity of the plants. Pb levels in pollution zones forming under the influence of the electric lamp plant were approximately 1000 ppm, significantly exceeding the maximum permissible concentration. Undoubtedly, Pb concentrations exceeding maximum permissible levels can be found in the air of these parts of the city. The combined impact of industrial plants and motor vehicles has led to an increase in Pb concentrations in the soil of Erevan to a level of more than 100 ppm on an area of 32 km^2. Vast city territories have Pb concentrations in the soil of over 500 ppm. Such a Pb concentration level is observed over an area of 13 km^2 in Erevan. According to the results of a special study conducted by the Institute of Applied Geophysics, the Pb level exceeded the maximum permissible concentration in 20–60% of the samples.

5
Hygienic Consequences of Environmental Pollution by Lighting Plants

The study of mercury distribution in the hair of Saransk children (about 650 children), conducted by Yanin (1992, 1996a), showed that Hg levels were less than 0.1 µ/g for more than 50% of children (background = 0.26 µ/g, critical level = <2 µ/g). Of the checked children 13% had Hg concentration levels 1.5 times higher than the background level; and 7%, 3 times higher. Fifty percent of the children had increased Pb concentrations (1.5 and more times higher than the background level). With 20% of the children, the maximum permissible level of Pb concentration was exceeded, with 5% the level was critical. The Cd level was by two to five times above the background level with 32% of the children. Seven percent of the children had a critical level of Cd. These data show that environmental pollution by Pb and Cd play the major role in the city. However, the fact of Hg accumulation in the biosubstrates of people is remarkable in itself.

In particular, a specific distribution of childrens' diseases was noted while analyzing distribution peculiarities of different childrens' diseases in Saransk (Table 10). As is shown in Table 10, city areas with increased Hg concentrations in the soil and air are characterized by the high frequency of such specific diseases as pyelonephritis and cystitis. These parts of the city are characterized by intensive metallic pollution. It would not be correct to connect the incidence of these diseases only with the adverse Hg environmental impact, but, at the same time, we cannot underestimate this possibility.

It should be noted, that children of electric lamp plant workers had higher mercury levels in their hair at an age of 5–7 years as compared to children whose parents had worked at other plants. This may be connected with the transport of Hg to homes on clothes, shoes, etc. This specific source of living accommodation pollution is of great hazard to children (Yanin 1996a). The probability of Hg input into the human body is increased due to non-premeditated and especially Hg spills in living areas and on public premises, which often happens in cities where Hg production is located.

The situation is more acute with workers of electric lamp plants who have direct contact with Hg. The study of Hg concentrations in the hair of SELZ workers exceeded the background level by 1.5 times with 50% of the workers and by 3 times excess with 30%. Sometimes the Hg level in the hair was close to the critical Hg level (<2 μ/g).

Some SELZ workers, especially women, had health problems typical of long-term Hg influence. Special checkups of the state of health during pregnancy of two groups of women, those working at the plant (experimental group) and women who did not (control group) showed that complications during pregnancy and birth were different for the two groups. However, the frequency of complications was higher in the experimental group. On average SELZ workers gave birth 2.6 h longer than the women of the control group. Late toxicosis occurred more often in the experimental group. Premature birth happened twice as often in the experimental group. They had evident anaemia two to four times more often. They also had changes in the peripheral state of blood (increased level of leukocytes and reduced erythrocyte sedimentation rate). The children of electric lamp plant workers lagged behind in mental and physical

Table 10. Distribution of pyelonephritis and cystitis in children in Saransk. (Yanin 1992)

City district	Incidence for 1000 children
Luminescent lamp plant impact zone	18
Central part	21
Other regions	0.9–4.8

development. Child hypotrophy and hypertrophy and the birth of twins have been registered more often in the experimental group of women.

6 Luminescent Lamp Utilization

The wide use of luminescent lamps at home and in industry does not exclude the probability of Hg input into the surroundings. This stipulates the necessity of utilizing these lamps and establishing special companies to register, collect and process mercury-containing lamps and other electrical engineering goods.

Companies involved in luminescent lamp utilization were set up in several industrial regions of Russia during the past several years. One of the first Russian regional demercurization stations operates in Novgorod with the support of the Mercury municipal ecological company. This station serves some cities of the country's northwest. The technology employed at this station is in no way inferior and even surpasses some similar international technologies. Such companies exist in Moscow, Mytishi, Dubna, St. Petersburg, Cheboksary, Kirov, Naberezhny Chelny, Nizhni Novgorod, etc., and in some other cities of the CIS (the Ukraine, Belorus, the Baltic states).

However, the absence of used mercury lamp stock-assessment and collection and the absence of a governmental policy and laws significantly impede the activities of these companies and do not favour the development of new demercurization centres. Public education activities are organized on a weekly basis for the public and for different organizations. There are no collection stations for used lamps, and in most cases, a mercury-containing lamp is thrown away as trash. This happens inadvertently, because of a lack of knowledge, or because there is no other way.

The total amount of annually utilized luminescent lamps in Russia and the CIS is not known. They constitute a small share of the total annual production and consumption. Thus, 600 000 luminescent lamps containing 60 kg of Hg were utilized at the largest mercury lamp-processing centre in Cheboksary (the Volga Region) in 1989 (Avrukh 1994). Annual utilization of luminescent lamps totals 300 000 in Novgorod (V.V. Bessonov, pers. Comm.). It may be assumed that the total amount of processed lamps does not exceed 10% of their annual consumption. The rest find their way to dump sites that are not designed for the utilization of such wastes. This leads to environmental pollution by this toxic metal.

7 Conclusion

The analysis of the data received shows that luminescent lamps are not the main source of mercury pollution of the environment on a global scale. According to our estimations, the global total multisource Hg emission into the environment totals 140–150 t. According to preliminary assessments, the annual Hg input into the environment totals approximately 10 t (Russia) and about 18 t (the CIS countries).

Modern luminescent lamp plants located in different cities of the CIS are the emission source of many chemical elements into the environment (Pb, Cd, Hg, Tl, Mo, W, rare earth elements, Zn, Ni, F, Sb, As, Li, Cu, etc.). The intensity and scale of mercury environmental pollution is often less than that of heavy metals that are used in production or are part of the raw materials. This fact should be taken into consideration while working out ecological programmes and conducting research work in the regions close to such plants. It should be remembered that any enterprise using Hg in its technological cycle is a source of increased hygienic and ecological hazard. The data on Hg accumulation in the human body and occupational diseases of workers confirm this fact. The quality of a city environment may be determined to a significant extent by the operation of large electrical engineering plants.

In general, when discussing ecological safety issues pertaining to the production and use of luminescent lamps one should distinguish the following: (1) ecological aspects connected with environmental pollution caused by the production and use of lamps; (2) problems of occupational hazards and industrial sanitation; (3) social and domestic aspects connected with the probability of inadvertent and premeditated input of Hg and its compounds and of other pollutants into private homes and public buildings, etc.; (4) technological problems of technology improvement, purification improvement, waste utilization and the development of utilization methods for used electrical engineering goods; (5) organizational problems of used goods collection. Public education activities aimed at explaining mercury and heavy metal pollution hazards when using luminescent lamps and other electrical engineering goods are of great importance.

Acknowledgements The present chapter was prepared with partial support of the Program on Peace and International Cooperation of the John D. and Catherine T. MacArthur Foundation.

References

Avrukh VS (1994) Waste mercury-containing lamps are an environmental hazard. In: Mercury hazards – the problem of the 20th century. The Foundation for Mercury Hazards, St. Petersburg, pp 38–44 (in Russian)

Burenkov EK, Yanin EP, Kizhapkin SA, et al. (1993) Ecological-geochemical assessment of the Saransk environment. IMGRE, Moscow (in Russian)

Dobryansky LA, Artemenko VM, Morozov VI (1992) Mercury in the environment of some Ukrainian towns. IGN ANU, Kiev (in Russian)

Fedorov VV (1969) Production of luminescent lamps. Energiya, Moscow (in Russian)

Fedorov VV (1992) Luminescent lamps. Energoatomizdat, Moscow (in Russian)

Kokinov AM, Fedorenko AS, Prytkov AA (1994) Methods of mercury hazard reduction in production and use of light sources. In: Mercury hazards – the problem of the 20th century. The Foundation for Mercury Hazards, St Petersburg, pp 79–82 (in Russian)

Kozhushko GM, Koryagin OG (1994) Solution methods of mercury problems in light sources of Ukraine. In: Mercury hazards – the problem of 20th century. The Foundation for Mercury Hazards, St Petersburg, pp 83–86 (in Russian)

Roizman VY (1994) Problems and solutions of mercury lamp demercurisation. In: Mercury hazards – the problem of 20th century. The Foundation for Mercury Hazards, St Petersburg, pp 44–45 (in Russian)

Rokhlin GN (1991) Gas-discharge light sources. Energoatomizdat, Moscow (in Russian)

Waymaus JF (1987) The state and perspectives of discharge light sources. Svetotekhnika 5:20–24 (in Russian)

Yagolnitzer MA, Obolensky AA, Babich VV, et al. (1993) The choice of an effective development strategy of a mining and smelting plant in new management conditions (modelling, imitation, decision-making, environmental protection). Novosibirsk (in Russian)

Yanin EP (1992) Mercury in the industrial city environment. IMGRE, Moscow (in Russian)

Yanin EP (1995) Mercury around a large chemical factory (central Kazakhstan). In: Regional and global mercury cycles: sources, fluxes and mass balances. NATO Advanced Research Worksh, 10–14 July 1995, Novosibirsk, Russia. Abstr Novosibirsk, p 39

Yanin EP (1996a) Specific mercury input source into living premises. In: Mercury. Complex safety system. The Foundation for Mercury Hazard, St Petersburg, pp 45–48 (in Russian)

Yanin EP (1996b) Ecological-geochemical aspects of luminescent lamp production and use. In: Mercury. Complex safety system. St Petersburg, pp 61–73 (in Russian)

Mercury in Gas and Oil Deposits

N.A. Ozerova, N.P. Mashyanov, Yu. I. Pikovsky, V.V. Ryzhov, A.E. Chernova, A.A. Ganeev, S.E. Sholupov, and L.A. Dobryansky

The main regularities of location of mercury-bearing gas and oil deposits are discussed. The mantle source of mercury in these deposits is debated. Correlative relations of mercury with other elements are described. Some evaluation examples of reserves of mercury in gas and oil deposits with high mercury content are given. Examples of regional and local prognosis of mercury-bearing deposits are shown.

1 Introduction

The study of mercury occurrences in oil and gas deposits was started in the 1960s. The paragenesis of mercury with hydrocarbons was not suggested earlier. Pioneer works in this direction were published by Bailey et al. (1961); Acterberg and Zaanen (1972); Morrison (1972); Tunn (1973, 1976); Dickenstein et al. (1973); Glogoczowsky et al. (1977). Our investigations in the areas of the former Soviet Union (fSU) are also related to this period: Saukov et al. (1972) and Ozerova et al. (1974). More intensive studies began to be produced in the 1980s and 1990s. Our study took place in different parts of the fSU, such as Central Asia, Kazakhstan, North Caucasus, Caspian Region, Ukraine, Kamchatka and Siberia. Work in Croatia was also carried out. We do not propose to make a detailed examination of mercury behaviour in gas and oil deposits in the current chapter. Our main task is to report the main regularities of mercury distribution in these deposits. The same applies to methods of mercury analysis: we give an account of only the principles of mercury determination and mention the appropriate papers for a fuller explanation of the methods.

2 Methods of Determination of Mercury

We used chemical methods of determination of mercury during the first period of investigation.

Gas was pumped through the barbater with a known mercury absorber, iodine solution in iodine potash, at 40–60 l/h. The volume of pumped air is about $n \cdot 100$

Environmental Science
Mercury Contaminated Sites (ed. by R. Ebinghaus et al.)

m^{-3} and more. Mercury was determined by turbocalorimeter with the help of the Polezhaev reactive.

This method of mercury determination in oil uses the combustion of oil in the air force with an oxidizer, granulated copper oxide, and with simultaneous capture of mercury in the refrigerator. Metallic mercury was determined by the turbocalorimeter in a kind of $CuHgI_3$ salt difficult to dissolve (for details see Ozerova et al. 1974; Saukov et al. 1972).

The chemical methods are fairly difficult.

The atomic-absorptional method using the Zeeman effect (atomic absorption spectrometer RA-915) was prepared and used for this reason. A spectrometer RA-915 has a background corrector itself: reduction of the initial resonance radiation up to 5% does not cause false response. Hydrocarbon gases were analyzed by this method later. However, the magnitude of background absorption could exceed the acceptable level during the analysis of samples with the complex matrix (oil and oil products). In order to reduce the magnitude of background absorption, a two-section atomizer was used. The sample is evaporated and mercury compounds are partially decomposed in the first step; then all mercury compounds are fully decomposed and the hydrocarbon matrix is combusted in the next step (Ganeev et al. 1995a).

Only atomic mercury is determined during the analysis of hydrocarbon gases. Experiments with pyrolisis (temperature level 750–800 °C in the reactor) and photolysis (temperature level approximately 50–60 °C in the reactor) were made to evaluate possible organic mercury compounds (HgR). The mercury input during the decomposition of its compounds could be more than 30%, as seen from Table 1 (Ganeev et al. 1995b).

Table 1. Hg concentrations in gases using photolysis and pyrolysis, 10^{-6} g/m^3

Gas deposits	Hg_{atom}	Hg from HgR	
		Photolysis	Pyrolysis
Oposhnya (Dnieper-Donetsk basin)	52.5 ± 0.3	2.1 ± 0.8	2.3 ± 0.5
Mirnenskoye	19.1 ± 0.3	5.8 ± 1.5	
(Northern Caucasus)	0.5 ± 0.1	< 0.1	

3 Main Regularities in the Distribution of Mercury-Bearing Gas and Oil Deposits

Mercury content in oils and gases varies broadly (Table 2). We have fewer analyses of condensates of hydrocarbon gases compared to those of oils and

Table 2. Hg concentrations in oil and gas deposits

Oil	$< 1 \times 10^{-6}$–2.1×10^{-3}%
Gas	1×10^{-8} g/m^3–14 mg/m^3

gases. Mercury content in condensates is $n \times 10^{-4}$ -1.5×10^{-3} g/l; but it is mercury in condensates that stipulates the reserves of mercury in the gas-condensate deposits. Metal mercury was revealed in transport tubes in some gas and gas-condensate deposits (Mirnenskoye deposit in the North Caucasus, Kultuk in the Middle Asia of the fSU).

Mercury in the hydrocarbon gases occurs in Precambrian, Caledonian and Hercynian tectonic structures and young folded belts of gas-oil and gas deposits. Mercury-bearing gases are localized in practically all the stratigraphic deposits of the Paleozoic, Mezozoic and Cenozoic where gas deposits exist. The lithological composition of the rock-collectors differs greatly (Ozerova and Pikovsky 1982; Ozerova 1986), so there is no stratigraphic or even lithologic control for gases with higher mercury component in the oil-gas area.

We analyzed the structural-geologic pecularities of all the known natural gas deposits with a mercury concentration of more than 1×10^{-6}g/m^3(Ozerova Pikovsky 1982; Ozerova 1986). Analysis shows that the localization of these deposits in the crossing zones of longitudinal deep faults, which are parallel to the young folded systems, by large transverse planetary lineaments is their general feature. Both lineament systems are, as a rule, fault zones thousands of kilometres long looking like graben-like depressions or uplifts in the sedimentary cover. These fault zones are surrounded by subparallel marginal faults, which are characterized by abruptly deep gradient of the crystallic basement and Moho discontinuity and modern tectonic activity. The latter is impressed in modern movements with two times and more higher heat flow, compared to nearby regions and hydrothermal activity including mercury etc.

This could be illustrated by the example of the Karpinsky lineament, well studied by us. It is distinctly seen in Fig. 1 that the mercury-bearing deposits of the European continent occur at the crossing conjunction of the Karpinsky lineament and two of the largest Rhine-Lybian and Arabian-African mantle lineaments (rift belts).

All these data infer the conclusion that the mercury content in natural gases is stipulated by rising of mercury-containing fluid fluxes from the Earth's mantle along the deep fault zones during their activization.

This clearly illustrated by the seismological sections through the Karpinsky lineament (Ozerova 1986). The gas deposits of the Altmark province in Germany with high mercury concentrations (average 3, up to 14 mg/m^3) are obviously related to the zone of deep mantle faults (Fig. 2). Localization of the gas deposits Oposhnya and Gadach of the Dnieper-Donetsk graben in the seismological sections are similar (Fig. 3). The highest mercury concentration, up to 11 mg/m^3 in the gas deposits in the fSU studied by us was revealed in the Oposhnya deposit.

The highest mercury potential, incidentally, was firstly prognosed for this deposit by us based on the localization of this deposit in the crossing zone of the Karpinsky lineament structures and the similar large zone of transverse deep faults (Fig. 4). Secondly, based on the activization of faults in the Alpine epoch (illustrated by submeridional river orientation along these faults, direct geophysical observations of the vertical and horizontal movements, strong emissions of gases in the mines of Krivoy Rog, Ukraine etc.). All these factors are favorable for the mercury rising from the Earth's depths (Ozerova 1986).

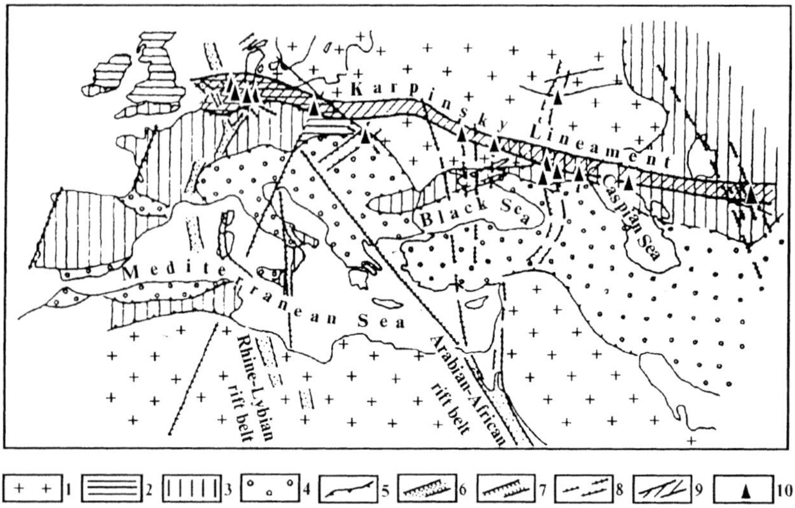

Fig. 1. Location of mercury-bearing gas and oil-gas deposists within the Karpinsky lineament 1–4 old areas of various age, *1* Precambrain; *2* Caledonian; *3* Hercynian; *4* Alpine; *5* boundaries between fold areas of various age; *6* continental rift belts (Rhine–Lybian and Arabian–African); *7* large zones of transverse dislocations; *8* transverse deep faults; *9* Karpinsky lineament; *10* gas deposits and occurrences with mercury content in gas over 1×10^{-6} g/m^3

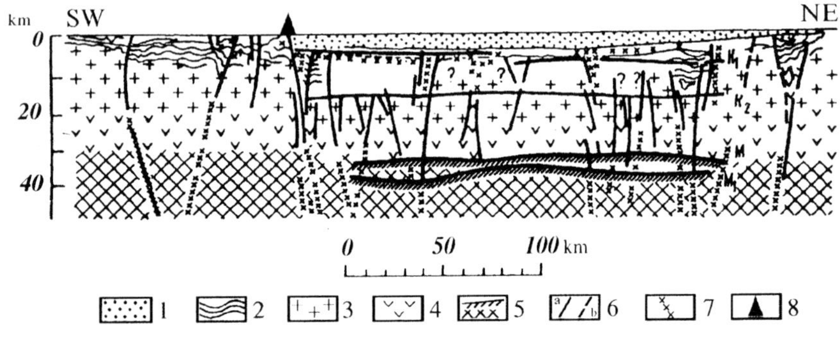

Fig. 2. Seismological section of the area of Zalzvedel-Pekenzen deposit (geotraverse TS-1) *1* sedimentary cover; *2* paleozoic folded basement; *3* "granitic" layer; *4* "basaltic" layer; *5* upper mantle; *6* main faults (a known, b proposed); *7* deep faults; *8* gas deposit Zalzvedel-Pekenzen, Altmark, Germany

Gas deposits of the Stavropol uplift area of the Karpinsky lineament (Mirnenskoye, Takhta-Kugulta deposits, etc.) could be used as another example. Here, the mercury content in gases is clearly correlated with an active tectonic life of the area in modern time – positive and negative tectonic movements, higher seismicity and large positive anomaly of the heat flow. Finally recent investigations for measuring mercury content in the near-surface atmosphere from a moving platform (in a car) by the authors (Ozerova and Mashyanov 1989)

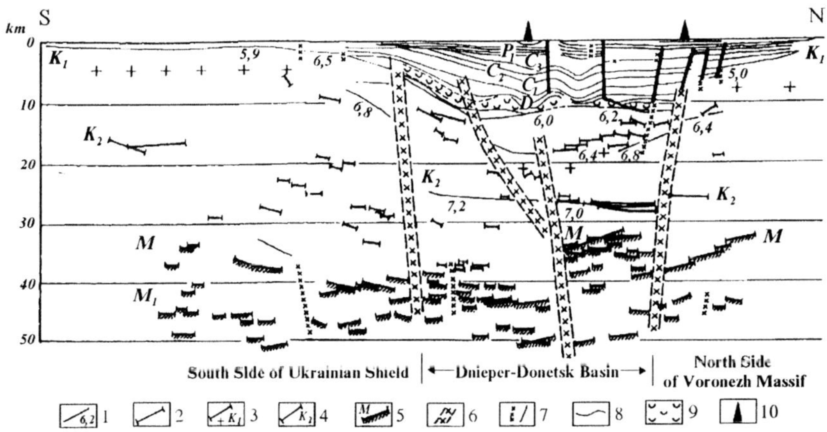

Fig. 3. Seismological section across the Dnieper-Donetsk basin and locatin of mercury-bearing gas deposits *1* efractive boundaries; *2* relfectory areas; *3* basement surface; *4* boundaries in the consolidated crust; *5* reflectory areas corresponding Moho discontinuity; *6* deep faults; *7* tectonic faults by seismic and geological data; *8* boundaries of stratigraphic complexes; *9* Devonian saliferous deposists; *10* mercury-bearing gas deposits Oposhnya and Gadach

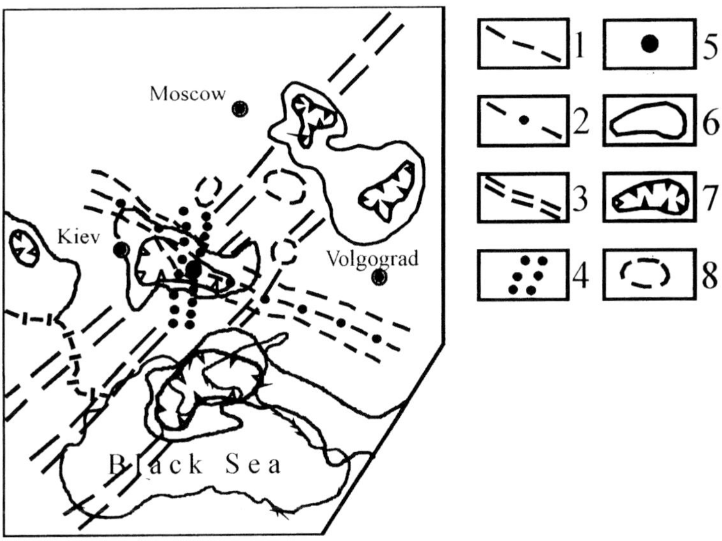

Fig. 4. Location of Hg-containing gas deposit Oposhyna. *1–4* deep faults of different direction; *5* Oposhnya; *6–8* isolines in proximity to strong earthquakes

showed the existence of long mercury anomalies above the faults in this part of the Karpinsky lineament.

We studied in more detail the Mirnensk ore deposit among those of the Stavropol uplift. Its high mercury potential is stipulated by the regional position– in the crossing conjunction of longitudinal faults of the southern flank of the

Table 3. Distribution of Hg in gases of Mirninskoye deposits (Northern Caucasus)

Age	$Hg, g/m^3$
Pg mkp_3	$5 \times 10^{-8} - 3 \times 10^{-7}$
K_1	$2 \times 10^{-7} - 4 \times 10^{-5}$

Manich dislocations and submeridional ones of the Trans-Caucasian lineament. Increase in mercury content with depth was revealed by gas sampling in the boreholes: the mercury content is lower in the Upper Paleogene deposits than in the Cretaceous ones, which are 2000 m lower (Table 3).

A linear zone of extremely high concentration also increasing with depth was revealed comparable to the described distribution of mercury (higher concentrations in the Cretaceous deposits and lower in the Paleogene). Thus, this is a linear penetrated zone, not shown on the geological map (Fig. 5). The revealed zone of higher penetration clearly corresponds to the structure of a large northwestern regional fault, as shown by the analysis of data received together with the regional geological materials.

The higher mercury content in gases in the deeper gas-bearing layers, described above (in the Mirnenskoye deposit) was also noted by us and in macroscale for the whole province. Thus, an obvious increase in mercury content at a depth from n × 100 m to 3 km (Table 4) is observed for the gas-oil deposits

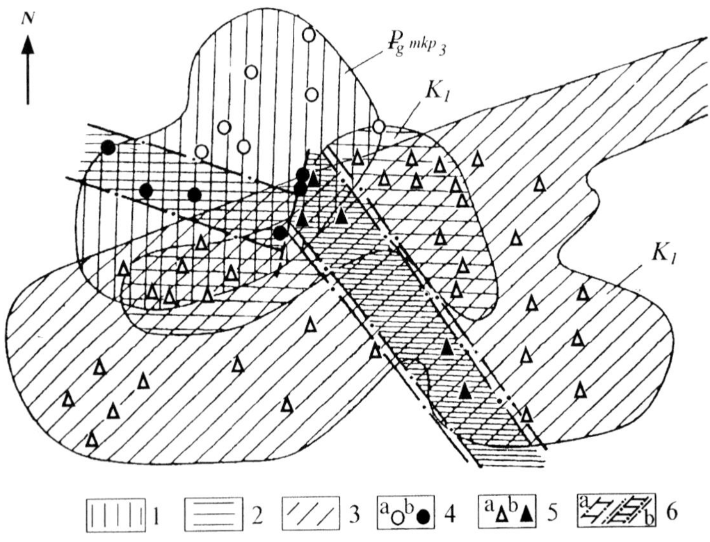

Fig. 5. Scheme of mercury-bearing Mirnenskoye gas deposit. Gas deposits: *1* Maykop deposits; *2,3* Lower Cretaceous deposits (*2* layer II, *3* layer IV); *4* wells of Maykop deposit with mercury content in gas: a non-discovered, *b 1–3*(conventional values); *5* wells of Lower Cretaceous deposits with mercury content in gas: less than 400 (conventional values), *a* and more than 400 (conventional values)-*b;* 6 zones of maximal concentrations of mercury in gases of Maykop deposit *a* and Lower Cretaceous deposits *b*

Table 4. Distribution of Hg in gases of the Middle-European platform depending on age and depth of gas-containing beds, 10^{-6} g/m^3

Depth of gas-containing beds (m)	Age				
	C	P	T	J	K
To–1000	–	–	–	–	15
From–1000 to –2000	–	10–90	–	15;100	20
From–2000 to –3000	–	180	15;30;200	–	–
> –3000	300;400	1000–3000	–	–	–

of the West European part of the Karpinsky lineament. This once more argues in favour of the deep genesis of mercury in gas-oil deposits and its rising from the depth.

Meanwhile, the mercury content is not stable in the wells in gas deposits while deposits, as a whole, retain their characteristics. This non-stability could be attributed to natural causes as well as technogenic ones and, preferably, to both (with varied dominance). A significant dispersion of mercury content was shown by many years of sampling in the Oposhnya deposit for years, days and hours (chemical determinations) and later constant determinations during the separate 24 h (atomic absorption method). It cannot be excluded that the seismoactive factor is the most important for the Oposhnya deposit (see Fig. 4, ore deposit situated in the epicentre of isolines concerning strong earthquakes). Correlation of mercury content in gases with seismic parameters corresponding in time in the seismic stations of the Ukraine is now under consideration.

What are the fellow travelers ("satellites") of mercury in degassing processes?

The correlation of higher mercury content with He anomalies is fairly notable. The geological conditions defining the occurrence of He anomalies are generally similar to those of mercury-bearing gases. The Karpinsky lineament with higher He-bearing along its length is a good illustration of the same correlation (Ozerova, Pikovsky 1982; Ozerova 1986, 1996). Mercury correlation with the "mantle helium" should be stressed in terms of the deep character of mercury degassing. Only a few data exist concerning this, but they are of special significance. A higher $^3He/^4He$ was discovered in the gases of the Stavropol uplift.

The relation of mercury to nitrogen, hydrogen and carbon dioxide is traced in some cases. Thus, nitrogen concentration is very high, 50–75, sometimes up to 99 vol% in the gas of ore deposits of Altmark (Germany) with a maximal mercury content in gas. A similar situation is also found in the ore deposits of Poland. Some German investigators (Dickenstein et al. 1973) concluded the deep nature of a considerable part of the nitrogen from these deposits based on the isotopic composition of nitrogen and geologic-geochemical characteristics of the gas. Distinct correlation between mercury concentration and carbon dioxide was observed by us in the Mirnenskoye deposit, Stavropol anticlise: mercury content is low, 5×10^{-8}–1.8×10^{-7} g/m^3 in hydrocarbonic gases of Upper Paleogene deposits where CO_2 is also fairly low, 0.06–0.24 vol%; where the CO_2 content is higher, 3.65–6.66 vol%, the mercury content is also higher, 2×10^{-7}–3.9×10^{-5}

g/m^3 (Ozerova and Mashyanov 1989; Ozerova 1986). Hydrogen was observed in mercury-bearing methane gases in Donbass (Donetsky basin).

The reserve of mercury in some hydrocarbon deposits corresponds to that of a mercury deposit and we stress once more that the noticeable part of mercury is in the condensate of gas in gas-condensate deposits. Our approximate calculation has shown that the reserve of mercury in most gas deposits, Groningen, Vustrov-Zalzvedel-Pekenzen (with fairly high mercury content) is assessed at several thousand tons, while the reserve of mercury in the oil field Cymrik, California (with a unique high mercury content in oils, up to 2×10^{-3}) is, at a minimum, 800 tons (Ozerova 1983, 1986).

While there is not enough material on mercury content in deposits of the world's different regions (varying strongly because of different geodynamic settings), it is only possible to calculate yearly mercury content approximately concerning reworking and using in the world's industry of oil, gas-condensate and gas. Thus, for oil and condensate yearly mercury content is tens or hundreds in sum and for gas, tons to tens of tons.

4 Practical Consequences

It is possible to discuss new regional structures with waiting mercury-bearing gas and oil deposits based on the characteristics and analysis of their regional geological position.

Such a structure, recommended by us 15 years ago as potentially mercury-bearing (Ozerova 1986), is the Pannonsk-Volinsky structure (Fig. 6). It belongs to the category of deep structures with a long history, being formed just in the Riphean. The Pannonsk-Volinsky depression is traced across the folded construction of the East Carpathians from the northeast to southwest for more than 900 km. It closes with the Riphean avlacogen of the East European platform in the northeast and the Solnoksk graben in the Baikal-Caledonian basement of the Large Hungarian depression in the southwest, and stretches in southwestern direction to Zagreb, Croatia. The Pannonsk-Volinsky depression is part of a large lineament, crossing the young folded system and old platform.

Fault activization of the Pannonsk-Volinsky depression in the Carpathians took place up to modern time. The mercury-bearing Bitkovsky oil deposit (1×10^{-4} Hg in oils) was influenced by it as well as hydrothermal processes forming the mercury deposits of the Vishkovsky ore group, the Uglyansky ore field and mercury-bearing polymetallic deposits – Beregovsky and the ore area Baya-Mare. Mercury-bearing gas deposits – Podravina gas field (1–2 mg/m^3 in gas (Spirič 1996) were discovered in the southwestern part of the structure, near Zagreb, that could well locate in the proposed mercury-bearing structure of the Pannonsk-Volinsky lineament. Separators for capture of mercury were made in some deposits. These are hopeful signs in the search for mercury-bearing gases in Hungarian territory.

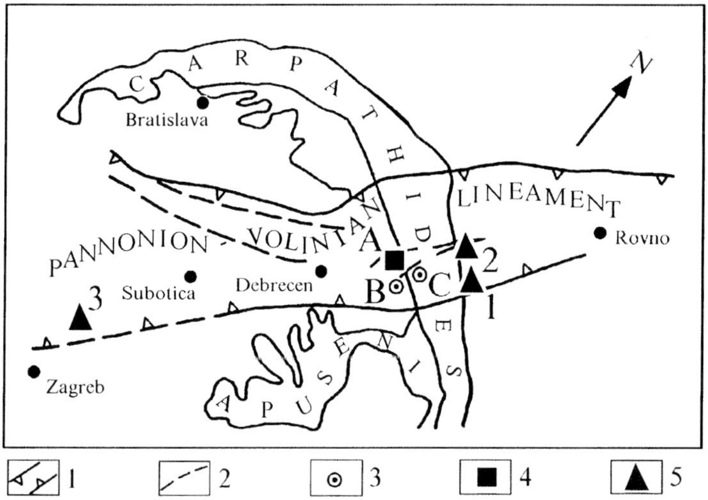

Fig. 6. Pannonsky–Volinsky lineament. *1* faults surrounding lineament; *2* faults; *3* mercury deposits: *B* Vishkovskoye; *C* Uglyanskoye; *4* Beregovskoye polymetallic deposit; *5* oil deposits: *1* Bitkovskoye: *2* Borislavskoye: *3* Molve

Mercury extraction and cleaning of gases and reworking products of gas and oil deposits with separating mercury as a fairly toxic element is too important for environmental protection. It is important even in cases when mercury is not an important industrial component: the huge amounts of world annual reworking gas and oil should be accounted.

Investigations of mercury are important from two aspects, as revealing the main regularities of its migration concerning hydrocarbons as a part of a large geochemical cycle of mercury migration and evaluation of possible mercury extraction or cleaning of gases and oil products for environmental protection (taking into account the huge scale of extraction of hydrocarbon gas and oil).

References

Achterberg A, Zaanen JJ (1972) Sporen kwik in het Groninger aardgas. Chem Weekbl 68:11–59

Bailey EH, Snavely PD, White DE (1961) Chemical analyses of brines and oil, Cymrik field, Kern County, California. Geol Surv Prof Pap D 424:306–309

Dickenstein GCh, Cluschko WW, Goldbecher K, Muller EP et al. (1973) Zum Auftreten von Quecksilber in Erdgasen am Beispiel der Rotliegenderdgaslagerstatten. Z Angew Geol 19:492–494

Ganeev AA, Pogarev SE, Ryzhov VV, Sholupov SE, Dreval TV (1995a) The new method for direct and express determination of mercury in oil, gas condensate and biological samples. Ecol Chem 4:122–126

Ganeev AA, Ryzhov VV, Maydurov AD, Mashyanov NR, Sholupov SE, Ozerova NA, Chernova AE (1995b) Direct definition of mercury-organic compounds in the natural gas and air with using of pyrolysis and photolysis. Vestnik S Peter Univ 4:73–80 (in Russian)

Glogoczowsky JJ, Karpinski T, Streetelski J, Zuk W (1977) Zwazki genetyczne miedzy naftowa gasem siemnym i substancya organiozna, rozproszona wutworach permu Monokyly Pezedsudeckiej. Nafta 2:37–42

Lubas J (1984) Geochemiczne aspekty wystepowania rteci w gazie ziemnym na Nizu Polskim. Nafta 40:371–375

Morrison J (1972) NAM recovers mercury produced with Dutch natural gas. J Oil Gas 70:72–73

Ozerova NA (1983) New mercury ore belt in Western Europe. Int Geol Rev 25:1095–1100

Ozerova NA (1986) Mercury in endogenic ore formation. Nauka, Moscow (in Russian)

Ozerova NA (1996) Mercury in geological sistems. In: Bayes W, Ebinghaus R, Vasiliev O(eds) Regional and global mercury cycles: sources, fluxes and mass balances. Kluwer, Dordrecht, pp 463–474

Ozerova NA, Mashyanov NR (1989) New in mercury-metry. In: Burenkov EK, Morozov VI (eds) Geochemical methods and scientific-technical progress in geological studying of the Earth's depths. Nauka, Moscow, pp 69–81 (in Russian)

Ozerova NA, Pikovsky Yul (1982) Mercury in hydrocarbonic gases. In: Geochemistry of ore-forming processes. Nauka, Moscow, pp 102–136 (in Russian)

Ozerova NA, Pikovsky Yul, Shikina ND, Gruzdeva MA, Fedin AD (1974) On mercury in oil and gas deposits. J Geol Ore Dep 4:85–91 (in Russian)

Ozerova NA, Mashyanov HR, Pikovsky Yul, Ryzhov VV, Ganeev AA, Sholupov SE, Chernova AE, Leontyev IA, Zherebtsov YuD, Dobryansky LA (1996) Mercury in gas and oil deposits. In: 4th Int Conf on Mercury as a global pollutant, 4–8 Aug 1996, Hamburg, p 336

Saukov AA, Aydinyan NKh, Ozerova NA (1972) Essay on geochemistry of mercury. Nauka, Moscow (in Russian)

Spiric Z (1996) Mercury removal from natural gas with sulphur impregnated activated carbon-our experience. Addition and Revised Abstracts 4th Int Conf on Mercury as a global pollutant. 4–8 Aug 1996, Hamburg

Tunn W (1973) Quecksilberspuren in Erdgasen aus deutschen Erdgasfelden und ihr Einfluss auf die Lufft.! Erdöl und Kohle-Erdgas - Petrochem. ver Brennst Chem 26:498–500

Tunn W (1976) Untersuchungen über Spurenelemente in Gasen aus deutschen Erdgas-und Erdölfeldeth. Ibid 29:1–211

II.b Mining

Atmospheric Mercury in Abandoned Mine Structures and Restored Mine Buildings at Mt. Amiata, Italy

R. Ferrara, B.E. Maserti, B. Mazzolai, F.Di Francesco, H. Edner, S. Svanberg, and E. Wallinder

Mt. Amiata, located in central Italy, is characterized by the presence of large deposits of cinnabar (mercury sulfide) from which mercury has been extracted since ancient Roman times. The Abbadia San Salvatore mining and smelting plant was the largest in the Amiata district until it was permanently shut down in 1982. About 2 million flasks of mercury (34.5 kg each) were produced in the 80-year period the plant was in use. At present, this zone is the site of abandoned mine structures, and in 1990 some buildings were restored and converted for commercial activities. A mercury flux of 100 g/h from the mine structures was estimated in July 1995 using the LIDAR remote sensing technique. Using point monitors, a map was made of the distribution of air mercury levels around these structures, showing very high values (1000–15 000 ng/m^3) near the old furnaces and cooling towers. The air mercury concentrations inside the restored mine buildings range from 200 to 2200 ng/m^3, with higher values in the winter months due to decreased ventilation.

1 Introduction

Until the 1980s, mining for mercury production showed strong growth due to the high market price of this metal (Nriagu 1979).

Large mining centers sprang up in cinnabar-rich areas characterized by geologic anomalies such as that in the Mediterranean basin, where over 500 000 tons of mercury were produced, approximately 50% of the world production (Bargagli 1990; Ortega Girones and Hernandez Sobrino 1992; Palinkaš et al. 1995).

Mercury prices began to decline in the 1970s when falling demand, due to emerging toxicological problems with mercury use and production and heightened public awareness with regard to environmental issues, caused many mines to close.

The mining and processing operations often left behind desolate landscapes. Huge deposits of roasted cinnabar resembling small hills often surround the abandoned mines and smelting plants.

The Mediterranean basin still contains the world's largest active installation, that of Almadén (Spain), while those at Idrija (Slovenia) and Mt. Amiata (Italy) have been closed.

At Idrija the landscape is returning to normal thanks to an ongoing cleanup program; only traces of the smelting plant remain. There are no banks of roasted cinnabar because, during the period the plant was in operation, processing residues were dumped into the Idrijca River, which empties into the Adriatic Sea (Kosta et al. 1978).

Environmental Science
Mercury Contaminated Sites (ed. by R. Ebinghaus et al.)

The situation at Mt. Amiata is different. Here, roasted cinnabar banks and abandoned mining and smelting complexes are still present. The largest plant in this district is located near the village of Abbadia San Salvatore. The works were shut down permanently in September 1982. Since these structures are located on the outskirts of the village, some of the outlying buildings were recently reconverted for commercial use.

In this first survey we studied air mercury levels inside the restored buildings, outside the buildings, and in the nearby abandoned smelting plant.

2 Study Area

The study area covers 150 000 m^2 and includes the zone of mining installations at Abbadia S. Salvatore and restored buildings (Fig. 1). Still present today are an

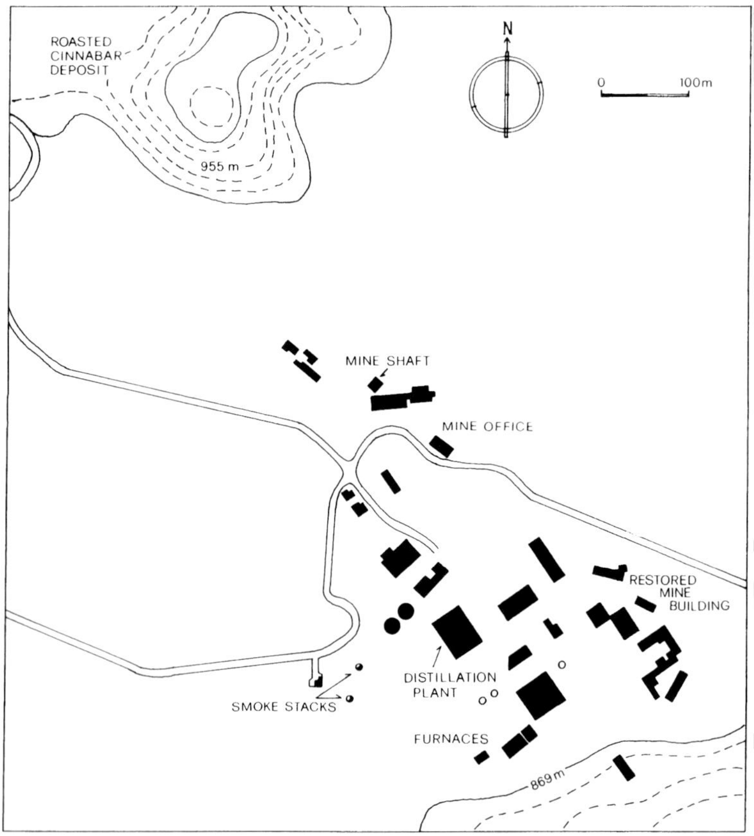

Fig. 1. Map of the Abbadia San Salvatore abandoned mine structures and restored mine buildings

elevator to reach the mine galleries (dug to a depth of 400 m), cinnabar crushing and sorting systems, a dryer, roasting furnaces, mercury condensation towers, a water purification plant, and all the buildings needed for bottling and storage. About 2 million flasks of mercury (34.5 kg each) were produced during the 80 years of activity (Bargagli 1990).

Inside the mine complex two types of furnaces are found, with condensation towers connected to two smokestacks: a modern, efficient Nesa furnace, and an older Cermak-Spirek furnace.

Some buildings located a few tens of meters from the old roasting ovens were converted to commercial activities such as the manufacture of leather bags, draperies, and so forth.

The remediation work basically consisted of completely replastering the walls and laying new flooring. One building was entirely rebuilt alongside the old ones, apparently with the partial use of materials from the primitive roasting furnace of the mine complex.

Figure 1 also shows the roasted cinnabar bank located a few hundred meters from the Abbadia mining structure. Formed by the accumulation of processing residues, the spoil bank covers an area of about 120 000 m^2.

3 Experimental

Air mercury concentrations were determined using point monitors based on preconcentration of the metal on gold traps and analysis, after thermal desorption, by atomic absorption spectrophotometry. Each determination is the result of two simultaneous samplings made by drawing in air at a rate of 1 l/min by means of a battery-powered diaphragm pump (Ferrara et al. 1992). The intake time was just a few minutes for the measurements made inside the buildings, whereas outdoor sampling times were 5–15 min. In the zone of the mine complex these times were further reduced to avoid collecting excessive amounts of mercury on the gold traps, which could result in problems of trap cleanliness and loss of instrument response linearity.

The measurements inside the buildings were continued over a 1-year period to check the effects of ambient temperature variation.

Determinations of the mercury flux from the mine complex were performed with the remote sensing technique known as LIDAR (light detection and ranging), developed at Lund University (Sweden). This laser spectroscopy technique, described by Edner et al. (1987), gives a three-dimensional picture of the distribution of mercury concentrations in the atmosphere; once wind speed and direction are known, it is possible to arrive at the amount of mercury emitted over time from a localized or diffuse source. The mobile LIDAR is housed in a Volvo F610 truck. Electric power for field work is supplied by a 20-kVA diesel generator installed in a trailer towed behind the truck.

Data from horizontal or vertical scans are presented as a two-dimensional plot with the concentration value indicated by a gray scale. The LIDAR system is placed in such a position that the plume can be scanned at an angle of about 90° to the wind direction. The flux is calculated from a vertical LIDAR scan by first integrating the measured concentrations over the area of the vertical plane mapped out by the scan. This integrated value is then multiplied by the wind speed perpendicular to the plane at the height of the plume's center of gravity. Flux measurements can be determined with a maximum accuracy of ± 20%.

4
Results and Discussion

On July 16 and 21, 1995 (T air = 30–36 °C), from 11:00 to 18:00 h, the LIDAR system was used to determine air mercury concentrations over the mine complex. The measurements were made from different positions to catch the plume at a more favorable angle with respect to the wind direction so as to obtain the most reliable mercury flux value possible. A mean value of 66 g/h was obtained.

Since part of the flux could not be measured due to the presence of obstacles in the zone scanned by the laser, it was estimated that the total mercury flux given off by the abandoned mine structures was approximately 100 g/h. This high value is comparable to the value measured for emissions from five geothermal power plants (120 g/h) located 4 km from Abbadia San Salvatore (Ferrara et al. 1994), and is much higher than the flux from all the roasted cinnabar spoil banks (a few grams per hour) present on Mt. Amiata (Ferrara et al., 1998).

Figure 2 shows an example of a horizontal section of the distribution of the air mercury concentration determined with the LIDAR system over the mine complex (07.16.95; time 17:30 h). Along the given measurement directions, mean mercury concentrations were evaluated as presented in the circles for the integration intervals marked along the lines. The length of the integration intervals was selected depending on the concentration level of mercury. In September 1990, Edner et al. (1993), using the same technique over the same mine complex (T air = 14 °C), measured mercury levels ranging from 10 to 340 ng/m^3.

Attempts to measure the flux over the Abbadia mine structure with the LIDAR technique in September 1994 (from the 19th to the 27th; T air = 10–30 °C) failed due to lack of wind. Analogous determinations performed over the smaller mine complex at Bagnore, also located in the Amiata district, showed emissions were 20 times higher in July than in September (Ferrara et al., 1998). In line with this finding, the mercury emission in the month of September from the Abbadia S. Salvatore mining installation should be approximately 5 g/h. Even lower values can be expected in the winter months, when temperatures average about 5 °C. In fact, mercury emissions such as those from mine structures and roasted cinnabar spoil banks are closely dependent on ambient temperature (Ferrara et al. 1997).

Although the LIDAR technique has shown itself to be particularly well suited for determining mercury fluxes, it usually cannot be used, due to the presence of

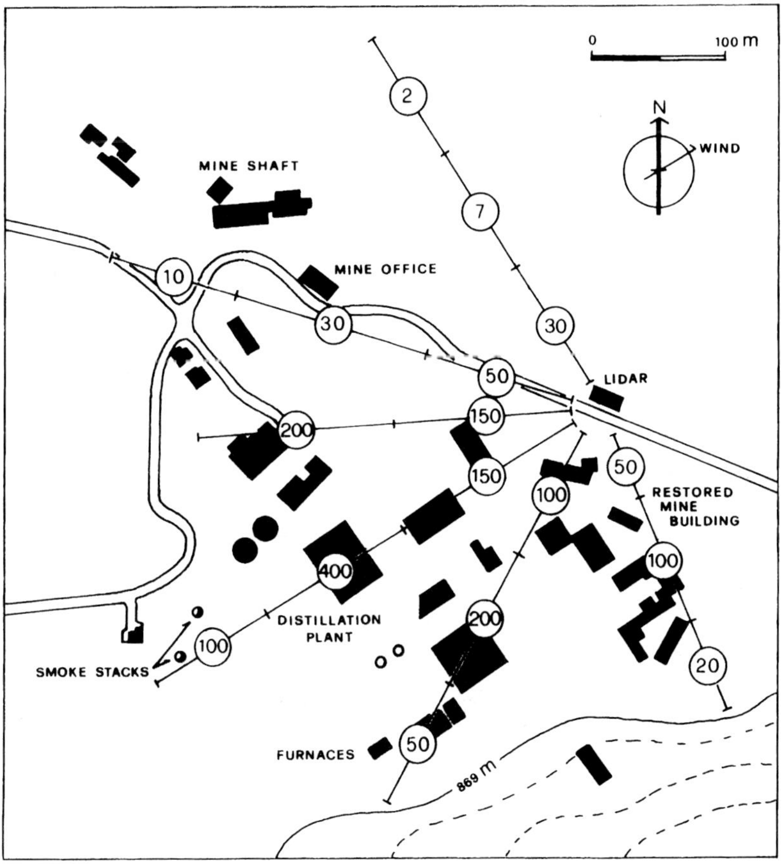

Fig. 2. Horizontal section of distribution of the air mercury concentration (ng/m^3) over the Abbadia S. Salvatore mine complex, measured with the LIDAR system (07.16.95; time 1730 h)

obstacles, to measure concentrations at ground level, that is, at a height where mercury is more likely to be taken in by people. It was therefore decided to measure the air mercury concentration in the zone of the mine structures by means of point monitors placed 1 m above the ground. The map in Fig. 3 shows the distribution of the mercury concentration determined with this technique in 34 sampling stations (St.) located in the area of the mine complex and outside the restored buildings. This distribution was determined on 25 June 1996, during a day with little or no wind and an ambient temperature of 20 °C. Extremely high values (>15 000 ng/m^3) were observed near the condensation towers of the old Cermak-Spirek furnaces (Fig. 3; St. 6). These high values are attributable to the metallic mercury with which the structure is permeated and to mercury accidentally spilled during the long period of mining activity. Part of the mercury seeped down into the earth, but little pools are still visible today on the floor of the structure.

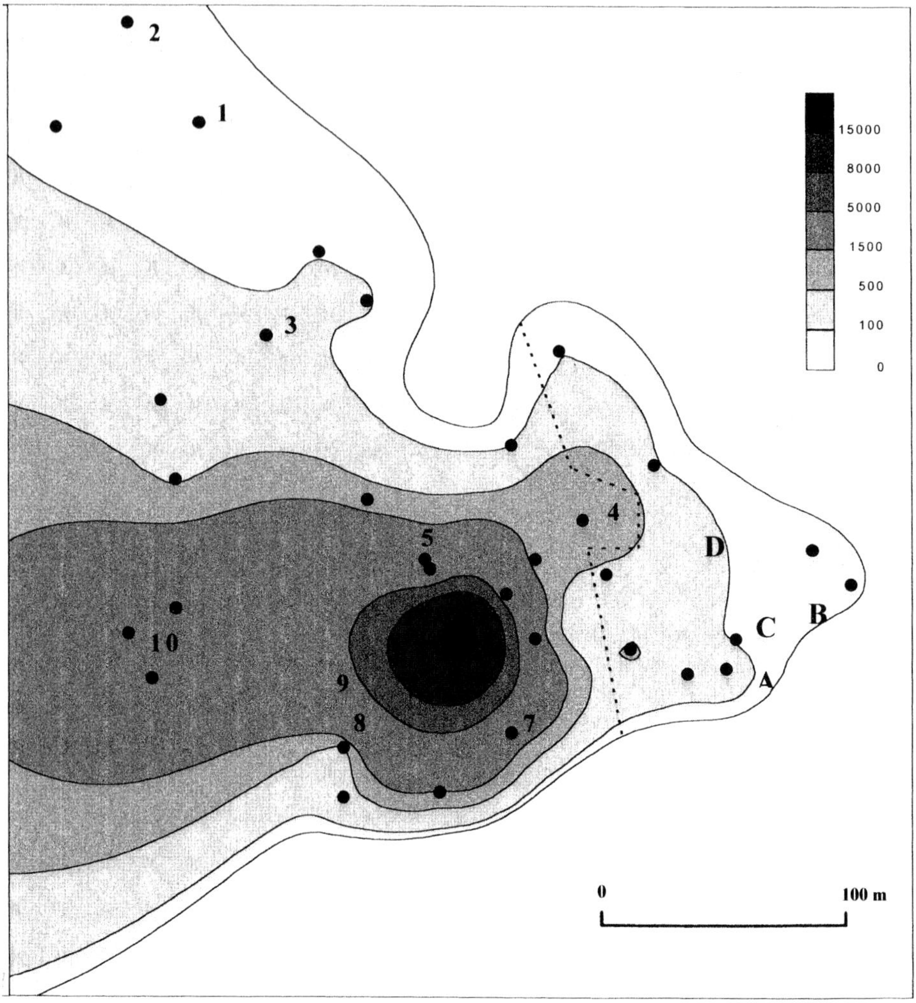

Fig. 3. Map of Hg (ng/m³) distribution in atmosphere, measured to a height of 1 m, in 34 sampling stations (•) at the abandoned mine structures and restored mine buildings of Abbadia S. Salvatore (Mt. Amiata, Italy; 25 June 1996). Mine structures (stations): *1* mine offices; *2* mine shaft; *3* wagons; *4* old furnaces; *5* dryer system; *6* cooling towers; *7* Goulder furnaces; *8* nesa furances; *9* nesa cooling towers; *10* staks. *Dotted line* denotes fence separating the mine structures from the restored mine buildings. Restored Buildings (stations): *A* bag production firm; *B* drapery; *C* wine cellar; *D* leather production firm

Near the modern Nesa furnaces (St. 8) and the related condensation towers (St. 9) the mercury concentrations were approximately 700–2000 ng/m³.

Again in Fig. 3 at station 4, an area now covered by a thick layer of grass, an atmospheric mercury concentration of 1700 ng/m³ was measured. The reason for this considerably high value is due to the fact that the first ovens installed for cinnabar roasting stood in this zone, and although they were razed many years ago, the ground is still contaminated with mercury. The datum reported above shows that even razing buildings is not really enough to ensure a cleanup of the contaminated area.

The mercury concentrations shown in Fig. 3, measured with point monitors, are much higher than the concentrations reported in Fig. 2, determined with the LIDAR technique. This is explained by the fact that with the point monitors the measurements were made 1 m above the ground, while with the LIDAR system they were generally made at a height of around 10 m, and mercury concentrations decrease exponentially with height (Kvietkus and Sakalys 1994). At present, the ground is the main source of the metal due to the widespread presence of the pools of mercury, mostly scattered around the condensation plant, and to the fallout of mercury-rich dust from the stacks throughout the long period of mining.

The broken line in Fig. 3 represents the fence separating the mine structures from the commercial activities initiated in 1990.

Outside the restored buildings, the air mercury levels range between 50 and 500 ng/m^3, with values essentially depending on the distance from the main polluting structures.

Concentrations inside the buildings range between 200 and 2200 ng/m^3. Values on the order of 1000–1500 ng/m^3 were measured in the drapery shop, a newly constructed building (cf. Table 1). These high levels confirm the use of materials from the razed furnaces to make the foundation. Environmental problems linked to the use of roasted cinnabar as a building material (in buildings, roads, paving for squares, and so on) have surfaced only recently.

The highest levels of atmospheric mercury were observed in the winter months, despite the fact that the temperatures inside the buildings (T = 18–20 °C), all of which are equipped with central heating, were lower than temperatures during the summer months (T = 27–30 °C). The cause for this trend is the decreased ventilation in the buildings during the winter. This tendency was confirmed by a mercury concentration value of over 10 000 ng/m^3 measured in a building used as a storehouse for bottled wine and opened only rarely.

5 Conclusions

In the summer months, the abandoned mine structures of Abbadia San Salvatore constitute one of the main sources of atmospheric mercury in the Mt. Amiata

Table 1. Air mercury concentrations (ng/m^3) inside the restored mine buildings

	Sept 1994	July 1995	Jan 1996	March 1996	June 1996
Leather production firm	400 ± 149	260 ± 10	350 ± 185	215 ± 9	210 ± 12
Bag production firm	700 ± 45	200 ± 22	1080 ± 550	780 ± 230	317 ± 20
Drapery shop		185 ± 80	1180 ± 163	570 ± 114	190 ± 60

mineralized district. For the first time, thanks to the use of the laser spectroscopy technique (LIDAR), it was possible to determine the emission flux, which proved to be 66 g/h (estimated value 100 g/h; July 1995). The results of this study show that remediation operations are absolutely necessary in the zone of the abandoned mine complex, particularly considering their position on the outskirts of the village. High air mercury concentrations (1700 ng/m^3) in the zones where the old furnaces once stood, now covered by thick vegetation, demonstrate the difficulty of designing a cleanup project that would be capable of significantly reducing the presence of the metal in the area.

The high air mercury concentrations (200–2200 ng/m^3) measured inside some restored buildings now used for commercial activities require a health inspection program that should include an initial survey of the workers and technical interventions on the buildings.

These results also show the poor effectiveness of the decontamination work done on the buildings and the complexity of the cleanup work on the abandoned mining complex.

At present, a remediation project is under study that calls for removing the top layer of earth (to a maximum depth of 7 m in the zones of the first roasting ovens) and transporting the material to a specially designed dump.

However, we feel that this remediation strategy might prove to be both costly and ineffective as mercury might be found in greater amounts deeper down due to seepage phenomena.

References

Bargagli R (1990) Mercury emissions in an abandoned mining area: assessment by epiphytic lichens. In: Cheremisinoff PN (ed) Encyclopedia of environmental control technology, vol IV. Gulf Publishing Company, Houston, Texas, pp 613–640

Edner H, Fredriksson K, Sunesson A, Svanberg S, Uneus L, Wendt W (1987) Mobile remote sensing system for atmospheric monitoring. Appl Opt 26:4330–4338

Edner H, Ragnarson P, Svanberg S, Wallinder E, Ferrara R, Maserti BE, Bargagli R (1993) Atmospheric mercury mapping in the cinnabar mine area. Sci Total Environ 133:1–15

Ferrara R, Maserti BE, Edner H, Ragnarson P, Svanberg S, Wallinder E (1992) Atmospheric mercury determinations by LIDAR and point monitors in environmental studies. In: Merian E, Haerdi W (eds) Metal compounds in environment and life, vol IV. Science and Technology Letters. Northwood, UK pp 29–38

Ferrara R, Maserti BE, De Liso A, Edner H, Ragnarson P, Svanberg S, Wallinder E (1994) Could the geothermal power plant at Mt. Amiata (Italy) be a source of mercury contamination? In: Watras CJ, Huckabee JW (eds) Mercury pollution integration and synthesis. Lewis, Boca Raton, pp 601–607

Ferrara R, Maserti BE, Andersson M, Edner H, Ragnarson P, Svanberg S (1997) Mercury degassing rate from mineralized area in the Mediterranean basin. Water Air Soil Pollut 93:59–66

Ferrara R, Mazzolai B, Edner H, Svanberg S, Wallinder E (1998) Atmospheric mercury sources in the Mt. Amiata area, Italy. Sci TOTAL ENVIRON 213:13–23

Kosta L, Ravnik V, Dermelj M, Pilhar B, Stegnar P, Byrne AR, Lokar J, Vakselj A, Novak J, Prosenc A (1978) Vestn Slov Chem Soc (SKD) Ljubljana 25:463–493

Kvietkus K, Sakalys J (1994) Diurnal variations in mercury concentration in the ground layer atmosphere. In: Watras CJ, Huckabee JW (eds) Mercury pollution integration and synthesis. Lewis, Boca Raton, pp 243–250

Nriagu JO (ed)(1979) Production and uses of mercury. In: The biogeochemistry of mercury in the environment. Elsevier/North-Holland Biomedical Press, Amsterdam, pp 23–40
Ortega Girones E, Hernadez Sobrino A (1992) The mercury deposits of the Almaden syncline, Spain. Chron Rech Min 506:3–24
Palinkaš LA, Pirc S, Miko SF, Durn G, Namjesnik K, Kapelj S (1995) The Idrija mercury mine, Slovenia, a semi-millennium of continuous operation: an ecological impact. In: Richardson M (ed) Environmental toxicology assessment. Taylor & Francis, London, pp 317–341

Mercury in the Town of Idrija (Slovenia) After 500 Years of Mining and Smelting

V. Miklavčič

Due to mining and smelting processes, and due to the naturally increased contents of mercury in soil, the wider surroundings of Idrija are contaminated with mercury. The Idrija ore deposit contains rocks enriched with cinnabar originating from periods up to Middle Triassic. Native mercury can be found primarily in Carboniferous clastic rocks which reach the surface in the western part of the ore deposit. In the 1970s, when the production of mercury reached its peak, the first intensive studies of the ecological cycle of mercury were made. New analytical techniques and improved methods for determining the total mercury and methylmercury contents were introduced in order to study the uptake and distribution of this element in the ground, air, water, plants, animals and man. After 20 years, systematic determinations of total mercury in environmental and biological samples were started again. In our work, total mercury in air, lichens sampled in situ, transplanted lichens, carrots and beans were determined, using gold amalgamation cold vapour atomic absorption spectrometry (CV AAS) after acid decomposition under pressure, or in the case of air measurements, the double amalgamation technique followed by CV AAS. Measured concentrations of total gaseous mercury in air varied from 10 to 277 ng Hg m^{-3}. The total mercury contents in lichens sampled in situ were in the interval between 3 to 23 µg Hg g^{-1} dry sample (d.s.). The total mercury content in carrot ranges from approximately 60 to 760 ng Hg g^{-1} d.s. The concentrations of total mercury in beans were in the interval between 2 to 23 ng Hg g^{-1} d.s.

1 Introduction

The discovery of one of the world's largest mercury deposits in 1490 led to the opening of the Idrija Mercury Mine, which, over a period of 500 years of continuous operation, experienced exceptional technical, technological and scientific development, which also left its mark on the town of Idrija. However, 500 years of mining activities produced harmful effects on the environment due to pollution with mercury and, partly, ionizing radiation from radon and its daughters. Over these 500 years, generations of Idrija miners extracted 107 000 tons of mercury (Hg) for commercial purposes. Taking into account all the losses, which were particularly excessive in the early periods of extraction, it may be estimated that the total quantity of extracted Hg surpassed 147 000 tons.

The Idrija Mine operated successfully until the global Hg crisis in the 1970s. In 1969 the results of many years of research on the influence of Hg and its compounds on the environment and humans were published in the United States (Jensen and Jernelov 1969; Jernelov 1969). Large quantities of Hg had been found in American rivers in the 1960s. In addition to the published findings, the toxicity of this element, particularly its organic compounds, was brought to light by the

Environmental Science
Mercury Contaminated Sites (ed. by R. Ebinghaus et al.)

mass poisoning, permanent injuries and deaths that occurred in Japan, Minamata (1956) and Niiagata (1965), and in Iraq (1972) as a result of the negligent use of Hg preparations. At the end of the 1960s, reports of increased Hg content in birds and some types of fish were published in Sweden (Johnels and Westermark 1969). The above-mentioned disasters and findings met with a worldwide response, resulting in limitation of the further production and broad industrial use of this unique liquid metal. The Hg crisis and falling demand (and price) led to a decision on the gradual shutdown of the Idrija Mine, which was passed by the Slovene Assembly in 1987. It was foreseen that the mine would be gradually shut down by 2006 and that monitoring of the consequences of the shutdown process would be conducted until 2009.

Mass poisonings with organic Hg compounds triggered numerous internationally coordinated studies of the harmful effects of Hg on the environment, its toxic effects on living organisms and the transformation of inorganic Hg into more toxic organic Hg forms, particularly methylmercury. In the early 1970s (1969–1974), when Hg production in the Idrija Mine was at its peak, the research group of the Department of Nuclear Chemistry at the J. Stefan Institute, headed by Prof. Lado Kosta, in collaboration with the Research Unit of the Idrija Mercury Mine, began intensive studies of the Hg ecological cycle in the Idrija region. These studies, which were undertaken as a response to the previously mentioned contamination with mercury compounds and based on the fact that Idrija and its immediate surroundings offered an ideal location for studies of this kind, involved monitoring of uptake, spread and transformation processes of Hg in the Idrija region, with the objective of determining the health hazards existing in the area due to the presence and harmful effects of Hg. These studies were of great importance at that time, which is evident from the fact that similar studies undertaken at the Almaden Mercury Mine, with the strong financial support of the University of Rochester, were in their initial phase in 1974.

The studies dealt with Hg circulation in practically all ecological systems and were complemented by the development and upgrading of analytical methods for determining submicrogram quantities of Hg in a broad range of samples from the area, as well as methods for tracing other Hg compounds. New techniques and improved methods for determining the total Hg content and methylmercury (MeHg) content were introduced in order to study the uptake and distribution of this element in the ground, air, water, plants, animals and man. The analytical methods used were based on neutron activation and volatilization of mercury under specific conditions (Kosta and Byrne 1969; Byrne and Kosta 1974), gas chromatography of volatile organic Hg compounds (Zelenko and Kosta 1973), and in the determination of mercury values in air by flameless atomic absorption (Kosta et al. 1974).

Mining activities, the roasting of ore and the use of ore residues, as well as the naturally increased Hg content in the ground have heavily contaminated Idrija and its surroundings with Hg. The Idrija ore deposit contains rocks enriched with cinnabar originating from periods up to Middle Triassic. Native Hg can be found primarily in Carboniferous clastic rocks which reach the surface in the western part of the ore deposit in the so-called Pront area (Mlakar 1974). The hazard of Hg contamination to humans and the environment increased during operation of the mine and smeltery. The gases emitted through the main exhaust stack continuously contaminated the air with Hg. According to some measurements and estimates, approximately 20 kg of Hg were emitted into the atmosphere daily in the form of vapor or aerosols (Kosta et al. 1974). Hg concentrations in air measured in the 1970s ranged from 0.1 to 10 μg Hg m^{-3} of air; the highest concentration in the air, found in the immediate vicinity of the smeltery, was 8.5 μg Hg m^{-3} of air. The Hg concentration in the town centre was 0.6 μg Hg m^{-3} of air and was 6 hundred times higher than at the reference locations (Kranj, Ljubljana), whereas Hg concentrations in uncontaminated locations did not exceed 0.001 μg Hg m^{-3} of air.

This type of pollution resulted in increased Hg values in plants and animal species in the Idrija area. Carrots from gardens located directly below the stack contained much more Hg (400 to 800 ng Hg g^{-1} of fresh sample) than those from the Pront area (50 to 100 ng Hg g^{-1} of fresh sample), where the ground is composed of Carboniferous shale containing native Hg. Results are shown in Table 1. The Hg content in other vegetables was much lower than in carrots, ranging from 10 to 50 ng Hg g^{-1} fresh sample (f.s.) below the stack and 1 to 10 ng Hg g^{-1} f.s. in the Pront area.

Hg content was also determined in the meat of animals most frequently consumed by the inhabitants of Idrija (rabbits, chickens). The values obtained for Hg content in muscle were generally higher in the case of animals from the area below the stack than those from the Pront area, but were relatively low in all cases, below 100 ng Hg g^{-1} f.s. (Stegnar 1980). Constant exposure to mercury was most evident in the highly increased Hg content in the tissues and organs of professionally exposed workers in the mine and smeltery. A very important finding was that of an approximately 1:1 molar ratio of selenium and mercury in the organs accumulating – and retaining, even after 20 years, – the highest Hg contents, i.e., the pituitary, thyroid, liver and kidney. Even the Hg concentrations found in other inhabitants of Idrija were an order of magnitude higher than those found in persons from unexposed areas (Kosta et al. 1975).

In 1977 the mine's operation was temporarily discontinued. The main benefit was reduced air and environmental pollution, which contributed to decreased Hg concentrations in living organisms in the area below the smokestack. However, measurements showed that pollution of the environment with Hg had not been completely eliminated. This was evident in the Hg concentrations found in carrots (average value was 500 ng Hg g^{-1}), which, after termination of the smeltery's operation, remained practically identical to those observed during its operation (average value was 488 ng Hg g^{-1}).

Table 1. Total mercury content in carrot and bean from two locations in Idrija town and reference location (village Ig near Ljubljana), in ng Hg g^{-1} of fresh sample (Byrne and Kosta 1970; Stegnar 1980)

Location Sample	Vicinity of smelting plant	Pront area	Reference location (Ig)
Carrot	440–826	48.8–102	0.35–0.70
Bean	3.6–21.9	1.2–1.3	

During the course of compiling a report entitled *Studies of the Degree of Environmental Pollution with Mercury in Idrija and its Surrounding* in 1980 (Stegnar 1980), it was established that mercury is a major polluter of the environment, where it undergoes transformation processes and appears in various chemical forms that are even more hazardous, particularly for humans. For this reason, researchers warned that special attention should be devoted to the selection of appropriate technology upon the reopening of mine and in seeking effective solutions for the rehabilitation of the environment.

The mine was reopened in 1983 and operated with minimum capacity until its operation was finally stopped in 1994. Today, after almost 500 years of operation, the Idrija Mine is in the final phase of its gradual shutdown. The time has come to focus all investigations on environmental issues in the Idrija region and to develop adequate solutions for its remediation.

In the period between 1983 and 1990, very few studies dealing with the problem of Hg in the environment were conducted in Idrija. The majority of studies undertaken in this period were dedicated to the exposed population of miners and their protection. However, a comprehensive study of the present state of the environment has not been made to this day. The situation changed after 1990, probably as a result of the increased awareness of environmental problems. A number of studies of Hg contamination of the environment and biological samples have recently been carried out in Idrija and its immediate surroundings (Miklavčič 1996).

2
Materials and Methods

In the period from 1991 to 1993 numerous measurements of total gaseous Hg in air were conducted using sampling, storage and final detection procedures specially developed and optimized for this purpose (Lupšina-Miklavčič 1994). Total gaseous Hg in air was determined by double amalgamation cold vapour atomic absorption spectrometry (CV AAS) after sampling on gold traps.

Epiphytic lichens have long been regarded as suitable tools for monitoring the relative levels of atmospheric pollution. They are largely dependent on wet and dry deposition from the atmosphere (Valenta et al. 1986; Brosset and Lord 1991; Iverfeldt 1991; Lindberg et al. 1991). In 1993 and 1994 the epiphytic lichens *Parmelia sulcata, Parmelia caperata* and *Hypogimnia physodes* were collected from six locations in the town of Idrija around the Mercury Mine at different distances from the smelting plant and mine shafts. The nearby town of Cerkno-Trebenče served as a reference location. The reference location (1) and sampling locations (2–7) are shown in Fig. 1.

In November 1993, branches over grown with lichens were transplanted from an area believed to be unpolluted with Hg (Trebenče, reference location) to three different locations in Idrija (A–C) and exposed for a period of 5 months. The locations were randomly selected, as the results of measurements of Hg

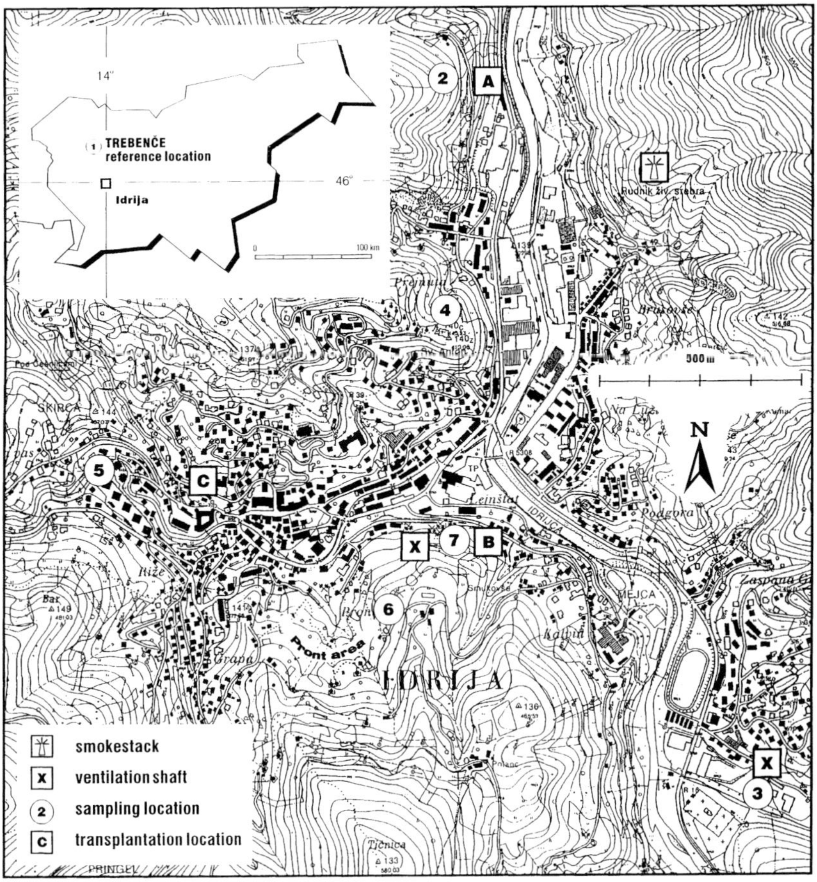

Fig. 1. Map showing the reference location, locations of sampling and transplantation areas of lichens and the mercury sources

concentrations in lichens sampled in situ were still unknown at the time. The locations of transplantation are also shown in Fig. 1.

Since in the 1970s (Stegnar 1980) researchers found that some plants (especially carrot) can bioaccumulate and asimilate Hg during their growth, we collected samples of carrot and bean from ten different locations in Idrija town, in the autumn of 1995, in order to compare the results obtained with those from the 1970s. Figure 2 shows the sampling locations of carrots and beans. Bean sample consisted of climbing (or pole) beans, and only the inner seed, not the pod, was analyzed. Carrots were carefully cleaned by slicing to remove outer contamination.

Total Hg in freeze-dried samples of lichens, carrots and beans was determined by gold amalgamation spectrometry (CV AAS) after acid decomposition under pressure (Horvat et al. 1991). Both analytical techniques are used for routine analysis of total Hg in the Laboratory of the Idrija Mercury Mine.

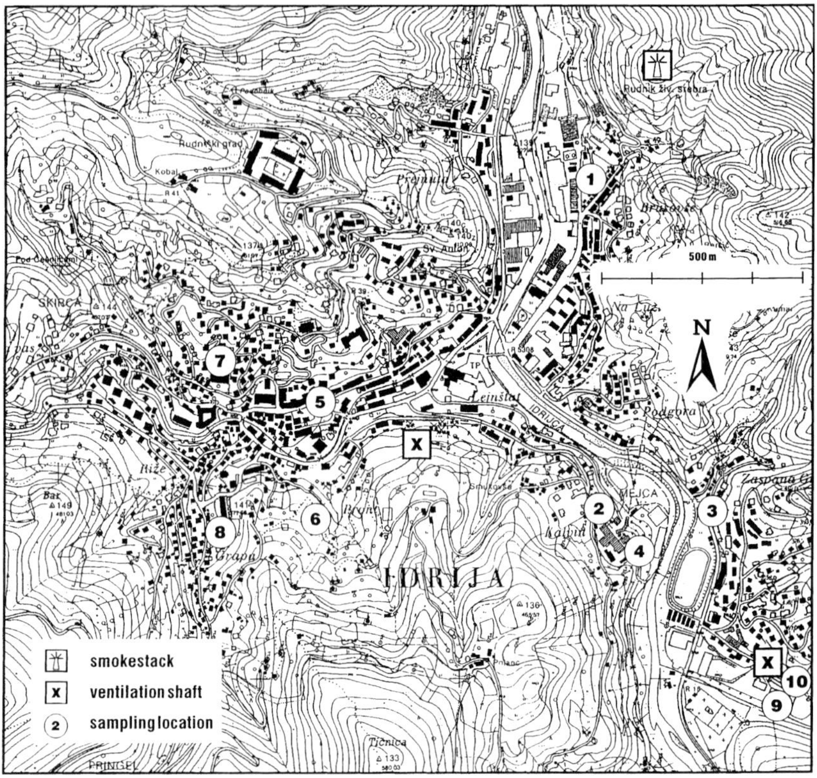

Fig. 2. Map showing the locations of the sampling areas of carrot and bean and the mercury sources

3 Results and Discussion

In the year 1992, 76 measurements of total gaseous Hg in air were performed at one location outside our laboratory (located close to the Idrija centre), three parallel samples for measurement (Lupšina et al. 1992). The duration of sampling was adapted to the weather conditions. Most of the measurements involved momentary concentrations of Hg in air, which means that the sampling period lasted a few hours at maximum. The air intake ranged from 10 to 60 $l\ h^{-1}$. Thus, the smallest and the largest volumes of air samples were 0.03 m^3 and 0.1 m^3, respectively.

The values of Hg concentrations in air obtained varied significantly, the lowest being 10 ng Hg m^{-3} (19 October 1992) and the highest 277 ng Hg m^{-3} (28 August 1992). The results of previous measurements of momentary Hg concentrations in air conducted in a natural environment have shown that Hg concentrations increase with increasing external air temperature, while wind reduces Hg concentration markedly, as is evident in Fig. 3.

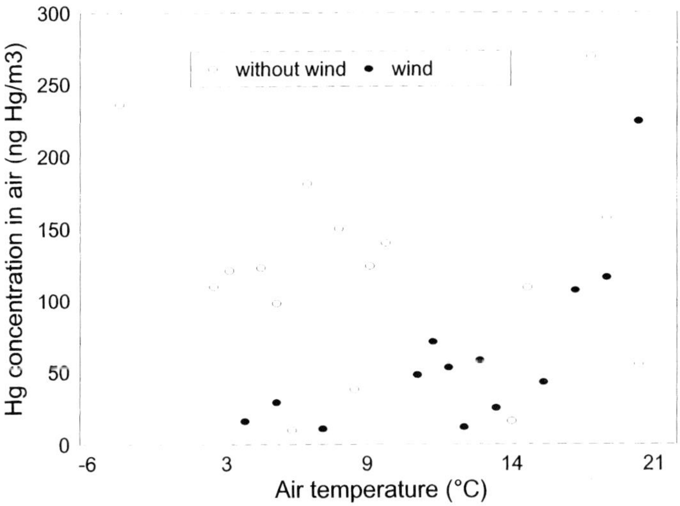

Fig. 3. Total gaseous mercury concentrations in air as a function of air temperature with or without influence of wind

These results obtained under different wind conditions point to the considerable influence of the wind on Hg concentrations in air. This indicates that the main source of Hg emission into the air is most likely the mine's ventilation shaft, whose effects on Hg concentrations in air is greatly reduced under favourable wind conditions. Measurements performed in calm weather gave an average Hg concentration of 135 ng Hg m^{-3} of air and those performed in windy weather an average Hg concentration of only 55 ng Hg m^{-3}.

As is evident from Fig. 4, the total Hg content in lichens sampled in situ ranges from approximately 3 to 23 μg Hg g^{-1} d.s. According to the lowest concentration of total Hg in lichens, found at the reference location Trebenče (0.23 μg Hg g^{-1} d.s.), the results obtained in Idrija were at least 10 times, even 100 times higher. These results are in accordance with previous data on Hg levels in air there (Lupšina et al. 1992). Lichens sampled in areas which are most affected by air from mine's ventilation shafts show high Hg contents. The highest levels were recorded in the northern part of Idrija town, probably due to the prevailing winds in the south-north direction down the valley of the river Idrija, spreading air-borne Hg, and because of the relief, as in the north of Idrija the valley is narrowing.

The results confirmed the excessive contamination of air and the environment in the Idrija region, which is nevertheless not uniformly distributed and is dependent on many factors, in particular natural and geographic ones. In our opinion, the most significant role is played by the air flow.

This study also included an attempt at active biomonitoring using transplanted lichens. We transplanted lichens from the reference location (Trebenče) at three different locations in Idrija town for a period of 5 months (from the beginning of November 1993 to the end of March 1994). The first location (designated A in

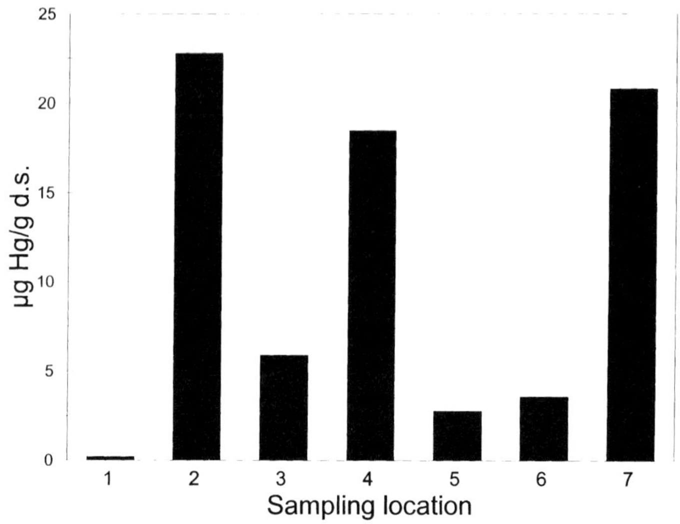

Fig. 4. Total mercury content in lichens sampled in situ

Fig. 2) was chosen north of Idrija where the basin is closing in, the second (B) was near one of the mine's ventilation shafts and the third location (C) was the Secondary School garden, quite far from major known pollutants in Idrija. A graphical presentation of Hg concentrations in transplanted lichens is given in Fig. 5.

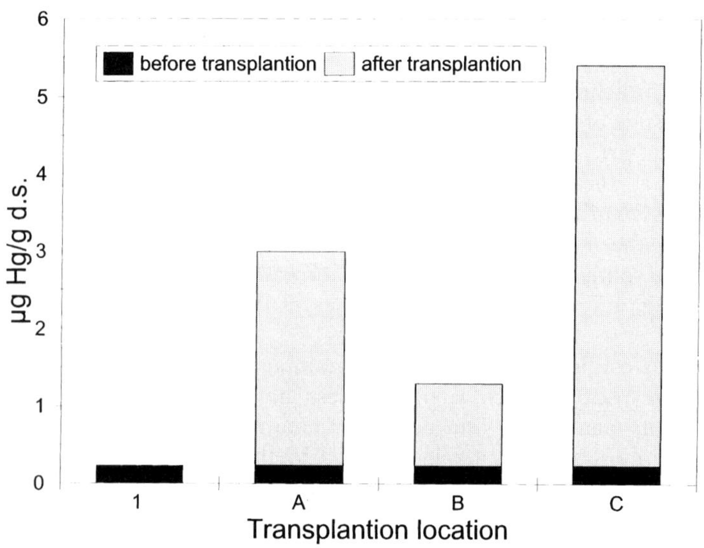

Fig. 5. Total mercury content in transplanted lichens

As is evident from Fig. 5, the content in lichens increased considerably during the exposure period at all three locations. The total Hg concentrations in lichens transplanted to the school garden (C) are significantly higher than those of the other environments. Probably the anomaly could be explained by the relief, since the garden is situated in a weakly ventilated valley. It is possible that polluted air collects here. Another source of Hg could be the Idrija Municipal Museum, which is located in the castle in the vicinity. In the past, this castle was used for storage of mercury. High values were also obtained in the northern mining area (A), which is in accordance with the concentrations of total Hg in lichens sampled in situ from this part. The Hg content in lichens from transplanted location B was the lowest, although we expected the highest. We discovered later that the location for transplantation was too close to the mine's ventilation shaft. The lichens were transplanted in a forest environment, which could reduce the dry and wet deposition.

It is known that biomonitoring using lichens excludes all momentary factors and integrates the level of air pollution during the entire exposure period. The results obtained confirmed the usefulness of epiphytic lichens for the routine monitoring of air pollution with Hg. The only precondition for the utilization of this method is the study of the natural geographical conditions in the Idrija region, particularly of air flows and the microclimate in order to enable the selection of suitable permanent locations for objective monitoring of air pollution with Hg over longer periods of time with the help of epiphytic lichens.

Samples of carrot and bean were collected in the autumn of 1995 from ten different locations in Idrija town. The sampling locations are shown on Fig. 2. Results for total Hg in carrot and bean are presented in Table 2. Total Hg was determined in dry samples.

The results for Hg content in carrot were in the interval between 60–760 ng Hg g^{-1} of dry sample. The highest value (760 ng Hg g^{-1} d.s.) was measured close to the smelting plant. High values (738 ng Hg g^{-1} d.s., 489 ng Hg g^{-1} d.s.) were found at some locations where we could not explain

Table 2. Total mercury content in samples of carrot and bean (collected in autumn 1995) in ng Hg g^{-1} of dry sample

Sampling location	Carrot	Bean
1	760 ± 38	18 ± 5
2	489 ± 69	
3	151 ± 18	6.6 ± 3.5
4	738 ± 54	
5	321 ± 39	
6	275 ± 29	23 ± 0.4
7	396 ± 30	
8	60 ± 9	2.0 ± 0.6
9	78 ± 19	6.4 ± 4.2
10	163 ± 10	4.0 ± 2.0

Results of three independent analyses are expressed as means ±95% confidence limits.

the results obtained. Close to the Municipal Museum, which is located in castle, we found high concentrations of mercury content too (396 ng Hg g^{-1} d.s.), which agrees with our measurements of mercury in transplanted lichens and measurements of mercury in air (Gosar et al. 1996). In the centre of the town the result obtained was quite high (321 ng Hg g^{-1} d.s.), which confirms the influence of the ventilation shaft located in the town centre. In the Pront area, where native mercury and cinnabar ore are present in the surface soil, we found 275 ng Hg g^{-1} of dry sample. We found that other ventilation shaft does not have a great influence. Determinations of mercury in carrot from this part of town (locations 9 and 10 in Fig. 2) did not show such high concentrations (78–163 ng Hg g^{-1} d.s.). The lowest mercury content (60 ng Hg g^{-1} d.s.) in carrot was found far away from the smelting plant and ventilating shafts. It is evident that there was no source of Hg, nearby.

We found beans at only six of the sampling locations. The concentrations of Hg in beans are 2 to 7 times lower than in carrot. The higest values were determined at the location close to the smelting plant (18 ng Hg g^{-1} d.s.) and Pront area (23 ng Hg g^{-1} d.s.). These are quite interesting results, because the concentration of Hg in carrot from these locations were very different. Probably the carrot reflects air pollution with Hg and the bean soil pollution. The Hg contents from other locations were in the interval between 2.0–6.6 ng Hg g^{-1} d.s. These results confirm the fact that carrot really can bioaccumulate and assimilate Hg during its growth. Probably it could distinguish between originating in air and that in soil.

4 Conclusion

If we compare our results obtained with those from the 1970s (Stegnar 1980), we can see that the smelting plant was really an important source of Hg pollution, and that the Pront area results reflect the fact that at this site Hg originates from the geological structure.

Generally, the Hg content in carrot from Idrija town are still very high. The Slovenian legislation allows on MAC of 100 ng Hg g^{-1} of dry sample and 20 ng Hg g^{-1} of fresh sample. It has been shown that only at two sampling locations were the concentrations under the MAC. In six locations the values obtained are higher than 200 ng Hg g^{-1} d.s.

From the results and measurements it is evident that much mercury is still present in the Idrija town environment in spite of the fact that most mercury production was ended in 1977. It is evident from trends found in this work that a longterm programme of environmental monitoring is required. Such a comprehensive survey should be carried out at intervals to follow declining trends in mercury levels, not only to safeguard man and the environment, but also to provide valuable data on the ecological cycle of mercury.

Acknowledgements The author is grateful to the Idrija Mercury Mine, which has provided the financial support for this research work and to all researchers who studied mercury or its influence in Idrija, both in the past and the present.

References

Brosset C, Lord E (1991) Mercury in precipitation and ambient air a new scenario, Water Air Soil Pollut 56:493–506

Byrne AR, Kosta L (1970) Studies on the distribution and uptake of mercury in the area of the mercury mine at Idrija, Slovenia, Yugoslavia. Vest Slov Kem Drus 17:5–11

Byrne AR, Kosta L (1974) Simultaneous neutron-activation determination of selenium and mercury in biological samples by volatilization. Talanta 21:1083–1090

Gosar M, Pirc S, Šajn R, Bidovec M, Mashyanov NR, Sholupov SE (1996) Distribution of mercury in the atmosphere over Idrija, Slovenia. Environ Geochem Health 19:101–110

Horvat M, Lupšina V, Pihlar B (1991) Determination of total mercury in coal fly ash by amalgamation cold vapour atomic absorption spectrometry. Anal Chim Acta 243:71–79

Iverfeldt Å (1991) Occurrence and turnover of atmospheric mercury over the nordic countries. Water Air Soil Pollut 56:251–265

Jensen S, Jernelov A (1969) Biological methylation of mercury in aquatic organisms. Nature 233:753

Jernelov A (1969) Conversion of mercury compounds. In: Chemical fallout, Chap 4. CC Thomas, Springfield, Illinois

Johnels AG, Westermark T (1969) Mercury contamination in the environment in Sweden. In: Chemical fallout, Chap 10. CC Thomas, Springfield, Illinois, pp 221–239

Kosta L, Byrne AR (1969) Activation analysis for mercury in biological samples at nanogram level. Talanta 16:1297–1303

Kosta L, Byrne AR, Zelenko V, Stegnar P, Dermelj M, Ravnik V (1974) Studies on the uptake, distribution and transformations of mercury in living organisms in the Idrija region and comparative areas. Vest Slov Kem Drus 21:49–74

Kosta L, Byrne AR, Zelenko V (1975) Correlation between selenium and mercury in persons exposed to inorganic mercury. Nature 254:238–239

Lindberg SE, Turner RR, Meyes TP, Taylor GE, Schroeder WH (1991) Atmospheric concentrations and deposition of Hg to a deciduous forest at walker branch watershed, Tennessee, USA. Water Air Soil Pollut 56:577–594

Lupšina V, Horvat M, Jeran Z, Stegnar P (1992) Investigation of mercury speciation in lichens. Analyst 117:673–675

Lupšina-Miklavčič V (1994) Determination of low level Hg concentrations in the air. In: Master's Thesis, Faculty of Chemistry, University of Ljubljana, Ljubljana (in Slovene)

Miklavčič V (1996) Mercury in the Idrija region. In: Meeting of researchers – Idrija as a natural and anthropogenic laboratory, mercury as a global pollutant, Abstracts, Idrija, 24–25 May 1996, pp 1–7

Mlakar I (1974) New aspects on the origin of the Idrija mercury ore deposit. In: 1st Int Mercury Cong, Abstracts, Barcelona, 6-10 May 1974, pp 157–159

Stegnar P (1980) Studies of the degree of environmental pollution with mercury in Idrija and its surrounding. Report on Environment contamination with mercury in Idrija and its surrounding, Jožef Stefan Institute, Ljubljana (in Slovene)

Valenta P, Nguyen VD, Nürnberg HW (1986) Acid and heavy metal pollution by wet deposition. Sci Total Environ 55:311–320

Zelenko V, Kosta L (1973) A new method for the isolation of methyl mercury from biological tissue and its determination at the parts-per-milliard by gas chromatography. Talanta 20:115–123

Evaluation of Internal Doses of Mercury at Intermittent Exposure to Elemental Mercury at the Mine in Idrija

A. Kobal, E. Nanut, M. Horvat, P. Stegnar, and A. Sabadin

Our experience suggested that the urine mercury concentration is not a valid indicator of individual internal doses received during intermittent type of exposure to elemental mercury. We decided to define the current internal dose of mercury through the analysis of total blood mercury. Thirty two miners were observed before and immediately after exposure to elemental mercury. Because the workers made use of respiratory protection, the residual external exposure was assessed with a personal sampler in 15 miners. Blood and urine samples of all examined miners were taken before and immediately after exposure for analysis of total mercury. Total mercury was determined in plasma, erythrocytes and also in the 18-h urine samples of 20 miners. The mercury concentration in all urine postshift samples was corrected to a urinary specific gravity of 1.024 and the postexposure 18-h urine samples of 20 workers was corrected to a urine volume of 1 ml/min. Mercury in the blood, urine and gold trap was determined by cold vapour atomic absorption spectrometry. A very good correlation was found between elemental mercury in inhaled air and the postexposure blood mercury concentration ($r = 0.96$, $P = 0.000$). A significant increase ($P < 0.001$) in blood mercury and urine mercury was established after exposure. The urine and blood mercury concentration after exposure was well correlated ($r = 0.68$, $P < 0.001$). No correlation was found between the blood mercury concentration before and after exposure ($r = 0.29$, $P > 0.05$). A significant increase ($P < 0.01$) in erythrocyte mercury and predominantly in plasma mercury was also established during exposure. The correlation between plasma and postshift urine mercury was higher ($r = 0.82$, $P = 0.000$) than between erythrocyte and postshift urine mercury ($r = 0.70$, $P = 0.002$), if the correction of urine mercury concentration to a urine volume of 1 ml/min was used. Our results confirm that blood mercury concentration is a valid and reliable indicator of current or recent individual exposure to elemental mercury at intermittent type of exposure. At this type and level of exposure, urine mercury in postshift samples, particularly if adjusted to a urine volume of 1 ml/min, reflect not only the accumulation of mercury from earlier exposures but partly also the current group level exposure.

1 Introduction

Our practical experience in assessing miners intermittently exposed to elemental mercury (Hg^{0}) in the Idrija Mercury Mine (Kobal 1991) and the experiences of other authors (Piotrowski et al. 1975; Nakaaki et al. 1975; WHO 1976) suggested that the urine mercury (U-Hg) concentration is not a valid indicator of individual internal dose received during exposure. Therefore, we decided to define the current internal dose of mercury (Hg) through the analysis of total blood mercury (B-Hg), erythrocyte mercury (E-Hg) and plasma mercury (P-Hg). We shall also study the relationship between B-Hg and U-Hg concentrations in postshift urine samples.

Environmental Science
Mercury Contaminated Sites (ed. by R. Ebinghaus et al.)

2
Study Design, Subject and Methods

Thirty two miners (16 mercury miners and 16 assistant miners) were observed before and immediately after exposure in the mine. Where native mercury was mined, the workers were exposed to Hg^{o} (Hg vapour and Hg liquid aerosols). Before our observations the miners and their assistants were not exposed to Hg^{o} for a period of 2 months (the half-life of U-Hg). The physical activity of the miners and their assistants was practically the same, and their loads varied from 5–9 Kcal per minute. During their work, all the miners regularly used air-purifying helmets with mercury-absorbing filters and anti-dust filters (AH6 Airstream Racal safety helmets with AS60502 filter). Because the workers made use of respiratory protection, the degree of current external exposure cannot be evaluated on the basis of Hg^{o} measurements in the working areas. In 15 miners the residual external exposure was assessed with a personal sampler positioned inside the helmet. Breathing zone air was continuously aspirated through a glass tube containing a gold trap where Hg^{o} was amalgamated and later analyzed by double amalgamation cold vapour atomic absorption spectrometry (CVAAS) (Horvat 1989). The Hg^{o} concentration in the air of working places was measured by means of the instant reading method (Mercury Vapour Indicator, MVI Shawcity, range 0.00–2.00 mg/m^3, sensitivity 1 µg/m^3, repeatability ±5%). Blood and urine samples of all examined miners were taken before and after exposure for analysis of total Hg. Blood postexposure samples were taken 1 hour postshift, urine postshift samples were collected in the first 2 h after the shift. Total Hg was determined in blood, plasma, erythrocytes and also in the 18-h urine samples of 20 miners. The 18-h urine samples included postshift urine (12:00–14:00 h), afternoon urine (14:00–22:00 h) and morning urine (10:00–18:00 h). The mercury concentration in all urine postshift samples was corrected to a urinary specific gravity (SG) of 1.024 [Barber and Wallis 1986; SG corrected concentration: $(C)_{Hg}$*(SGo-1)/(SG-1); $(C)_{Hg}$ = measured concentration, SG = 1024, SG = measured]. The mercury concentration in the postexposure 18-h urine samples of 20 workers was corrected to a urine volume (UV) of 1 ml/min (Araki et al. 1986; UV-corrected concentration: Ui*Vib, Ui = non-adjusted concentration, Vi = urinary flow rate (ml/min), b = the value specific for U-Hg equal to 1). U-Hg was analyzed on the day of sampling.

Mercury in the blood and urine was determined by reduction-amalgamation CVAAS and by neutron activation analysis (NAA). The limit of B-Hg detection by CVAAS was 0.1 ngHg/g in a 0.2 g sample, with a coefficient of variability (CV) from 3 to 10%. The limit of U-Hg detection was 0.1 ng/g in a 0.5 g sample, CV varied (depending on the concentration) from 5 to 10%. The limit of B-Hg detection by NAA was 0.1 ng/g, with CV dependent on the Hg concentration in the sample, generally about 5% (Kosta and Byrne 1969). The quality control program for analysis of total Hg included the certified reference materials, comparison with NAA results and the results of other laboratories. The statistical evaluation of data was performed using a paired t-test, Pearson's correlation coefficient and linear regression analysis.

3
Results

The Hg^0 concentration in the air of working areas in the pit was on average 0.22 ± 0.23 mg/m^3. The Hg^0 concentration in inhaled air, measured for 9.0 ± 5.5 consecutive days in 15 miners using a personal sampler, was on average 0.13 ± 0.15 mg/m^3. A very good correlation was found between Hg^0 in inhaled air and the B-Hg concentration ($N = 15$, $r = 0.96$, $P = 0.000$, $y_{(B\text{-}Hg)} = 349.77x^* + 9.72$; *the regression coefficient is significant at $P < 0.001$) during exposure in the mine (Fig. 1).

A significant increase ($P < 0.001$) in B-Hg and U-Hg was established during exposure (Table 1). No correlation was found between the B-Hg concentration before and after exposure ($r = 0.29$, $P > 0.05$). The U-Hg and B-Hg concentration after exposure (Fig. 2) were well correlated ($N = 30$, $r = 0.68$, $P < 0.001$, $y_{(U\text{-}Hg)} = 1.35x^* + 14.31$; $^*P < 0.001$). The standard error of U-Hg prediction was high ($se_y = \pm 7.7$). A significant increase ($P < 0.01$) in E-Hg and predominantly in P-Hg was also established after exposure. Immediately after exposure, 30% of Hg was found in the packed cell volume (PCV) and 70% in plasma (Fig. 3). The average ratios (ratio measured directly) of erythrocyte to plasma mercury in the same volume unit before and immediately after exposure are 0.92 and 0.72, respectively. The average values of mercury in 18-h, postshift urine samples do not differ significantly from those of the morning samples. However, the average values of U-Hg in postshift samples adjusted to UV of 1 ml/min are statistically significantly higher than the average values of U-Hg in 18-h urine samples (Fig. 4). The interindividual variability in the UV-adjusted U-Hg values of postshift samples is higher (CV = 99%) than in the values adjusted to

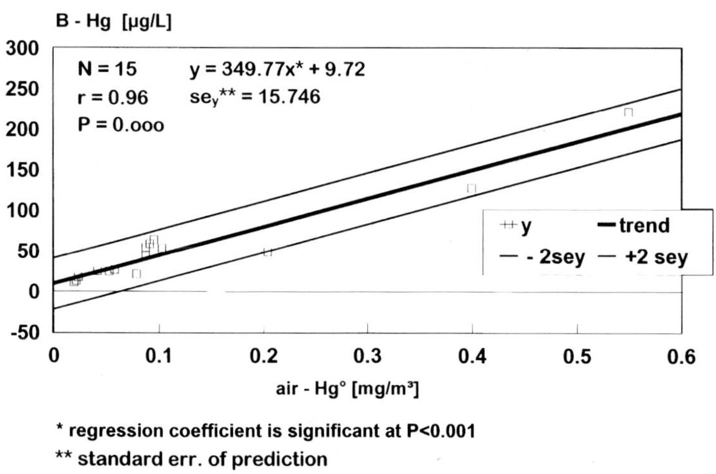

Fig. 1. Correlation between individual daily levels of mercury in inhaled air and those in blood samples taken 1 h after exposure

Table 1. *t*-test and correlation between B-Hg and U-Hg concentrations before and immediately after exposure

t-test[a]	B-Hg (μg/l) before-after	U-Hg (μg/l)[b] before-after
N	29	32
x	6.6–47.7	40.6–68.7
SD	2.0–45.2	26.3–51.7
P	0.000	0.000
Correlation		
N	29	32
r	0.29	0.68
P	>0.05	0.000

[a]Paired value
[b]Postshift samples corrected for SG: 1.024.

SG (CV = 73%). The correlation between P-Hg and postshift U-Hg was higher ($r = 0.82$, $P = 0.000$, $y = 1.357x^* + 23.79$; $^*P < 0.001$) than between E-Hg and postshift U-Hg ($r = 0.70$, $P = 0.002$, $y = 2.862x^* + 0.965$; $^*P < 0.001$), if the correction of U-Hg concentration to a UV of 1 ml/min was used (Fig. 5).

4 Discussion

The absence of any correlation whatsoever between the blood mercury concentration before and after exposure, at significant increase ($P < 0.001$) of

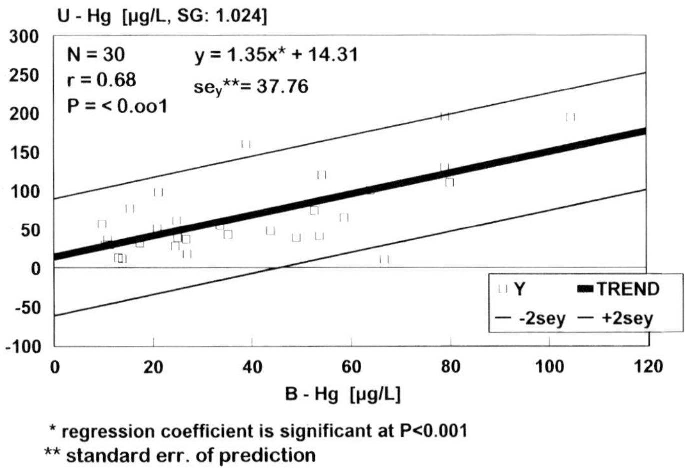

Fig. 2. Correlation between U-Hg and B-Hg concentration in samples after exposure period

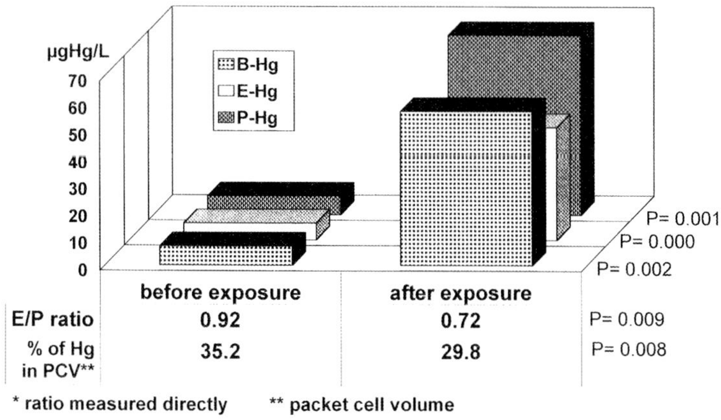

Fig. 3. *B-Hg*, E-, P-*Hg* and E/P *Hg* ratio in miners ($N = 18$) before and after exposure period

B-Hg during exposure, indicates that the blood mercury concentration depends solely on the current absorption of Hg^0. This is also in line with the results of our earlier studies of blood mercury kinetics (Kobal 1991), in which the Hg elimination from the blood was described by a two-component model and an exponential decrease: $B\text{-}Hg_{(T)} = 43.6\ e^{-0.117T} + 8.2\ e^{-0.0036T}$. The half-life of the fast part of excretion (up to day 14) is about 5 days and the slow part about 28 weeks (Fig. 6). Our findings correspond with the results published by Cherian et al. (1978) and some subsequent studies (Barregård et al. 1992; Sällsten et al. 1993).

The ratio of erythrocyte to plasma mercury is higher ($E/P = 0.92$) at lower B-Hg (7 ± 1 μg/l) than at higher B-Hg concentrations (57.0 ± 53.4 μg/l) directly after increased absorption of Hg^0 ($E/P = 0.72$). A relatively greater accumulation of mercury in plasma was observed during exposure. In the case of

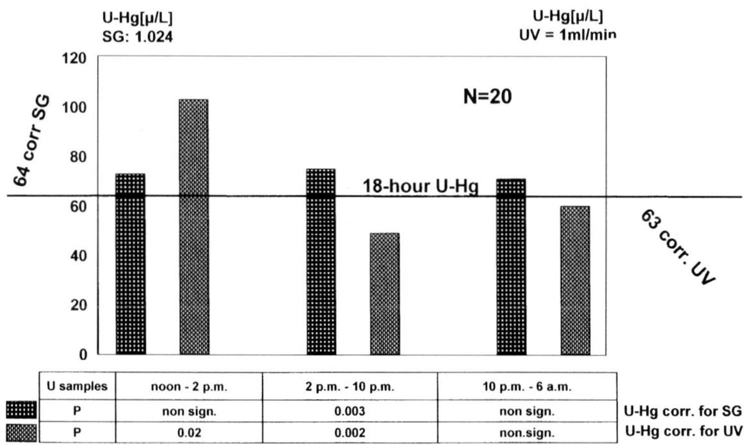

Fig. 4. Differences between 18-h U-Hg sample and postshift, afternoon, morning U-Hg samples corrected for SG (1024) and UV (1 ml/min)

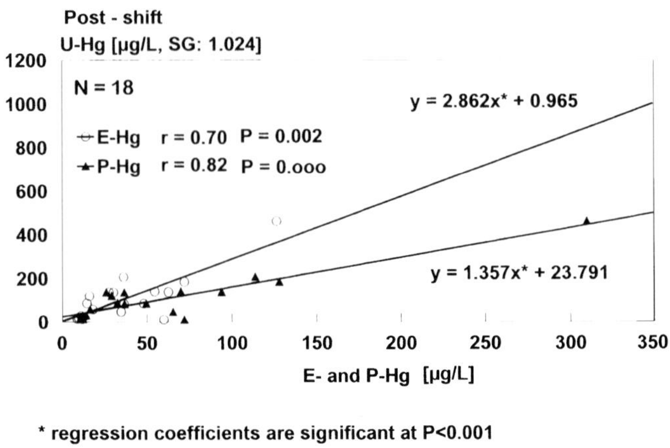

Fig. 5. Correlation between U-Hg in postshift samples corrected to UV (1 ml/min) and E-Hg, P-Hg at miners in the postexposure period

simultaneous occupational exposure to Hg^0 and exposure of methylmercury (Me-Hg) from nutrition, the distribution of inorganic mercury between erythrocytes and plasma is even more in favour of the latter (Suzuki et al. 1970). The results of our studies (Horvat and Kobal, unpubl.), conducted on 18 females who live in an environment not exposed to Hg^0 and do not eat seafood, showed that, at very low total blood mercury values (in a range of up to 2 ng/g), the average E/P ratios of Me-Hg and inorganic mercury were 5.3 and 0.91, respectively. Our results suggested that the distribution of mercury among erythrocytes and plasma is relatively uniform at lower inorganic mercury concentrations in blood; the relative increase in inorganic Hg in plasma after exposure to Hg^0 is evidently the result of increased absorption.

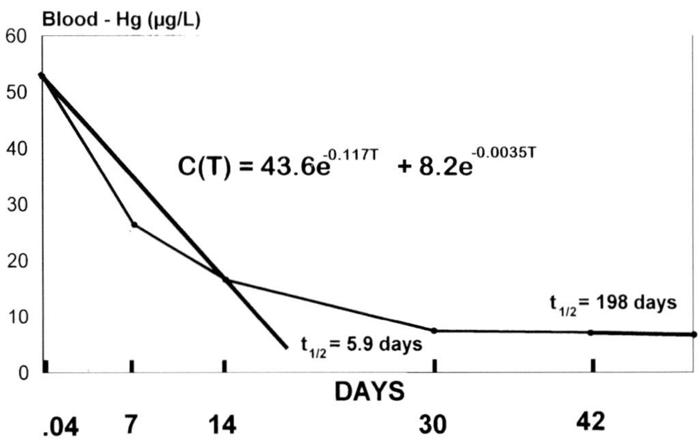

Fig. 6. Elimination kinetics of B-Hg after exposure

The correlation of Hg^0 content in inhaled air with blood mercury is exceptionally high and statistically significant ($r = 0.96$, $P = 0.000$). The regression analysis showed that the ratio of Hg^0 concentration in inhaled air to total blood mercury (μg/m^3:μg/l) at an air mercury level of 50 μg/m^3 and exposure of 4 h/day is 1:0.54, which partly corresponds to the ratio 1:0.49 obtained by Schäller and Triebig (1984), but not so well to the ratio 1:0.45 obtained by Roels et al. (1987) in workers who were long-term exposed to Hg^0 for about 8 h per day. On the basis of these relationships obtained in our study, it can be concluded that intermittent exposure (4 h per day) to a Hg^0 concentration in air of 50 μg Hg/m^3 results in a B-Hg concentration of 27 μg/l. At this level of external exposure the B-Hg concentration obtained from Roels et al. (1987) was 22.6 μg/l. Despite the shorter daily exposure of the miners observed, the B-Hg concentration is slightly higher, which is probably related to their higher pulmonary ventilation (about 27 l of air per minute) during their work in the mine.

The level of correlation of blood and urine mercury concentrations in miners prior to exposure is in accordance with the results of Lauwerys and Buchet (1973), who found the correlation at low blood mercury values (10 μg/l) to be low ($r = 0.36$, $P < 0.05$, $N = 40$). Our postexposure results are very similar to the findings of Schäller and Triebig (1984), who obtained a similar correlation between blood and urine mercury in workers immediately after exposure to Hg^0. Our results also correspond to those of Nakaaki et al. (1975), as the mercury concentration in urine increases significantly only during exposure to higher concentrations.

Based on individual values obtained under our investigative conditions, it was shown that the ratio between B-Hg (μg/l) and U-Hg (μg/l SG: 1.024) is 1:1.87 (for external exposure to a Hg^0 concentration of 50 μg/m^3 – indirectly evaluated data). On the basis of this ratio we can conclude that a B-Hg concentration of 27.21 μg/l results in a U-Hg concentration of 51 μg/l (SG: 1.024). Roels et al. (1987) reported a ratio between B-Hg (μg/l) and U-Hg (μg/g creat) of 1:2.7, which indicates higher values of U-Hg. The difference in the blood-urine mercury ratio between the two investigations is in our opinion mostly related to the "type of exposure". Roels et al. (1987) observed workers with long-term uninterrupted exposure; on the other hand, we observed workers with long-term, but interrupted and intermittent exposure. The ratio between air-Hg (μg/m^3) and U-Hg (μg/g creat) obtained by Roels et al. (1987) is 1:1.2. If we calculate our results (from the relationship between air-blood and blood-urine mercury) the ratio between air-Hg (μg/m^3) and U-Hg (μg/l SG: 1.024), is 1:1, which is expected for the intermittent type of exposure in our study.

The regression analysis of postshift U-Hg concentrations and B-Hg ($y = 1.35x^* + 14.3$; $^*P < 0.001$) showed a relatively high standard error of U-Hg prediction ($se_y = 37.7$) from the blood mercury concentration, which clearly indicates a high interindividual variation in Hg elimination in urine, depending not only on the current exposure, but also on the accumulation of mercury in the kidneys in earlier exposures.

The results obtained indicate that the urine samples collected immediately postshift with the values adjusted to UV (1 ml/min), which were very good correlated to plasma mercury, have a greater potential also for the estimation of current group level exposure. Our results suggested that the average mercury concentrations in the morning urine samples of the observed miners can replace the 18-h (or 24-h) collection of urine. These findings correspond with the circadian rhythmicity of the urinary mercury excretion established by Vokac et al. (1980). Our results are not entirely comparable with the findings of Roels et al. (1987), as our case evidently involves a different type of occupational exposure to Hg^0, and the U-Hg concentration in postshift samples was not corrected to urinary creatinine. Both creatinine and specific gravity adjustments of U-Hg concentration seem to be affected by urine volume (Araki et al. 1986), which is very important in the evaluation of U-Hg concentration in urine samples immediately after the hard work load.

5 Conclusion

Our results confirm that postexposure B-Hg concentration is a valid and reliable indicator of individual internal doses received during the current and recent intermittent type of exposure to various Hg^0 concentrations in air. Owing to the fast elimination of Hg from blood, more attention should be focused on the blood sampling time. Immediately after exposure, U-Hg seems to be predominantly dependent on the P-Hg level. At this type and level of exposure, U-Hg in postshift samples, particularly if adjusted to a UV of 1 ml/min, reflect not only the accumulation of mercury from earlier exposures but also the current group level exposure. In individual evaluations the variation of U-Hg excretion in postshift samples should be taken into account. The results of a Health Safety Programme for miners intermittently exposed to Hg^0 in the Idrija mine (Kobal 1994; Kobal and Dizdarević 1996) and applied over the last 25 years, confirm the results obtained in our study.

In further validation of the biological threshold limit value for mercury concentration in blood and urine in workers intermittently exposed to Hg^0 we propose to take into account also the ratio between air-blood and blood-urine mercury concentrations and the high interindividual variation in mercury elimination in urine obtained in our investigation.

References

Araki S, Aono H, Murata K (1986) Adjustment of urinary concentration to urinary volume in relation to erythrocytes and plasma concentrations: an evaluation of urinary heavy metals and organic substances. Arch Environ Health 41:171–177

Barber TE, Wallis G (1986) Correction of urinary mercury concentration by specific gravity, osmolality and creatinine. J Occup Med 28:354–359

Barregård L, Sällsten G, Schütz A, Attewell R, Skerfving S, Järvholm B (1992) Kinetics of mercury in blood and urine after brief occupational exposure. Arch Environ Health 47:176–184

Cherian MG, Hursh JB, Clarkson TW, Allen JA (1978) Radioactive mercury distribution in biological fluids and excretion in human subjects after inhalation of mercury vapour. Arch Environ Health 33:109–114

Horvat M (1989) Development and study of analytical methods for determination of low level mercury concentration and its application in analysis of biological and other environmental samples. Ph. D. Thesis University of Ljubljana, Slovenija

Hursh JB, Clarkson TW, Cherian MG, Vostal JV, Maille RV (1976) Clearance of mercury (Hg197, Hg203) vapor inhaled by human subjects. Arch Environ Health 31:302–309

Kellershohn C, Comar D, Poec C LE (1965) Determination of the mercury content of human blood by activation analysis. J Lab Clin Med 66:169–176

Kobal AB (1991) Occupational exposure to elemental mercury and its influence on mercury in blood, erythrocytes, plasma, exhaled breath and urine, and catalase activity in erythrocytes. Ph. D. Thesis University of Ljubljana, Slovenija

Kobal AB (1994) Quicksilber aus Idrija–historisch und aktuell–eine arbeitsmedizinische Betrachtung. Zentralbl Arbeitsmed 44:200–210

Kobal AB, Dizdarević T (1996) The health safety programme for workers exposed to elemental mercury at the mine in Idrija. 4th Int Conf on Mercury as a global pollutant, Congress Centre Hamburg, Germany, 4–8 Aug 1996, Abstracts p 16

Kosta L, Byrne AR (1969) Activation analysis of mercury in biological samples at nanogram levels. Talanta 16:1297–1303

Lauwerys R, Buchet JP (1973) Occupational exposure to mercury vapors and biological action. Arch Environ Health 27:65–68

Nakaaki K, Fukobari S, Tada (1975) An experimental study of inorganic mercury vapour exposure. J Sci Lab 12:705–716

Piotrowski J, Trojanowska B, Mozilnicka EM (1975) Excretion kinetics and variabilty of urinary mercury in workers exposed to mercury vapor. Int Arch Occup Environ Health 35:245–256

Roels H, Abdeladim S, Ceulemans E, Lauwerys R (1987) Relationships between the concentrations of mercury in air and in blood or urine in workers exposed to mercury vapour. Ann Occup Hyg 2:135–145

Sällsten G, Barregärd L, Schütz A (1993) Decrease in mercury concentration in blood after long-term exposure: a kinetic study of chloralkali workers. Br J Ind Med 50:814–821

Schäller KH, Triebig G (1984) Personenbezogene Probennahme von Quecksilberdämpfen am Arbeitsplatz. Arbeitsmed Sozialmed Präventivmed 19:289–292

Suzuki T, Miyama T, Katsunuma H (1970) Mercury contents in the red cells, plasma, urine and hair from workers exposed to mercury vapour. Ind Health 8:39–47

Vokac Z, Gundersen N, Magnus P, Jebens E, Bakka T (1980) Circadian rhythnicity of the urinary excretion of mercury, potassium and catecholamines in unconventional shift-work systems. Scand J Work Environ Health 6:188–196

World Health Organization (1976) Environmental health criteria 1:Mercury. WHO, Geneva

Mercury in Terrestrial Food Webs of the Idrija Mining Area

A. Gnamuš and M. Horvat

To evaluate the degree of contamination with mercury and its transfer in food webs in the active mercury mining area of Idrija, concentrations in soil profiles, plants and animal tissues were analyzed during the period from 1990 to 1996. Total mercury (T-Hg) and methylmercury concentrations (Me-Hg) were determined in body tissues and fur of roe deer, *Capreolus capreolus* L. ($n = 20$), their diet (composite plant samples of a minimum of 40 plant species), soil samples, and in tissues of their predators such as the wolf, *Canis lupus* L. ($n = 2$), and lynx, *Lynx lynx* L. ($n = 8$). The study was focused on the wider area of the active mercury mine in Idrija, the abandoned mercury mining area in Podljubelj, where mining and smelting ceased in 1900, and five non-polluted locations in Slovenia. Special attention was paid to proper sample collection, preparation and storage. The concentration levels of total and methylmercury in roe deer tissues are correlated with those in their composite vegetal diet, and soil samples from the polluted area. A correlation with the distance from the mercury source was observed (T-Hg: $r^2 < 0.62$; Me-Hg: $r^2 < 0.73$). Herbivore tissues which originate from the active mining area contain up to 350 times higher concentrations of mercury than controls. The tissues of old animals mostly contain high Hg concentrations compared to the corresponding tissues of young animals. However, in the Idrija area the relationship with the age of roe deer was less distinctive due to the influence of the particular spatial exposure gradient. Total mercury concentrations in roe deer organs descend as follows: kidney >> liver > fur > pineal gland >> suprarenal gland > spleen > masseter muscle > lung > pituitary gland > eye sections > sections of the central nervous system (CNS). A large portion of Hg (20 to 85%) in animal tissues from the contaminated area was present as methylmercury (except in the liver and kidney), despite the fact that inorganic Hg and elemental Hg are the primary sources of pollution in the area. The percentage of Me-Hg in composite plant samples and soil profiles was very small (0.14 to 2%). Our data indicate that in the terrestrial environment, roe deer tissues can be successfully used as an indicator of mercury contamination, transfer and accumulation processes. Furthermore, a positive correlation between mercury concentrations in certain roe deer and human tissues from residents of the mining area suggests that roe deer tissues may be successfully used as biomonitors and thus as an early warning of elevated mercury concentration levels in man. However, mercury levels in selected predator species do not follow locally enhanced Hg concentrations in their prey (roe deer) or in vegetation and soils from the polluted area. Since the predators selected occupy larger territories and their prey contains different mercury levels, they are considered less suitable for monitoring of locally enhanced Hg contamination.

1 Mercury in the Terrestrial Environment

Due to its high toxicity, persistence, biogeochemical cycling, widespread distribution throughout the global environment, accumulation and concentration in food chains, mercury is receiving a great deal of interest. Although Hg is a natural component of the Earth's crust, its concentrations in the biosphere have increased significantly during the last few decades due to anthropogenic

Environmental Science
Mercury Contaminated Sites (ed. by R. Ebinghaus et al.)

emissions. Moreover, approximately two-thirds of the total world's current yield of Hg has been produced in the past century (Andren and Nriagu 1979). In recent years, worldwide fossil-fuel combustion fluxes with global effect, and a few mines and smelters with very distinctive local effects produced the majority, i.e. about 60%, of the current annual Hg input to the atmosphere (Nriagu and Pacyna 1988; Baeyens et al. 1996). Ultimately, various transport and scavenging processes return the majority of airborne Hg to the terrestrial environment close to the pollution source (Lindberg et al. 1979; Baeyens et al. 1996). Deposition of the element in surface soils and its reemission induce elevated levels on and in plants, which subsequently results in accumulation of the metal in herbivorous and further in carnivorous animals or man.

Approximately two-thirds of the world's mercury resources are found in the Mediterranean basin area (Bargagli et al. 1986), Idrija (Slovenia) being one of the most notable (Fig. 1). In our study, a systematic sampling survey in the Idrija cinnabar (HgS) ore deposit area was designed in order to evaluate the distribution pattern and transformation rate of the principal Hg species (Hg^{0}, Hg^{2+} and Me-Hg), their mobility and availability to plants and other terrestrial organisms living in the area known as one of the largest natural Hg accumulations in the world. Although the study deals with the heavily polluted mining area, we believe that the basic principles from this study go beyond the particular geographical site and may be used for comparisons with any other site polluted with mercury. Naturally, each site of contamination has its own

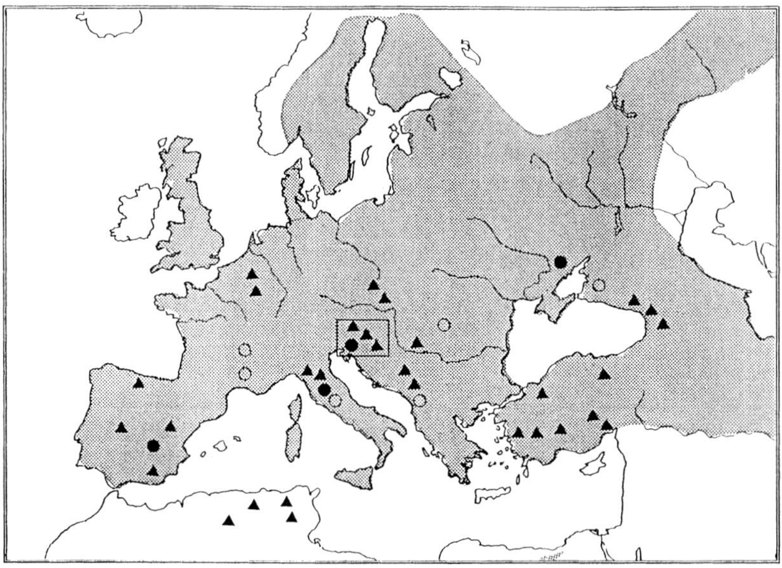

Fig. 1. Geographical distribution map (area in *square* enlarged in Fig. 2): area of the present distribution of roe deer (*Capreolus capreolus L.*) in Eurasia (After Kurt 1970). ●, ▲, ○ Locations of active and inactive mercury mines and deposits rich in Hg in Eurasia (courtesy of Mt. Amiata Mining Company, after UNEP/FAO/WHO 1987). ●, Very productive mines; ▲, previously productive mines; ○, presence of mercury

particularities. To reveal them and to enable comparisons with other heavily polluted sites in the world, some efforts were made to describe the soil types and the parameters controlling their metal capacity, the composition of parent materials, as well as vegetation types in the area surveyed.

2
Mercury in the Study Area

The Idrija cinnabar ore deposit with an estimated thickness of 800 m is sited in a Paleozoic and Trias fracture and covers an area of approx. 350 000 m^2, situated beneath the town of Idrija and its surroundings. The Idrija mine is the world's second largest mercury mine, with continuous mining, smelting and refining operations for over 500 years. So far, over 12.7 million tons of Hg-ore were excavated and the Idrija mercury mine contributed about 13% of the world Hg stock – over 108 000 tons of commercial mercury, with a total of mined Hg of over 145 000 t (courtesy Idrija Mining Company). In the last 5 centuries, over 37 000 t of Hg has been officially discharged to the environment due to the inefficient smelting technology; the actual figures of Hg emissions must have been much higher. Thus, the estimated average annual anthropogenic release into the air from the Idrija mining region would be approx. 70 t of Hg, which represents almost 2% of the annual amounts of Hg released on the global scale by human activities, according to recent literature estimations (Nriagu and Pacyna 1988; Fitzgerald 1995). The mine's average annual production in the third quarter of this century was between 400 and 500 t of commercial Hg/year, but was reduced in the last few years from 1986 to only about 50 t of Hg per year. In 1990 the Idrija mine produced only about 25 t/year, production has been reduced even more since and the mine is now being finally closed since the summer of 1996 (courtesy Idrija Mining Company). A geographical map of Slovenia showing the origin of the samples is presented in Fig. 2. Approximately 40 km west of Slovenian's capital of Ljubljana, Idrija is situated in a climatic transitional region between the Alps, Karst and Adriatic with a prevailing continental temperate climate and slight sub-Mediterranean and sub-montane influence. Geographically, Idrija is located in very heterogeneous terrain at an altitude of approx. 330 m in the relatively narrow valley of the river Idrijca, surrounded by hills of an average altitude between 600 and 900 m (see Fig. 3a). The main Hg deposits, mine and smelter are situated in the middle of the densely populated valley of the town of Idrija and it is, therefore, subjected to natural and strong anthropogenic Hg emissions. A semimillenium of mining resulted in a multilevel network of together over 700 km of longitudinal shafts beneath the town's surface, down to –38 m below sea level (courtesy Idrija Mining Company). The mine still represents 12% of the world's Hg reserves, or around 4 million tons of cinnabar ore containing 0.1 to 0.33% of Hg (Drovenik 1992). Idrija Hg-bearing minerals contain 0.1 to 86.2% of Hg, but the richest Idrija ore deposits with average Hg contents between 25 and 30% were excavated in the mine's early production in the 16th century. Ancient mineral paragenesis in the Idrija ore deposit area

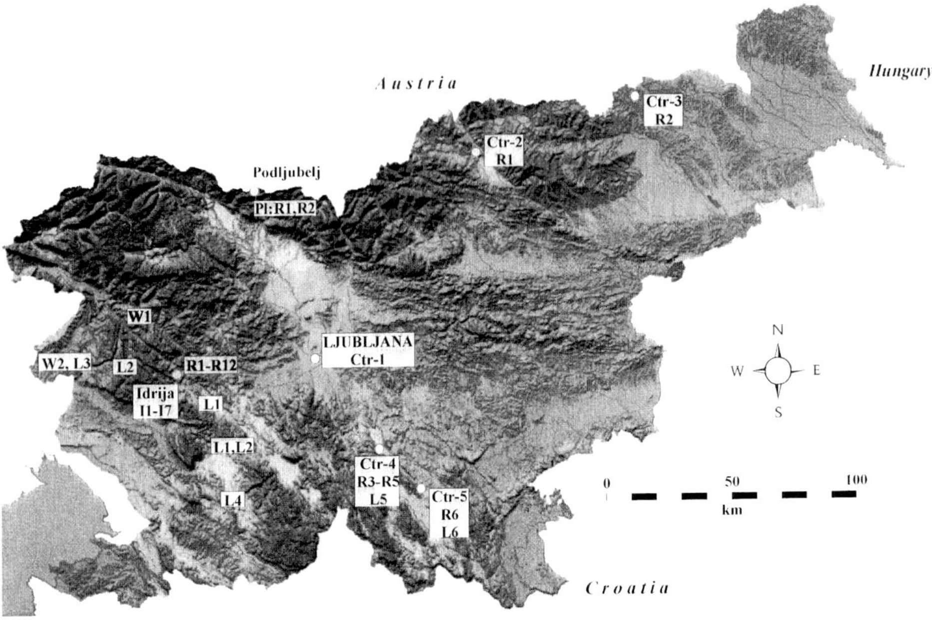

Fig. 2. Geographical relief map of Slovenia showing the origin of the samples. *I–1* to *I–7* Sampling areas in the district of the active Idrija mercury mine; *R1* to *R12*, *L1*, *L2*, *W1*: exposed animals – wider area of the Idrija mercury mine; *Pl*: sampling areas in the district of the long-abandoned mercury mine in Podljubelj: *R1*, *R2*: roe deer from the Podljubelj area; *Ctr–1* to *Ctr–5* control sampling areas: *R1* to *R6*: roe deer controls; *L1* to *L6* lynx controls; *W2* wolf control

resulted in the production of considerable quantities of native elemental Hg, which is a unique geological phenomenon.

The climatic transition and distinctive relief of the area result in a relatively high annual precipitation, mostly disposed through the spring and late autumn, and frequent mists. Average daily Hg concentrations in air in the town centre are 200 to 400 ng Hg/m^3 (range 10 to 16 000 ng Hg/m^3 Scholupov and Mashyanov, unpubl. data), and strongly depend on weather conditions (Lupšina-Miklavčič 1994). In the wider Idrija surroundings, average T-Hg air concentrations range between 25 and 75 ng Hg/m^3 (Scholupov and Mashyanov, unpubl. data). General wind conditions in the Idrija valley are very complex, due to the heterogeneous orography, causing air turbulence with winds frequently changing direction at approx. every 50 m in the town centre. Therefore, air turbulence results in circulation of lower air masses inside the valley, causing strong deposition of

Fig. 3a-b. Geographical relief of the Idrija region **a** Digital relief model of the Idrija area showing sampling polygons *I-Sm*, *I-1* to *I-7* and average T-Hg concentrations in soil samples collected in the period from 1990 to 1996, ranged in four clusters relating mercury concentrations **b** Digital relief model of the Idrija area showing sampling polygons *I-1* to *I-4* grouped into *Zone A*, characterized by considerably high Hg concentrations, and polygons *I-5* to *I-7* grouped into *Zone B*, characterized by moderately elevated Hg concentrations; *R1* to *R12*: Roe deer exposed

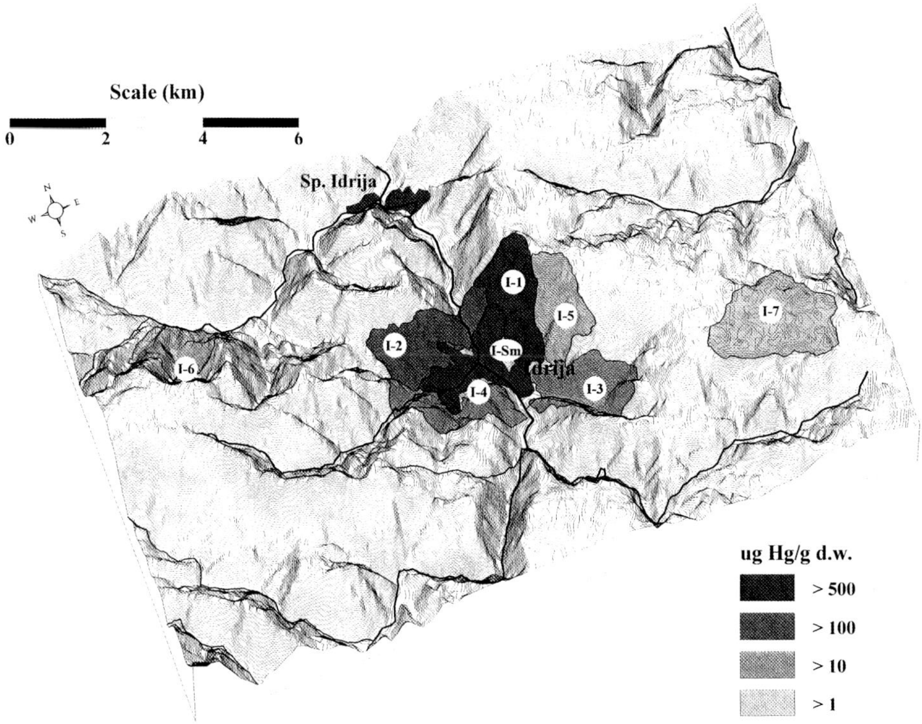
Scale (km)
0 2 4 6
N
W E
S
Sp. Idrija
I-1
I-5
I-7
I-2
I-Sm
Idrija
I-6
I-4
I-3
ug Hg/g d.w.
> 500
> 100
> 10
> 1

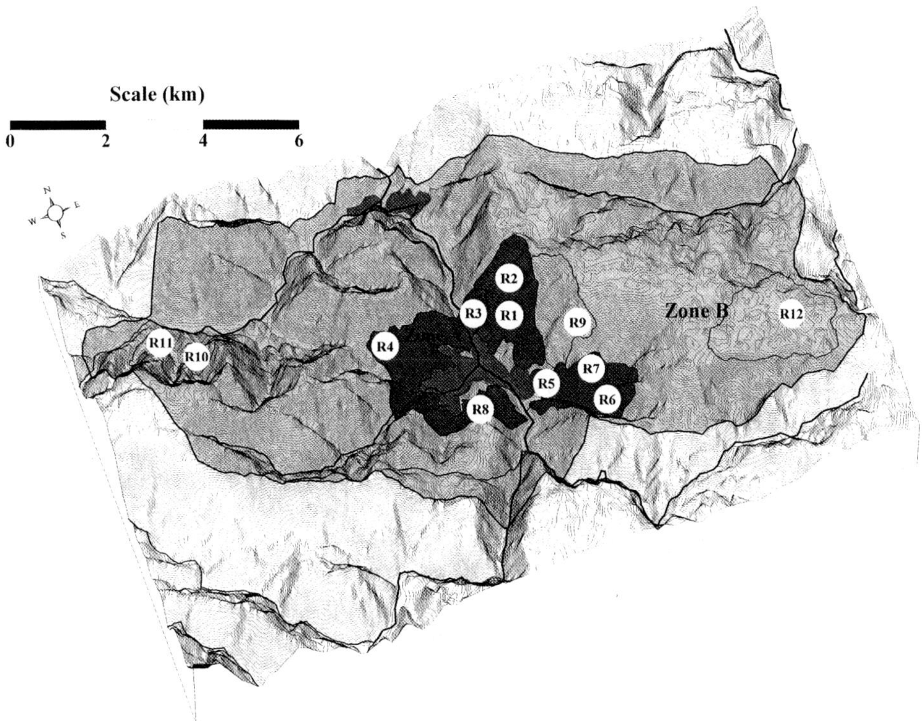
Scale (km)
0 2 4 6
N
W E
S
R2
R3
R1
R9
Zone B
R12
R4
R11
R10
R7
R5
R6
R8

airborne mercury inside the closed narrow valley area and the slopes of the surrounding hills. Otherwise moderate winds in the upper air layers of the Idrija area flow predominantly from the south-southeast, and also periodically from the north (Petkovšek 1980), causing the highest Hg deposition rates in the prevailing wind directions.

However, the Idrija Hg deposits with the mine-smelter complex represent one of the biggest point source contributions of Hg to the atmosphere. The area of the town and surrounding hills is densely covered with numerous tailing sites from the past. According to some estimates, an area of approx. 90 km^2 out of 425 km^2 of the Idrija community has been used for deposition of waste ignition residues from smelting. It is also subjected to severe contamination by airborne Hg from the mercury mine ventilation shafts, and periodically by flue gases from the smelting works. Since the mine with its ventilation shafts, smelting complex and neighbouring tailings sites, as well as the area especially rich in geological Hg deposits (i.e. Hg-bearing rocks) named Pront, are all positioned in a relatively closed valley separated by less than 1 km, the area may be regarded as a uniform point source of pollution. These characteristics and a high rate of the ground Hg contamination distinguish the whole area as a unique natural laboratory for the investigation of biogeotransformations and food-chain transfer of mercury in the terrestrial environment.

3 Experimental

3.1 Sample Collection and Sampling Procedures

Collection of samples in the Idrija district started in 1990 as part of a long-term complex study using environmental and biota samples to reveal the extent of contamination and transformations of Hg deposited in the area heavily affected by naturally and anthropogenically increased Hg concentrations. In this biogeochemical and ecological survey the wider mine and smelter area was taken into consideration. To evaluate contamination with mercury in the heavily polluted natural terrestrial environment, forest soil profiles, composite plant samples and tissues of animals that graze on selected plant species were selected. Since it is planned that the mine and smelter will be permanently closed within the next few years, the opportunity has been chosen to determine the consequences of a semi-millenium of continual mercury ore processing. One of the aims of our survey was to record the ultimate state as regards contamination of the surrounding environment with the objective of observing the following restoration processes in the affected area. In addition, our objective was to estimate the fate and significance of Hg accumulated in the soils based on the bed-rock parent characteristics and individual soil types, and to determine the factors that influence the transfer of Hg to plants. For comparison with the heavily polluted

Idrija area, similar samples were collected from the abandoned Hg mining area in Podljubelj and five control areas without mercury contamination (see Fig. 2). In the present chapter, sampling procedures are described, the existence of some biogeochemical mechanisms related to the soil-plant-animal Hg transfer under natural conditions of terrestrial environment are suggested and analytical data from the 6-year survey are summarized and discussed. Some aspects related to the mercury concentrations in vegetation and wildlife as subjects in the food-chain transfer and biotransformations of mercury in the terrestrial environment have been presented recently (Gnamuš et al. 1995; Gnamuš 1996). A thorough statistical evaluation, further discussion and final conclusions will be presented when other environmental quality data have been collected and analyzed.

3.1.1
Soil Samples

Nine equal-sized forest sampling sites of approx. 1.5 km^2 were selected (Fig. 2: I-1 to I-7, Pl, Ctr-1). Seven of them cover the Idrija district (Fig. 3a: I-1 to I-7). In order to reveal the extent of intensive anthropogenic contamination, soil samples were collected in the immediate vicinity of the active smelter complex of the Idrija Hg mine (Fig. 3a: I-Sm) and the abandoned smelter complex of the Podljubelj Hg mine (Pl-Sm). However, these two sampling polygons were much smaller than the remaining sampling sites.

Four polygons (I-1 to I-4) were near the Idrija town centre:

I-1 covers the northeastern suburbs of Idrija including the heavily contaminated area in the vicinity of the smelter complex,
I-2 covers the northwestern and northern suburbs,
I-3 covers the southeastern suburban forest edges near the town centre in the vicinity of the mine ventilation shafts, and
I-4 covers the southern and southwestern suburbs of Idrija.

The other three sampling polygons were selected in the wider Idrija area:

I-5 at approximately 2 km to the east-northeast,
I-6 at approx. 4 km to the northwest, and
I-7 at approx. 5 km to the southeast.

The control sampling site (Ctr-1) was located 40 km away in the forest near Ljubljana Zoo (see Fig. 2). In addition, corresponding samples were collected in the abandoned mercury mining area in Podljubelj (Fig. 2: Pl), where mining and smelting ceased in 1900. Although the production of the Podljubelj mercury mine was much smaller compared to the Idrija mine, the aim of collecting the samples in its vicinity was to reveal the extent of contamination approx. 100 years after cessation of production. A standard soil sampling procedure (Rump and Krist 1992) was used. The depth of forest soil important for Hg distribution studies is considered to be 0 to −20 cm (Padberg 1991; Padberg and Stoeppler 1991). Since the forest soils in the studied areas are rather shallow, only soil profile blocks of depth from the surface (0 cm) to approx. −10 or −15 cm were found. In order to obtain representative samples from each location, forest soils were collected from 25 randomly chosen sampling sites within each sampling area. Each subsample was collected with undisturbed stratification, and was packed and sealed in a polyethylene bag as a soil profile block of different depth depending on the soil type. In the laboratory, soil samples were coarsely subdivided into two horizons regarding the qualitatively estimated content of organic matter. The topsoil horizon was classified as the humus layer including litter (A-horizon and O-horizon) with average depth from the surface (0 cm) to approx. −5 cm (alterations in depth were approx. 2 cm in various soil types or subsamples), and the bottom soil horizon was classified as the deeper soil layer from a depth of approx. −5 to −15 cm (alterations in depth were approx. 5 cm in various soil types or subsamples). Composite samples were prepared in the laboratory. Due to the expected high concentrations of Hg in soil, samples were prepared in a separate laboratory, away from that used for preparation of plant and animal samples. Samples were air-dried for 4 days at approximately 20 °C. Bigger particles, e.g. stones, were removed, and then the samples were homogenized and ground in an agate mortar. During the whole sample preparation procedure, special precautions were necessary to avoid cross-contamination of samples. Therefore, homogenizing was started with control samples, followed by less contaminated and continued with the more contaminated samples. The homogenizer was thoroughly cleaned after each sample. For each sample, the first aliquot ground was discarded, and then the bulk of the sample was ground. Homogenized samples were stored in plastic containers, sealed in polyethylene bags, and stored in a refrigerator until further processing. An aliquot taken for determination of moisture content was dried at 85 °C to constant weight. All data are expressed on a dry weight basis.

3.1.2
Plant Samples Corresponding to Roe Deer Diet

Composite plant samples corresponding to roe deer diet consisted of a minimum of 40 species of herbaceous plants (whole plants) and small twigs with foliage of woody plants, with the approximate classification of composition and percentages shown in Table 1. The most abundant and preferred plant species included in the composite plant samples from the Idrija district are listed in Table 2. Since the number of species comprised in roe deer diet represent a very large part of

Table 1. Approximate composition of average composite plant samples representing roe deer diet in the investigated areas

Classification of vegetation types included in roe deer diet		Summer diet (%)		Autumn Diet (%)	
Woody plants	Deciduous plants		30		35–30
	Chamaephytes and climbers	~50	20	~65	25
	Conifers		0		5–10
Herbs		~50		~35	

Table 2. List of main plant species included in the composite plant samples representing roe deer diet collected through a number of sampling exercises in the period from 1990 to 1996 with range of percentages of each particular species in whole sample

Main species included in composite plant samples of roe deer diet	Summer diet range (%)	Autumn diet range (%)
Acer campestre + *A. pseudoplatanus*	~5–10	~7–15
Salix caprea + *Salix* spp.	~2–8	~3–12
Quercus sessiliflora	~4–5	~5–8
Carpinus betulus	~4–6	~6–10
Populus tremula	~3–7	~4–9
Fagus sylvatica	~4–6	~5–9
Sorbus aria + *S. aucuparia* + *S. torminalis*	~2–10	~3–14
Ulmus glabra	~2–4	~3–5
Corylus avellana	~1–2	~3–5
Other: *Pyrus, Prunus, Rosa, Rhamnus, Tilia*, etc.	~5	~5
Rubus idaeus + *R. saxatilis agg.*	~7–18	~10–25
Vaccinium myrtillus	~1–2	~2–5
Picea abies + *Abies alba*	0	~5–10
Fragaria vesca	~1–7	~35
Oxalis acetosella	~2–8	~2–6
Trifolium spp.	~1–14	~1–11
Gentiana asclepiadea	~4–6	~1–3
Mercurialis perennis	~2–6	~1–4
Cyclamen purpurascens	~1–5	~1–4
Omphalodes verna	~1–2	~1–2
Salvia glutinosa	~6–8	~2–3
Other: *Potentilla, Euphorbia, Mellitis, Viccia*, etc.	~10	~4–7

the vegetation in the surveyed areas, composite plant samples may be regarded also as vegetation samples. The samples were collected in the same sampling areas as the soil profile samples within an area of approximately 1.5 km^2 at each location: I-1 to I-7, Ctr-1 and Pl (see Fig. 2). In addition, corresponding composite plant samples were collected in the smaller polygons of the immediate vicinity of the active smelter complex of the Idrija Hg mine (Fig. 3a: I-Sm) and the abandoned smelter complex of the Podljubelj Hg mine (Pl-Sm).

Vegetation samples were collected twice a year, in summer and late autumn in the period from 1990 to 1996, except at the polygons Pl and Pl-Sm, where the sampling collection was performed since the autumn 1995. Samples collected in summer covered the climax of foliage growth and the estimated maximum of aerial contamination by airborne Hg species (highest dry deposition). Autumn samples were collected at the end of the vegetation season, when wet deposition is highest due to the high autumn precipitation. Final results for the mercury content of each year's vegetation samples are expressed as an average of the summer and autumn sampling exercises, and the mercury contents in vegetation from different years were compared. Composite plant samples of approx. 2 to 3 kg (fresh weight) were collected in large sackcloth bags inside larger polyethylene bags to avoid conditions favourable for possible biotransformation of mercury during sampling. All vegetation samples were collected on the same day, so that the influence of changing weather conditions or other factors affecting dry and wet deposition of mercury between sampling at different locations was minimized. The sampling days were chosen randomly and were within the same week every season and every year, provided that 5 or 6 days before there was no larger rainfall. At all sampling sites, as well as herbs, only small twigs and leaves of woody plants positioned in the 0 to 1.8 m sphere were collected. The samples were transported to the analytical laboratory within the same day of sample collection. Each composite plant sample was immediately sorted by species, and fresh weight of each species was determined. To match the percentage of individual plant species listed in Table 2, the new composite sample with a defined percentage of individual species was composed. The samples were not washed or otherwise treated, since the objective was to determine the whole budget of plant Hg contamination and to estimate the mercury intake of herbivorous animals as they consume plants directly – without pretreatment. After weighing, plants were cut into smaller segments. Composite plant samples were then lyophilized in a Christ Alpha 1-4 freeze drier at −50 °C and approx. 1.5 Pa for 3 days, starting from the controls to the most contaminated. After lyophilization, samples were weighed again to determine the wet/dry ratio and ground in a Brabender mill. During the whole sample preparation procedure, special precautions were necessary to avoid cross-contamination of samples. Powdered samples were subsequently transferred into paper bags, packed into plastic dishes sealed inside polyethylene bags and stored in a refrigerator until further processing. Moisture content was determined on a separate subaliquot (not the one taken for analysis) by drying at 85 °C for 24 h. All results regarding plant material are expressed on a dry weight basis.

3.1.3
Animal Samples

Twenty roe deer (*Capreolus capreolus* L.) of different age and sex were sampled at random during the hunting seasons 1990 to 1996 (Table 3). Twelve roe deer originated from localities I-1 to I-7 (Fig. 3b: R1–R12) around Idrija, whereas control roe deer R1–R6 were collected at four non-polluted localities Ctr-2 to Ctr-5 elsewhere in Slovenia, and the other two roe deer were collected in the abandoned Podljubelj mining area (Pl:R1, R2; see Fig. 2). Also tissue samples of the predator species lynx (*Lynx lynx* L.; $n = 8$) and of two wolves (*Canis lupus* L.) were collected from the southern to northwestern parts of Slovenia (L1–L6, W2), some of them relatively close to the contaminated Idrija area (L1, L2, W1). Tissue samples were carefully removed from the body and stored in separate polyethylene bags less than 20 min after the animal died. In the field samples were labelled, placed into polyethylene bags and temporarily stored at +4 °C in a refrigerator, transferred to the laboratory and deep frozen (−20 °C). Before further processing, brain, eyes, pituitary, pineal gland, suprarenal gland, masseter muscle, lungs, liver, spleen, kidney and fur were dissected. Twenty-five different roe deer tissues were selected to establish Hg distribution and accumulation in the body:

Table 3. Roe deer identification numbers and area codes of their origins (I-1 to I-7, Pl, Ctr-2 to Ctr 5), sex and age structure of animals used for Hg analyses, and description of their sites of origin

Roe deer (R)/ Polygon code	Sex	Age	Origin – description
Idrija – active mercury mining area			
Zone A			
R 1 / I-1	♀	5–6 years	Idrija – near current Hg smelter chimney
R 2 / I-1	♀	3–4 years	Idrija – near current Hg smelter chimney
R 3 / I-2	♀	3–4 years	Idrija-N (Prejnuta) – near abandoned Hg smelter
R 4 / I-2	♀	1 year	Idrija-N (Češnjice) – W of abandoned Hg smelter
R 5 / I-3	♀(infant)	8 months	Idrija-S (Cegovnica) – near mine ventilation shaft
R 6 / I-3	♂	1 year	Idrija-S (Ljubevč) – 0.5 km E of mine ventilation shaft
R 7 / I-3	♀	5 years	Idrija-S (Ljubevč) – 0.5 km E of mine ventilation shaft
R 8 / I-4	♂	4–5 years	Idrija-S (under Polanc) – 1 km S of Hg smelter chimney
Zone B			
R 9 / I-5	♀(infant)	6 months	Gore above Idrija – 2 km E of Hg smelter chimney
R 10 / I-6	♀	3–4 years	Sr. Kanomlja – 4 km NW of Idrija
R 11 / I-6	♀(infant)	8 months	Sr. Kanomlja (Brdo) – 4 km NW of Idrija
R 12 / I-7	♀	6–8years	Dole above Idrija – 5 km SE of Idrija
Podljubelj – abandoned mercury mining area			
R 1 / Pl	♀	10–12 years	Podljubelj – 2 km N of the abandoned Hg mine
R 2 / Pl	♀	8–10 years	Podljubelj – 3 km N of the abandoned Hg mine
Controls – four different locations in Slovenia (Ctr-2 to Ctr-5)			
R 1 / Ctr-2	♀(infant)	5 months	N Slovenia: Slovenj Gradec (NE), Kremžar vrh
R 2 / Ctr-3	♀	1 year	NE Slovenia: Zg. Kungota – 10 km N of Maribor
R 3 / Ctr-4	♀(infant)	8 months	S Slovenia: Ribnica (N), Mala gora-Vel. Poljane
R 4 / Ctr-4	♀	2 years	S Slovenia: Ribnica (N), Mala gora-Vel. Poljane
R 5 / Ctr-4	♀	10 years	S Slovenia: Ribnica (N), Mala gora-Vel. Poljane
R 6 / Ctr-5	♀	8–10 years	S Slovenia: Kočevje (NE), Stari log – Šenberg

1. Brain (encephalon): cortex cerebelli, nucleus dentatus, temporal cortex cerebri, nucleus ruber, nucleus amigdaloides, hypocampus, hypothalamus, thalamus and medulla oblongata.
2. Eye: cornea and iris, lens crystallina, corpus vitreum, chorioidea and sclera, retina and eye muscles.
3. Endocrine glands – pituitary (hypophysis), pineal gland (epiphysis) and suprarenal gland (glandula suprarenalis).
4. Chewing muscle (Musculus masseter).
5. Heart (Myocardium).
6. Lungs.
7. Liver.
8. Spleen.
9. Kidney.
10. Fur from between the ears and eyes.
11. Jaws were dissected for subsequent age determination by comparison of the dental condition of our samples with a jaw sample collection.

Results on six to ten roe deer tissues are selected for presentation in this study: temporal brain cortex, pituitary, masseter muscle, liver, kidney and fur; in some figures lungs, spleen, myocardium and fat tissue concentrations are added. From the predator species the corresponding tissues were collected for comparison, except for the brain tissues, endocrine glands and fur, which were not available.

Soft tissues were macerated on sterilized glass plates using a clean scalpel. A homogeneous mass was obtained and aliquots of the mass sampled for total and Me-Hg determination. Fur samples were cut into small pieces with a pair of scissors. Laboratory tools and ware needed for preparation of samples were washed with an aqueous detergent solution, steamed in hot nitric acid vapour, rinsed with double distilled Hg-free water and dried. To avoid changes in the tissue water content the processing of the homogenized animal tissues were carried out immediately after maceration.

3.2 Determination of Total Mercury (T-Hg)

3.2.1 *Soil Samples*

T-Hg in soil profile samples was determined by cold vapour atomic fluorescence spectrophotometry – CV AFS. An aliquot of homogenized soil profile sample (100 to 300 mg) was digested with 10 ml of an acid mixture (HNO_3:H_2SO_4, 7:3 v/v) in a closed Teflon digestion vessel (~30 ml) at 70 °C for 4 h. After the digestion, 1 ml of strong oxidizing solution was added (1 ml of BrCl) and the sample was diluted to 25 ml with MilliQ water. An aliquot of the digested sample was then transferred to the reaction vessel containing 20–30 ml of MilliQ water with 5 ml of HCl and 0.5 ml of $SnCl_2$ to reduce Hg^{2+} to Hg^0. The Hg vapour was then swept from the solution with nitrogen at 450 ml min^{-1} through a soda lime trap onto gold-coated silica sand. The mercury collected on the trap was then released into the CV AFS analyzer (Brooks Rand: Model-2) by heating (600 °C) the trap for 2 min. Peak area integration was used for quantification of Hg (Liang and Bloom 1993). Some of the soil profile T-Hg contents were also measured by neutron activation analysis, NAA (Byrne and Kosta 1974).

3.2.2 *Plant and Animal Samples*

An aliquot of homogenized tissue samples, 150 to 300 mg (dry weight for plant tissues and wet weight for animal tissues) was carefully weighed and introduced into a 15-ml Pyrex ampoule and 2 ml of 70% HNO_3 (Merck-MOS Selectipur) added. Ampoules were sealed and heated at 90 °C for 24 h. The mineralized samples were diluted with double distilled water (DDW) and the total mercury (T-Hg) content in aliquots was determined by gold amalgamation cold vapour atomic absorption spectrometry (CV AAS; Horvat et al. 1986) using an LCD Milton Roy Mercury Detector (Model 1255) and computer-guided reduction/amalgamation software.

3.3 Determination of Monomethylmercury Compounds (Me-Hg)

3.3.1 *Soil Samples*

The technique for determination of monomethylmercury compounds in soil samples is described in detail elsewhere (Horvat et al. 1988, 1993; Liang et al. 1994a,b). It consists of water vapour distillation, followed by aqueous phase ethylation, precollection on a Carbotrap or Tenax at room temperature and separation of ethylated species on a GC column (a Chromosorb W-AW-DMCS packed column, 40 cm long, i.d. 2 mm, operated isothermally at 110 °C). Separated mercury species were then transformed into elemental Hg by heating at 900 °C and swept into an atomic fluorescence detector (CV AFS, Brooks Rand: Model-2). Peak heights were used for calculations.

3.3.2 *Plant and Animal Samples*

A detailed description and a flow-chart of the quantitative method for determination of monomethylmercury compounds in biota samples by CV AAS is given elsewhere (May et al. 1987). The method consists of HCl leaching, followed by ion exchange separation (Dowex 1X8, 100–200 mesh) of organic and inorganic Hg, and conversion of organic Hg into inorganic by UV irradiation. An aliquot of the sample was then transferred to the reduction cell and mercury quantitatively determined by gold amalgamation CV AAS (LDC Milton Roy Detector). In principle, this method allows only the quantification of total organic mercury compounds. However, through the comparison of different separation and detection techniques, it is known that when analyzing biota samples the total organic mercury concentrations correspond to the concentration of monomethylmercury only. In addition, the accuracy of the results was verified by analyzing certified reference materials (CRMs).

3.4 Analytical Quality Control

Quality control of the results for total and monomethylmercury analyses in samples was performed by analysis of certified reference materials, CRMs. Relative standard deviations of replicates in all samples analyzed was from 3 to 8% of the mean. All samples were analyzed in duplicate (or triplicate), and three blank samples were run with each batch of analyses. The CRMs were also analyzed in duplicate together with each batch of samples, and data quality control charts were systematically recorded.

3.4.1 *Soil Samples*

For verifying the accuracy of total and monomethylmercury measurements in soil samples, an IAEA reference material IAEA-356, trace elements and monomethylmercury in polluted marine sediment, was used. This is at present the most suitable reference material of composition similar to soil that is also certified for monomethylmercury (Horvat et al. 1994).

3.4.2 *Plant and Animal Samples*

Analytical quality control of the results for total and monomethylmercury in plant and animal samples was performed by analysis of CRMs obtained from the National Research Council of Canada (NRCC): DORM-1, dogfish muscle; DOLT-1, dogfish liver; and TORT-1, lobster hepatopancreas. The same CRMs were used to verify the accuracies of total and monomethylmercury determinations in plant and animal samples. Although these CRMs are marine animal matrices, they are at present the only available reference materials of biotic origin with certified values for monomethylmercury that are comparable to the values found in the samples from the area heavily polluted by mercury.

4 Results and Discussion

4.1 Soil Samples and Composite Plant Samples Corresponding to Roe Deer Diet in the Idrija Area

The area surrounding Idrija is mostly covered with temperate continental deciduous climax forestal vegetation – the main tree species is beech forming various communities (lowland submontane *Hacquetio-Fagetum*, thermophilic *Ostryo-Fagetum* and the mesophilic *Omphalodo-Fagetum* or *Abieti-Fagetum Dinaricum*), while at higher altitudes the area is covered with montane beech forest (*Lamio orvale-Fagetum*). The soils of the surveyed area developed mainly on carbonate rocks and are rather shallow. In the higher parts and slopes rendzina is the predominate soil type, but it is supplemented with deposits of brown carbonate soils on limestone and dolomite, and eutric cambisols on different carbonate and noncarbonate rocks (conglomerate, brecchia and schists). To a smaller extent, dystric cambisols and dystric ranker on the Permo-Carboniferous schists and grits appear in locations east of Idrija, where the vegetation also changes to acidic fir forest (*Galio rotundifolii-Abietetum*).

During the year the composition of plant species in roe deer stomach varies (Klötzli 1965; Adamič 1990). Also, the roe deer's energy consumption changes considerably during the year. The energy consumption is the highest during the summer-autumn period, is reduced drastically due to the lack of food in winter and increases very little until late spring. Therefore, the summer composite vegetation samples, collected to reflect roe deer food at that time, contained a higher percentage of herbs and twigs of deciduous trees and a small percentage of conifer twigs compared to the late autumn composite vegetation samples (Tables 1 and 2). The mercury contents in vegetation samples from different years were compared. Plant species in roe deer diet were determined using analysis of the stomach content, and hypothetical browsing preferences were determined for the area around Idrija (Adamič 1990).

No significant differences of total and methyl mercury were found on comparing air dried (clean room, 4 days, room temperature) and freeze-dried plant samples. This observation is not in agreement with the observations of La Fleur (1973), who claimed that lyophilization of certain biological samples without sulphydryl groups may result in decomposition of up to 50% of Me-Hg to inorganic mercury. In order to avoid possible contamination during air drying, freeze drying was chosen.

The highest vegetation mercury concentrations occur in the heavily contaminated location of the smelter complex (I-Sm), where the true extent of the heavy anthropogenic close-range contamination in the Idrija area is revealed. In this location Hg values of 45 to 65 µg/g d.w. in the composite plant samples and up to 900 µg/g d.w. in soils were found (see Fig. 3a). The extent of the heavy anthropogenic close-range contamination is even more evident from the mercury concentrations in soil profiles and grassland vegetation from the hill in the immediate vicinity of the smelter chimney outlet, where T-Hg concentrations reach 2500 to 3000 µg/g d.w. (Gnamuš et al. 1996). Me-Hg concentrations in soil and grassland vegetation from the top of the hill in the immediate vicinity of the smelter chimney outlet reach very high values of 400 to 600 ng/g d.w. in soil samples, and about 300 to 400 ng/g d.w. in the grassland vegetation (Gnamuš et al. 1996).

4.2 Roe Deer as a Bioindicator of Mercury Contamination in the Terrestrial Environment

Roe deer have a number of advantages, such as their defined food chain, high abundance and wide distribution in Europe and Eurasia (see Fig. 1, shaded area). Each animal ranges over a restricted area and their distinctive territoriality is expressed over the majority of their adult life. Their tissues are relatively easy to obtain due to overpopulation caused by anthropogenic elimination of predators such as the wolf and lynx in most European forests. Therefore, the population density of roe deer in the monitored area is large enough that culling of a few animals per year does not threaten the species with extinction (Gnamuš et al. 1991). Their long life-time enables us to obtain chronological information about the concentration levels, behaviour and chemical species of mercury in naturally exposed animals. As they are vigorous feeders, the uptake of the element from food is of importance and high accumulation of contaminants in their tissues is found (Borg et al. 1969; Hoellerer and Coduro 1977; Krynski et al. 1982; Lutz 1985; Frøslie et al. 1986; Babinska-Werka and Czarnowska 1988; Holm 1993). Reviewing the data obtainable for terrestrial wild mammals (Wren 1986), relatively high concentrations of mercury were established in roe deer tissues in various studies, despite the position of roe deer at the bottom of the terrestrial food webs as herbivores. Additionally, it is possible to study certain processes in vivo, such as possible biomethylation in the gastrointestinal tract. Another very important feature is that these animals have certain characteristics close to those of humans – both are placental mammals with a similar life range. Roe deer often live very close to human residences, especially in small towns, closely surrounded by woodlands and grasslands, such as Idrija. Our results from the wider Idrija region and their comparison with former data from the same areas (Byrne et al. 1971; Kosta et al. 1972, 1974a) confirm the feature that long-term anthropogenic intervention in Idrija mostly prevents discrimination between the natural and anthropogenic contributions of Hg in the surrounding environment. As concerns animal tissue Hg concentrations, the predominant role of anthropogenic pollution is suggested, although no pre-anthropogenic environmental data from the area are of course available for comparison. In this respect, comparison with the data measured in corresponding samples from the Podljubelj area (see Fig. 2), where Hg mining was abandoned approx. a century ago, is instructive.

4.3
Do Biota Reflect a Mercury Exposure Gradient Related to the Distance from the Source of Contamination?

A summary of T-Hg concentrations in composite soil profiles from the polygons I-Sm and I-1 to I-7 in the Idrija area is presented graphically in Fig. 3a. The original results on soil T-Hg content in the Idrija area were first clustered in the four concentration ranges, following airborne contamination from the smelter chimney and geological Hg deposits (see Fig. 3a). However, biota mercury concentrations in the Idrija area do not follow such distinctive geological spatial contamination pattern. Especially in roe deer tissues from the polygons I-1 to I-4 in the wider smelter/mine area, i.e. from forestal town suburbs of Idrija, mercury concentration differences as concerns the distance from the source of pollution were not so distinctive. Therefore, in order to attain a better insight into the complex features of mercury contamination in the affected area, the results obtained from individual animals, as well as from composite plant and soil samples were grouped into three differentiated contamination zones. Thus, related to the distance from the smelter as the primary source of mercury contamination in the active mining area, three contamination clusters were formed (see Fig. 3b).

As shown in Fig. 3a,b, the confined small area in the immediate vicinity of the smelter with highest Hg values was identified as the smelter zone (identification code: I-Sm). The polygons I-1 to I-4 lie in the wider smelter/mine area and were grouped into Zone A, which is characterized by considerably high Hg concentration levels in soil profiles and biota samples. The Hg concentrations in samples obtained from the areas I-5, I-6 and I-7 were grouped into Zone B characterized by moderately elevated Hg concentrations. The same principles of grouping were employed for all types of samples from the investigated areas.

A summary of the original analytical results as T-Hg and Me-Hg concentration ranges in various samples grouped into contamination clusters is presented in Table 4. The actual values regarding composite plant samples and roe deer tissues in the clusters from the areas investigated are shown in Tables 5 and 6, respectively.

Except for the elevated levels in soil samples, Hg concentrations in composite plant samples and roe deer tissues from the abandoned Podljubelj mining area (identification code: Pl) are comparable to the controls. Therefore, values from the Podljubelj area are not presented separately.

T-Hg and Me-Hg concentrations in biota samples (selected roe deer tissues and the composite plant samples representing their diet) as a function of the distance from the Idrija mercury mine/smelter complex are presented graphically in Fig. 4a,b. Evidently, the concentrations of mercury in all samples analyzed decrease with the distance from the Idrija mine/smelter complex. As the concentrations of mercury from the heavily contaminated sampling sites are

Table 4. Summary of the concentration ranges of T-Hg and Me-Hg in soil profiles and biota samples, collected in the period from 1990 to 1996

Sampling area	Sample	N	Total Hg[a] (μg/g) Range	Me-Hg[a] (μg/g) Range	Me-Hg (%) Range
Smelter zone	Soil	12	800–925	0.047–0.064	0.005–0.007
	Humus	12	770–900	0.066–0.078	0.008–0.009
	Vegetation	12	45–65	0.065–0.097	0.14–0.16
	Roe deer tissues	–	–	–	–
Zone A	Soil	48	9–640	0.036–0.060	0.009–0.47
	Humus	48	8–500	0.030–0.071	0.015–0.59
	Vegetation	48	1.5–45	0.025–0.090	0.17–1.7
	Roe deer tissues[b]	200	0.02–75	0.010–0.700	22–70[c]
Zone B	Soil	36	1.5–4	0.004–0.035	0.03–0.90
	Humus	36	0.8–3	0.003–0.030	0.35–0.95
	Vegetation	36	0.3–0.7	0.005–0.015	1.8–2.1
	Roe deer tissues[b]	100	0.006–5	0.002–0.360	35–85[c]
Controls					
Ctr-1	Soil	12	0.43–0.70	0.0008–0.001	0.14–0.18
Ctr-1	Humus	12	0.44–0.76	0.0008–0.001	0.13–0.19
Ctr-1	Vegetation	12	0.10–0.14	0.0017–0.0021	1.5–1.6
Ctr-2 to Ctr-5	Roe deer tissues[b]	150	0.001–0.02	0.0003–0.003	13–49[c]

N = number of samples analyzed.

[a] All concentration values d.w. except for roe deer tissues (f.w.)!

[b] 25 different tissues for each animal.

[c] Except for kidney with 0.25–0.74% and liver with 8.5–11%.

Table 5. Average T-Hg and Me-Hg concentrations in the composite plant samples representing roe deer diet (composed of over 40 plant species) from the investigated areas clustered into contamination zones

Sampling area	N	Total-Hg (μg/g d.w)			Me-Hg (ng/g d.w.)			Me-Hg (%)	
		Average	SD	Range	Average	SD	Range	Average	Range
Smelter zone	12	52.9	7.25	45.3–62.7	80.3	13.4	65.4–96.7	0.15	0.14–0.16
Zone A	48	14.4	2.57	1.55–42.9	41.8	6.74	23.8–90.3	0.84	0.17–1.67
Zone B	36	0.45	0.031	0.277–0.680	8.63	0.63	5.0–14.0	1.89	1.79–2.06
Controls	12	0.122	0.0121	0.110–0.137	1.87	0.16	1.7–2.1	1.54	1.50–1.57

N = number of samples analyzed.
Note: The original results are clustered in the four contamination zones (Smelter zone, Zones A, B, Controls) related to the distance from the source of pollution in Idrija, and were collected through a number of sampling exercises in the period from 1990 to 1996 (see also Fig. 3a,b)

several orders of magnitude higher than in the Idrija's wider surroundings and control areas, values are represented on a logarithmic scale. Each value in the graphs represents an average Hg concentration obtained for a certain type of sample through a number of sampling exercises in the period from 1990 to 1996. Also, correlations between mercury values in biota tissues and the distance of their site of origin from the Hg pollution source are presented. A comparison with average Hg concentration values in biota samples from the control areas (Ctr-1) and the abandoned mining area in Podljubelj (Pl) for the composite plant samples, and Ctr-2 to Ctr-5 as well as Pl for the roe deer tissue samples, was made to evaluate the size of contamination in the active mining area.

In the active mining area of Idrija, mercury concentrations in composite plant samples representing roe deer diet, and in roe deer body tissues reflect a mercury exposure gradient related to the distance from the source of contamination. Of course, mercury values in biota correspond by certain factors to those in soil profiles from each particular area (see Table 4; original numerical data in Gnamuš et al. 1996). As in soil profiles, the highest mercury values in the active Hg mining area were found in composite plant samples of the extremely polluted smelter zone (I-Sm), followed by polygon I-1 inside Zone A and considerably less contaminated Zone B. The same is true for roe deer tissues. Unfortunately, no roe deer tissue samples could be obtained from the smelter zone, since there are no roe deer living in this small polygon.

The mercury values from the Idrija area were compared to the controls (Ctr-1 to Ctr-5). Average vegetation T-Hg concentrations in the immediate vicinity of the Idrija smelter complex (smelter zone) were approx. 4× higher than those measured in Zone A, approx. 120× higher than Hg values measured in the samples from Zone B, and approx. 430× higher than the controls. Average T-Hg concentrations from the distinctively contaminated Zone A differ from the less contaminated Zone B by a factor of approx. 30× for vegetation samples, and approx. 2.5 to 5× for the various roe deer tissues. Differences in average T-Hg concentration levels between the Zone A and the controls follow factors of approx. 120× for vegetation samples, and approx. 10 to 65× for the various roe deer tissues. Average T-Hg concentrations in Zone B differ from the controls by

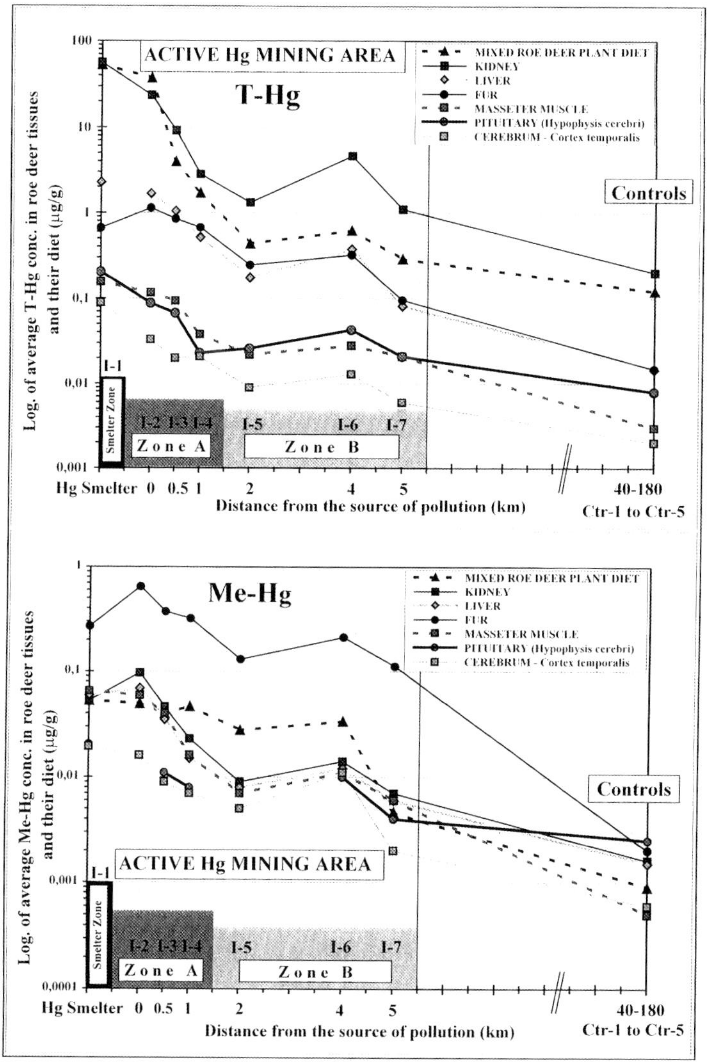

Fig. 4. Summary of the results for T-Hg (**a**) and Me-Hg (**b**) concentrations obtained from the analysis of the composite plant samples and selected roe deer tissues, related to the distance from the source of mercury contamination. Note that Hg concentration values are presented on a log scale. All concentration values f.w. except for mixed plant diet samples (d.w.)

factors of approx. 4× for vegetation samples, and approx. 3 to 20× in the various roe deer tissues. From these differentiation factors it can be stated that concentration differences between particular areas in the heavily polluted Zone A and less polluted Idrija surroundings clustered in Zone B were smaller in roe deer tissues and larger in the composite plant samples representing roe deer diet. In contrast, the concentration differences between samples from the moderately polluted environment and the controls were relatively greater in roe deer tissues, and relatively smaller in vegetation samples. Mercury concentrations in roe deer tissues from the affected Idrija mining area are approximately 1 or 2 orders of magnitude higher than those of the controls, depending on the distance of their site of origin from the smelter (see Fig. 4).

A comparison of Hg content between various roe deer tissues and their composite plant diet collected within the clustered contamination zones A and B in the Idrija area and the control areas showed some interesting features. In Zone A roe deer brain tissue Hg levels were approx. 80 to 200 times lower than in the composite vegetation, except for the sampling area I-1 with approx. 1000× higher vegetation values. In contrast, roe deer kidney Hg levels from this cluster were approx. 1.5 times higher than the composite roe deer diet. Again, the exception was sampling area I-1 with a reversed animal-to-plant ratio. In Zone B brain Hg levels were approx. 50 times smaller, and kidney Hg levels were approx. 3 to 7 times higher than in the composite vegetation samples of the area. Mercury concentration ranges for other roe deer tissues in relation to the Hg values in the composite plant samples representing roe deer diet deviate from the accumulation patterns found in kidney and brain, and show animal- to -plant ratios between the two extremes. In the contaminated area roe deer liver and fur Hg concentration levels were usually approx. within the same range as in their vegetation diet. For the controls, Hg values in various roe deer tissues were approx. 15 to 60 times lower than in the composite vegetation samples, and fur Hg concentrations deviated from the expected approximate equality with vegetation Hg levels, being about 8× lower. The trends in Hg content of the composite plant samples representing browsed roe deer diet and selected roe deer tissues as a function of the distance from the source of Hg pollution are indicated by the lines between mercury values (see Fig. 4) and resemble those in other environmental samples (Kosta et al. 1974b; Hess 1993; Gnamuš et al. 1996). Evidently, mercury values decrease with the distance from the mercury source, except in the sampling area I-6. Despite being reduced, Me-Hg concentrations in composite plant samples and roe deer tissues still reveal deviations in the sampling area I-6. The deviations in the sampling area I-6 show that distance from the pollution source cannot be the only measure of pollution monitoring. The actual pollution must be accessed through a more complex approach. Considering the distance of the sampling area I-6 from the pollution source, enhanced T-Hg concentrations established in the I-6 top soil (Gnamuš et al. 1996), and expressed in biota samples, may be influenced by the prevailing winds which sweep airborne Hg westerly from the heavily contaminated mine-smelter area. Since the sub-soil Hg content disagrees with deviating elevated values in the top soil (Gnamuš et al. 1996) and biota from the area, it is very likely that the enhanced mercury concentrations in area I-6 may be related to enhanced airborne Hg contamination rather than to some undiscovered bed-rock mercury deposits in the area. The elevated Hg contents in the area could arise due to a higher Hg deposition related to airborne pollution from the Idrija smelter chimney. However, since Hg concentrations in soil and plants have been changing little during the past 5 years, despite the drastic reduction of mercury production in the smelter, the elevated Hg contents in samples from sampling area I-6 are most probably related to one or several of the many undiscovered tailing sites from the past, which are spread around Idrija. Because of the high market value of mercury, in the mine's early days ore was frequently stolen from the mine to produce the precious metal in improvised low recovery smelter works built in the hills around Idrija.

Deviations in vegetation and animal T-Hg values related to the distance from the source of Hg pollution are much smaller than in the soil profiles (Gnamuš et al. 1996). The same is true for Me-Hg concentrations.

In general, the concentration differences in the composite plant samples are expected to be small due to the relatively uniform distribution of selected plant species over the sampling areas. The sampling site particularities are even less expressed for roe deer due to the fact that those terrestrial mammals range over territories with different directions of exposure. Thus, biota display more representative data and the impact of possible locally enhanced soil Hg concentration at certain sampling sites is less expressed due to their integrating action.

Comparison of Me-Hg concentrations in the three contamination zones of Idrija area and controls, relating the environmental and biota sample concentration levels, show differences much smaller than those observed for T-Hg. Following the same territorial sample grouping as for T-Hg, Me-Hg concentrations in various types of samples are within the same order of magnitude (Table 4, Fig. 4), which suggests different features of Me-Hg accumulation incomparison to inorganic Hg forms. Average Me-Hg concentrations in vegetation of the heavily contaminated smelter zone (I-Sm) are about 10% higher than in the polygon covering north-eastern suburban Idrija area (I-1), which comprises the wider smelter area (Fig. 3a), but are approx. 25–50% higher than the average values for those samples obtained from the areas I-2 to I-4 clustered in Zone A (see Fig. 4b). Similar trends were discovered in the soil profiles from the region (Gnamuš et al. 1996). Environmental and biota samples from Zone A of the Idrija area contained higher Me-Hg concentrations than the samples from Zone B, but the differences were less significant than in T-Hg concentrations (see Table 4). Me-Hg concentrations in the distinctively contaminated Zone A were approx. 4 times higher for vegetation samples and about 1.5 to 4.5 times higher for the various roe deer tissue samples, compared to the less contaminated wider Idrija surroundings (Zone B). On comparing Me-Hg concentrations in biota samples from the distinctively contaminated Zone A and controls, values approx. 20 times higher were found in vegetation samples, and approx. 20 to 210 times in most roe deer tissues from Zone A. The exceptions were pituitary glands in which Me-Hg contents were only 3 times higher than in controls. Average Me-Hg concentrations in the wider Idrija surroundings (Zone B) differ from the controls by factors of approx. 5 times for vegetation samples, and approx. 10 to 80 times for various roe deer tissues, except in the pituitary gland with approx. 3 times higher values. Average Me-Hg concentrations in the immediate vicinity of the smelter complex (smelter zone) are about twice as high as those measured in vegetation samples from the Idrija forestal suburbs, approx. 8 times higher than in the wider Idrija surroundings, and approx. 40 times higher than in controls. However, mixed roe deer diet and roe deer tissues consistently reflect the fluctuations in Me-Hg concentrations and T-Hg levels related to the distance from the pollution source.

From the data presented in Fig. 4 it is evident that the Hg concentration differences in vegetation reflect the microterritorial fluctuations in terrestrial Hg contamination better than roe deer tissues, whereas roe deer reflect the macro-territorial exposure gradient more efficiently. Since the living ranges of selected roe deer specimens were probably very seldom restricted to the size of a particular sampling area (I-1 to I-7) only, roe deer tissue Hg concentrations usually represent a measure of the contamination which is not entirely dependent on the particular limited sampling location, as in soil profiles and plant samples. However, the sampling areas I-1 to I-7, used for the collection of soil profiles and vegetation samples, had to be restricted to a size of approx. 1.5 km^2 for operational reasons and to standardize the sampling conditions. The living range of a particular specimen is also the main reason why roe deer is a more useful bioindicator than various predator species as concerns monitoring the exposure gradient of strong airborne Hg contamination in Idrija, which features distinctively local character. Good agreement between environmental and biota mercury concentration fluctuations related to the distance from the source of Hg pollution shows that roe deer effectively take over the Hg levels from their environment and efficiently reflect the Hg contamination pattern of the area in which they live. Roe deer tissues from the contaminated region expressed a large distinction in Hg contents compared to the controls. These data indicate that roe deer tissues may be successfully used as an indicator of Hg contamination in the terrestrial environment.

By comparing the data obained from roe deer tissue analysis to the Hg concentration values measured in various environmental samples from the area, we may conclude this section by stating that roe deer as a representative of the terrestrial wildlife species sufficiently reflect the mercury exposure gradient related to the distance of the site of their origin from the contamination source. In addition, vegetation samples reflect the territorial Hg exposure gradient. The use of bioindicators is very important for Hg risk assessment since the plant and animal tissue Hg values represent biologically available Hg, whereas ambiental monitoring considering Hg in air or in soil profiles reflects external contamination mainly. Together they give us a fairly accurate picture of the size and extent of contamination in a certain area.

4.4
The Concentrations of Total Mercury (T-Hg) and Methylmercury Compounds (Me-Hg) in Roe Deer Tissue Samples and Their Browsed Diet from the Heavily Polluted Area

As established by earlier studies concerning plant samples from the examined area (Furlan and Kosta 1972; Stegnar 1973) and other Hg-contaminated areas (Huckabee et al. 1983; Bargagli et al. 1986; Barghigiani et al. 1989; Ferrara et al. 1991), various terrestrial vascular plant species from the same sampling site show important differences concerning tissue Hg concentrations. Regarding different accumulation rates of terrestrial plant species related to the local/individual factors (e.g. soil types, texture and properties, microenvironment, exposure of the sampling microlocation and of the plant specimen, individual physiological demands and characteristics, etc.) it is advisable for biomonitoring to use a multispecies composite mixed plant samples of a defined individual species percentage composition. Biomonitoring using such composite samples instead of single plant species samples is much more representative for the contaminated area examined. Moreover, a higher representativeness as concern Hg concentrations can be obtained, since the variety of over 40 plant species usually represent a very large portion of a certain area's forest vegetation cover. However, the main concern of the terrestrial vegetation mercury content in this article was related to the Hg transfer from the soil and airborne Hg contamination into food chains. The contamination of vegetation due to aerial Hg deposition and root uptake results in adsorption/absorption of the metal on/in plant leaves, by which the extent of Hg accumulation in the first link of the terrestrial food chain is achieved. Thus, Hg accumulated in terrestrial vegetation is transferred to the primary consumer's link (herbivorous animals) in terrestrial food webs, in this study represented by roe deer.

Mercury concentrations in roe deer tissues are strongly dependent on the chosen tissue (Table 6). Different T-Hg contents of various animal tissues are related to their metabolic activity and function, and are frequently relatively high (see Fig. 4 deviations in Hg values between various roe deer tissues obtained from each particular sampling area). In this section mercury values for each body tissue of the particular roe deer from the contamination zones A and B of the polluted Idrija area are summarized separately according to the order of the Hg concentrations measured. For all sampling sites an identical order of tissues was discovered, as presented below. In brackets ranges of Hg concentration values expressed on f.w. (fresh weight) basis are presented. The approximate overall concentration ranges of mercury (T-Hg) in all examined tissues and organs of roe deer from the Idrija district (Zones A and B) follow the order:

kidney (1–75 μg/g) >> liver (0.1–1.7 μg/g) > fur (0.1–1.3 μg/g) >> pineal gland (0.02–0.3 μg/g) >> masseter muscle (<0.15 μg/g) > suprarenal gland (<0.12 μg/g) > spleen (~ 0.11 μg/g) > *lungs* (~ 0.1 μg/g) > pituitary (0.02 – 0.1 μg/g) > eye sections (<0.1 μg/g) > sections of the CNS (<0.03 μg/g).

The concentrations of T-Hg in roe deer tissues from the Idrija mining area were about 100 times higher than those in the controls. The same principles of roe deer tissue ranking were used for Me-Hg tissue accumulation also. For most of the selected roe deer tissues except fur, Me-Hg concentrations in the Idrija area were within 10 and 100 ng/g f.w., while in fur the average Me-Hg concentrations were up to 5 times higher than in other tissues and organs. In brackets, Me-Hg concentration ranges expressed on f.w. basis are presented. The corresponding sequence of roe deer tissues ranked by descending Me-Hg concentrations differ from the one obtained for T-Hg tissue values in the Idrija district and follow the order:

fur (100–500 ng/g) > kidney (10–100 ng/g) > liver (10–70 ng/g) > masseter muscle (10–60 ng/g) > other tissues incl. endocrine glands, spleen, lungs, eye sections and sections of CNS (5–15 ng/g).

Table 6. Average T-Hg and Me-Hg concentrations in roe deer tissue samples from the investigated areas clustered into contamination zones

Group/Tissue	N	Roe Deer							
		Total-Hg (ng/g f.w.)			Me-Hg (ng/g f.w.)			Me-Hg (%)	
		Average	SD	Range	Average	SD	Range	Average	Range
Idrija-Zone A: R1-R8									
Cerebrum Cortex temp.	8	26.1	8.95	15.2–40.3	11.9	5.33	4.12–20.1	45.1	27.1–64.8
Pituitary	5[a]	114	54.2	66.9–204	14.8	4.86	11.1–20.3	12.8	9.1–19.4
Masseter muscle	8	85.1	37.6	27.5–143	33.9	20.6	15.6–65.0	41.8	18.5–63.0
Liver	7	1184	866	324–2268	42.5	18.1	22.2–68.5	5.24	1.88–10.3
Kidney	7	22 775	18 388	2977–56 168	54.6	27.6	22.7-97.3	0.47	0.092–0.99
Fur	8	834	295	517–1266	389	179	165–715	46.8	28.8–69.3
Idrija-Zone B: R9-R12									
Cerebrum Cortex temp.	4	10.3	3.73	5.97–14.7	7.02	4.32	2.10–11.7	62.9	35.2–81.2
Pituitary	4	28.2	10.1	20.9–43.1	10.0	2.07	8.28–12.7	36.9	29.5–41.6
Masseter muscle	4	27.5	7.39	21.3–36.6	12.0	4.40	6.97–16.0	44.6	31.3–70.5
Liver	4	276	266	73.9–664	10.1	3.60	6.18–14.4	5.59	1.73–8.36
Kidney	4	3351	1500	1207–4639	13.0	6.92	7.34–22.6	0.43	0.24–0.61
Fur	4	247	171	100–487	167	130	66.4–358	67.4	45.5–84.0
Controls: R1-R6									
Cerebrum Cortex temp.	6	1.74	0.16	1.57–1.98	0.79	0.33	0.28–1.23	45.4	17.8–72.4
Pituitary	6[a]	4.88	3.54	1.27–10.1	1.74	0.96	0.44–2.55	60.5	18.6–80.6
Masseter muscle	6	2.93	1.25	1.45–5.01	1.05	0.74	0.44–2.32	30.7	14.4–56.0
Liver	4	14.8	3.03	10.4–17.0	1.49	0.41	0.91–1.84	10.0	8.75–11.4
Kidney	5	154	73.1	81.3–224	1.18	0.40	0.80–1.66	0.98	0.38–1.55
Fur	6	21.6	13.1	6.76–48.2	8.93	9.60	1.34–22.0	42.4	12.6–96.9

N = number of samples analyzed.

[a]Number of samples analyzed for T-Hg contents equals number of samples analyzed for Me-Hg contents except for Zone A – Pituitary where N = 3 and Controls-Pituitary where N = 4 for Me-Hg content analysis!

Note: Results are clustered in the contamination zones (Zones A, B, Controls) related to the distance from the source of pollution in Idrija, and were collected through a number of sampling exercises in the period from 1990 to 1996 (see also Fig. 3b).

The differences in Me-Hg concentrations between various roe deer tissues were smaller than differences in the T-Hg values (see Fig. 4a,b). It should be noted that in some cases only one or two samples from the particular polygon were compared to the others, and may not be representative. Due to a small amount of pituitary tissue sample (usually the total tissue mass was less than 150 mg) in some cases only T-Hg was determined (see Table 6: Pituitary N*).

Me-Hg ranges in selected animal tissues from the Idrija district as a percentage of overall Hg contents in descending order were as follows (see Table 6: Zones A and B):

fur (~30–85%) > brain sections (30–80%) > masseter muscle (~20–70%) > pituitary (<40%) > other tissues incl. some endocrine glands, spleen, lungs, eye sections (<20%) > liver and blood (~1–10%) >> kidney (~0.09–1%).

Comparison of T-Hg concentrations in roe deer tissues and their browsed diet from the areas investigated showed some important features (see Fig. 4). Mercury concentrations in most roe deer tissues are much lower than the values in browsed roe deer diet of a particular sampling area. From Fig. 4 it is also evident that Hg concentrations in various roe deer tissues differ significantly. Thus, it is difficult to decide which animal tissue would be most appropriate for comparison with their composite plant diet and environmental concentrations, since mercury contents in some roe deer tissues were much smaller than in diet, and in other tissues the ratio was reversed. The lowest animal tissue T-Hg contents were found in brain tissues and were approx. 50 to 1000 times lower compared to their diet browsed from the areas from which the roe deer originated. The highest animal tissue values were found in kidneys, with approx. 1.5 to 7 times higher Hg content than in the composite plant samples representing roe deer diet. The exception was sampling site I-2 where the average Hg concentration in roe deer kidney samples was approx. 40% lower, and the brain where Hg levels were approx. 1000 times lower than in vegetation diet. The difference probably arose due to a disagreement between the anticipated range territories of roe deer with identification Nos 3 and 4 (see Fig. 3b), and the sampling area I-2 used for collecting browsed roe deer diet. Namely, the range territories of these two roe deer specimens probably did not coincide with the wider smelter area in the northeastern part of Idrija from which composite plant samples representing roe deer diet were collected. By comparing the data on Hg concentration levels in roe deer tissues and their diet from area I-2, it appears that roe deer in the wider smelter area seemed to avoid the vicinity of the smelter complex, since their tissue Hg concentrations deviate from the expected values.

4.5 Is There Any Evidence of Hg Accumulation and Biotransformation Processes Occurring in Roe Deer Tissues?

It is difficult to establish trends in accumulation and biotransformation of mercury in roe deer tissues from the heavily polluted Idrija area, since a relatively small number of animals was examined so far. Despite roe deer being an abundant mammal species in the polluted Idrija region, it is relatively difficult to obtain a large number of animal tissues from the polygons where Hg values in environment are heavily enhanced. As roe deer population occupying territories in the heavily polluted town's suburban area inside the Idrija valley is small, the culling of the animals must be reasonable and proportionate to the wider examined area, so that the normal population structure is preserved (Gnamuš et al. 1991).

Despite elemental Hg being the main Hg form introduced into the environment, and inorganic Hg compounds representing the majority of the form found in the contaminated terrestrial environment of Idrija, large Me-Hg values and percentages were measured in roe deer tissues compared to its concentrations in the air, soil profiles and roe deer plant foodstuffs from the area. Large portions of Hg were present as Me-Hg in the majority of selected roe deer tissues except lungs, spleen, liver and kidney. Often Me-Hg concentrations in roe deer tissues are of the same order of magnitude as in the composite vegetation samples, despite T-Hg concentration values in the latter are being much higher (see Tables 4,5,6). Thus, average concentrations of Me-Hg in the contaminated soil profile samples from the Idrija area account only for up to 0.04% of T-Hg (range 0.005 to 0.9%; Gnamuš et al. 1996), and in vegetation samples around 0.27% of T-Hg (range 0.14 to 2%), whereas Me-Hg concentrations in most roe deer tissues from the contaminated area varied from 20 to 90% of T-Hg – for example in roe deer muscle it represents up to 80% (range 22 to 80%). The means of the Me-Hg percentage values for ten different animal tissues compared to the percentages of Me-Hg in soil profiles, plants and human tissue samples from the Idrija area and unpolluted areas of Slovenia are presented in Fig. 5a,b. Relatively low percentages of Me-Hg in roe deer tissues obtained from the contaminated area were found in blood, tissues of the internal organs like lungs, spleen, liver and kidneys, and eye sections, where mean Me-Hg values varied from 0.08% in kidney cortex up to 15% in lungs and eye sections with the exception of *Lens crystallina* with the Me-Hg percentage between 40 and 50%. Therefore, we believe that a selective accumulation of particular Hg species occurs in the target tissues of mercury accumulation at the level of the primary consumers.

Due to the low percentage of Me-Hg in composite vegetation foodstuffs of roe deer, and the high Me-Hg percentage in most of the roe deer tissues, a complex sequence of mechanisms involving processes of selective uptake and elimination, transformation and retention must take place in the metabolism of these herbivorous animals. Through these processes relative elevation of roe deer tissue Me-Hg content compared to their plant diet occurs. Most probably, the primary reason for enhanced Me-Hg values in roe deer tissues may be selective retention of methylated Hg species from the vegetation diet in the body tissues of the herbivore.

Indeed, high Me-Hg values in tissues of roe deer living in the contaminated area confirm the widely accepted feature that in animal tissues Me-Hg is especially accumulated. It is believed that accumulation and bioconcentration processes are expressed at different metabolic rates in each specimen, and therefore tissue Me-Hg concentrations varied between specimens living in the contaminated area. However, transformations of mercury inside the animal tissues must not be neglected. The relatively high Me-Hg values in the target tissues of mercury accumulation suggest that in addition to preferential accumulation and retention, at least part of the organic mercury may arise from metabolic biotransformation processes i.e. in-vivo methylation of Hg. The most favourable factors for in-vivo methylation of Hg are present in the animal's gastro-intestinal tract. Factors such as the deer's nutritional habits, suitable pH and temperature in the rumen of herbivores, high concentrations of inorganic mercury present in the plant foodstuffs, the presence of CO_2, CH_4, and important microorganisms suggest the existence of biomethylation processes. Of course, the Me-Hg concentration level in each tissue is a result of methylation and demethylation processes, which additionally complicate the issue.

It is suggested that both various accumulation patterns and important transformation mechanisms depend on the different toxicological and pharmacological features of the particular Hg species involved and individual characteristics of the animals. A special study is anticipated which will be devoted to solving problems dealing with Hg transformation and transport processes in wildlife metabolism, and in the polluted forestal terrestrial ecosystem as a whole. Also, the significance of some metabolic processes inside the gastrointestinal tract of ruminants for the biotransformation rate of Hg in herbivore tissues will be established and evaluated.

4.6
Is Wildlife Particularly Vulnerable as Concerns Environmental Hg Contamination?

As stressed in the WHO-IPCS Environmental Health Criteria (WHO 1976, 1989, 1990, 1991) and some other reviews (Lindqvist 1991; O'Brien et al. 1993), only a few studies have been published on truly wild, non-laboratory terrestrial mammals in relation to their exposure to mercury and its compounds in the natural environment. From such studies some important information was gained with respect to mercury toxicity. In one of the first studies performed on wild mammals (Aulerich et al. 1974), it was shown that mink exposed to enhanced concentrations of inorganic Hg in their diet (as in the case of natural exposure in the Idrija area) showed no apparent clinical signs of poisoning, whereas the toxic effects of Me-Hg were found to be much worse. However, from the small number of studies concerning mercury concentrations in terrestrial wildlife and the food chain transfer of the element, such as that of Bull et al. (1977), it was proposed (WHO 1976, 1989, 1990, 1991) that exposure via soil, soil water and through terrestrial food webs should be emphasized, rather than through drinking water and inhalation pathways. These later were found to be of less importance for the ecotoxicology of terrestrial organisms, even though heavily polluted areas could be an exception. The results of our present study confirm the above (see Table 4 and Fig. 4). There is a strong correlation between the mercury concentration in the mixed plant diet samples and mercury concentration values in roe deer tissues. Thus, the vigorous vegetation feeding characteristics of herbivorous

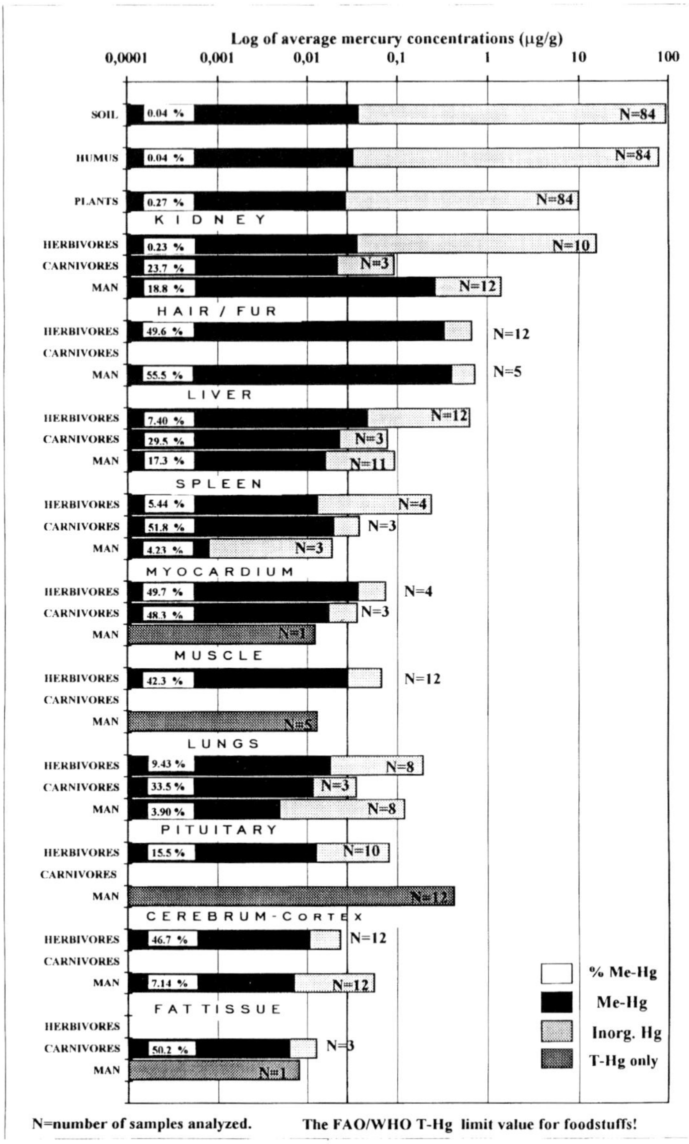

Fig. 5 a The average mercury concentrations in various samples of terrestrial food webs collected in the active mining area of Idrija during 1990–1996.

animals suggest the paramount role of Hg intake through the diet in the polluted environment.

Unfortunately, no toxicological or pharmacological data are available for roe deer. Despite the well-known long-term Hg retention in brain and pituitary, the mercury concentrations in these tissues obtained from roe deer living in the polluted area were low compared to most other tissues examined. The Hg levels

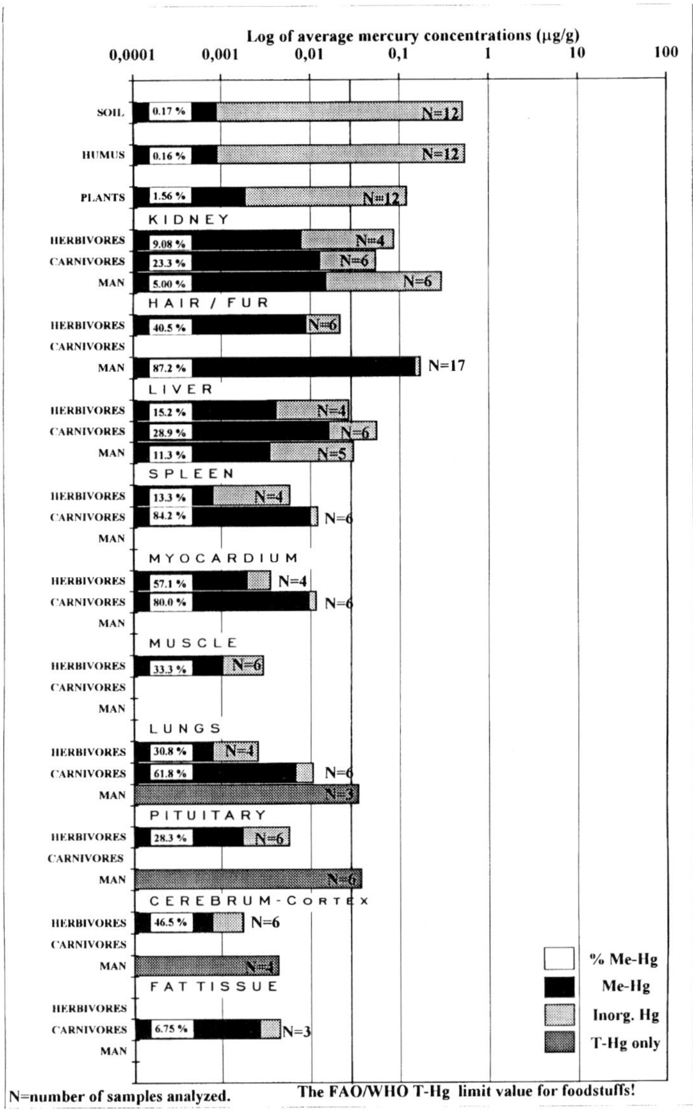

Fig. 5 b The average mercury concentrations in various samples of terrestrial food webs collected in unpolluted *areas* in Slovenia during 1990–1996. The FAO/WHO T-Hg limit value for foodstuffs is indicated (0.03 µg/g fresh wt.)

in muscle, which represents a tissue whose Hg absorption/retention is much more dependent on the metabolic rate and thus subject to frequent changes, were much higher than in the above mentioned brain and endocrine tissues, typical "hot spots" of Hg retention in man. The reasons for this discrepancy are still unknown and should be physiologically assessed by a separate study. Liver and kidney represent organs having a leading role in metabolic modifications and transformations of various Hg species entering the animal body and their enhanced Hg concentrations were expected. It is important to note that T-Hg concentrations in kidney cortex of exposed roe deer reached up to 75 mg Hg/g f.w. and were an order of magnitude higher than Hg values in liver. It is expected that the processes of partial conversion of Me-Hg compounds, i.e. demethylation and elimination of Hg are carried out in these two organs. On the other hand, depending on the metabolic rate and concentrations and species of Hg present, they may transfer some Hg to other organs, causing potential adverse effects in cytogenetic, physiological, neurological or metabolic processes. Considering high Hg concentrations in fur, a possible mode of Hg elimination through the seasonal replacement of fur is suggested. Thus, fur may have an important role in long-term balancing of the Hg body burdens in areas contaminated by Hg. Considering its high Hg content, the possible contribution of airborne Hg deposition to the fur surface should be taken into account and its role and size in the overall fur Hg concentrations elucidated. It is well known that elemental Hg^{o}, which is the most abundant Hg form in Idrija air, does not accumulate in fur related tissues, e.g. hair (Chatt and Katz 1988), but since the fur of the animals was not washed prior to analysis, the possible contribution of surface contamination to the T-Hg fur concentrations must not be neglected.

4.7
Is It Possible to Explain the Observed Age and Sex-Related Hg Concentration Fluctuations?

The relationship between Hg content in tissues and the age of the organism was studied in roe deer. If roe deer are to be considered reliable bioindicators of environmental Hg fluctuations, the alterations of the Hg distribution patterns due to age and sex should be studied, beside the response to Hg distribution from the spatial contamination pattern. Positively correlated mercury values in roe deer and environmental samples with distance from the Hg pollution source were already established (see Table 4 and Fig. 4; Gnamuš et al. 1995, 1996). Good agreement between Hg concentrations in various roe deer tissues and the environmental data was shown.

To minimize interferences with the spatial contamination pattern in the polluted Idrija area, the age and mercury content correlations of the animals within each of the contamination zones (Zones A, B) or even within each of the sampling polygons (I-1, I-2,) should be compared (see Table 3 and Fig. 3a,b). In this way, the anomalies in Hg pollution between the sampling areas could be greatly reduced. However, a relatively good age-related accumulation pattern was observed in control animals only.

We must be aware of the fact that the number of roe deer (n = 20) examined so far is relatively small and that deviations from the expected trends might be related to the individual differences between particular specimens. In the distinctively contaminated Idrija area the concentrations of mercury in the limited number of roe deer tissues obtained so far did not reflect the age of the animals, probably because of the influence of the severe spatial exposure gradients. Additional samples, especially from the heavily polluted Zone A inside the Idrija valley, must be surveyed for more reliable results. However, as concerns Idrijan roe deer (see Table 3: Zone A), positive age/T-Hg correlation was observed in lungs ($r^2 = 0.73$), but correlations in other tissues were not statistically significant.

Sex differences cannot be observed from the data collected so far, since all the culled animals, except two (one roebuck and one roecalf – see Table 3: R6 and R8), were females.

To establish a relationship between the Hg content in tissues and the age/sex of roe deer in this heavily polluted area, further studies with a greater number of specimens are required.

4.8 Do Wildlife Reveal the Toxic Effects of Enhanced Hg Concentration Values in the Heavily Polluted Environment?

So far, no visual damage was observed on the macromorphological level in roe deer tissues despite some very high Hg concentrations, which greatly exceed the maximum levels accepted under FAO/WHO regulations for food substances. Since no apparent clinical signs of poisoning were observed either in plants or in roe deer from the heavily polluted area in Idrija, we can confirm the evaluation of the effects of Hg in the terrestrial environment given in the WHO-IPCS Environmental Health Criteria (WHO 1989): "Acute effects would not be seen in terrestrial plants growing in natural soils, nor in terrestrial birds or mammals". Nevertheless, a variety of the stochastic effects related to the enhanced presence of Hg in the Idrija environment may occur. They may result in various potentially dangerous effects which are very difficult to establish and evaluate.

In order to identify any tissue damage it would be necessary to perform histological and cytogenetic studies. On the other hand, a semimillenium of intensive exploitation of mercury, resulting in highly elevated Hg concentrations in the Idrija valley environment, may represent a period long enough to enable organisms living in the heavily affected area to develop effective phylogenetically conditioned mechanisms preventing adverse effects.

It is difficult to establish whether any harmful vegetational responses to the heavy Hg pollution in Idrija area exist since no visual adverse effects were observed which could be directly related to the highly elevated Hg concentrations. Perhaps the only visual sign of heavy pollution might be the absence of the majority of plant species forming roe deer diet from the vegetation cover in the most heavily contaminated Smelter Zone. However, this may be primarily the consequence of enhanced SO_2 and flue gas emissions from the smelter chimney, as well as the soil depth and characteristics.

It is important to stress that most forms of mercury are highly toxic to the animals, which is not the case for plants (WHO 1989), except at very high concentrations. Long-term retention and subsequent toxicity of Hg to plants may also be prevented by effective defence mechanism such as efficient Hg elimination through litterfall. In our present study, Hg concentration levels in plants were higher than in most animal tissues from the contaminated area. Apparently, relatively effective mechanisms for Hg detoxification and elimination were developed through the phylogenetic development of the organisms in the heavily polluted area. However, since the higher plants and animal indicators show no toxic effects to relatively high mercury concentrations, a comparative study focused on detoxifying mechanisms in plants (including the metal complexing peptides called phytochelatins) and similar mechanisms (metallothioneins) in animals could be instructive.

4.9 Wildlife as an Early Warning for Human Health Hazards of Hg Contamination?

Comparison of the present results on roe deer tissues with data on the Idrijan and control human populations in the 1970s and 1980s shows relatively good agreement (Table 7; Byrne et al. 1971; Kosta et al. 1972, 1974a,b,c; 1975; Zelenko 1986; Horvat et al. 1991). Although the most extensive study on humans was performed about 25 years ago, when the mercury mine and smelting plant discharged several tens of kilograms of Hg per day into the air, the former data is comparable with those in the present study, as shown in Table 7. The exceptions are pituitary and brain cortex with significantly lower mercury concentrations in roe deer compared to the values measured in these tissues of the Idrijan human residents. It is possible that these high concentrations of mercury in human brain and endocrine system are the consequence of the high physiological and metabolic activity of these tissues in man. To elucidate this problem, a special study should be designed. However, most of the present roe deer tissue Hg concentrations are even higher than those in comparable human tissues obtained 15 to 25 years ago (Byrne et al. 1971; Kosta et al. 1972, 1974a,b,c; 1975; Zelenko 1986; Horvat et al. 1991) (see also Fig. 5a: herbivores vs. man). These comparisons indicate that roe deer tissue concentrations may be used as an early warning of elevated Hg tissue values in humans. If any cytogenetic or histological changes in selected roe deer tissues are found, we may predict similar adverse effects of high Hg concentration would be found in human tissues. Further investigations and comparative studies of the present situation in humans is needed to confirm the above and to indicate which roe deer tissues are the most suitable for comparison and for mercury intoxication risk assessments in humans.

4.10 The Prey:Predator Relationship as a Reflection of Territorial Hg Contamination

A summary of the average T-Hg and Me-Hg concentrations in various samples from the terrestrial environment in the Idrija area and unpolluted control areas

Table 7. Comparison of the average T-Hg and Me-Hg concentrations in selected human and roe deer tissues from the Idrija area and in controls

Origin and tissues analyzed	Total-Hg (ng/g f.w.)				Me-Hg (ng/g f.w.)			
	N	Average	SD	Range	N	Average	SD	Range
Man								
Idrija								
Brain	21	55.7	80.0	0.4–300	9	7.2	7.6	2.0–26
Hair	5	731	736	160–1950	8	406	220	150–840
Pituitary	24	1043	3114	2.2–15 400	–	–	–	–
Kidney	23	834	1372	33.8–5500	7	39.2	79.9	33.6–220
Lungs	8	137	85.7	42–271	–	–	–	–
Liver	11	92.5	43.1	41–184	6	16	9	1.8–28
Muscle	5	12.9	7.6	4.9–24	–	–	–	–
Controls								
Brain	4	4.3	1.8	2.5–6.7	–	–	–	–
Hair	17	171	84.9	80–350	17	150	81	80–340
Pituitary	6	36.9	23.7	7–63.6	–	–	–	–
Kidney	6	179	157	11–365	–	–	–	–
Lungs	3	34.5	15.3	18.4–48.8	–	–	–	–
Liver	5	26.8	12.3	10.0–39	2	3.5	0.7	3.0–4.0
Muscle	–	–	–	–	–	–	–	–
Roe deer								
Idrija								
Brain-cortex	12	22.6	16.1	5.7–62.3	12	10.7	5.28	2.0–19.8
Fur	12	639	382	105–1270	12	323	186	66..4–715
Pituitary	10	81.6	67.0	20.9–204	5	12.7	4.54	8.5–20.3
Kidney	10	15 563	18 818	1207–56 168	8	35.9	32.9	7.3–97.3
Lungs	8	193	124	87.6–332	4	17.1	6.6	13.5–27.0
Liver	12	637	694	74.0–2268	8	46.6	29.7	6.2–85.3
Muscle (*M. masseter*)	12	67.0	40.6	210–143	10	28.3	21.6	5.8–65.0
Controls								
Brain-cortex	6	1.74	0.16	1.57–1.98	6	0.79	0.33	0.28–1.23
Fur	6	21.6	13.1	6.8–48.2	6	8.9	9.6	1.34–22.0
Pituitary	6	4.88	3.54	1.27–10.1	4	1.74	0.96	0.44–2.55
Kidney	4	154	73.1	81.3–224	4	1.18	0.40	0.80–1.66
Lungs	4	2.58	0.99	1.22–3.48	4	0.80	0.31	0.51–1.22
Liver	4	14.8	3.03	10.4–17.0	4	1.49	0.41	0.91–1.84
Muscle (*M. masseter*)	6	2..93	1.25	1.45–5.01	6	1.05	0.74	0.44–2.32

Note: Results of the human tissue Hg concentration analysis from the Idrija area and of human controls from Slovenia are summarized by the present authors from several articles published in the 1970s and 1980s.

in Slovenia is shown in Fig. 5a,b. As mercury concentration differences in various samples cover several orders of magnitude, values are represented on a logarithmic scale. For comparison, data on the available Idrijan human tissues (Fig. 5a) and available human tissue Hg concentrations from unpolluted regions in Slovenia (Fig. 5b) are added, as man is the ultimate consumer in the food webs.

From the bars in Fig. 5, Hg transfer and transformation rates in different trophic levels of the food web may be established and comparisons of the prey vs. predator tissue mercury concentrations can be made (herbivores vs. carnivores).

In the specific case of the Idrija region, where a strong local Hg pollution pattern is expressed, roe deer as large terrestrial herbivorous mammals proved to be very efficient bioindicators of Hg contamination. As discovered from our investigation, roe deer reflect locally enhanced Hg contamination in the Idrija area more distinctively compared to their raptors. Despite the fact that the two selected predator species prey on roe deer primarily and are directly above roe deer in the food web, the data on Hg concentrations in selected tissues of lynx and wolves from the wider Idrija area show some deviations from the expected prey-to-predator pollutant magnification (see Fig. 5a: herbivores vs. carnivores).

The main reason for the unsuitability of large carnivorous predator species for monitoring the specific Hg pollution pattern in the Idrija region, compared to their prey, appears to be their large territorial range. As established before (see also Figs.3a,b), the Hg pollution in the Idrija area is strongly influenced by specific geographical relief and is spatially restricted. The individual territory of the lynx or the wolf is usually much larger than the individual herbivore's foraging area. It is also substantially larger than the heavily contaminated region in the Idrija area. Thus, individuals of the predator species only seldom plunder the most exposed prey (roe deer) and rarely contact the heavily affected area itself. Therefore, the results obtained from the chosen predator species in the Idrija area do not follow the general prey-to-predator relationship ratios for Hg concentrations which should be expressed in more static food webs in the area. However, the characteristic prey-to-predator pollutant magnification was observed in unpolluted areas of Slovenia affected by the moderate global Hg pollution only, since territorial distribution of the pollutant in these areas was less influenced by geographical or other factors (see Fig. 5b: Herbivores vs. Carnivores).

The specific accumulation and distribution patterns in animals at different trophic levels of the food webs will be revealed through further studies including other animals from the studied areas. Also, the importance of mercury transfer to the higher links in terrestrial food webs of the contaminated areas and man will be elucidated.

5 Conclusions and Recommendation

- A semi-millenium of anthropogenic cinnabar ore processing in Idrija has caused severe contamination of the natural compartments and biota in the area. Mercury concentration levels in environmental and biological samples from the Idrija area mostly greatly exceed the maximal allowed concentrations set by the Steering Committee for the Protection and Management of the Environment and Natural Habitats (CDPE) for European soil resources, and FAO/WHO for food substances (see Fig. 5a,b: The FAO-WHO T-Hg limit value for foodstuffs!).

- Environmental and biological data obtained from the studied areas in the Idrija district were compared to enable a critical evaluation of the environmental changes in mercury concentrations 5 to 10 years after cessation of intensive mercury production in the mid-eighties. Despite drastic reduction from approx. 500 t/year in 1986 (several tens of kg of Hg emitted daily from the smelter chimney), to about 25 t/year in 1990 and to almost complete closure in the past 5 years, the mercury values in the samples analyzed do not follow such a trend. Concentrations of mercury are higher that one would expect from production trends alone. The results of the study reflect the tendency of Hg species to accumulate and be retained in environmental matrices and biota, and its capability to transform into biologically available forms in the natural environment. This process poses a potential contamination hazard for all living organisms in the area, especially in the immediate vicinity of the mining/smelting complex where concentration levels are the highest. These features suggest that gradual closure of the mine and smelting plant does not instantly solve the problem of elevated Hg concentrations in biota inhabitating the surrounding environment, a fact which is also to be expected from the persistence of the contaminant itself.
- The study demonstrated a useful way of estimating the current degree of mercury and methylmercury accumulation in a particular polluted terrestrial environment through a complex ecological survey, which includes analyses of some environmental compartments (soil profiles) and wildlife food webs.
- Exposure gradients were recorded in both animal and plant indicators, as well as in soil profiles when approaching the source of contamination.
- It was demonstrated that roe deer could successfully be used as indicators of mercury contamination in terrestrial ecosystems, and their advantages and disadvantages in relation to their actual bioindicative value were explored and described. Considering their high abundance and wide distribution in Europe and Eurasia, the positive correlations with the environmental Hg pollution and preliminary data on the relationship between Hg content in tissue and the age of the organisms, roe deer appear to fulfill all the conditions necessary for their use as biomonitors.
- The small number of the animals examined prevents drawing general conclusions as concern the age/sex and mercury content correlations. Additional roe deer specimens are required to discover the possible role of the sex of the animals in Hg accumulation trends and to extend their age structure. They would also enable a thorough statistical evaluation of the results to be made.
- Despite the relatively high concentrations of Hg in some roe deer tissues, no visual macro-morphological abnormalities were discovered. To uncover possible adverse effects of life-time exposure of animals to elevated Hg concentrations in the Idrija area, histological and cytogenetic tests are required. The specific tissues which accumulate the most Hg were found and selected for future cytogenetic and histological studies. The relative accumulation rates for total and methylmercury concentrations were

established for different tissues of roe deer, and the relationship to their diet was found.

- Some efforts were made to establish the role of herbivores in transferring various mercury species from plant foodstuffs into the higher links of the terrestrial food web. The importance of herbivore Hg transformation processes for accumulation in consumers of the higher trophic levels, including man, will be assessed through further studies. The data from the study should be complemented with a larger number of herbivores, more individual raptors in the present food web in the polluted area, and also biota from other terrestrial and aquatic food webs, in order to demonstrate their interdependence.
- Since the potential harmful effects of Hg for some larger terrestrial wild mammals in the natural environment affected by anthropogenic Hg emissions are comparable to those for the exposed human population in the area animals may serve as appropriate indicators of Hg contamination.
- Our data on Hg concentrations in soil samples, composite plant samples corresponding to roe deer diet, roe deer tissues as well as the tissues of their raptors, provide sufficient data to confirm roe deer tissues as indicators of mercury contamination, transfer and accumulation in the terrestrial environment. Furthermore, the good correlation between mercury levels in roe deer and some tissue analyses of the mining area's human residents illustrates that tissues of roe deer may be successfully used as an early warning of possible health effects of mercury on humans.
- Comparison of the present results on Hg concentration values in the roe deer tissues with the data on similar tissues obtained from some postmortem human autopsy cases of the Idrija residents in the early 1970s shows similar mercury concentrations, despite the passage of more than two decades.
- The mercury contamination problem of the area is complicated due to the interdependence of a variety of abiotic and biotic factors influencing the transformation processes of the element in biogeochemical cycles, the occurrence and intensity of its accumulation and transfer of mercury compounds in the food webs. Further analysis and Hg speciation in the tissues of animal indicators within the complex ecological monitoring programme will be needed to elucidate accumulation patterns and important transformation mechanisms dependent on the different toxicological and pharmacological features of the Hg species involved.
- Meaningful biomonitoring programs should be implemented in the areas where distinctive local contamination exists. It is of special concern to control heavy local pollution, to evaluate data acquired, and to determine possible toxic effects of long-term Hg exposure on the organisms living in their natural habitats. The fate of mercury in the heavily polluted environment should be monitored to prevent possible harmful effects to humans. In particular, it would be important to conduct additional studies on the parameters that influence the mobility and transformation of mercury under various environmental conditions.
- It is suggested that careful trend biomonitoring of mercury and methylmercury compounds should be continued in order to observe the natural

restoration processes in the affected environment after complete cessation of Hg production. Some efforts to expand these studies in the Idrija area should be made to reveal the mechanisms of biogeochemical cycling, transformation and biomagnification processes in this unique natural and anthropogenic laboratory.

- Further studies, including environmental matrices such as soils, waters, flood sediments, and biota samples will be necessary to control environmental Hg fluctuations and transformations over the next decade. They should enable the extent and velocity of the restoration proceses to be established and gradual diminution of environmental and biota tissue Hg concentrations which should follow the final closure of the mercury mine/smelter complex in Idrija. In this respect, additional samples from the Podljubelj area, where Hg mining was abandoned approximately a century ago, were collected for comparison and are presently awaiting analysis.

Acknowledgements The authors are grateful to the J. Stefan Institute, Dept. of Environmental Sciences, Laboratory for Radiochemistry in Ljubljana, where most of the analytical work was carried out, particularly Ms. Ana Prosenc, who carried out analyses by NAA, to Dr. Anthony R. Byrne for discussion and suggestions, to Prof. Peter Stegnar and Dr. Borut Smodiš for enabling cooperation between J. Stefan Institute, Dept. of Environmental Sciences and Dept. of Biology, University of Ljubljana, to Dr. Rober Šajn from Institute for Geology, Geotechnics and Geophysics, Ljubljana for preparing the digital relief model of the Idrija area, the staff of the Idrija Hunter's Association, especially to Mr. E. Miklavčič for collecting the animal samples from the Idrija area, to Prof. M. Adamič from Biotechnical Faculty, Dept. of Forestry for cooperation and helpful assistance, to Prof. J. Lorger from the Veterinary Faculty for help in precise dissection of brain tissue from the animals, and to Prof. T. Pardubsky and Prof. A. Bidovec from the Veterinary Faculty for consultations and help in collection of samples. The authors would also like to thank Mr. M. Zupan from Biotechnical Faculty, Dept. of Agronomy for help at soil types determination in the areas surveyed. Special thanks are due to Dr. Lian Liang of the IAEA-MEL, Monaco, for her help in carrying out the analyses of total and methylmercury in soil samples. The research was partly sponsored by the Rector's Foundation at the University of Ljubljana and Slovenian Scientific Foundation.

References

Adamič M (1990) Food habits as an element of the management of ungulate populations with emphasis on the red deer (*Cervus elaphus* L.). Ph D thesis. Univ of Ljubljana, Ljubljana, Slovenia (in Slovene, English Summary)

Andren AW, Nriagu JO (1979) The global cycle of mercury. In: Nriagu JO (ed) The biogeochemistry of mercury in the environment. Elsevier-North Holland, Biomedical Press, Amsterdam, pp 1–21

Aulerich RJ, Ringer RK, Iwamoto S (1974) Effects of dietary mercury on mink. Arch Environ Contam Toxicol 2:43–51

Babinska-Werka J, Czarnowska K (1988) Heavy metals in roe-deer liver and alimentary tract and their content in soil and plants in central Poland. Acta Theriol 33:219–230

Baeyens W, Ebinghaus R, Vasiliev O (Eds.) (1996) Global and regional mercury cycles: sources, fluxes and mass balances. Kluwer, Dordrecht, pp 1–563

Bargagli R, Barghigiani C, Maserti BE (1986) Mercury in vegetation of the Mount Amiata area (Italy). Chemosphere 15:1035–1042

Barghigiani C, Siegel BZ, Bargagli R, Siegel SM (1989) The contribution of mercury from thermal springs to the environmental contamination of Mt. Amiata. Water Air Soil Pollut 43:169–175

Borg K, Wanntrop M, Erne B, Hanko E (1969) Alkyl mercury poisoning in terrestrial Swedish wildlife. Viltrevy 6:301–370

Bull KR, Roberts RD, Inskip MJ, Goodman GT (1977) Mercury concentrations in soil, grass, earthworms and small mammals near an industrial emission source. Environ Pollut 12:135–140

Byrne AR, Kosta L (1974) Simultaneous neutron activation determination of selenium and mercury in biological samples by volatilization. Talanta 21:1083–1090

Byrne AR, Dermelj M, Kosta L (1971) A neutron activation study of environmental contamination and distribution of mercury in animals and fish. In: Nuclear techniques in environmental contamination. STI/PUB/268 IAEA, Vienna, Austria pp 415–425

Chatt A, Katz SA (1988) Hair analysis – applications in the biomedical and environmental sciences. VCH., Weinheim, Germany

Drovenik M (1992) Kovinske mineralne surovine Slovenije. (Metal-containing mineral ores in Slovenia.) Onesnaževanje in varstvo okolja. Zavod za Tehnično Izobraževanje, Ljubljana, Slovenia pp 31–36 (in Slovene)

Ferrara R, Maserti BE, Breder R (1991) Mercury in abiotic and biotic compartments of an area affected by a geochemical anomaly (Mt. Amiata, Italy). Water Air Soil Poll 56:219–233

Fitzgerald WF (1995) Is mercury increasing in the atmosphere? The need for an atmospheric mercury network (AMNET). Water Air Soil Pollut 80:245–254

Frøslie A, Haugen A, Holt G, Norheim G (1986) Levels of cadmium in liver and kidneys from Norwegian cervides. Bull Environ Contam Toxicol 37:453–460

Furlan J, Kosta L (1972) The mercury contamination and the content of other essential elements in native plants in a contaminated area at Idrija. Zb Bioteh Fak Univ.Ljublj. Kmetijstvo 19:45–50 (in Slovene, English Abstr)

Gnamuš A (1996) The use of biological indicators in monitoring the pollution of air with mercury in the broader Idrija region. In: Miklavčič V, Kobal AB (ed) Idrija as a natural and anthropogenic laboratory – mercury as a global pollutant. Idrija mercury mine (RŽS Idrija), Slovenia, pp 34–41

Gnamuš A, Horvat M, Stegnar P (1991) Exposure to, and bioaccumulation of mercury in deer in the surroundings of Idrija. IJS Technical Report DP-6200, Jožef Stefan Institute, Dept. of Nuclear Chemistry, Ljubljana, Slovenia (in Slovene, English Summary)

Gnamuš A, Horvat M, Stegnar P (1995) Der Quecksilbergehalt von Rehwild und geästem Laub für die Bewertung der Umweltbelastung im Bergbaugebiet von Idrija – eine Fallstudie aus Slowenien. Z Jagdwiss 41:198–208

Gnamuš A, Horvat M, Stegnar P (1996) Biomonitoring of mercury contamination in the Idrija mining area. IJS Technical Report DP-7403, Jožef Stefan Institute, Dept of Environmental Sciences, Ljubljana, Slovenia

Hess A (1993) Verteilung, Mobilität und Verfügbarkeit von Quecksilber in Böden und Sedimenten am Beispiel zweier hochbelasteter Industriestandorte. PhD Thesis. Ruprecht-Karls Universität, Heidelberg, Germany (in German)

Hoellerer G, Coduro E (1977) On heavy metal contamination in venison. Z Lebensm Unters Forsch 163:260–263 (in German)

Holm J (1993) Investigation of roe-deer – criteria for use as a bioindicator in specimen banking. Sci Total Environ 139-140:237–249

Horvat M, Zvonarič T, Stegnar P (1986) Optimization of a wet digestion method for the determination of mercury in blood by cold vapour atomic absorption spectrometry (CV AAS). Vestn Slov Kem Druš 33:475–487

Horvat M, May K, Stoeppler M, Byrne AR (1988) Comparative studies of methylmercury determination in biological and environmental samples. Appl Organomet Chem 2:515–524

Horvat M, Begič I, Prosenc A, Smrke J, Konda D, Byrne AR, Stegnar P (1991) A study of some trace elements in human hair and the foeto-placental system in women from the central Slovenian region. In: Aitio A, Aro A, Jarvisalo J, Vainio H (eds) Trace elements in health and disease. The Royal Society of Chemistry, Thomas Graham House, Science Park, Cambridge CB4 4WF

Horvat M, Bloom NS, Liang L (1993) Comparison of distillation with other current isolation methods for the determination of methyl mercury compounds in low level environmental samples, part 1. Sediments. Anal Chim Acta 281:135–152

Horvat M, Mandič V, Liang L, Bloom NS, Padberg S, Lee YH, Hintelmann H, Benoit J (1994) Certification of methylmercury compounds concentration in marine sediment reference material, IAEA-356. Appl Organomet Chem 8:533–540
Huckabee JW, Sanz Diaz F, Janzen SA, Solomon J (1983) Distribution of mercury in vegetation at Almaden, Spain. Environ Pollut (Ser A) 30:211–224
Klötzli F (1965) Qualität und Quantität der Rehäsung (in Wald und Grünlandgesellschaften des nördlichen Schweizer Mittellandes). Veröffentlichungen des Geobotanischen Institutes der Eidg. Techn. Hochschule. Hans Huber, Bern, (in German)
Kosta L, Byrne AR, Stegnar P, Zelenko V (1972) Uptake of mercury by plants and its distribution in living organisms in an environment with increased concentration of this element. In: Radiotracer studies of chemical residues in food and agriculture. IAEA-PL-469/5, Vienna, Austria pp 47–59
Kosta L, Byrne AR, Zelenko V, Stegnar P, Dermelj M, Ravnik V (1974a) Studies on the uptake, distribution and transformations of mercury in living organisms in the Idrija region and comparative areas. Vestn Slov Kem Druš 21:49–76
Kosta L, Zelenko V, Stegnar P, Ravnik V, Dermelj M, Byrne AR (1974b) Fate and significance of mercury residues in an agricultural ecosystem. In: Isotope tracer studies of chemical residues and the agricultural environment. IAEA–PL–512/30, Vienna, Austria pp 87–102
Kosta L, Zelenko V, Ravnik V, Levstek M, Dermelj M, Byrne AR (1974c) Trace elements in human thyroid with special reference to the observed accumulation of mercury following long-term exposure. In: Comparative studies of food and environmental contamination. IAEA, Vienna, Austria pp 541–550
Kosta L, Byrne AR, Zelenko V (1975) Correlation between selenium and mercury in man following exposure to inorganic mercury. Nature 254:238–239
Kurt F (1970) Das Rehwild. Bayerische Landes-Verlag, München, Germany (in German)
Krynski A, Kaluzinski J, Wlazeiko M., Adamowski A (1982) Contamination of roe deer by mercury compounds. Acta Theriol 27:499–507
LaFleur PD (1973) Retention of mercury when freeze-drying biological materials. Anal Chem 45:1534–1536
Liang L, Bloom NS (1993) Determination of total mercury by single-stage gold amalgamation with cold vapour atomic spectrometric detection. J Anal Atom Spectrom 8:591–594
Liang L, Horvat M, Bloom NS (1994a) An improved speciation method for mercury by GC/CVAFS after aqueous phase ethylation and room temperature precollection. Talanta 41:371–379
Liang L, Bloom NS, Horvat M (1994b) Simultaneous determination of mercury speciation in biological materials by GC/CVAFS after ethylation and room-temperature precollection. Clin Chem 40:602–607
Lindberg SE, Jackson DR, Huckabee JW, Janzen SA, Levin MJ, Lund JR (1979) Atmospheric emission and plant uptake of mercury from agricultural soils near the Almaden mercury mine. J Environ Qual 8:572–578
Lindqvist O (1991) Mercury in the Swedish environment. Water Air Soil Pollut (SpecTssue) 55:1–261
Lupšina-Miklavčič V (1994) Determination of total mercury in the air. M Sc Thesis. Univ of Ljubljana, Ljubljana, Slovenia (in Slovene)
Lutz W (1985) Ergebnisse der Untersuchungen von Rehen (*Capreolus capreolus* L.) und Hasen (*Lepus europaeus* Pallas) auf Schwermetalle und chlorierte Kohlenwasserstoffe in Nordrhein-Westfalen. Z Jagdwiss 31:153–175 (in German)
May K, Stoeppler M, Reisinger K (1987) Studies in the ratio total mercury/methylmercury in the aquatic food chain. Toxicol Environ Chem 13:153–159
Nriagu JO, Pacyna JM (1988) Quantitative assessment of worldwide contamination of air, water and soils by trace metals. Nature 333:134–139
O'Brien DJ, Kaneene JB, Poppenga RH (1993) The use of mammals as sentinels for human exposure to toxic contaminants in the environment. Environ Health Perspect 99:351–368
Padberg S (1991) mercury in terrestrial ecosystem. Investigations of transport and transformation mechanisms. Ph D Thesis. Univ of Tübingen, Tübingen, Germany (in German, English Summary)
Padberg S, Stoeppler M (1991) Studies of transport and turnover of mercury and methylmercury. In: Merian E, Haerdi W (eds) Metal compounds in environment and life. vol 4. Interrelation between

chemistry and biology. Science and Technology Letters, Northwood, UK, and Science Reviews, Wilmington, Delaware, USA, pp 329–340

Petkovšek Z (1980) Vetrovne razmere v Idriji. (Wind conditions in Idrija.) FNT Technical Report, Univ of Ljubljana, Ljubljana, Slovenia (in Slovene)

Rump HH, Krist H (1992) Laboratory manual for the examination of water, waste water and soil. 2nd edn.VCH, Weinheim, Germany

Stegnar P (1973) Mercury uptake and its distribution in plants and animals. M Sc Thesis. Univ of Ljubljana, Ljubljana, Slovenia (in Slovene)

UNEP/FAO/WHO (1987) Assessment of the state of pollution of the Mediterranean Sea by mercury and mercury compounds. MAP Technical Reports Series No 18, UNEP, Athens, Greece

World Health Organisation (1976) Mercury. WHO, UNEP Environmental health criteria 1, Geneva, Switzerland pp 1–131

World Health Organisation (1989) Mercury – Environmental aspects. WHO-IPCS Environmental health criteria 86, Geneva, Switzerland pp 1–115

World Health Organisation (1990) Methylmercury. WHO-IPCS Environmental health criteria 101, Geneva, Switerland pp 1–144

World Health Organisation (1991) Inorganic mercury. WHO-IPCS Environmental health criteria 118, Geneva, Switzerland pp 1–147

Wren CD (1986) A review of metal accumulation and toxicity in wild mammals I. Mercury. Environ Res 40:210–244

Zelenko V (1986) Porazdelitev metil-Hg v človeških možganih. (Distribution of methylmercury in the human brain.) IJS Technical Report DP-4454, Jožef Stefan Institute, Dept of Nuclear Chemistry, Ljubljana, Slovenia (in Slovene)

Vegetation on Contaminated Sites near an Hg Mine and Smelter

V. Banásová

1 Introduction

Former mercury mines and smelter are situated in Rudnany (east Slovakia).

Mine spoil heaps are composed of the stones which had no value for processing in the past. However, the soil which developed there contains higher amounts of Hg. The main pollution source of Hg contamination was a smelter plant complex situated at the mouth of the Rudnianska dolina valley. According to Bobro (1989) and Bobro et al. (1992), compounds of S, Hg, Cu, Sb, Bi, As, Mn, Zn, Pb and Fe were emitted.

The vegetation polluted by mercury was studied by many authors. The majority of these focused on mercury accumulation in the plants. Fungi and lichens (Lodenius 1979), water plants (Lodenius 1980) and other plants (e.g. Weaver et al. 1984; Koski et al. 1988) were the objects of such studies.

The first data on vegetation structure in the surroundings of the old smelter complex situated on the Rudnany village margins were published by Hajdúk (1961). Pišút (1962) published a bioindicator study based on lichens from these sites. The same author (Pišút 1984) published a comparative study of these places after 20 years. He noticed that in the surroundings of the smelter no epiphytic lichens occurred. The paper by Hauskrecht and Hajdúk (1966) also includes mercury levels in the plants of *Allium montanum*, whose population size increased considerably, for a short time, in the smelter surroundings. These papers deal with the situation in the surroundings of the old smelter complex on the Rudnany village margins, which was still operating until the 1970s. Subsequently, a new smelter was built outside Rudnany. Kaleta (1982) observed the vegetation changes in its surroundings after 10 years. Maňkovská (1984) presented mercury levels recorded in soils and spruce needles. The impact of emissions from the smelter on the population of *Coronilla varia* were studied by Banásová and Holub (1993). Mercury concentrations in different plant species were presented by Banásová et al. (1993) and Banásová and Holub (1994), and lichens by Lackovičová et al. (1994).

The aim of this chapter is the evaluation of:

- Contamination of soils and plants by Hg, Cu, S.
- Characteristics of the vegetation on Hg mine spoil heaps.
- Influence of an Hg smelter on plant populations and plant communities.

Environmental Science
Mercury Contaminated Sites (ed. by R. Ebinghaus et al.)

2 Methods

Investigations were carried out under field conditions on mine spoil heaps and in the vicinity of the smelter.

2.1 Mine Spoil Heaps

The cover of plant species was estimated using methods of phytosociological relevés according Braun-Blanquet (1964). The soil samples were taken from the rhizosphere (0–5 cm), oven-dried and sieved through a 2-mm sieve. Total N was analyzed using Kjehldahl, oxidizable carbon Cox using the Walkley-Black methods. Plant and soil samples were oven-dried and ground for analyses of heavy metals. Total Hg content was determined using cold vapour analysis (CVAA). Total Cu in soil and plants (dry matter) was analyzed after wet digestion (concentrated nitric and sulphuric acids 3:2) using flame AAS.

2.2 Smelter

The study of *Calamagrostis epigeios* populations was carried out during smelter operations.

The *Calamagrostis epigeios* populations were analyzed on seven plots of 1 m^2 at different distances from the smelter. Within each plot the following determinations were carried out:

- total Hg content in soils using CVAA analysis,
- total above ground biomass of *C. epigeios*/plot, biomass of *C. epigeios* panicles/plot,
- number of panicles,
- size and weight were measured on 30 panicles on each plot.

Vegetation structure of plant communities was studied on the transect from the smelter in the direction of prevailing winds in the time, after smelter activity stopped.

The study of plant communities was carried out on five sites at different distances from the smelter. On each site the phytosociological relevés were recorded according to Braun-Blanquet (1964).

Soils samples from a depth of 0–5 cm and samples of above ground parts of plants were taken on the sites where vegetation relevés were recorded. The Hg concentrations were measured using CVAA analysis and copper by AAS. Sulphur content was determined by LECO analyzer, while pH (H_2O) was measured by potentiometer.

3
Results and Discussion

3.1
Hg Mine Spoil Heaps

The exploitation of mercury in this region dated since the 18th century (Kodera 1990). In the surroundings of the mine there are small, unreclaimed heaps. The soils developed several years after dumping and mosaic vegetation cover was established by colonization from native local plant species.

3.1.1
Characteristics of Soils

The substrate of mine spoil heaps is composed of stones. The spoil heaps have a graveley texture, so the plants growing there are deficient in water and nutrients (0.65–2.28% C_{ox}, 0.1–0.19% N). The input of heavy metals from the weathering spoil material to the soil has been recorded. According to our analyses, the soil contains high concentrations of metal residues. For example, Hg concentration ranged from 72.7–4700 mg/kg and Cu from 40.0–11 300 mg/kg. Soil established on the heaps has an alkaline reaction, with pH ranging from 7.3–8.1.

3.1.2
Contamination of Plants

Notable contamination of aboveground plant parts by mercury, copper and sulphur was found on study heaps. The aboveground parts of dominant plant species were analyzed.

Higher amounts of Hg and Cu occurred in *Agrostis stolonifera* than in *Silene vulgaris*. Mercury concentration reached 11.2 mg/kg in the aboveground parts of *Agrostis stolonifera* and reached 7.44 mg/kg in *Silene vulgaris*. Concentrations of Cu were 87 mg/kg in *Agrostis stolonifera* and 18.5 mg/kg in *Silene vulgaris*.

3.1.3
Characteristics of Vegetation

Colonization of spoil heaps by higher plants depends on soil properties such as weathering, chemical changes and microclimate (Schubert 1954; Ernst 1965). One of the basic conditions is the ability of plant species to develop metal resistance (Ernst 1990).

Plants colonies on Hg mine spoil heaps are very patchy. Vegetation cover is a mosaic, depending on the weathering and development of soils. High percentages of barren soil occur there. Species such as *Agrostis stolonifera*, *Cardaminopsis arenosa* and *Silene vulgaris* dominate. Other plants, such as *Hypericum perforatum*, *Lotus corniculatus*, *Dianthus carthusianorum* and *Pimpinella saxifraga*, constituted only a minor part of the stand. Shrubs, *Sambucus racemosa* and *Salix capraea*, occur in the margin of old heaps with a low cover.

Agrostis stolonifera, which also grows on the heaps after Hg mining, is also a frequent colonizer of toxic mining spoil and can adapt to the poor soils and to the high concentration of metals (e.f. Schubert 1954; Ernst 1974; Banásová 1976).

Some plants growing on Hg mine spoil heaps react to the extreme ecological conditions by creating red spots on the leaves *(Hypericum perforatum, Agrostis stolonifera, Pimpinella saxifraga, Silene vulgaris)* or black apical necrotic spots on the leaves of shrubs *(Sambucus racemosa, Salix capraea)*. The reduction of the number of flowers in blossoms to one or two was noticed in *Lotus corniculatus* (Fig. 1). Retarded plant growth was found in *Plantago lanceolata* (Fig. 2).

3.2 Hg Smelter

The Hg smelter operating since 1970 provided an opportunity to follow the stages of damage to surrounding ecosystem in a short time. Ten years after starting the smelting process, trees began dying (Kaleta 1982) and herb layer cover increased. After about 20 years of smelter work operations, we recorded greater changes in

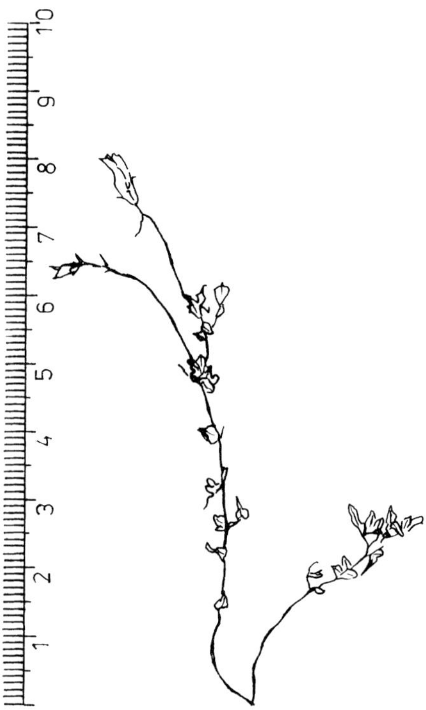

Fig. 1. Reduction of inflorescence in *Lotus corniculatus* growing on Hg mine spoil heaps

Fig. 2. Reduction of growth in *Plantago lanceolata* growing on Hg mine spoil heaps

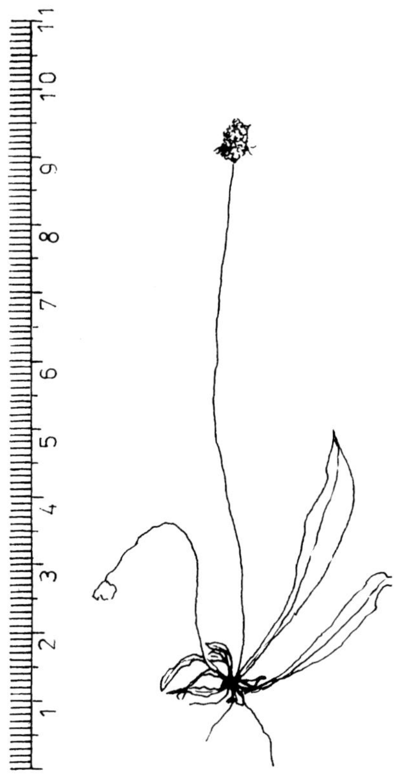

the original stands. Some species, e.g. *Calamagrostis varia*, were extinct, having been replaced by the more tolerant *Calamagrostis epigeios*. The originally abundant species had become rare, e.g. *Coronilla varia* (Banásová and Holub 1993), and the expansive and tolerant species *Calamagrostis epigeios* and *Sambucus ebulus* had spread out.

3.2.1
Contamination of Soils

The main emissions were Hg, Cu and other metals, plus SO_2 gas emissions. Dispersion of emissions depends on landscape configuration and prevailing winds. The Stožky hill is situated on the opposite side of the smelter complex. The distribution of emissions was different not only in horizontal, but also in vertical distances (Banásová and Holub 1993). According to our measurements, the highest concentration of metal and sulphur in soils was on the northern slopes of the hill, just opposite the smelter, mainly in their lower and middle parts, approximately up to the level of the chimney heights. Toxic agent content decreases on hilltops and with increasing distance from the smelter (Fig. 3). The parent rocks are limestone and dolomite in this region, so soils have an alkaline

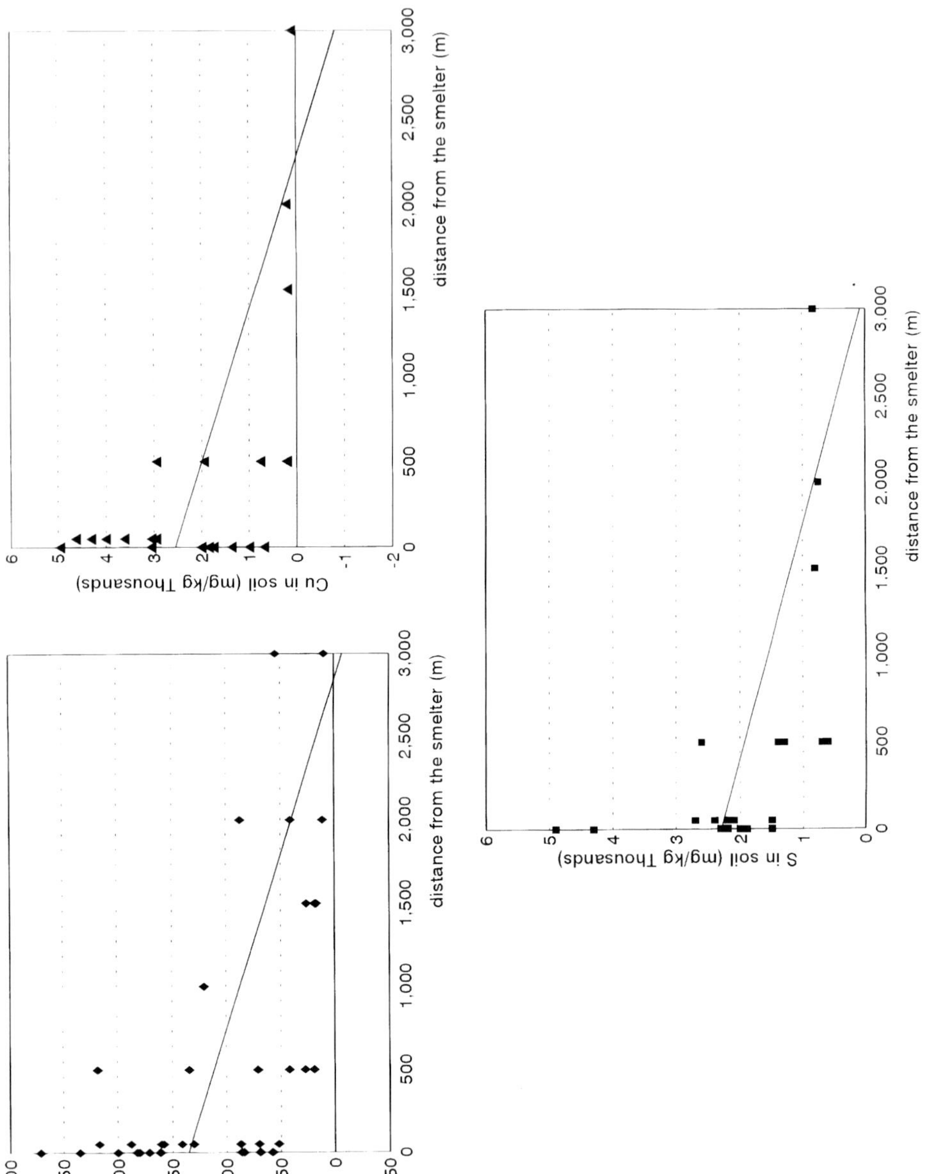

Fig. 3. Distribution of Hg, Cu and S in soils along the transect from the smelter

pH (7.4–7.9). Due to SO_2 emissions, soil pH decreased to extreme values (4.6–4.0) in the surroundings of the smelter.

3.2.2
Contamination of Plants

Notable contamination of aboveground parts of plants by mercury, copper and sulphur was found. The aboveground parts of trees, shrubs and forbs growing in the surroundings of the smelter were analyzed. High amounts of mercury and copper were found in the leaves of birch (*Betula pendula*) and in the aboveground parts of *Calamagrostis epigeios* and *Coronilla varia* (Fig. 4).

3.2.3
The Influence of the Hg Smelter on Plant Populations

The impact of air pollution on *Coronilla varia* populations (Banásová and Holub 1993) near the Hg smelter was studied. A reduction in leaf size of *C. varia* and fruit (loment) size, and total absence of blossoms were observed on highly contaminated sites near the smelter. The vitality of a *C. varia* population, at a

Fig. 4. Maximum total Hg concentration in aboveground part of plants growing around the smelter

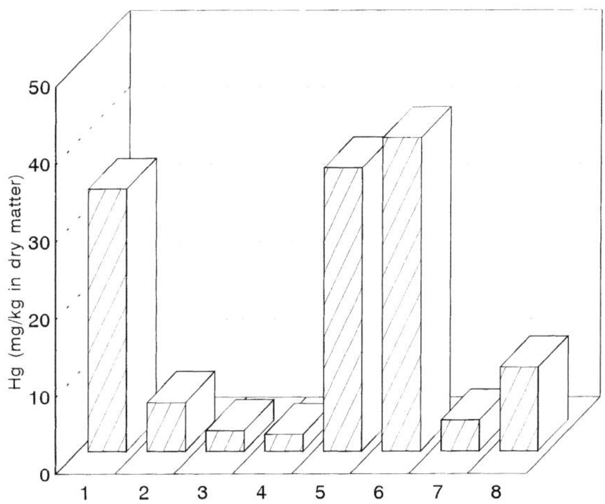

1. Betula pendula
2. Picea abies
3. Sorbus aucuparia
4. Rubus idaeus
5. Calamagrostis epigeios
6. Coronilla varia
7. Senecio fuchsii
8. Hypericum perforatum

distance of more than 1200 m from smelter, was still good. Leaf size was almost equal to that of the control. Plants on this site flowered and produced seeds.

Deforested sites with changed soil chemistry are occupied by the invasive plant species *Calamagrostis epigeios* with 80–100% cover. The cover of *C. epigeios* stand was reduced to 50% on the sites with the highest contamination. *C. epigeios* creates monotonous cover on the more contaminated areas. The negative impact of Hg smelter operation occurred also in *C. epigeios*, although it belongs to the tolerant and common plant species. A notable decrease in plant height occurred on high contaminated sites. Development of panicles in the population was influenced by pollution. On highly contaminated soils the frequency of inflorescence decreased (Fig. 5). The panicle length showed no great differences between plants growing on low or high contamination (Fig. 5). Sizes of panicles are probably determined on a genetic basis and are practically the same near to and at great distances from the smelter (Fig. 5).

The impact of high contamination occurs also in the aboveground biomass values. A moderate stimulatory effect is shown by low Hg concentration, but increasing Hg concentration decreased total aboveground biomass. (Fig. 5)

Weight of panicles decreased on highly contaminated soils. The decrease in panicle weight occurring on less contaminated soils is probably affected by the competition of higher cover of other plants.

The total biomass of the aboveground parts of *C. epigeios* decreased with increasing Hg content (Fig. 5).

The damage symptoms showed some *C. epigeios* plants growing in the communities near the smelter chimney. Deformed panicles, less seed production, and bleaching are observed there (Fig. 6).

3.2.4
The Influence of the Smelter on Plant Communities

The original forests died out and were replaced by herbaceous vegetation in the surroundings of the smelter. Some of the species belonging to the original forest communities, such as *Senecio fuchsii* and *Rubus idaeus*, occurred also in the degraded stands (Fig. 7). Both the cover of this species and species richness decreased with increasing contamination (Fig. 8). *Calamagrostis epigeios* becomes the dominant species and creates monotonous cover, lacking any other species. Diversity of plant cover increases where emission influence is slight, due to invading species, i.e. *Sambucus ebulus*, or plants of original rock and forest communities.

Depending on the distance from the smelter, degree of contamination and landscape configuration, the following degradation stages have arisen:

First Stage. Beyond a distance of 3 km from the smelter, original stands of *Picea excelsa*, *Fagus sylvatica* and *Abies alba* occur. The first damage symptoms were visible on spruces and firs. The pines (*Pinus sylvestris*) did not manifest

Fig. 5. Number, length, biomass of *Calamagrostis epigeios* panicles, and total above-ground biomass of *C. epigeios* in relation to Hg contamination of soil around the Hg smelter

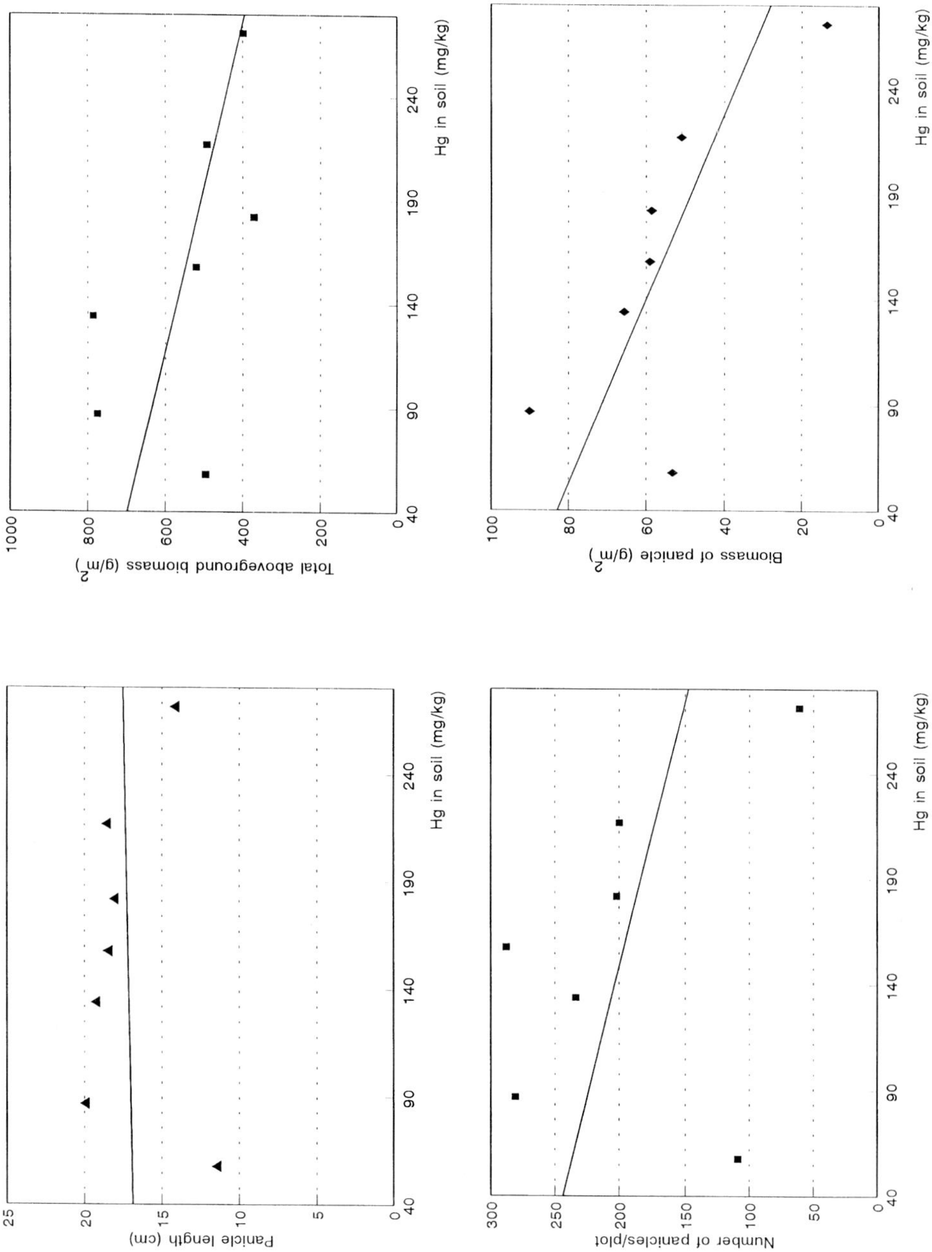
Total aboveground biomass (g/m2)
1000
800
600
400
200
0
Hg in soil (mg/kg)
40
90
140
190
240
Biomass of panicle (g/m2)
100
80
60
40
20
0
Hg in soil (mg/kg)
40
90
140
190
240
Panicle length (cm)
25
20
15
10
5
0
Hg in soil (mg/kg)
40
90
140
190
240
Number of panicles/plot
300
250
200
150
100
50
0
Hg in soil (mg/kg)
40
90
140
190
240

visible symptoms of damage. The tree, shrub and herb layers were normally developed. The shrub layer consisted mostly of *Sorbus aucuparia*, *Coryllus avellana*, *Sambucus nigra* and *Rubus idaeus*. The herb layer with a dominance of *Senecio fuchsii* did not manifest visible damages. *Calamagrostis epigeios* did not occur in this stand. The floristic composition was close to the spruce forests in wider surroundings (c.f. Hajdúk 1963). The average number of species was 27 and that of the woody plant species 8.5. Mercury levels in the soil fluctuated from 9.1 to 54.3 mg/kg.

Second Stage. The damage was much more obvious at 2 km from the smelter works. Rapid decrease in tree vitality occured. The lower twigs of spruces lost their needles, their crowns were strongly opened and the surviving needles were necrotic. The seedlings from natural regeneration were dwarfed, the increments were inhibited and bizzare malformations occurred. The cover of *Picea excelsa* was reduced and replaced by a higher cover of *Sorbus aucuparia* (Fig. 6). In general, the cover of the tree layer was reduced. The shrub layer was well developed.

An obvious decrease in vitality was observed in *Rubus idaeus*. It reached maximally 80 cm, did not branch, and its fructification was reduced. Due to the forest opening, the herb layer cover increased. The common species in the herb layer were *Sesleria albicans*, *Festuca pallens*, *Brachypodium pinnatum* and others. *Senecio fuchsii*, the plant of original forest communities, was still present, but its cover was reduced (Fig. 6). Patches of *Calamagrostis epigeios* appeared in the stands. The average number of species was 10.5 and that of the woody plant species 3.

Mercury levels in the soil ranged from 40.8 to 87.0 mg/kg. pH was 7.0.

Third Stage. The contamination of soil by mercury increased at the distance of about 1 km from the smelter works and reached about 120 mg/kg. Soil pH (7.53) was still slightly alkaline. A sudden decrease in the number of species was recorded. The tree layer is missing in the stands. The conifers disappeared totally from the stands and the deciduous trees like *Betula verrucosa* reached only the height of the shrub layer and occurred sporadically there. *Rubus idaeus* had only simple, unbranched shoots and reached a height of only 10–30 cm. They were sterile, without flowers and fruits, and had necrotic leaves. Cover of *Senecio fuchsii*, a species of the original forest community, was low (Fig. 6). Plants like *Origanum vulgare*, *Campanula carpatica* and *Coronilla varia* survived in the herb layer. *Calamagrostis epigeios* become a dominant species, but apical necroses occurred sometimes on its leaves. The undecomposed litter from the preceding year formed a layer about 2 cm thick.

Fourth Stage. A simple community was developed on large areas in the front of the smelter works. The tree and shrub layers did not occur and the herb layer consisted only of monocoenosis of *Calamagrostis epigeios* (Fig. 6). Its populations exhibited a reduced vitality, manifested by a large number of sterile individuals and by necroses on the leaves.

Sporadic tufts of *Festuca duriuscula* and *Gallium mollugo* and sterile individuals of *Coronilla varia* occurred. The undecomposed detritus formed a layer of 10–15 cm. Small leaves and shortened branch axes in birches formed bizzare broom-shaped trees. The average number of species was 3.1. Mercury level in the soil was very high and reached 160–200 mg/kg. Soil pH decreased rapidly to 4.2.

Fig. 6. Deformity of *Calamagrostis epigeios* panicle growing around the Hg smelter. *Right* normal developed panicle

Fifth Stage. The last degradation stage close to the smelter works was characterized by the occurrence of bare land. The area of bare land reached ca. 8 m^2, occurring among the patches of monocoenoses of *Calamagrostis epigeios*. The surfaces are covered by a 15–20-cm-thick layer of undecomposed dead plants from preceding years. The soil contamination was high, the mercury levels reached 200–235 mg/kg. Due to the SO_2 emissions, soil pH reached 4.0, although the parent rock is limestone.

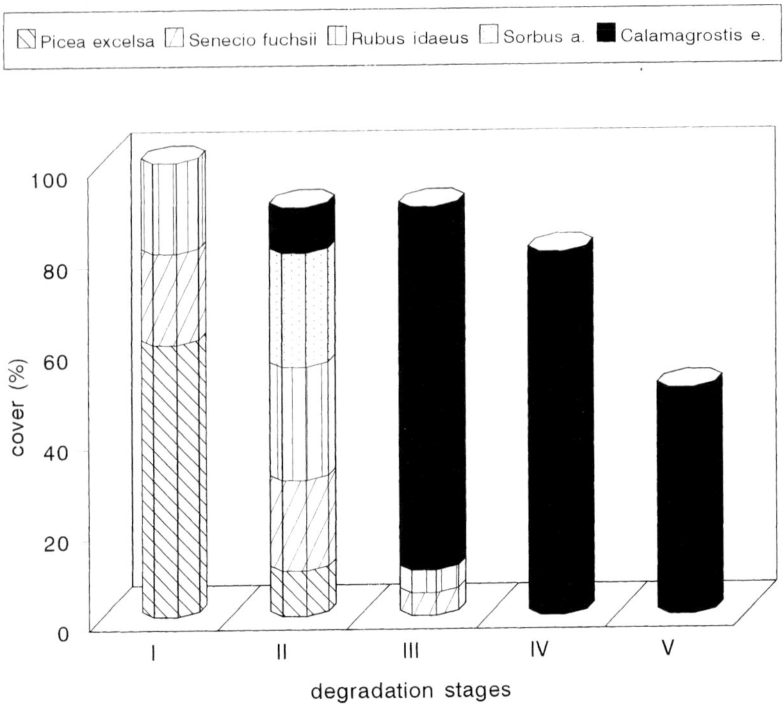

Fig. 7. Cover of main plant species in degradation stages of plant communities in the surroundings of the Hg smelter

4
Conclusion

Extreme ecological conditions on the spoil heaps pose problems for plant growth. The vegetation is patchy, with plants growing in the fissures between stones. The plants which can grow on the Hg spoil heaps belong to the group of plants with wide ecological amplitude. They can survive in conditions with deficiency of water and nutrients, and in high concentrations of Hg and Cu. These include *Agrostis stolonifera* and *Cardaminopsis arenosa*, *Silene vulgaris* species in these areas. Ruderal plant species do not occur very often.

A secondary effect of the metallurgical processing of raw materials is contamination of the environment by emissions and subsequent changes in vegetation (e.g. Hajdúk 1961; Kaleta 1982, 1991; Schubert 1985; Kontrišová 1980). The stand structures change rapidly (Hajdúk 1963, Kowarik and Sukopp 1984 etc.).

The chemical parameters, such as high concentrations of mercury and other metals including SO_2, were changed the ecosystem around the smelter. The *Coronilla varia* and *Calamagrostis epigeios* plants and populations showed

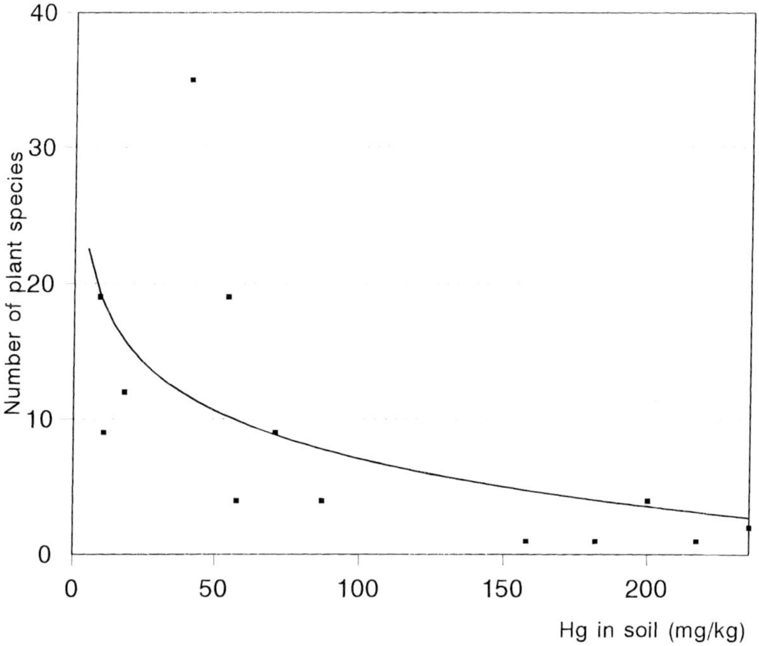

Fig. 8. Plant species richness in relation to Hg contamination of soil in the surroundings of the Hg smelter

morphological changes and a reduced degree of vitality and reproductive potential due to the impact of the smelter. The vegetation around the smelter is characterized by species-poor stands with prevailing *Calamagrostis epigeios*. The soil pH decreased to 4.0 in the most contaminated sites and bare land occurred between the stands of *Calamagrostis epigeios*. Occurrence of bare land was recorded at first in 1993, 20 years after the start of smelter operations.

The impact of the smelter on the vegetation was a result of the synergistic effects of pollution by Hg, Cu, sulphur dioxide and other pollutants. The extinction of the original phytocoenoses and development of the degradation stages around the smelter complex near Rudnany have been caused by emissions. Both qualitative (the structure of plant communities) and quantitative parameters (number of species) have been changed in the impacted area due to pollution.

Acknowledgements This research was supported by the Slovak Academy Sciences, Project No. 4105 Plant communities of Slovakia, which is gratefully acknowledged.

I would like to express my thanks to Dr. Z. Holub for help during the investigations in the field and for constructive comments.

References

Banásová V (1976) Vegetation of copper and antimony mine spoil heaps. Biol pr 22, 1, VEDA Bratislava, 109 pp (in Slovak)

Banásová V, Holub Z (1993) Response of *Coronilla varia* L. populations to environmental pollution by mercury and other pollutants. Biológia (Bratislava) 48:403–406

Banásová V, Holub Z (1994) The impact of the exploatation and processing of mercury ores on the vegetation. Proc Mineral Resources, Environment and Health, Ostrava, pp 11–14 (in Slovak)

Banásová V, Holub Z, Zeleòáková E (1987) The dynamics structure and heavy metal accumulation in vegetation under long-term influence of Pb-Cu immissions. Ekológia (ČSSR) 6:101–111

Banásová V, Pišút I, Holub Z (1993) The contamination of soils and vegetations by mercury and other elements from emissions of the smelter in Rudnany. Čistota ovzdušia 23:267–271 (in Slovak)

Bobro M (1989) Mine and smelter activity on the middle part of Spiš region and the occurrence of the pollution. Rudy 9:249–252 (in Slovak)

Bobro M, Zaic J, Merva M, Bálintová M, Hančuľák J, Leško O (1992) Concentration of mercury in both the workplace and environment of treatment complex iron mines in Rudnany. Proc Environ Miner Proc Ostrava:84–89 (in Slovak)

Braun-Blanquet J (1964) Pflanzensoziologie. Springer, Wien New York

Ernst W (1965) Ökologisch-soziologische Untersuchungen der Schwermetall-Pflanzengesellschaften Mitteleuropas unter Einschluss der Alpen. Abh Landesmus Naturkd Münster Westfalen, 27, p 1

Ernst W (1974) Schwermetallvegetation der Erde. Gustav Fischer, Stuttgart

Ernst WHO (1990) Mine vegetation in Europe. In: Shaw AJ (ed) Heavy metal tolerance in Plants. Evolutionary aspects. CRC Press, Boca Raton pp 22–37

Hajdúk J (1961) Quantitative and qualitative changes in the phytocoenoses in consequence of smelter works emissions. Biológia (Bratislava) 16:404–418 (in Slovak)

Hajdúk J (1963) Florographic characteristics of the Galmus region. Biol Pr, pp 1–52 (in Slovak)

Hauskrecht I, Hajdúk J (1966) Gehalt an Hg und anderen Elementen im Berglauch *Allium montanum* F.W. Schmidt, der um den Exhalationsquellen des Eisenwerkes in Rudnany wächst. Biológia (Bratislava) 21:, 676–680

Kaleta M (1982) Forest ecosystems in the surroudings of the smelter in Slovakia. Čistota ovzdušia 5:1–7 (in Slovak)

Kaleta M (1991) Long-term influence of smelter work emissions in Rudnany on the vegetation. Čistota ovzdušia 21:239–244 (in Slovak)

Kaleta M, Banásová V (1992) The influence of emissions from the smelter works in Rudnany, Krompachy amd Nižná Slaná on the vegetation and soils. Proc Ochrana a tvorba životného prostredia v najvýznamnejších sídelných aglomeráciách ČSFR, Košice, pp 109–116 (in Slovak)

Kodera M (ed) (1990) Topographic mineralogy of Slovakia. VEDA Bratislava (in Slovak)

Kontrišová O (1980) Meadow communities in an area influenced by fluorine-type emissions (žiar basin). Biol Pr, VEDA Bratislava 26, 2, 160 pp (in Slovak)

Koski E, Venäläinen M, Nuorteva P (1988) The influence of forest type, topographic location and season on the levels of Al, Fe, Zn, Cd and Hg in some plants in southern Finland. Ann Bot Fenn 25:365–370

Kowarik I, Sukopp H (1984) Auswirkungen von Luftverunreinigung auf die spontane Vegetation (Farn- und Blutenpflanzen). Angew Bot 58:157–170

Lackovičová A, Martiny E, Pišút I, Streško V (1994) Element content of the lichen *Hypogymnia physodes* and spruce needles in the industrial area of Rudnany and Krompachy (NE Slovakia). Ekológia (Bratislava) 13:415–423

Lodenius M (1979) Mercury content of *Hypogymnia physodes* and pine needles affected by a chloralkali works at Kunsankoski, SE Finland. Ann Bot Fenn 16:7–10

Lodenius M (1980) Aquatic plants and littoral sediments as indicators of mercury pollution in some areas in Finland. Ann Bot Fenn 17:336–340

Mankovská B (1984) The effects of atmospheric emissions from the Krompachy, Nižná Slaná, Rudnany iron ore mines on forest vegetation and soils. Ekológia (ČSSR).

Pišút I (1962) Bemerkungen zur Wirkung der Exhalationsprodukte auf die Flechtenvegetation in der Umgebung von Rudnany (Nordostslowakei). Biológia (Bratislava) 17:481–484

Pišút I (1984) Die epiphytische Flechtenflora in der Umgebung der Ortschaft Rudnany (Nordostslowakei). Zb. Slov Nár Muz Prír Vedy 30:27–37
Schubert R (1954) Die Schwermetallgesellschaften des östlichen Harzvorlandes. Wiss Z Martin-Luther Univ Halle-Wittenberg Math Naturwiss Reibe 3:51–70
Schubert R (1985) Bioindikation in terrestrischen Ökosystemen. Gustav Fischer, Jena, 283 pp
Weaver RW, Melton JR, DeShin Wang, Duble RL (1984) Uptake of arsenic and mercury from soil by bermudagrass *Cynodon dactylon*. Environ Pollut 33:133–142

Modern Mercury Contamination from Historic Amalgamation Milling of Silver-Gold Ores in the Carson River, Nevada and Jordan Creek, Idaho: Importance of Speciation Analysis in Understanding the Source, Mobility, and Fate of Polluted Materials

P.J. Lechler

Approximately a century after peak mining and milling activities at two historic mining camps in the western United States, Hg pollution persists in the Carson River, Nevada, and Jordan Creek, Idaho. Active channel sediments in these streams are still enriched in Hg at levels of 2 to 4 orders of magnitude above regional background.

The dominant form of Hg is still elemental Hg, the form historically released to the environment (although some Hg sulfide was released from natural mineralization in the Jordan Creek watershed). However, in more-anoxic reservoir sediments and deep soil horizons at the Carson River Superfund Site, significant proportions of Hg sulfide have been formed where total sulfur exceeds 0.05 to 0.10 weight percent. Eh-pH diagrams describing Hg and S forms in a variety of surficial environments agree with the Hg speciation results for elemental and sulfide species obtained in this study using a combination of published methods.

1 Introduction

Prior to the introduction of cyanidation technology at the beginning of the 20th century, mercury-amalgamation of precious metal ores was common practice throughout the western United States. Amalgamation is the alloying and collection of fine gold-silver (Au-Ag) particles in puddles, droplets, or coatings of mercury (Hg). The Hg is then collected and heated to evaporate the Hg away from the precious metals; the Hg is recondensed in a retort and saved for reuse. Some loss of Hg (and associated precious metals) is attendant in the many steps in this metallurgical process.

This western US mining era began with the discovery of gold in California in 1848 and the Californian gold rush of 1849 (Fisher and Holmes 1968). Placers from the Comstock lode in Nevada were discovered in 1859 and pan amalgamation of these ores, mainly between 1860 and 1895, released approximately 6.4×10^6 kg of Hg to the Carson River/Lahontan Reservoir system (Smith 1943; Ansari 1989). It is also estimated that 3×10^6 troy ounces of Au and 64×10^6 troy ounces of Ag were lost along with the Hg. These tremendous amounts of Hg, Au, and Ag now reside in mill tailings and the channel sediments and flood plain deposits of the Carson River, largely along the 110-km stretch between Carson City and Fallon, Nevada. Approximately 100 mills processed Comstock ore at one time or another during the peak mining period. Recognition of the magnitude of this pollution led to the

Environmental Science
Mercury Contaminated Sites (ed. by R. Ebinghaus et al.)

establishment of this area as a US Environmental Protection Agency Superfund Site in 1990 (S. Hogan, pers. comm.).

Placers from the Silver City/Delamar area (also known as the Owyhee Country) of southwestern Idaho were discovered in 1863 and led to the pan-amalgamation milling of lode ores in the Silver City/Delamar areas, and the release of Hg to Jordan Creek, until the beginning of the 20th century (Wells 1983). Jordan Creek flows out of the Owyhee Range in southwestern Idaho and into the Owyhee River in southeastern Oregon. After being dammed to form the Owyhee Reservoir in Oregon, the river flows into the Snake River, which is dammed to form Brownlee Reservoir. Both fluvial systems are still contaminated with Hg, with concentrations in the range of 1 to 10 µg/g total Hg in active channel sediments of the Carson River, up to 100 µg/g in Lahontan Reservoir, and up to 70 µg/g in Jordan Creek. All of these creeks, rivers, and reservoirs in Nevada, Idaho, and Oregon contain at least some fish which are contaminated with Hg.

Both mining camps were enriched in Ag relative to Au, and the Ag was largely present as the sulfide or sulfosalt, which led to the use of the Washoe pan process for the effective recovery of Ag from these ores (Smith 1943; Wells 1983). The Washoe pan process was developed on the Comstock as a variation of the Mexican Patio Process for the extraction of Ag from complex ores. It involved the treatment of finely crushed (stamp-milled) ore slurry in a large metal pan which was heated with steam to speed up the process. The Ag sulfides or sulfosalts were decomposed through the addition of large amounts of salt (NaCl) which dissolved perhaps 60% of the Ag as a Ag chloride complex. Elemental Hg was added to the slurry to amalgamate free Au, electrum (Au-Ag alloy), and elemental Ag which was reduced from its chloride complex through the addition of iron nails to the mixture.

The form of Hg released to the environment by miners in both areas was mainly elemental Hg, although some Hg sulfide was released from natural mineralization in the Jordan Creek watershed as well. Eh-pH diagrams indicate that the form of Hg stable in these surficial environments is elemental Hg, and Hg speciation analyses performed as described below indicate that the form of Hg present in these systems is still dominantly the elemental form. Exceptions to this exist, however, as explained later. The watersheds of these systems still contain subaerial mine and mill tailings contaminated with Hg (up to 4600 µg/g total Hg in tailings in the Carson River watershed, for example) which will remain a source of Hg to these fluvial systems, ensuring their continued pollution with Hg for a very long time. Thousands of tons of mine and mill wastes are present on the flood plains of both watersheds; however, the fine-grained mill tailings typically generated by historic stamp mills are much more abundant in the Carson River watershed, Jordan Creek being more impacted by coarser-grained mine wastes. Niriagu (1994) estimated that 63380 tons of Hg may have been released from 1820 through 1900 as a result of amalgamation milling of precious metal ores throughout the United States, and Basham et al. (1996) concluded that this was probably a reasonable estimate based on their development of a database of historic amalgamation sites in the US.

The chemical form of Hg present in contaminated systems is important from the perspective of bioavailability of Hg to humans and the ecosystem. Thus, the chemical form of Hg in sediments and soils is an important determination, but there is currently little consensus on the best methodology for determining the distribution of Hg species in soils or sediments. This contribution reports on the distribution of Hg and its chemical form in sediments of the Carson River and Jordan Creek, and soils from the Carson River Superfund Site and nearby areas. While the quality assurance of speciation methodology is difficult (Wallschlager et al. 1996), compatibility of the speciation results reported here with theory (Eh-pH diagrams), other geochemical features of the systems, and systematic, geochemically logical spatial variations in speciation, impart confidence to the techniques employed.

2 Methods

2.1 Field Methods

Sediments were collected from the upper 10 to 20 cm of the active channel of Jordan Creek, the Carson River, and the bed of Lahontan Reservoir. Soils and tailings were collected as integrated samples from discrete subsurface horizons, defined by visually determined media heterogeneities such as changes in grain size, texture, color, and degree of induration. The samples were placed in plastic bags and refrigerated until analysis. Wet samples were used for all analyses, while moisture content was determined on a separate sample split; all analytical results were corrected to a dry weight basis (105 to 110 °C).

Five sediment samples from the active channel of the Carson River, and one from the bed of Lahontan Reservoir, were collected from more-or-less evenly spaced sites to investigate spatial variations in the chemical speciation of Hg during downstream transport (Fig. 1). In the same way, nine samples were collected every 1.6 km down Jordan Creek from just below Silver City to the base of the Owyhee Range, 14.5 km downstream (Fig. 2).

The upstream sample site in the Carson River is located above the point at which mercuriferous tailings are judged to be entering the river and therefore constitutes a background sample representative of the concentration and speciation of Hg in uncontaminated sediments in this area. The next four samples were collected from 8 to 58 km downstream in the river and the final sample was collected from Lahontan Reservoir, 12 km more distant (Fig. 1). No such background sample was collected from Jordan Creek because of logistical difficulties, although the upstream sample below Silver City is substantially less contaminated than downstream samples. This is compatible with most of the amalgamation milling of ores being performed below Silver City (Wells 1983).

Soils in the western Nevada study area are typical semiarid soils characterized by poor A and B soil-horizon development. Often, a layer of caliche is developed

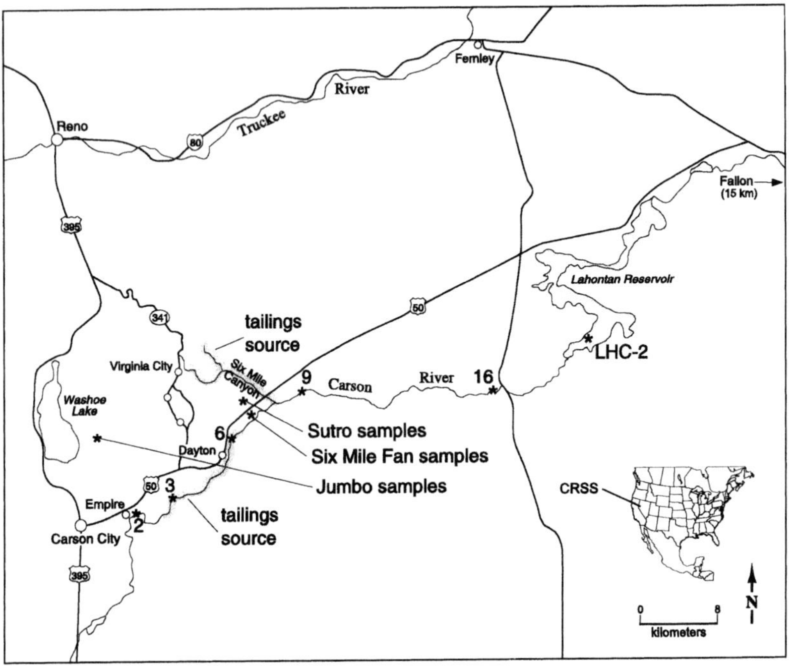

Fig. 1. Location map of the Carson River Superfund Site in west-central Nevada, USA. Carson River sample locations *2*, *3*, *6*, *9*, and *16* are shown, as is the location of sample LHC-2 in Lahontan Reservoir. Soil sampling locations from *area 1* (Sutro and Six Mile Fan locations) and *area 2* (Jumbo location) are also shown

to a greater or lesser extent several centimeters below the surface. Soil samples were collected from hand-dug trenches as "channel" samples from discrete pedologic horizons as described above. Samples were obtained from two principal areas: (1) two sites on the strongly contaminated alluvial fan at the mouth of Six Mile Canyon and (2) two sites along Jumbo Grade, a small, historic gold mining area peripheral to the main Comstock District (Fig. 1).

2.2 Analytical Methods

Gold and silver were determined on aliquots taken from the solutions generated in step 1 of the sequential extraction series described below. Silver was determined directly by flame atomic absorption spectrometry using D_2 background correction. Gold was determined by graphite furnace atomic absorption spectrometry after extraction into MIBK and back-washing of iron with 10% (v/

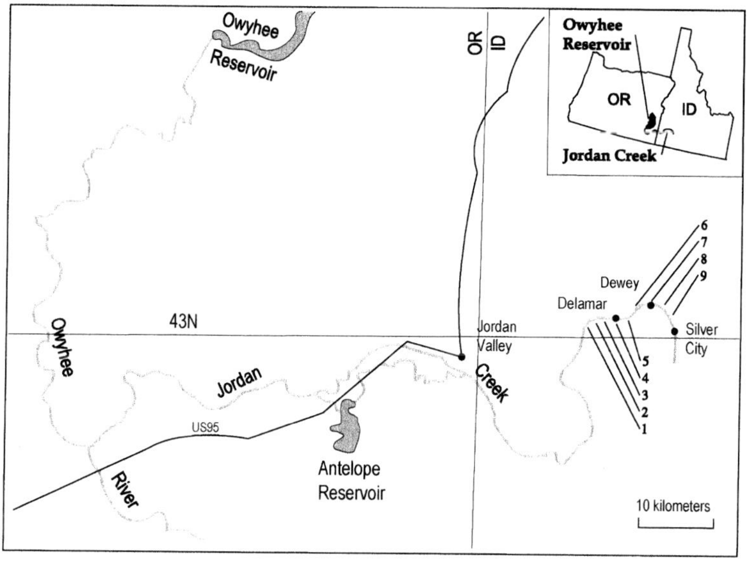

Fig. 2. Location map of Jordan Creek in southwestern Idaho and southeastern Oregon. Jordan Creek sample locations in the Silver City–Delamar area of historic mining are shown

v) HCl. Total sulfur was determined by the classic $BaSO_4$ precipitation technique after sample digestion with HNO_3 and $KClO_3$ (Scott 1939).

Quality assurance of these data is tied to standards from the US National Institute of Standards and Technology and the US Geological Survey, replicate determinations, and round-robin interlaboratory comparisons (Lechler and Desilets 1991). Precision and accuracy of total Hg determinations is better than ± 10% of the amount present; NIST-8407 (50 ± 2 µg/g certified) was found to contain between 49.9 and 53.0 µg/g total Hg using the methodology described below; spike recovery was better than 97%; all samples are run in triplicate. Precision of the individual speciation analyses described below is better than ± 10%, and generally better than ±5%, of the amount present; accuracy of the individual speciation analyses is currently indeterminate.

2.3 Mercury Speciation Method

After a review of the literature covering the determination of various species of Hg in solid samples, the following sequential extraction method (Table 1) was established (based partially on the work of Eganhouse et al. 1978; Di Giulio and Ryan 1987; Quevauviller et al. 1992; Biester 1994):

Table 1. Mercury speciation procedure

Mercury species	Procedure
1. Total Hg	Place a 10 g sample in a 250-ml beaker, add 5 ml HNO_3 and 15 ml HCl, cover with a watch glass, and gently boil for 45 min. Dilute to 50 ml with 3 N HCl and determine total Hg by cold vapor atomic absorption spectrometry (CVAAS)
2. Elemental Hg	Place two 10 g samples in two 250-ml beakers in an oven for 48 h at 180 °C [the work of Biester (1994) clearly indicates that 180 °C is required to liberate elemental Hg from some soils]. Place petri dishes containing activated charcoal over the uncovered beakers, covering approximately ½ of each beaker's opening. This minimizes transfer of Hg vapor from one sample to another during heating and captures Hg vapor that would otherwise be released to the laboratory environment (this charcoal is subsequently disposed of with other hazardous laboratory wastes). Determine the Hg content of one of the heated samples and compare this concentration to the total Hg content of the sample; the difference is elemental Hg. The second heated sample is saved for the determination of other Hg species in the sample, starting at step 3
3. Exchangeable Hg	Using the second, heated sample from step 2, add 50 ml 0.5 M $MgCl_2$, mix at room temperature for 2 h, let stand for 60 min, decant 40 ml, centrifuge, and determine exchangeable Hg by CVAAS. Wash residue with 50 ml of deionized water, let stand for 60 min or until clear and pour off wash solution; save residue for step 4
4. Strongly bound Hg	Using the sample residue from step 3, add 50 ml 0.5 N HCl, mix at room temperature for 2 h, let stand for 60 min, decant 40 ml, centrifuge, and determine strongly bound Hg by CVAAS. Wash residue with 50 ml of deionized water, let stand for 60 min or until clear and pour off wash solution; save residue for step 5
5. Organic Hg	Using sample residue from step 4, add 25 ml 0.2 N NaOH, mix at room temperature for 2 h. Add 25 ml 4% (v/v) CH_3COOH (glacial acetic acid), mix at room temperature for 2 h, let stand for 60 min, decant 40 ml, centrifuge, and determine organic Hg by CVAAS. Wash residue with 50 ml of deionized water, let stand for 60 min or until clear and pour off wash solution; save residue for step 6
6. Hg sulfide and residual Hg	Using sample residue from step 5, add 25 ml saturated sodium sulfide (Na_2S) solution, mix at room temperature for 24 h; centrifuge and decant clear supernatant solution, add 25 ml DI water, mix, centrifuge, and decant, twice. Determine residual Hg on residue (by running total Hg on this portion). Subtract elemental, exchangeable, strongly bound, organic, and residual Hg from total Hg; difference is Hg sulfide

Definitions:

Elemental Hg is that fraction of the total Hg which is volatile upon heating as described above. Mercuric chloride will contribute to the elemental Hg number if it is present in significant quantities and should first be removed by leaching with deionized water and quantified if desired. Mercuric chloride is present at less than 1% of the total Hg in Carson River and Jordan Creek samples

Exchangeable Hg is that portion of the total Hg which can be exchanged off of active sites in soils or sediments (e.g., clays, Fe and Mn oxyhydroxides, and organic matter)

Strongly-bound Hg reflects Hg from several soil/sediment components such as Fe and Mn oxyhydroxides, a portion of the organically bound Hg which is exchanged to solution through

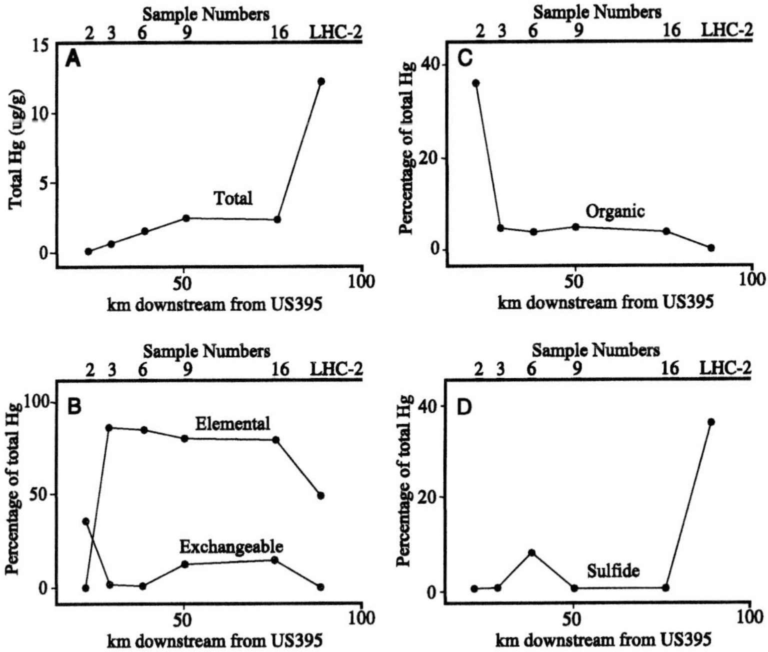

Fig. 3A–D. Distribution of total Hg (A) and elemental Hg (B), exchangeable Hg (B), organic Hg (C), and Hg sulfide (D) in Carson River and Lahontan Reservoir sediments

3 Carson River Superfund Site

3.1 Mercury in Sediments

Total Hg contents of Carson River sediments show a steady increase downstream as contaminated materials enter the active channel, mainly from tailings being eroded from banks as the river meanders through its contaminated floodplain

---

protenation of organic sites, and exchange from mineral surface sites where Hg is strongly sorbed. This makes interpretation of this "species" particularly ambiguous and difficult

Organic Hg is that Hg which is released for analysis upon digestion of organic acids with NaOH and organic bases with CH_3COOH

Hg sulfide is that portion of the total Hg soluble in saturated Na_2S solution and may be cinnabar, metacinnabar, or amorphous HgS

Residual Hg is that fraction of the total Hg which is not soluble and available for analysis in solution with other than hot aqua regia

Although methyl Hg is important to environmental investigations of Hg pollution, it is not determined in the method described here. Methyl Hg has been shown to be a very small fraction of the total Hg in Carson River sediments (Bonzongo et al. 1996)

(Fig. 3A). The proportion of elemental Hg in these samples starts near zero in the upstream sample at 22 km, and increases to greater than 90% at 29 km where the influence of mercuriferous mill tailings is manifested (subaerial tailings at millsites generally contain >95% elemental Hg (P.J. Lechler, unpubl. data)). Thereafter, the proportion of elemental Hg decreases slightly downriver, before decreasing sharply in the reservoir sample (Fig. 3B). In the Carson River samples the proportion of exchangeable Hg varies inversely with respect to elemental Hg and then both species decrease sharply in the sample from Lahontan Reservoir (Fig. 3B). This relationship is thought to reflect the slow dissolution of elemental Hg with readsorption of the released Hg onto clays, organic matter, and Fe and Mn oxyhydroxides which are then deposited in downstream, lower-energy environments (Lechler and Miller 1993). This interpretation is supported by the observation that the ratio of Hg to Au in the upper 10 to 20 cm of central channel bed sediments increases downstream as Hg is selectively dissolved out of upstream amalgam particles and becomes sorbed to fine-grained materials which are deposited in downstream areas (documented in Fig. 4.). This interpretation is further supported by the data of Bonzongo et al. (1996), who show that the dissolved fraction of Hg in upstream waters is approximately 50% of the total Hg, but this dissolved fraction quickly decreases to approximately 5% or less of the total Hg in downstream water.

Organic Hg distribution is similar to exchangeable Hg (Fig. 3B,C). While the upstream sample at 22 km contains very low total Hg (29 ng/g), the majority of this Hg is associated with organic matter and exchangeable particle sites in the sediment. Five percent or less of the Hg in the downstream samples is related to the small amount of organic matter present (0.1% or less organic carbon (J.R. Miller, pers. comm.)).

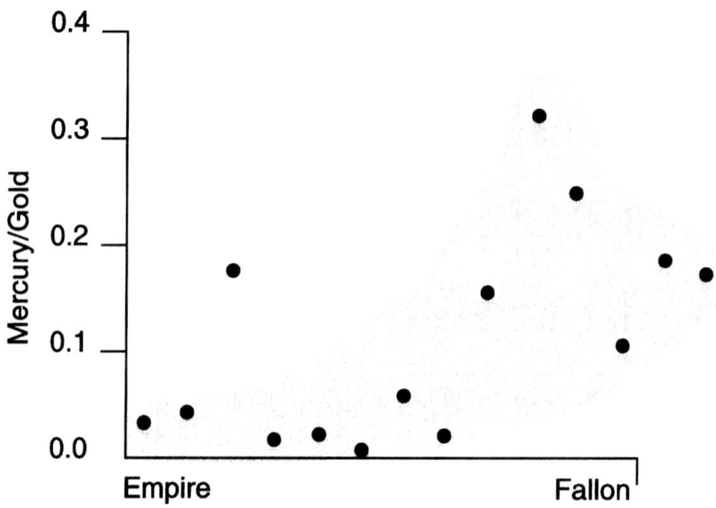

Fig. 4. Mercury/gold ratios in Carson River sediments from Empire (upstream) to Fallon (downstream) showing general increase in a downstream direction

The proportion of residual Hg in these samples is generally less than 10% except in the sample from Lahontan Reservoir which contains 36% sulfide Hg (Fig. 3D). This Hg sulfide is probably produced in situ, possibly where sulfate-reducing bacteria increase the activity of sulfide-sulfur in the sediments (Jorgensen 1983). It is reasonable that the low energy, more-anoxic reservoir environment would be conducive to higher reduced-sulfur activity (Jorgensen 1983).

3.2 Mercury in Soils

3.2.1 *High-Sulfur Area*

Area 1 (soils contaminated with Comstock tailings) is characterized by relatively high total Hg concentrations (approximately 1 to 160 μg/g) and high total sulfur (0.05 to 1%). Area 2 (soils contaminated with Jumbo tailings) media contain

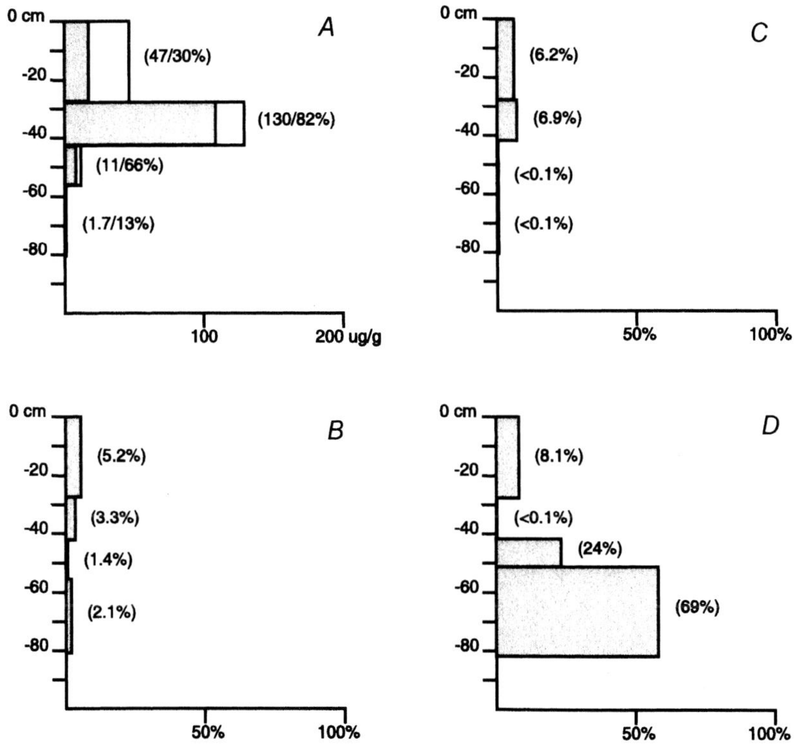

Fig. 5A–D. Distribution of total and elemental (*shaded*) Hg (**A**), exchangeable Hg (**B**), organic Hg (**C**), and Hg sulfide (**D**) in Six Mile Fan soil horizons. *Numbers in parentheses* are **A** μg/g total Hg/% of total Hg which is in elemental form; *B–D* % of total Hg which is in particular species form significantly lower total Hg (approximately 0.02 to 5 μg/g) and total sulfur (0.02 to 0.03%) levels.

Results of Hg speciation analysis of soil profile samples from one site in area 1, referred to here as Six Mile Fan, are depicted in Fig. 5A–D. Figure 5A shows that the horizon containing relatively pure mill tailings from 28 to 42 cm deep has both the highest total Hg and the highest percentage of elemental Hg in the profile. This is compatible with elemental Hg being the form released to the environment during the historic Comstock mining period. Both total and elemental Hg decrease in the lower two samples as the proportion of sulfide Hg increases (Fig. 5A,D). In this environment this trend is interpreted as reflecting the transformation and fixation of Hg as HgS in the lower, more reducing soil horizons. This interpretation is further supported by additional observations from area 2, a low sulfur environment, where such high sulfide Hg proportions in deep soil horizons are not observed (discussed below). Exchangeable Hg is similar in distribution to organic Hg in that it is a small proportion in all horizons but is highest in the shallowest samples (Fig. 5B,C).

The second sampling site in area 1, referred to here as Sutro, shows a distribution of total and elemental Hg similar to the SMF site (Fig. 6). The soil horizon visually identified as buried tailings has the highest total Hg and a high percentage of elemental Hg. The deepest soil sample has both much lower total Hg and, as at the SMF site, a low percentage of elemental Hg. All Hg species were

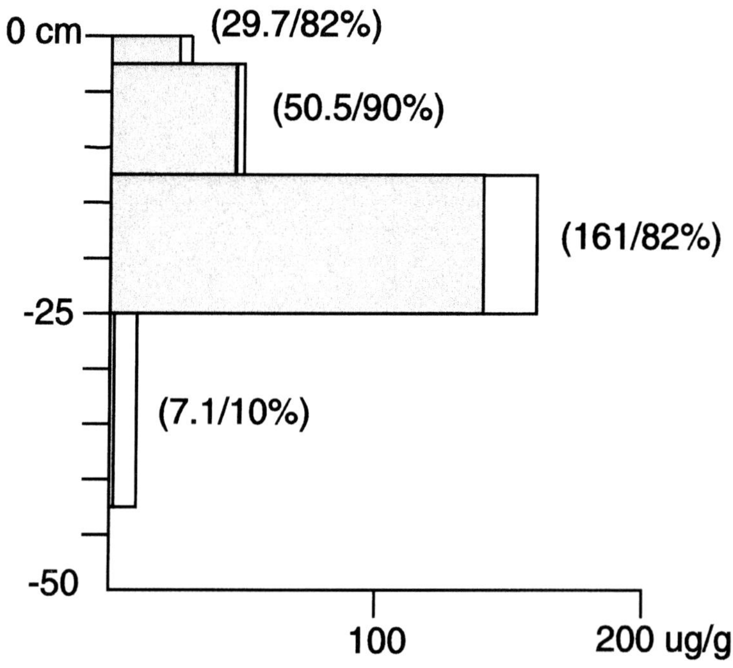

Fig. 6. Distribution of total Hg and elemental Hg (*shaded*) in Sutro location soil horizons. *Numbers in parentheses* are μg/g total Hg/% of total Hg which is in elemental form

not determined at this site, but this pattern is compatible with the bulk of the remaining Hg being present as HgS as it was at the other high-sulfur, SMF site, and as is common in other mining-contaminated areas like this (P.J. Lechler, unpubl. data).

3.2.2
Low-Sulfur Area

In the Jumbo Grade area (area 2), total Hg and total sulfur are both much lower in tailings and soils. Figure 7 shows the distribution of total and elemental Hg in a soil profile near an abandoned millsite. In the middle three soil horizons the proportion of elemental Hg is very high, compatible with this being the form of Hg released to the environment during historic gold milling in this area and the visual identification of tailings in these horizons.

In the deepest soil horizon, background concentrations are observed for total Hg (typically 10 to 30 ng/g in this area), and elemental Hg is undetectable. Elemental Hg is not the expected form for Hg in trace amounts in uncontaminated media (Tunnel 1969). Rather, Hg is more likely to be present in small amounts as ionic Hg in normal rock-forming minerals (typically present at approximately 100 ng/g or slightly higher (Tunnel 1969) but is not quantitatively extracted from these silicates by the total Hg procedure used here), to be strongly associated with organic matter, or to be sorbed to clays and Fe and Mn oxyhydroxides. The near-surface sample contains lower elemental Hg as well,

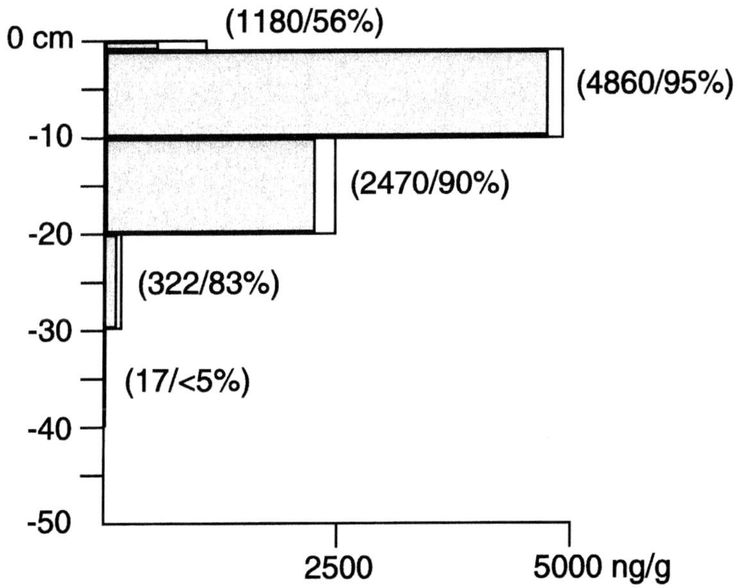

Fig. 7. Distribution of total Hg and elemental Hg (*shaded*) in Jumbo millsite soil horizons. *Numbers in parentheses* are ng/g total Hg/% of total Hg which is in elemental form

consistent with observations from area (1), indicating higher proportions of other Hg species in near-surface horizons where Hg probably becomes more strongly associated with organic matter in weakly developed A-horizons or is converted to Hg salts in the alkali-rich surface layers.

Compatible with the above observations for the contaminated millsite location are the distributions of Hg present in a relatively uncontaminated, background site uphill, but downwind, from the millsite location. At the background site, Hg concentrations reflect approximate regional background concentrations and the proportions of elemental Hg are low or undetectable. The exception is the near-surface sample which contains both higher total Hg and relatively high elemental Hg for this location (Fig. 8). This pattern of Hg distribution is compatible with atmospheric Hg vapor or particulate deposition from the upwind, contaminated millsite. Dispersion of Hg-contaminated particulates by eolian processes has been observed elsewhere in the Carson River Superfund site on the scale of 1 or more km from source tailings (see, for example, Lechler 1993; Lechler and Miller 1993).

The other contaminated site in area 2 contains mill tailings transported approximately 1 km downdrainage from the millsite location. Here, soil development is very weak due to frequent mixing/deposition of sediments during storm-driven runoff, and all horizons show high proportions of elemental Hg (Fig. 9). No contaminated soil samples from the low-sulfur area 2 exhibit the low elemental Hg or high sulfide Hg speciation pattern seen in deep soils from the high-sulfur area 1.

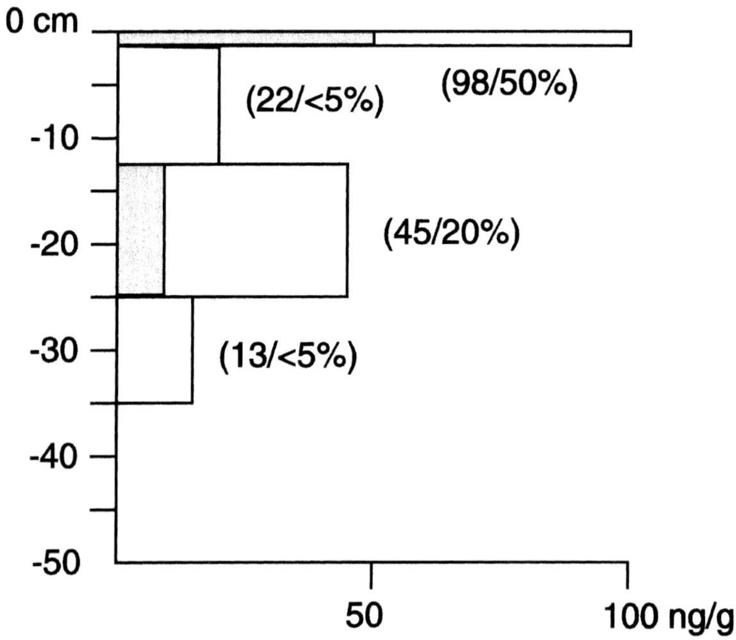

Fig. 8. Distribution of total Hg and elemental Hg (*shaded*) in Jumbo background soil horizons. *Numbers in parentheses* are ng/g total Hg/% of total Hg which is in elemental form

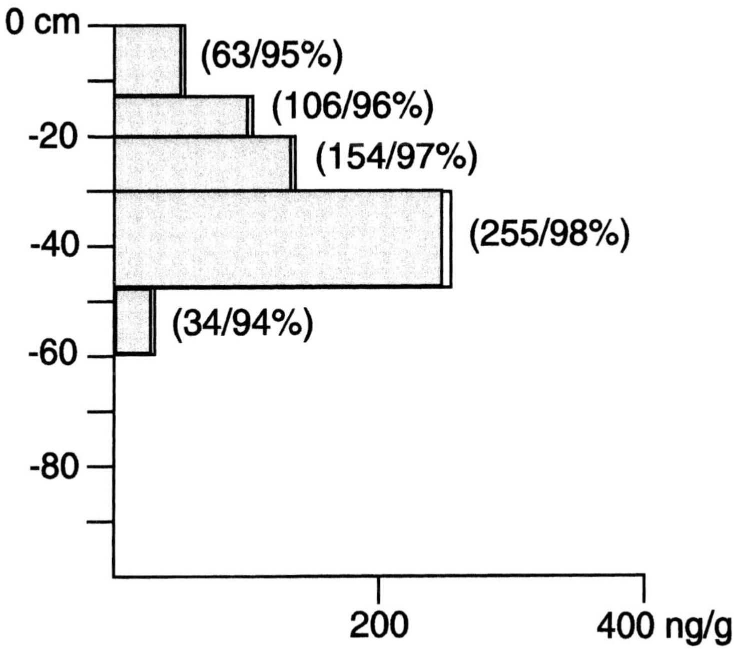

Fig. 9. Distribution of total Hg and elemental Hg (*shaded*) in Jumbo transported soil horizons. *Numbers in parentheses* are ng/g total Hg/% of total Hg which is in elemental form

4 Jordan Creek, Idaho

4.1 Mercury in Sediments

The main historical centers of mining activity along Jordan Creek were near the Dewey Mine at Happy Camp and just above the Delamar Mine at Discovery Bar (Wells 1983). Figure 10 shows that the main areas of Hg contamination, as judged by the highest Hg concentrations in active channel sediments, are in samples from these two locations. The lowest total Hg concentration is in the most-upstream sample, just below Silver City, where little milling of ores was done.

Exchangeable Hg is highest in the three samples immediately below Silver City, the Dewey Mine, and the Delamar Mine, suggestive of Hg being released partly in solution with the subsequent adsorption to fine, reactive particles in the sediments as occurs in the Carson River (Fig. 11). Strongly bound Hg shows a similar pattern downstream of Silver City and the Delamar Mine, but only weakly so proximal to the Dewey Mine. Organic Hg accounts for only 1% or less of the total Hg in Jordan Creek samples.

The other significant chemical form of Hg in Jordan Creek sediments is Hg sulfide. Natural Hg mineralization is known from volcanic rocks and epigenetic

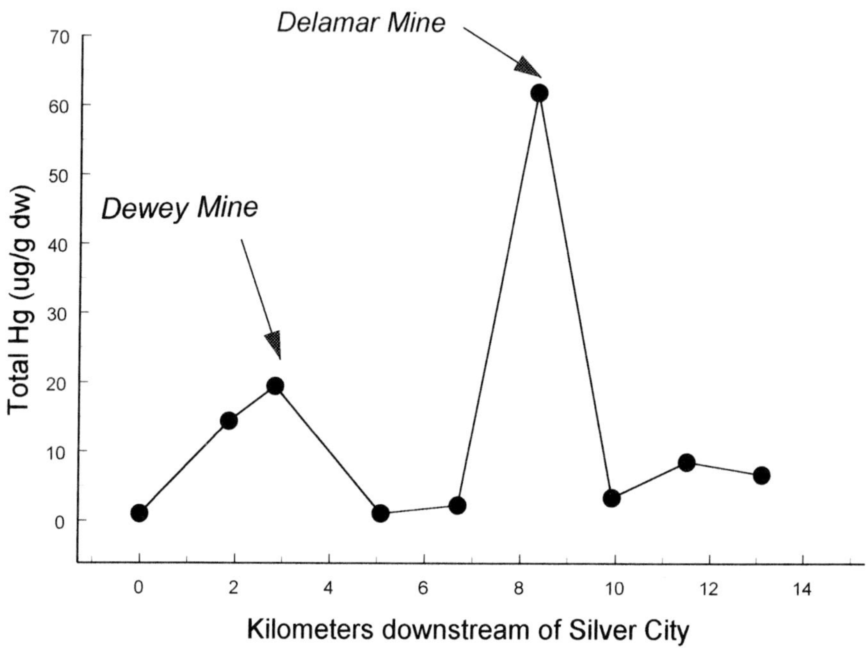

Fig. 10. Total Hg concentrations in Jordan Creek sediments, southwestern Idaho

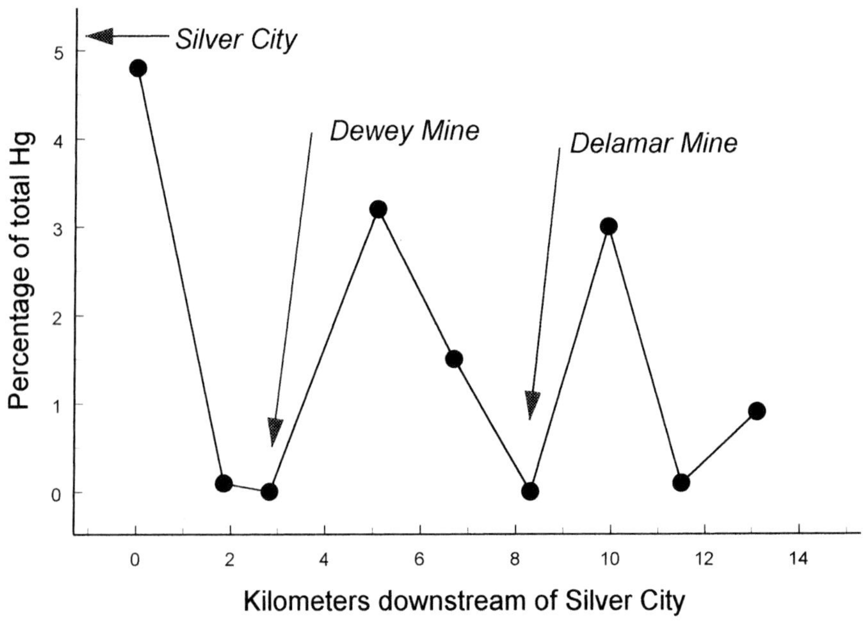

Fig. 11. Proportion of total Hg which is present in an exchangeable form in Jordan Creek sediments

mineralization in the Jordan Creek watershed (Williams 1974). Mineralization in a tributary which enters Jordan Creek between the Dewey and Delamar Mines supplies HgS-enriched sediments to Jordan Creek. This input of HgS is reflected in a high proportion of HgS in Jordan Creek sediments relative to elemental Hg where this tributary enters (Fig. 12). This peak in HgS in the sediments decays systematically downstream in the next four samples (approximately 6 km) as the effects of this tributary input are diluted by sediments low in HgS.

5
Discussion

It is apparent from the above descriptions of distribution and speciation of Hg in soils and sediments from sites contaminated by historic amalgamation milling, that high concentrations of Hg can persist in surficial environments for at least decades, probably centuries, and perhaps millennia. Moreover, the elemental form of Hg released during historic milling is the same form of Hg that dominates in these media today in the studied localities. Exceptions to this are in areas where the Hg is moved by surficial processes to more anoxic sites where its form may change at least partially to HgS (in less arid environments the Hg

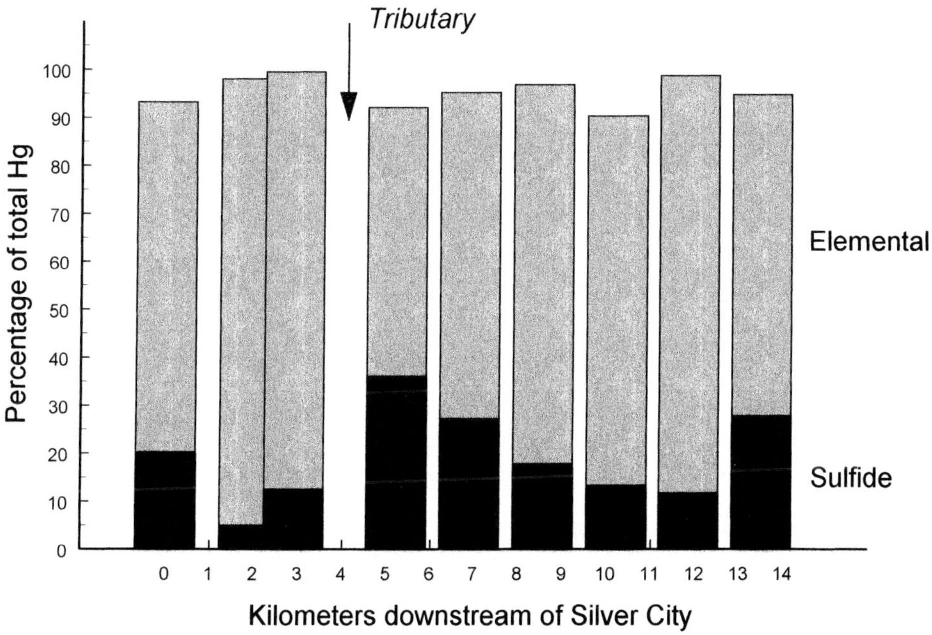

Fig. 12. Proportion of total Hg which is present as elemental Hg (*gray*) and as Hg sulfide (*black*) in Jordan Creek sediments

probably becomes strongly associated with organic matter as well). This transformation appears to be restricted, as would be expected, to areas where the sulfur content of soils or sediments is enriched (requiring, it appears, a minimum of 0.05 to 0.1% by weight of labile total sulfur).

These systematics are predictable and substantiated with Eh-pH diagrams which describe the behavior of Hg and S in surficial environments (Figs. 13, 14). Surficial soils, substantially in contact with the atmosphere in the CRSS area, are characterized by Eh-pH conditions approximately represented by A (sulfur-rich) and a (sulfur-poor) circles in Figs. 13 and 14. Anoxic conditions prevailing in deep soils are represented in Figs. 13 and 14 by circles B (sulfur-rich) and b (sulfur-poor). Thus, it should be expected that the elemental Hg released by past milling operations would indeed be stable in these near surface environments and should persist largely in the elemental form for any amount of time that the Hg remains stored in these soils, except where reducing conditions develop in sulfur-rich environments. This is what the Hg speciation analysis performed by the technique described above reports for the CRSS soils and the CRSS and Jordan Creek sediments, adding credibility to these experimental results.

The more acidic conditions measured for soils from high sulfur areas are also to be expected because the sulfides released to these soils and sediments along with the Hg will weather to form small amounts of ferric sulfate and sulfuric acid, driving the pH lower. These lower pH conditions also act to move the high sulfur soil and sediment conditions into the stability field of HgS.

Figure 13 also shows that under special surficial conditions of high chloride activity in which the stability field of $HgCl_2$ would expand into the area of circle A, an amount of soluble mercuric chloride would be expected to form. The high solubility of this form of Hg would cause it to be leached and removed from

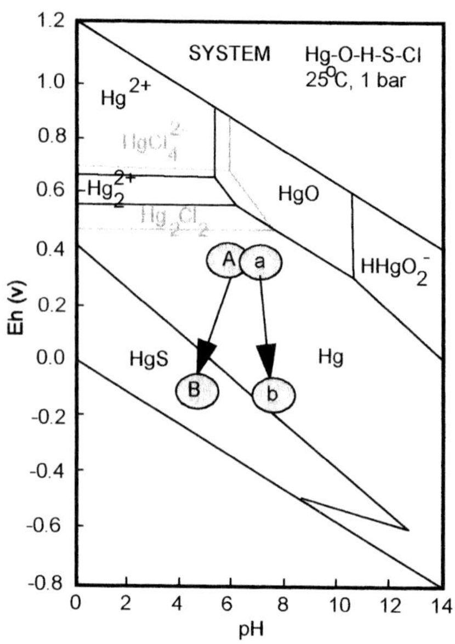

Fig. 13. Eh-pH diagram for the system Hg–O–H–S–Cl at 25 °C and 1 bar (activities: Hg = 10^{-8}, Cl = $10^{-3.5}$, S = 10^{-3}). Circle *A* indicates approximate conditions for near-surface, high-sulfur soils (elemental Hg stable); circle *a* indicates approximate conditions for near-surface, low-sulfur soils (elemental Hg stable); circle *B* indicates approximate conditions for buried, high-sulfur soils (Hg sulfide stable); circle *b* indicates approximate conditions for buried, low-sulfur soils (mainly elemental Hg stable). (Brookins 1988; Henderson 1982)

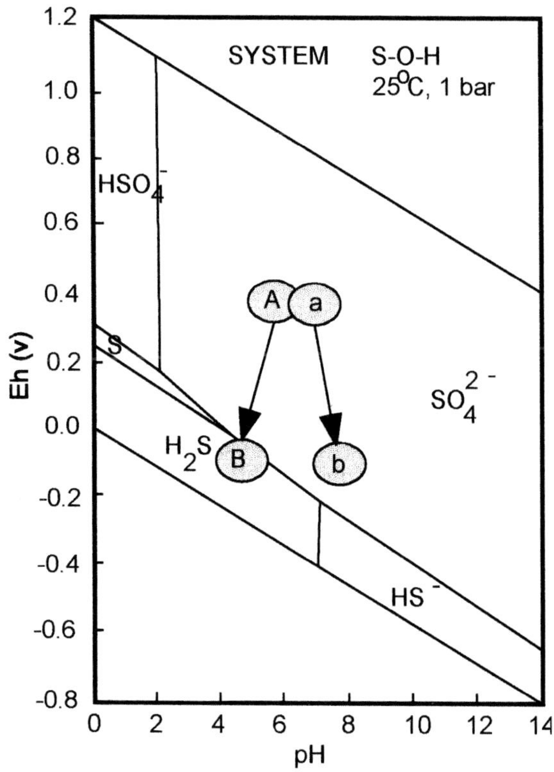

Fig. 14. Eh-pH diagram for the system S–O–H at 25 °C and 1 bar (activity of S = 10^{-3}). Circle *A* indicates approximate conditions for near-surface, high-sulfur soils (sulfate stable); circle *a* indicates approximate conditions for near-surface, low-sulfur soils (sulfate stable); circle *B* indicates approximate conditions for buried, high-sulfur soils (sulfide stable); circle *b* indicates approximate conditions for buried, low-sulfur soils (sulfate stable) (Brookins 1988; Henderson 1982)

surficial soil horizons, explaining why high concentrations of mercuric chloride have rarely been measured at the CRSS (P.J. Lechler, unpubl. data).

The Eh-pH diagrams shown in Figs. 13 and 14 reflect the distribution of Hg species at equilibrium for the P-T-X conditions for which the diagrams were drawn. Equilibrium may not be attained for some reactions in surficial and near-surface geologic systems, changing the shapes of the stability fields somewhat. The diagram also does not account for microbial activity and the effects of organic matter, both of which may alter the distribution of Hg species to some degree, especially with respect to methyl Hg. However, in arid to semiarid environments such as those described here, these effects do not appear to be significant.

6
Conclusions

Soils and sediments which are impacted by historic amalgamation milling operations in the western US are typically contaminated with total Hg in the range of 0.1 to 300 μg/g. The Hg released to the environment was mainly in the elemental form, the form which still dominates in the sediments today. Where these materials are transported from near-surface horizons largely in equilibrium with atmospheric oxygen to more anoxic, reducing environments in reservoirs or deep soil horizons, a significant proportion of HgS is formed. This occurs where total, labile sulfur exceeds 0.05 to 0.1 wt%.

These observations come from the analysis of the speciation of Hg in these soils and sediments by the methodology described above. As pointed out by Wallschlager (1996), quality assurance of speciation methodologies through the fortification of natural samples with discrete Hg species is impractical at best. The concurrence of stability fields of different Hg species on Eh-pH diagrams with the results of the chemical speciation analysis lends support to the validity of this methodology. Moreover, the geochemically reasonable shifts in observed speciation of Hg in these soils and sediments from one type of environment to another further supports the conclusions of the speciation methodology.

Mercury contamination from historic amalgamation milling of precious metal ores which contain associated sulfide minerals may be less bioavailable than Hg from ores essentially devoid of sulfides. Because of the relatively high solubility of most sulfide minerals in surface waters, most gold placer deposits (deposits found in unconsolidated alluvium) contain few sulfide minerals. Thus, Hg pollution resulting from the amalgamation milling of placer gold may generally be more harmful in terms of higher solubility, mobility, and bioavailability of the released Hg than from high-sulfur lode mining sites (mineralization formed in hard rock deposits, with the precious metals often being accompanied by sulfide minerals such as pyrite) because the low-S placer environments are not conducive to the fixation of Hg as relatively insoluble HgS.

References

Ansari MB (1989) Mines and mills of the Comstock region, western Nevada. Camp Nevada Mono 8, Univ. Nevada Press, Reno, pp 87–94

Basham EL, Lechler PJ, Miller JR (1996) Database of sites of historic mercury amalgamation milling of gold and silver ores in the United States. Abstracts Mercury Global Pollutant, 4–8, Aug 1996, Hamburg, Germany, p 141

Biester H (1994) Möglichkeiten der Anwendung eines temperaturgesteuerten Pyrolyseverfahrens zur Bestimmung der Bindungsform des Quecksilbers in Böden und Sedimenten. Heidelb Geowiss Abh 75, PhD Diss, Ruprecht-Karls-Univ, Heidelberg, 156 pp

Bonzongo JC, Heim KJ, Warwick JJ, Lyons WB (1996) Mercury levels in surface waters of the Carson River-Lahontan Reservoir system, Nevada: influence of historic mining activities. Environ Pollut 92 (2):193–201

Brookins DG (1988) Eh-pH diagrams for geochemistry. Springer, Berlin Heidelberg New York, 176 pp

Di Giulio RT, Ryan EA (1987) Mercury in soils, sediments, and clams from a North Carolina peatland. Water Air Soil Pollut 33:205–219

Eganhouse RP, Young DR, Johnson JN (1978) Geochemistry of mercury in Palos Verdes sediments. Environ Sci Technol 12 (10):1151–1157

Fisher V, Holmes OL (1968) Gold rushes and mining camps of the early American West. Caxton Printers, Boise, 466 pp

Henderson P (1982) Inorganic geochemistry. Pergamon, New York, 276 pp

Jorgensen BB (1983) The microbial sulfur cycle. In: Krumbein WE (ed) Microbial geochemistry. Blackwell, Oxford, pp 91–124

Lechler PJ (1993) Mercury vapor sampling at the Carson River Superfund Site. Proc Heavy Metals Environ, Toronto, Canada, pp 377–380

Lechler PJ, Desilets MO (1991) The NBMG standard reference material project. Nev Bur Mines Geol Newsl 10:1–2

Lechler PJ, Miller JR (1993) The dispersion of mercury, gold, and silver from contaminated mill tailings at the Carson River Mercury Superfund Site, west-central Nevada. Proc Perspectives Env Geochem Tropical Countries, Nov 29–Dec 3, Niteroi, Brazil, pp 433–436

Nriagu JO (1994) Mercury pollution from past mining of gold and silver in the Americas. Sci Total Environ 149:167–181

Quevauviller P, Donard OFX, Wasserman JC, Martin FM, Schneider J (1992) Occurrence of methylated tin and dimethyl mercury compounds in a mangrove core from Sepetiba Bay, Brazil. App Organomet Chem 6:221–228

Scott WW (1939) Standard methods of chemical analysis, vol I. Van Nostrand, Princeton, 1234 pp

Smith GH (1943) The history of the Comstock, 1850–1920. Univ Nev Bull 37:41–47

Tunnel G (1969) Mercury minerals, abundance in rock-forming minerals. In: Wedepohl KH (ed) Handbook of geochemistry. Springer, Berlin Heidelberg, New York 80-D-1-3

Wallschlager D, Spengler M, Desai MVM, Carvalhinho-Windmoller C, Wilken R-D (1996) How the matrix dictates the distribution of major inorganic mercury species in soils and sediments. 4th Int Conf Mercury as a Global Pollutant, 4–8 Aug, 1996, Hamburg

Wells MW (1983) Gold camps and silver cities. Idaho Bur Mines Geol Bull 22, 165 pp

Williams DR (1974) The influence of the physical environment on the problem of mercury contamination in the Jordan Creek drainage, Idaho. Abstract Geol Soc Am Cordilleran Sect Meet, Las Vegas, p 275

Gold Mining in Siberia as a Source of Mercury Contamination of the Environment

T.G. Laperdina, M.V. Melnikova, and T.E. Khvostova

The results of research on estimation of mercury contamination in the areas impacted by gold mining in Siberia – the oldest gold mining region of Russia – were considered. Investigations were carried out at the sites of gold ore and placer deposits in the Krasnoyarsk Krai, the Republic of Buryatia, the Republic of Yakutia, the Chita Oblast and the refining plant in Novosibirsk. Mercury concentrations in the environmental compartments of the industrial areas and background zones ranged from <0.001 to 4.0 mg/kg for rocks and minerals; from <1 to 183 ng/m^3 for atmospheric air; from 5 to 250 000 ng/m^3 for industrial air; from 0.013 to 18.9 mg/kg for soils; from <5 to 5000 ng/l for dissolved mercury and from <5 to 27 800 ng/l for particulate mercury of natural and industrial waters; from 0.008 to 54.2 mg/kg for bottom sediments. The influence of severe landscape climatic conditions of Siberia with massive or island spreading of permafrost on the extent of mercury contamination in gold mining areas was considered.

1 Introduction

Siberia is the oldest gold mining region of Russia. The first domestic gold was produced in Zabaikalye (Transbaikalia) from silver Nerchinsk Ores in 1719 (Ignatkin 1994). The intensive gold mining began later, when placer and ore gold deposits were discovered. In the middle of the 19th century gold mining in Russia ran to 26 t a year (40% of the world's total production), most of the gold being mined in Siberia. Up to 1917, a total of 2754.17 t of gold had been mined, 73.3% of this amount in Siberian mines (Foss 1963; Vyazelshchikov and Paritzky 1963). At present, the significant decrease in gold mining and the change in its structure are observed in the country due to complex economic conditions and transformation of territorial border lines. Thus, gold mining expected for 1996 accounts for 60 tons, that is less by 20%, than in 1995 and by 50% than in 1990. The proportion of ore gold is decreasing steadily, due to great investments in its mining, especially in working of new ore deposits and exhaustion of the richest ores in the old mines. For the period from 1966 to 1990 ore gold mining decreased by 15.4% and amounted to 30.293 t in 1990 (24.3% of the total amount of gold production). Ore gold mining has been decreasing in recent years even more rapidly. Thus, the corporation Zabaikalzoloto produced a total of 12.543 t of gold in 1990 and only 4.871 t (38.8%) of ore gold (Pluteshko 1993). In 1995 the total production in the Chita Oblast accounted for 8.729 t and the part of ore gold was only 14%.

Environmental Science
Mercury Contaminated Sites (ed. by R. Ebinghaus et al.)

The mining of placer gold resources at benches, along and under the riverbed, which are located, as a rule, in the valleys of small and medium-sized rivers is accompanied by irreversible landscape degradation and extreme deterioration of natural water streams, with transformation of their hydrologic, hydrochemical and hydrobiological characteristics. Recultivating measures at mined-out sites do not result in restoration of valuable vegetation species, or quantitative and qualitative composition of water biota in broken water streams. Traditional technologies of placer gold mining in Russia (dredges and high-pressure hydraulic giants) are, as a rule, ecologically dangerous, yielding a great deal of wastewaters with a high content of suspended clay matter. In extreme situations (rainfall, floods) emergency discharges of wastewaters from settling tanks are not uncommon. They result in silting of riverbeds, contamination of bottom sediments and water phase with the toxic metals that are constituents of gold-bearing ores (arsenic, antimony, mercury, lead, zinc, bismuth). This leads to a decrease in quantity and variety of hydrobiota and deterioration of food chains. Hydraulic methods of gold mining are either forbidden or under strict ecological control in highly developed countries. Thus, for example, in California, USA, hydraulic methods were forbidden as far back as 1884, and in Alaska they are used, but in accordance with the criteria established by the Federal Environmental Protection Agency and Alaska Department of Environmental Conservation (Yeend 1991; Young 1993). In Russia, despite the growth of the placer gold part of the total production of gold, the control to implement environmental protection measures in gold mining sites is, unfortunately, weakened. This is one of the reasons that placer gold mining activity has moved to remote, hard-to-reach areas, where the transport system is not sufficiently developed.

2
Sources of Mercury Contamination of the Environment Due to Gold Mining in Siberia

Mining of ore and placer gold is, as a rule, accompanied by environmental mercury contamination of differing intensity. The main sources of mercury emission to the environment are:

- Elevated mercury levels in the rocks and ores of gold deposits and mercury release during ore and gravel processing, and also its transition into an active migratory state in the processes of mercury-bearing rock oxidation at sites broken by mining.
- Direct use of mercury in the process of gold amalgamation of gold-bearing raw materials and its technological losses due to the use of technically imperfect equipment (incomplete mercury vapour condensation in the process of amalgamated gold burning, incomplete mercury trapping in the process of technological gas purification during forced ventilation, mercury losses with technological waters during amalgamation).

- Accumulation of a large amount of concentration tailings with elevated mercury contents and its emission into the atmosphere, surface water streams or groundwaters due to the processes of rock oxidation, tailings drainage by atmospheric precipitation etc.
- Reprocessing of gold bearing concentration tailings with elevated mercury contents, which is accompanied by mercury transition into an active migratory state.
- Reworking of industrial gold placers mined out previously by the dredge method with the use of sluiced amalgamation and transition of a great deal of mercury into an active migratory state, which was buried under the layer of bottom sediments in the old dredge pits in anaerobic conditions at temperatures near zero in a "preserved" state.
- Refining of gold alloy with natural or technogenic (amalgamation use) elevated mercury contents at refining plants, which is accompanied by vapour mercury emission into the atmosphere along with technological gases.

3
Amalgamation of Gold-Bearing Ores, Gravel and Concentrates

The method of gold amalgamation based on a selective wetting of native gold particles with mercury and their weighting is the oldest technology in gold extraction from ores. In Russia, the method of amalgamation of gold-bearing ores came into use on an industrial scale after 1870 (Roslyakov and Kirillova 1995) and was for a long time an obligatory component in traditional schemes of gold-bearing raw material concentration.

Amalgamation methods are classified into two groups (Vyazelshchikov and Paritzky 1963):

- Internal amalgamation, when gold trapping occurs inside crushing devices, simultaneously with rock crushing and native gold particle release from the mineral shell (in crushers and crusher-roll mills).
- External amalgamation, when gold trapping takes place outside crushing devices (on sluices, in amalgamators).

Originally, internal amalgamation was widely used in processing gold-bearing ores and gravels, mined by underground and open methods, and external sluice amalgamation was used in dredge technology. These methods were accompanied by great specific mercury expense, with great irreversible losses, and they were the most harmful for the health of the workers. Maximum mercury expense occurred in the process of sluice amalgamation, when mercury (up to 420 kg) was poured over the riffles of gold trapping sluices. In this case, mercury expenses accounting for 40–130 mg per 1 m^3 of gravel and 3–10% of metal were additionally extracted. The mercury losses at sluice amalgamation were great and could account for 55–70 mg/m^3 of the processed gravels (Zamyatin et al. 1973). The mercury content in dredge tailings could reach 0.2 g/m^3 due to mechanical losses (Ministry Cabinet of the USSR 1988). During internal amalgamation in barrels and mills, the correlation between mercury and extracted gold depends on gold particle size and accounts for (4–10):1.

Considerably smaller mercury losses take place in amalgamation of schlich concentrates, although mercury expenses here are also significant. For example, when gold was previously amalgamated in barrels, mercury expenses accounted for 10–15 g/g of gold (Vyazelshchikov and Paritzky 1963). At present, mercury expenses for amalgamation at the stage of finishing of gravity concentrates are considerably smaller and can account for 100–300 g/kg of gold.

According to the data of the corporation Zabaikalzoloto, approximately 2 t of mercury to 8 t of all processed gold arrived at gold-mining plants of the Chita Oblast up to 1989 annually, which corresponded to its annual irreversible loss and emission to the environment. It should be noted that in processing gold-bearing raw materials, concentration methods of cyanidation and flotation are widely used. Therefore the specific weight of mercury expenses per 1 kg of gold was greater than by a simple division of mercury expenses by total gold production, extracted by amalgamation, cyanidation and flotation (2 t per 8 t or 250 g of mercury per 1 kg of gold). It is impossible to make more accurate calculations due to limited and sometimes contradictory information.

Incomplete burning of amalgams causes mercury content in concentrates sent to refining plants to account for 0.2–4.0 g/t (Ministry Cabinet of the USSR 1988). As a result, mercury pollution of the environment and industrial spaces occurs not only in the areas impacted by gold-mining activities, but also in the areas near refining plants. According to the official data, in the USSR up to 1988 the total irreversible losses of metallic mercury at dredges and washing devices reached 6 t/year, at gold dressing plants near 3 t/year (Ministry Cabinet of the USSR 1988).

In the context of high mercury contamination of traditional gold-mining sites, cases of mercury poisoning, increasing morbidity of the attendants, the Chief Department on Precious Metals and Diamonds of the Cabinet of Ministers of the USSR prohibited gold amalgamation at dredges and washing devices from January 1, 1989, and from January 1, 1990 at gold extraction plants (GEP) and gold dressing plants (GDP) (Ministry Cabinet of the USSR 1988). However, in spite of the official prohibition, the use of amalgamation in gold mines still persists to the present, especially at the stage of gravity concentrate finishing. Thus, for example, there is evidence (Koledneva 1994) that when amalgamation was used in one of the small mines of the Chita Oblast amalgam burning was performed using the most primitive methods – on an open fire without mercury vapour condensation due to the lack of necessary equipment. Since mercury expenses from the process of gold amalgamation is not centrally accounted, it is impossible to estimate accurately annual mercury emission to the environment in the mining and processing of gold-bearing raw materials. However, the increasing use of mercury by small mines appears to be inevitable at present.

4
Investigations on Estimation of Mercury Contamination in Gold-Mining Areas in Siberia

There are few data in the literature on the results of investigations into mercury contamination of gold-mining areas in Siberia (Table 1, Fig. 1, Bodienkov 1984; Rukavishnikov 1984; Taisayev 1991; Laperdina et al. 1995, 1996; Roslyakov and Kirillova 1995; Tupyakov et al. 1995) due to a state restriction on publishing information on gold mining, and also due to a previous lack of an instrumental and methodical base, which would have allowed comprehensive investigations of the necessary sensitivity and accuracy. Determination of background mercury concentrations in the environmental compartments (and especially in natural waters) has become possible in Russia only during the past 8–10 years in connection with increasing attention to mercury as one of the most toxic pollutants and the development of highly sensitive and specialized equipment and methods.

The most detailed and comprehensive investigations on the estimation of mercury contamination in the gold-mining areas of Siberia were carried out in the Chita Oblast in 1991–1994 (Laperdina et al. 1995, 1996; Tupyakov et al. 1995). Mercury concentrations in background areas and zones impacted by some mines, located in different landscape and climatic zones, from the northern highland-taiga with mari landscape (very swampy hummocky zones with shallow lying permafrost) to highland-steppe were studied: Uryum and Tungir (Tungir-

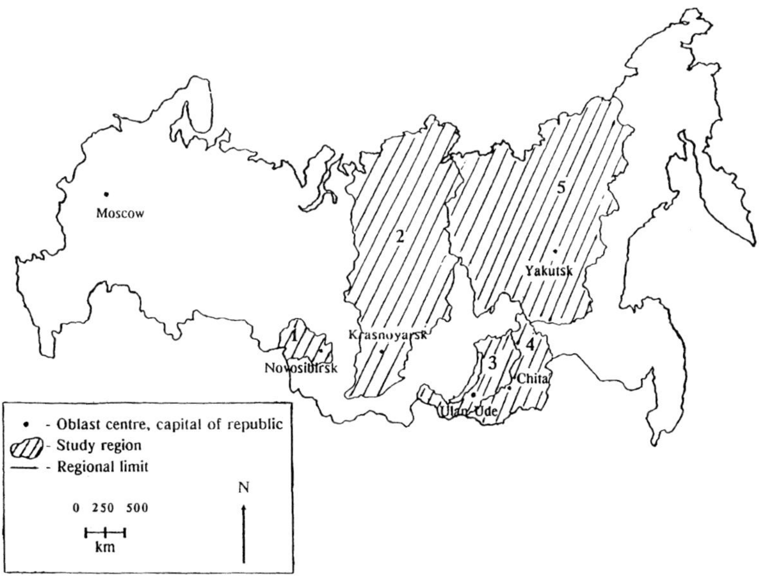

Fig. 1. Map of the studied regions in Siberia. *1* Novosibirsk Oblast; *2* Krasnoyarsk Krai; *3* Republic of Buryatia; *4* Chita Oblast; *5* Republic of Yakutia

Table 1. Mercury concentrations in the environmental compartments of gold-mining areas and background zones

Sampling sites	Concentration ranges of mercury								
	Rock, mineral (mg kg^{-1})	Eluvium (mg kg^{-1})	Atmos. air (ng m^{-3})	Industrial air (ng m^{-3})	Soil (mg kg^{-1})	Water			Bottom sediment (mg kg^{-1})
						Dissolved, (ng l^{-1})	Particulate		
							(ng l^{-1})	(mg kg^{-1})	
Chita Oblast (Laperdina et al. 1995, 1996)									
Background	0.001–0.028	0.013–0.033	<1–11		0.013–0.14	<5–15	<5–380	0.4–112	0.008–0.11
Industrial	<0.001–0.307	0.020–0.780	<1–183	5–250 000	0.015–3.59	<5–5000	<5–27 800	0.2–1860	0.010–54.2
Republic of Buryatia (Irokendinskoe deposit) (Taisayer 1991)									
Industrial									1.0–5.0
Krasnoyarsk Krai (Boguny deposit) (Roslyakov and Kirillova 1995)									
Background									0.034–0.18
Industrial									0.035–3.0
Novosibirsk Oblast (refining plant) (Roslyakov and Kirillova 1995)									
Background					0.03–0.08			0.04[a]	
Industrial					0.03–18.9			6.8[a]	
Republic of Yakytia (Bodienkov 1984; Omelchenko 1985, Rutavishnikov 1984)									
Background	0.01–0.04								
Industrial	0.15–4.0			100–230 000					

[a]Snow.

Olyokma raion), Ksenyevsky (Mogocha raion), Baleyzoloto (Baley raion) and Lyubov (Kyra raion). A total of 105 measurements of the mercury content in atmospheric air and the air of the mining site were performed; 56 samples of rocks and minerals, 80 samples of soils, 170 samples of natural and technogenic waters and 122 samples of bottom sediments were analyzed. Besides, selective analyses of the mercury content in 21 samples of vegetation, fish, and meat of wild animals were performed.

The research was carried out to estimate mercury contamination along the dispersion water flux in Irokendinskoe ore gold mines of Buryatia (Taisayev 1991), located in a bald peak-taiga landscape of fragmented highlands of the cryolite zone. At this site the main source of mercury and gold flux is a tailings dump of GEP, which has been in operation for 25 years. Flood erosion of dams of the tailings dump resulted in riverbed silting and formation of highly contrasted lithochemical mercury (1–5 mg/kg) and gold (0.5–4 mg/kg) flux. Besides, ore heaps and surface and underground mining driftage are also sources of mercury and gold. The clearly defined biogenic anomaly of gold and silver was also fixed at the head (500–600 m) of the technogenic flux (thicket, willow, Dauric larch, water mosses). However, such a biogeochemical anomaly was not revealed for mercury, on which basis the author concluded a poor bioassimilability of mercury in plants.

In 1993 in the Krasnoyarsk Krai, research was carried out to estimate mercury contamination of bottom sediments of the Boguny river in the area impacted by the gold-sulphide-quartz Boguny deposit, mined with the help of drifts and pits during World War II (Roslyakov and Kirillova 1995). Despite the fact that mining operations were stopped here about 50 years ago and since then this deposit has not been exposed to any technogenic activity using mercury, bottom sediments of natural water streams are still characterized by elevated mercury contents, exceeding background values by 8.6 times.

The research carried out in the area impacted by the refining plant in Novosibirsk have shown that highly contrasted technogenic mercury and lead haloes were formed in the soils around the plant. These metal contents in solid phase of snow were also investigated (Roslyakov and Kirillova 1995).

The problems of negative impact of metallic mercury vapours on the health of the people employed in the processes of gold concentration with the use of sluice amalgamation at dredges and processing of shlich concentrates at GEP and GDP in the eastern regions of the country have been studed by Russian medical scientists and hygienists for a long time since 1935 (Bodienkov 1984; Rukavishnikov 1984). It was determined that the processes of sluice amalgamation at dredges and external amalgamation at GEP are accompanied by very high mercury concentrations in the air of working places, exceeding the average shift maximum permissible concentration (MPC) by tens of times. This is detrimental to the health of the attendants, affecting mainly respiratory, urogenital, nervous systems, sense and digestive organs and liver. One should emphasize an increasing morbidity of women associated with pathology of female reproductive organs. Most workers employed in amalgamation processes were women.

5
Mercury Contents in the Environmental Compartments of Background Areas and Zones Impacted by Gold-Mining Activities

5.1
Rocks and Minerals

The Clarke of mercury in the crust, according to different data, ranges from 0.03 to 0.08 mg/kg, in rocks of various compound from 0.00021 to 0.40 mg/kg (Saukov 1975; Ozerova 1986; Ovchinnikov 1990). Mercury is an effective geochemical indicator of gold-bearing deposits and for some of them the second significant admixture (after silver) in native gold in both occurrence and concentration (0.1–5.0 g/kg; Stepanov and Moiseyenko 1993). Mercury haloes border ore bodies all along their length, increasing in thickness and contrast from lower to upper levels. Haloes are much thicker than gold ore bodies and extensive in the supra-ore part. Mercury concentrations in the haloes of near-surface and more subsurface deposits may reach 10–20 mg/kg, more common by 0.1–0.3 mg/kg (Roslyakov and Kirillova 1995). Therefore, industrial mining of ore gold deposits is, as a rule, accompanied by an admixture of mercury emission from bedrocks, gold-bearing ore and its reprocessing products and redistribution of emitted mercury between the environmental compartments, which leads to an increase in mercury concentration level in comparison with the natural background content (Table 1).

For placer gold deposits, the data on mercury haloes are inadequate to establish clear geochemical peculiarities. For some placers in Buryatia, however, the elevated mercury content (1.6–112 mg/kg) was also registered in hypergenic gold (Karmanov 1985).

In Siberia in the areas of traditional gold mining, a huge amount (millions of tons) of rocks and bedrocks heaps, concentration tailings with elevated mercury contents, have been accumulated at present. Problems of mercury emission from rocks of heaps and tailings due to physicochemical, chemical and microbiological processes, which take place during oxidation of rocks, their destruction, secondary mineral formation, rainfall drainage and also intensity and dynamics of migration processes in the atmosphere, soils, surface water streams and underground waters remain unstudied. The almost universal presence of permafrost and a seasonal frozen state make it difficult to transfer rules established for areas with warmer climate to Siberia, which is characterized by diversity of climate and landscape conditions. The results of the studies carried out in Zabaikalye (Laperdina et al. 1995, 1996), Yakutia (Omelchenko 1985) and Gorny Altai (Anoshin et al. 1995) have shown that decreased mercury contents in frozen rocks of gold ore deposit pits (Yakutia) due to active cryogenic processes, as well as the enrichment of oxidized metasomatically altered rocks with mercury are observed in the mining areas.

The increase in mercury content on the average by two to three times in oxidized rocks (0.013–0.033 mg/kg) in comparison with original rocks (0.005–

0.028 mg/kg) is observed in the background areas, heaps of open pits. More substantial (ten and more times) enrichment of weathering products with mercury (0.65–0.78 mg/kg) is registered in the heaps of shafts and drifts (Laperdina et al. 1995, 1996). Most likely, the cause is mercury degassing from original ores and rocks raised to the surface and its accumulation in secondary clay minerals. Besides, additional enrichment of friable sediments may be determined by an upward gaseous mercury migration through the zones of tectonic faults and underground driftages and pits, and also by mercury absorption from atmospheric precipitation. Accumulation of mercury in clay sediments, which can strongly bind metal, assists in some localization of mercury pollution, especially in zones of permafrost. Therefore, the methods suggested to estimate mercury emission to the environment due to gold mining, which lie at about 100 kg mercury/t of gold mined or 1–3 t of mercury from 1 million t of mining heaps actively migrate into the environment (Roslyakov and Kirillova 1995) we consider insufficiently substantiated. Mercury contamination of the areas broken by mining operations is, as a rule, of local and less intensive character than predicted by these methods.

5.2 Atmospheric Air and the Air of Technogenic Zones

The background mercury concentrations in the air of Siberia account for 1–11 ng/m^3 and depend on the vicinity of tectonic fault zones and also on the elevated mercury geochemical background near mercury, ore gold and some other deposits (Laperdina et al. 1995, 1996; Malgin and Puzanov 1995).

Mercury contamination of the atmosphere in the gold-mining areas is caused by two factors: a slight increase in mercury background levels due to processing of gravels and rocks with elevated mercury contents, and a greater increase due to the direct use of metallic mercury in concentration processes. The extent of air contamination is determined by intensity and duration of mercury use, concentration technology and equipment used (Table 1). Besides, landscape and climatic conditions determining the intensity of geochemical and biochemical processes, causing mercury redistribution between the environmental compartments, play an important role.

As the results of the study carried out in Zabaikalye have shown, the use of metallic mercury in concentration processes at GEP and GDP leads to a contrasting (25–183 ng/m^3), steady and, as a rule, local air contamination of these sites. In this case, contamination sources are gas wastes due to incomplete mercury condensation in the process of amalgam burning, mercury degassing from former and present tailing dumps and contaminated soil surface. Thus, near the Baleyzoloto GDP a local halo of gaseous mercury (25–40 ng/m^3) was registered, whose localization is most likely caused by the use of an effective system of vapour mercury condensation. Unlike the Baleyzoloto mine, in the zone impacted by the Lyubov mine, a total increase in mercury content in the atmosphere of both industrial (16–183 ng/m^3) and residential (12–152 ng/m^3) areas was observed. Maximum value was found at the former tailings dump of the GEP.

The results of measurements at dredges have shown that the mercury vapour content near the gravity concentration board during the amalgamation process is relatively low, and accounts for 1800–2000 ng/m^3, if amalgamation is performed at the stage of schlich concentrate, finishing in a non-isolated ventilated place (Laperdina et al. 1995, 1996). The mercury content in the air of dredges increases sharply during the process of sluice amalgamation and may reach enormous values in the isolated dredge compartments, exceeding MPC of mercury vapours for industrial sites (5000 ng/m^3) by 7.4–10 (Bodienkov 1984) or by 20–50 times (Laperdina et al. 1995, 1996) under different technological and climatic conditions. The level of mercury vapour content in the air of GEP and GDP depends on the amalgamation technology applied, the volume of processed ore, gravel and concentrates, the insulation of working places, the availability of a powerful ventilating system with condensation of mercury vapours, and it can vary from 50–1000 ng/m^3 at GEP of the Baleyzoloto mine (T.G. Laperdina, unpubl. data) to 1000–9000 ng/m^3 (Rukavishnikov 1984). With the use of external amalgamation at GEP and GDP the mercury vapour concentrations in the air of working places, especially in the process of mercury application and amalgam removal, may exceed MPC values by tens of times. Ineffective refining of gas wastes with high vapour mercury content (230 000 ± 45 000 ng/m^3) from burning furnaces causes a high level of mercury contamination in the atmospheric air (Rukavishnikov 1984).

Elevated mercury concentrations in the air (11–43 ng/m^3) were found also in areas of the gold reception centre (GRC) and were associated with mercury pollution of the industrial buildings and the soils surrounding them. Refining of gold-bearing concentrates and alloys at refining plants in Novosibirsk and Krasnoyarsk is accompanied by the evolution of technologic gases with high mercury concentrations, that leads to the environmental contamination of the territory around the plants due to incomplete mercury vapour condensation. Thus, gaseous mercury plays a significant part in the contaminated soils surrounding the refining plant in Novosibirsk (Roslyakov and Kirillova 1995).

5.3 Soils

The mercury content in the soils of background areas of Siberia ranges from 0.01 to 0.33 mg/kg (d.w.) (Anoshin et al. 1995; Laperdina et al. 1995, 1996; Malgin and Puzanov 1995), where the elevated values are registered in the soils near mercury, ore gold and some other deposits and over a large tectonic fault zone.

As the results of the studies carried out in gold-mining areas have shown (Table 1), high mercury contamination of the soils (1.24–3.59 mg/kg) is characteristic only for objects directly using metallic mercury (GDP, GEP, GRC). Thus, maximum mercury concentration in the soil (3.59 mg/kg) was registered at the Kavykta site, in the area of the former GRC, where amalgamated gold burning had been performed previously. The mercury contamination halo is not large here. The mercury content in the soil at approximately 100 m from the GRC was 0.037 mg/kg. Similar local soil contamination (2.41–2.91 mg/kg), observed in the area of the Itaka settlement near the GRC building, resulted from the use of metallic mercury, its careless storage and transportation. Elevated mercury concentrations (1.24 mg/kg) were also detected in sandy soil in the area of the Baleyzoloto GEP, exceeding the background values of the town of Baley by 20 times.

In the area of the Lyubov mine, which is characterized by the maximum air contamination among other settlements, the highest level of mercury concentration in the soil surface (0.03–0.57 mg/kg) was registered with an average of 0.20 mg/kg. A threefold increase in mercury content level was observed in cultivated soils in comparison with the background values.

Very intensive soil contamination was determinated around refining plants in Novosibirsk (Roslyakov and Kirillova 1995). A contrast technogenic mercury halo, extended according to the wind and occupying an area of about 1 km^2, was formed around the plant. Mercury concentrations in the soils are high and lie in the range of 0.03–18.9 mg/kg, average 0.5 mg/kg. Values exceeding MPC (2.1 mg/kg) are registered. Mean background mercury content in the soils of Novosibirsk accounts for 0.08 mg/kg and for soils of the Novosibirsk Oblast 0.03–0.04 mg/kg. In conjunction with mercury, high lead concentrations are registered in the soils around the refining plant: 10–478 mg/kg, average 140 mg/kg, local city background value 41 mg/kg and regional 24 mg/kg. In solid snow fraction, sampled in the area of soil anomaly, the mercury content reaches 6.8 mg/kg, and lead content up to 100 mg/kg, when local background accounts for 0.04 and 15 mg/kg, respectively. Hence, refining plants are significant sources of mercury and other toxic metal contamination of the environment.

Soils strongly bound mercury due to complexing with humus functional groups, and half-life of such complexes is estimated at 250 years (Trachtenberg and Korshun 1988). Thus, mercury pollution of soils is of a stable durable character and may result in pollution of the atmosphere, waters, plant and animal biocenosis, thus affecting the health of man.

5.4 Natural and Technogenic Waters

Mercury contamination of natural and industrial water streams and reservoirs in the areas impacted by processing of placer and ore gold deposits may be caused by: (1) direct emission of metallic mercury into the water in the process of sluice amalgamation; (2) contaminated wastes of GDP and GEP; (3) emission of dissolved and suspended mercury forms from processed gravels and rocks; (4) precipitation from the atmospheric air in the form of vapour, aerosols, dust and rainfall. High mercury contamination of water objects may take place in the process of erosion of tailing dumps and heaps of GDP and GEP with flood water and rainfall (Table 1).

Dissolved forms of mercury have low migratory ability due to their active sorption on suspended particles of clay and organic materials. Therefore, for the correct estimation of mercury contamination extent in water objects, levels of mercury concentration in dissolved and suspended condition have to be determined.

As the results of the study carried out in Zabaikalye have shown (Laperdina et al. 1995, 1996), in water streams of the background area the most commonly encountered concentrations of dissolved mercury range within a narrow interval: <5–10 ng/l. However, increased background value of mercury concentrations in underground and surface waters are registered in the areas of traditional gold mining, at sites without direct technogenic impact. Thus, the dissolved mercury concentrations in the underground waters used for drinking in some settlements of the Mogocha raion exceeded local background and accounted for 50–80 ng/l. This increase may result from the long-term use of metallic mercury during gold processing near these settlements and also from the geological peculiarities of the areas.

Mercury contamination of surface waters with technogenic waters may cause a dangerous deterioration in water quality, which is strongly regulated by the State fish-breeding criteria. According to these criteria, the increase in mercury concentrations in the areas of technogenic load must not exceed the background level of more than 10 ng/l. The exceeding of this criterion (10–40 ng/l) was observed in some small and medium-sized rivers (Cherny Uryum, Bugarikhta, Khangaruk).

The elevated concentrations of dissolved mercury (10–130 ng/l) are characteristic for dredge pits and, to a lesser extent, for settling tanks. However, high turbidity of technogenic waters determines the predominant content of suspended mercury. Thus, for example, concentrations of dissolved and suspended mercury accounted for 130 and 8830 ng/l, respectively, in one of the settling tanks of the Kavykta site.

The content of suspended mercury forms in the background water streams ranged in rather wide limits of volume (<5–380 ng/l), and specific (0.4–112 mg/kg d.w.) concentrations. The water objects, bearing technogenic load are characterized by a wider range of suspended mercury contents: volume concentrations from 5 to 27 800 ng/l; enrichment with mercury from 0.2 to 1860 mg/kg; turbidity in the settling tanks reached up to 12 070 mg/l. Positive correlative relations were established between volume concentrations of suspended mercury and water turbidity and also between volume and specific concentrations of mercury in suspended matter. The strength of correlation (0.41–0.97) is different and depends on both natural and technogenic factors. As the results of the analyses have shown, the enrichment of fine clay fractions with mercury is by five times higher in comparison with coarse fractions, in which the mercury content is close to Clarke values.

The maximum content of suspended mercury forms was found in the filtering reservoir of the GDP of the Baley mine 27 800 ng/l, that is higher by more than 3 orders of magnitude than the background level. In this case, suspended matter is extremely enriched with mercury, 321 mg/kg. Such an enormous value representing a dissolved mercury concentration of 5000 ng/l, indicates very high contamination in this reservoir. In the well located at several tens of metres from the gold dressing plant and used for the industrial water supply, the dissolved mercury concentration accounts for only 10 ng/l, and volume and mass concentrations of suspended mercury for 110 ng/l and 22.9 mg/kg, respectively. These values are an order of magnitude higher than similar concentrations in background underground waters. Hence, concentration of suspended mercury forms and enrichment of suspended matter with the metal are better indicators of water reservoir contamination than the content of dissolved mercury.

5.5 Bottom Sediments

One of the most dangerous effects caused by mining of gold placer deposits is a significant increase in turbidity of natural water streams, resulting in silting of riverbeds and deterioration of feeding base of fisheries as well as a steady mercury contamination of bottom sediments of industrial water reservoirs. During emergency discharges of waste waters from such reservoirs or flood erosion of dams and dikes, a potential danger of mercury contamination of natural water streams is created.

Mercury concentrations in bottom sediments in background waters Siberia range within narrow intervals from 0.01 to 0.18 mg/kg (Anoshin et al. 1995 Laperdina et al. 1995, 1996; Roslyakov and Kirillova 1995). Hence, the background water streams of the studied areas fall in the category of unpolluted.

As the results of the studies carried out in the gold mining areas of the Chita Oblast (Laperdina et al. 1995, 1996; Tupyakov et al. 1995), Buryatia (Taisayev 1991) and the Krasnoyarsk Krai (Roslyakov and Kirillova 1995) have shown that the mercury content in bottom sediments of technogenic water streams and reservoirs ranges widely from 0.01 to 54.2 mg/kg (Table 1). The most intensive mercury contamination of bottom sediments is observed in the area of the Lyubov mine. High concentrations of mercury (0.43–54.2 mg/kg) were found in silts of the former tailings dump of the gold-extracting plant. The elevated mercury content (0.42–1.29 mg/kg) was found in bottom sediments of the river Khangaruk, subjected to a vigorous influence of the tailings dump of the GEP. The former dredge pit, located in the bed area of this river, is also characterized by a high mercury content (2.83 mg/kg).

The study of bottom sediments of technogenic mercury flux formed by the tailings dump of the GEP (Buryatia) has shown that dam break of the tailings dump has resulted in the formation of highly contrasted and extended (more than 12 km) lithochemical mercury (1–5 mg/kg) and gold (0.5–4.0 mg/kg) flux due to mechanical migration. Accumulation of technogenic materials, silting of boulder and cobble bed, formation of technogenic dune ranges and bars took place over the first 3 km along the river bed. Further, transportation of technogenic materials prevails and silting of boulder and cobble bed are observed at some places. In the period of low water, the floodplain of the river and, in some places, of the river bed are covered with a crust of technogenic silt.

Similar steady, contrast and extended (about 17 km) mercury dispersion flux (0.04–3.0 mg/kg) is registered in the Krasnoyarsk Krai in the Bogunay river, eroding mine heaps. Gold mining was stopped 50 years ago but the contrast of technogenic mercury flux is still retained. Mercury content in bottom sediment of the river, 7 km upstream from the mine, varies between 0.03 and 0.18 mg/kg, average 0.064 mg/kg, and in technogenic flux average value accounts for 0.555 mg/kg, that is 8.6 times higher than background concentrations.

In the area of the Baleyzoloto mine the maximum mercury concentration in bottom sediments (3.24 mg/kg) was observed in fine-grained sandy ground on the bottom of filtering reservoirs of the GEP. The elevated concentrations were also found in bottom sediments of some open pits. Their enrichment with mercury accounts for the average of 0.62 mg/kg, which is approximately ten times higher than the level of mercury concentration in background water reservoirs. Similar results were obtained for the background and industrial water objects in the area of the Bukuka tungsten mine. Hence, mining and processing of rocks with elevated mercury content, characteristic of ore gold deposits and some other hydrothermal deposits, lead to an increase in mercury content, approximately by an order of magnitude, in bottom sediments of water systems which are not subjected to a direct mercury load.

If sluice amalgamation was not used in processing placers by the dredge method, the mercury content in bottom sediments of open pits, as a rule, is not high. Similar results are also registered in settling tanks at sites of hydraulic works. Thus, in the area of the Ksenyevsky mine, the average value in bottom sediments of technogenic reservoirs, except for a single abnormal value (0.130 mg/kg) accounts for 0.027 mg/kg. The average mercury concentration in bottom sediments of settling tanks at sites of hydraulic works in the Tungir-Olyokma raion, except for a single abnormal value (0.131 mg/kg), accounts for 0.030 mg/kg.

6 Influence of Climatic Factors on the Extent of Mercury Contamination Due to Gold Mining in Siberia

The most complete and comprehensive estimation of the extent of mercury contamination due to gold mining in Siberia was performed in some areas of Zabaykalye; therefore estimation of climatic factors influence was carried out on the basis of the results obtained from gold mining areas of Zabaykalye.

In spite of the long-term history of gold mining in Zabaikalye and previous intensive use of mercury amalgamation for concentration of gold-bearing raw materials, comparatively few areas of the region are characterized by a high mercury contamination. One of the basic reasons has to do with the severe climatic conditions of Zabaykalye, marked by an extremely continental climate with circumzero or negative average annual temperatures, deep seasonal freezing of soils and grounds, massive or island spreading of permafrost, the short period of open water in natural water streams and virtually total freezing of some of them in winter. These conditions produce an inhibiting effect on biochemical and geochemical processes of dissolving of metallic mercury which is transported into water systems. Dredge processing of placer deposits, as a rule, is carried out in insulated reservoirs with derevation of the basic river bed.

Previously, with the use of sluice amalgamation, metallic mercury settled down on the bottom of the reservoir and was buried under the layer of processed gravels. Thus, a significant amount of metallic mercury may be found under the layer of bottom sediments and grounds in anaerobic low-temperature conditions in the areas of mined-out placers, and presents a severe hazard to the environment in the reworking process of such technogenic deposits.

Low temperatures of waters in natural reservoirs and streams even in summer and total freezing of some of them in winter reduce the probability of mercury methylation processes and its emission into the water phase due to the development of heterotrophic microorganisms in the benthonic water layers and bottom sediments. The probability of biogenic mercury methylation and its accumulation in trophic chains of water ecosystems may increase only in shallow lakes, dredge pits and settling tanks.

The role of vegetation in the biochemical mercury cycle under severe climate conditions with spreading of permafrost will differ from areas with warm climate, due to a short vegetation period, lesser growth rates and accumulation of foliage, nutrition with waters overlying permafrost, which are predominantly formed by precipitation. Particular distinctions must be observed in the areas of bald peaks, tundra and mari landscapes, characterized by poor vegetation, chiefly consisting of mosses, lichens and low forest. The tundra zone is characterized by simple ecosystems and short trophic chains, therefore mosses and lichens, which can accumulate mercury and other heavy metals from soils and precipitation, will prevail in the food of mammalia. Thus, the elevated mercury concentrations in the meat of wild deer and Manchurian deer 0.039 and 0.035 mg/kg, respectively, exceeding the MPC for meat products (0.03 mg/kg) have been registered (T.G. Laperdina, unpubl. data).

The processes of freezing and thawing of rocks, soils, bottom sediments, water phase of rivers, lakes and industrial water objects are accompanied by a number of complex chemical, physicochemical and physical processes, leading to the transformation of mercury forms and their redistribution in the environment. Thus, it was determined that frozen rocks of the ore gold deposits in Yakutia (Omelchenko 1985) and some gold placers in Mogocha raion (Laperdina et al. 1995, 1996) are characterized by a decreased mercury content, resulting from the possible process of transference and concentration of mobile absorbed mercury forms on the interface "ice water" during rock freezing. The processes of rock weathering, which are mostly caused by temperature variations and water freezing and thawing in rock fractures, are accompanied by the formation of friable metasomatically altered products, which can be enriched with mercury. It was determined that in weathering products of rocks the mercury content may increase tenfold and more in comparison with the original rocks. As a result, mercury contamination of the environment in Siberia due to technogenic emission of mercury may be of a more local character in comparison to regions of warm climate.

Mercury contamination of the environment due to the use of primitive equipment for gold amalgamation in countries with tropical climate (Brazil, Tanzania, Indonesia and Vietnam) has reached an enormous extent at present. Thus, the mercury content in bottom sediments of gold-mining areas accounts for 5–129 mg/kg, mercury concentrations in suspended matter reaching 8.6 mg/l (Pfeiffer et al. 1989). Mercury contamination of soils is also registered. In such conditions the role of vegetative compartments in the biogeochemical mercury cycle is very important. Mercury accumulates and migrates throughout biotic components of ecosystems.

As a rule, such a high extent of mercury contamination of the environment in the areas impacted by gold-mining plants is not registered in Zabaikalye. Besides climate factors, this is most likely explained, by less mercury expenses in the process of concentration, a more developed concentration technology and stricter regulations established by State nature protection committees.

A high level of mercury contamination of the environment is registered in the area affected by the Luybov mine, according to the mercury content in bottom sediments of natural water streams (0.42–1.29 mg/kg), which is close to mean concentrations (1.6 mg/kg), observed at dredge placer workings in South America. In other mines studied, the contamination extent is lower and at present not dangerous, though in occasional cases an excess of the MPC for soils was recorded.

Massive or island spreading of permafrost on the Earth forms about a quarter of the land area. However, specific mercury behaviour and transformation in these severe climatic conditions are virtually not studied. As the results of our studies have shown, the extent of mercury contamination in the areas impacted by the gold-mining plants of Zabaikalye is significantly lower than in gold-mining countries with tropical climate. Therefore, investigations on temperature correction of chemical and biochemical processes, intensivity determining quantitative parameters of biogeochemical mercury cycle need to develop global-scale models of mercury emission, migration and balances.

7
Conclusion

As the results of the studies have shown, the contrast and localization of mercury contamination of the environment in gold-mining areas in Siberia due to the use of amalgamation technology mainly depend on duration and intensity, and also on natural landscape and climatic conditions. The most sensitive indicators of mercury pollution are atmospheric air, soil and bottom sediments of natural and technogenic reservoirs and streams.

In the context of the official prohibition of amalgamation use, lack of official centralized supplies of metallic mercury to gold mines and also toughening of fines for supplies of gold concentrates with an increased mercury content to refining plants, mercury expenses in amalgamation of gold-bearing raw materials have sharply reduced. The main mercury users are small scale mines not possessing enough finances to buy expensive enriching equipment and forced to use a cheap amalgamation method of gold extraction. At present, the proportion of such mines in the total amount of gold mines, for example in the Chita Oblast, is less than 10%. Because of lack of official data and according to a very approximate estimation, the amount of mercury used in gold mining in Siberia now accounts for about 10% of the annual total gold production. For example, in 1992 the total gold production in Siberia accounted for 124 53 tons, so mercury expenses could account for about 12 tons, half of this amount as emission into the environment.

A high level of mercury contamination of the environment was registered in technogenic and residential areas impacted by the Luybov mine and at other sites with direct use of metallic mercury. Hence, it is necessary to draw up a programme to reveal such highly contaminated sites and restore mercury-contaminated soils with developed techniques. At present, the use of mercury at gold-mining plants is illegal and uncontrolled. To prevent an increase in mercury contamination of the environment in gold-mining areas, it is necessary to prohibit mercury use or legalize objects where amalgamation of concentrates is performed and to establish regular control over the extent of mercury concentration in gas and water wastes and tailings dumps.

References

Anoshin GN, Malikova IN, Kovalev SI, Androsova NV, Sukhorukov FV, Tsibulchik VM, Shcherbov BL (1995) Mercury in the environment of Southern West Siberia. Chem Sustainable Dev 3:63–104

Bodienkov GA (1984) Labour hygiene in gold mining and concentration with a dredge method. In: Manyashin YA (ed) Problems of labour hygiene and professional pathology in gold mining in the east of the country. Ministry of public health, Moscow, pp 20–28 (in Russian)

Foss GV (1963) Gold. Gosgeoltechizdat, Moscow (in Russian)

Ignatkin YA (1994) Along wild steppes of Zabaikalye. Chita Publishing House, Chita (in Russian)

Karmanov NS (1985) Geochemical peculiarities of native gold on evidence of spark mass spectrometry of the ore sites of Eastern Siberia. Author's abstract. IGG, Novosibirsk (in Russian)

Koledneva N (1994) The tricks of mercury. Zabaikalsky rabochy 11:1 (in Russian)

Laperdina TG, Tupyakov AB, Yegorov AI, Melnikova MV, Askarova OB, Banshchikov VA, Khvostova TE, Tsybikdorzhiev Z, Bochko OK (1995) Environment pollution by mercury in the zone of influence of transbaikalian gold mining plants. Chem Sustainable Dev 3:53–62

Laperdina TG, Melnikova MV, Khvostova TE (1996) Mercury pollution of the environment due to gold mining in Zabaikalye. In: Baeyens W, Ebinghaus R, Vasiliev O (eds) Proc NATO ARW, Regional and global mercury cycles: sources, fluxes and mass balances. Novosibirsk, Russia, 10–14, July 1995. Kluwer, Dordrecht, pp 415–427

Malgin MA, Puzanov AV (1995) Mercury in soils, surface and soil air of the Altai–Sayan mountainous country. Chem Sustainable Dev 3:145–157 (in Russian)

Ministry Cabinet of the USSR (1988) The order of the Chief Department on precious metals and diamonds Moscow. No. 124 of 29.12.1988. On the prohibition of mercury (amalgamation) in the technological processes during concentration of gold-bearing ores and sands (in Russian)

Omelchenko MM (1985) Mercury at gold ore deposits of Yakutia and its role in geochemical prospectings. Abstract. LGI, Leningrad (in Russian)

Ovchinnikov LN (1990) Applied geochemistry. Nedra, Moscow (in Russian)

Ozerova NA (1986) Mercury and endogenous ore deposition. Nauka, Moscow (in Russian)

Pfeiffer WC, Lacerda LD, Malm O, Souza CMM, Silveira EG, Bastos WR (1989) Mercury concentrations in inland waters of gold mining areas in Rondonia, Brazil. Sci Total Environ 87/88:233–240

Pluteshko VP (1993) Gold mining in Russia in 1966–1990 and prospects of its development. AO Geoinformmark, Moscow (in Russian)
Roslyakov NA, Kirillova OV (1995) Environmental contamination by mercury owing to gold production in Russia. Chem Sustainable Dev 3:41–52
Rukavishnikov VS (1984) Sanitary estimation of labour conditions and health of attendants at gold extraction plants. In: Manyashin YA (ed) Problems of labour hygiene and professional pathology in gold mining in the east of the country. Ministry of public health, Moscow, pp 29–38 (in Russian)
Saukov AA (1975) Geochemistry. Nauka, Moscow (in Russian)
Stepanov VA, Moiseyenko VG (1993) Geology of gold, silver and mercury. Dalnauka, Vladivostok (in Russian)
Taisayev TT (1991) Technogenic dispersion flux of gold and mercury in bald peak-taiga landscapes. DAN USSR 371:719–722 (in Russian)
Trachtenberg IM, Korshun MN (1988) Mercury and its compounds. In: Philov VA (ed) Dangerous chemical substances. Inorganic compounds of the elements of 1–4 groups. Handbook. Khimiya, Leningrad (in Russian)
Tupyakov AV, Laperdina TG, Egorov AI, Banshchikov VA, Melnikova MV, Askarova OB (1995) Mercury concentration in the abiogenic environmental components near gold and tungsten mines and concentration complexes in the Eastern Transbaikalia. Water Resour 22:179–186 (in Russian)
Vyazelshchikov VP, Paritzky ZN (1963) Handbook on processing of gold-bearing ores and placers. GONTI, Moscow (in Russian)
Yeend W (1991) Gold placers of the Circle District, Alaska – past, present, and future. US Geol Surv Bull 1943:1–42
Young JE (1993) For the love of gold. Worldwatch May–June:19–26
Zamyatin OV, Lopatin AG, Sannikova NP, Chugunov AD (1973) Concentration of gold-bearing gravels and conglomerates. Nedra, Moscow (in Russian)

II.c Emissions and Atmospheric Dispersion

Estimation of Gaseous Mercury Emissions in Germany: Inverse Modelling of Source Strengths at the Contaminated Industrial Site BSL Werk Schkopau

O. Krüger, R. Ebinghaus, H.H. Kock , I. Richter-Politz, and C. Geilhufe

Anthropogenic emission sources of gaseous mercury at the contaminated industrial site BSL Werk Schkopau have been determined by measurements and numerical modelling applying a local dispersion model. The investigations are based on measurements from several field campaigns in the period of time between December 1993 and June 1994. The estimation of the source strengths was performed by inverse modelling using measurements as constraints for the dispersion model. Model experiments confirmed the applicability of the inverse modelling procedure for the source strength estimation at BSL Werk Schkopau. At the factory premises investigated, the source strengths of four source areas, among them three closed chlor-alkali productions, one partly removed acetaldehyde factory and additionaly one still producing chlor-alkali factory have been identified with an approximate total gaseous mercury emission of lower than 2.5 kg/day.

1 Introduction

In recent years, considerable progress has been made in specifying the anthropogenic sources of atmospheric mercury (EPRI 1994). For all Europe the annual anthropogenic emission of mercury has been estimated to be about 726 t, originating from 928 sources (Axenfeld et al. 1991). The dominating source categories in Europe are fossil fuel combustion and chlor-alkali plants. Waste incineration and non-ferrous metal smelting contribute less than 10%. Emissions from the former German Democratic Republic (GDR), including the different species of elemental mercury, divalent inorganic mercury and particulate mercury, accounted for more than 40% of the European total. The contribution of the GDR originated from a few relatively small but in most cases highly industrialized areas. According to Hellwig and Neske (1990), extremely high amounts were emitted in the region Halle/Leipzig/Bitterfeld, due to both burning of lignite coal in power plants without flue gas desulurization equipment and high losses of mercury from the chlor-alkali factories.

An important emitter of air pollutants in this region was the former Chemische Werke Buna (now Buna Sow Leuna Olefinverbund GmbH; hereafter referred to as the BSL Werk Schkopau) located near Halle/Saale. Before 1989, the company produced per year 1000 kt calcium carbide, 300 kt acetylene, 420 kt chlorine, 300 kt polyvinylchloride (PVC) and 120 kt synthetic caoutchouc. As a consequence of the reunification of Germany, about 50 production installations which were considered to be ecologically harmful were closed at the BSL Werk

Environmental Science
Mercury Contaminated Sites (ed. by R. Ebinghaus et al.)

Schkopau in the period of time between 1990 and 1992. Among them were three chlor-alkali production works (hereafter denoted L66, I54 and H56), one factory still producing chlor-alkali (denoted P159) and one acetaldehyde factory (denoted F44). In these factories, mercury was used as an electrode for chlorine and sodium hydroxide production and as a catalyst for acetaldehyde production. All production sites consist of partly removed buildings or buildings with openings. Since these buildings are continuing to release gaseous mercury into the surrounding atmosphere, presently, the main problem related to mercury at BSL Werk Schkopau is the redevelopment of this contaminated area. Furthermore, the variations of mercury concentrations in the surrounding air and the potential danger this posed for human health were unknown. Therefore, in order to obtain an implication about the amount of mercury in air an important task was to estimate the quantity of mercury emissions originating from specified sources at the factory premises.

This chapter focuses on the actual situation in 1994 at the industrial site BSL Werk Schkopau located near Halle/Saale. Emission source strength estimations based on measurements and model calculations are presented. The following section describes the measurement method and the quantities of atmospheric gaseous mercury measured at the BSL Werk Schkopau. After a brief description of the local dispersion modelling, an inverse modelling procedure which was applied to estimate source strengths of mercury emissions is explained. Model experiments due to the applicability of the inverse modelling method at the individual site BSL Werk Schkopau are discussed, and the results of the source strengths of gaseous mercury emissions are presented. The conclusions are outlined in the final section.

2
Measurements of Gaseous Mercury at the BSL Werk Schkopau

In order to determine mercury concentrations in air at the BSL Werk Schkopau, the analytical methods described in Ebinghaus et al. (1995) were applied. During the field campaigns total gaseous mercury (TGM) was collected on gold-coated glass beads. Figure 1 depicts the method of sampling. Basically, two quartz tubes of 0.4 cm internal diameter were packed with glass beads (100 mesh) to a length of approximately 1.5 cm, which results in an active sampling surface of 15–20 cm^2 per trap. To prevent contamination, a third tube which was fitted with a gold/platinum gauze was placed between the adsorber tubes and the pump. The ambient air was drawn through a 0.5-cm quartz wool plug before passing through the adsorber tubes. Approximately 10 l of air was collected at a flow rate of 20–30 l/h. After sampling, the tubes were closed with plastic caps and stored in a firmly sealed glass container. To prevent contamination during storage, 1 g of silver wool was kept in the container to trap gaseous mercury diffusing into it.

All samples were analyzed using cold vapour atomic fluorescence spectroscopy (CVAFS) with the two-step amalgamation technique (Fitzgerald and Gill 1976).

Fig. 1. Schematic sampling train used for total gaseous mercury

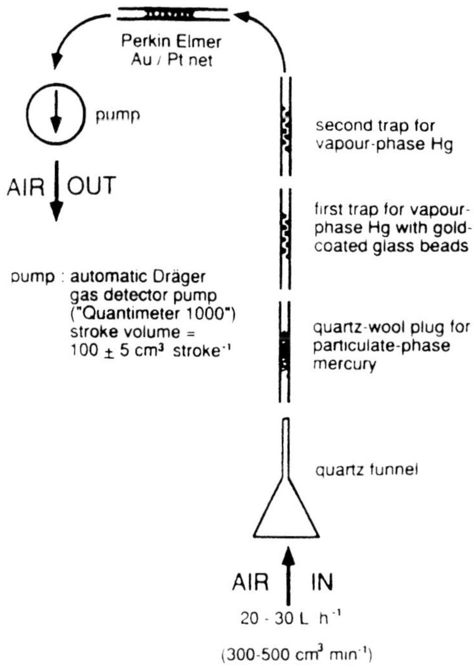

Calibration was carried out by injecting mercury-saturated air into the analytical column with a gas-tight syringe.

As the main emission sources of total gaseous mercury three chlor-alkali productions units (L66, I54/H56) as well as the acetaldehyde production unit (F44) including their contaminated surrounding areas were considered (Fig. 2). While earlier clearly increased concentrations were measured at hall F44, where mercury was used as a catalyst for acetaldehyde production, considerably high values of TGM were detected in the air at the chlor-alkali productions I54/H56 and L66. In these factories, elemental mercury was used as an electrode during electrolysis. Figure 3 shows L66, which was one of the halls for chlorine production. About 100 electrolysis cells for chlorine and sodium hydroxide production were in operation in hall L66. Each of these iron-made cells contained 1 t of elemental mercury as an electrode material (Fig. 4). During operation the approximate temperature inside the cells was between 80 and 90 °C, which caused the evaporation of considerable amounts of mercury.

In October 1993, more than 2 years after production was stopped, indoor concentrations of the halls of maximum 270 μg/m³ TGM were measured in hall F44. Inside the chlor-alkali production facilities of I54, values of 100 μg/m³ were detected. Close to heavily contaminated soil surfaces peak values of ambient air concentrations of about 100 μg/m³ were observed frequently.

Due to the specifics of the production pathways at the BSL Werk Schkopau, it was assumed that elemental mercury was the major species emitted into the atmosphere. Therefore, besides elemental mercury, other species which might

Fig. 2. The factory premises of BSL Werk Schkopau located near Halle/Saale, Germany. The entire BSL Werk Schkopau covers an area of approximately 4 km^2 with an adjacent waste disposal in the north-west of an additional size of 3 km^2

also be of interest with regard to different behaviours in the atmosphere, i.e. gaseous $HgCl_2$ and dimethylmercury, were not separately measured during the field experiments. However, both species contributed to the TGM value which was measured.

The intention of the in-situ measurements was to obtain information concerning the horizontal distribution of gaseous mercury 0.5 m above ground level. The area within a maximum distance of 5000 m from the sources was of main interest. All detections were made within the expected plume of mercury which was developed in the main direction of the wind shaped by the wind shear and temperature profiles.

In order to obtain data for different atmospheric conditions during the sampling period from December 1993 to June 1994, the measurements were carried out on an event basis. The reason for this was to include the influence of seasonal and temperature-related variations. Table 1 shows typical measurements

Fig. 3. The production site L66 of the BSL Werk Schkopau in 1994, four years after the closure of the former Chemische Werke Buna. The hall dimensions are a length of 162 m, a width of 37 m and a height 13 m

of total gaseous mercury in the surroundings of the BSL Werk Schkopau. It can be summarized that outside the factory premises, the maximum concentrations of higher than 500 ng/m^3 were measured in downwind direction at a distance of 1000 m from the sources. In upwind direction, values close to the background concentration for this region of 2 ng/m^3 were measured (Ebinghaus et al. 1995, Ebinghaus and Krüger 1996).

Fig. 4. Electrolysis cells used for the chlor-alkali production in hall L66

Table 1. Measurements in the surroundings of the factory premises of BSL Werk Schkopau in mean distance from the source areas L66, 154/H56 and F44

distance (m) / TGM (ng/m3)

wind-direct.	location : **N**	location : **NO**	location : **O**	location : **SO**	location : **S**	location : **SW**	location : **W**	location : **NW**
	dist. / TGM	dist. / TGM	dist. / TGM	dist. / TGM	dist. / TGM	dist. / TGM	dist. / TGM	dist. / TGM
N	500 / 6,5			1500 / 128	800 / 530	1200 / 18		
N	1000 / 8,5	.			800 / 450	1200 / 35		
N					800 / 415	1200 / 40		
N					1000 / 35	2000 / 129		
N					1000 / 43	2000 / 16		
N					1000 / 83	2000 / 17		
N					1000 / 95	2000 / 18		
N					1000 / 83	2000 / 15		
N					1000 / 261	2500 / 17		
N					1000 / 210	2500 / 28		
N					1000 / 152			
N					1200 / 26			
N					1200 / 300			
N					1200 / 193			
N					1500 / 235			
N					2000 / 27			
N					2000 / 20			
N					3000 / 91			
N					6000 / 11			
S	1500 / 63	1000 / 8,8	1000 / 33	1000 / 4,8	1000 / 6,0		2000 / 6,5	1000 / 8,5
S	4000 / 6,8	2500 / 5,5					2500 / 8,2	3500 / 4,4
S								4000 / 4,5
SW		1000 / 62	5000 / 7,0		1500 / 32			1000 / 63
SW		2000 / 56			3500 / 5,3			
SW		3000 / 72			4000 / 7,5			
SW		3500 / 10						
SW		4000 / 44						
SW		5000 / 2,6						
SW		12000 / 3,2						
SW		15000 / 2,4						
W		1000 / 161	800 / 41					
W		1500 / 5,7	800 / 25					
W		7000 / 5,4	800 / 35					
W			1000 / 66					
W			2500 / 6,3					
W			10000 / 4,0					
W			12000 / 6,0					
NW			500 / 8,8	1200 / 38	1200 / 20			
NW			1000 / 11	1500 / 46				
NW			1500 / 23	1500 / 78				
NW			5000 / 4,6	1500 / 39				
NW				1500 / 7,9				
NW				2000 / 9,0				
NW				4500 / 3,8				

3 Dispersion Modelling

The local atmospheric transport of gaseous mercury originating from emissions at the BSL Werk Schkopau was calculated for distances of a maximum of 10 km. A numerical model for simulating transport from individual point sources was applied. Earlier results of this transport model (Eppel et al. 1991) compared well with measurements within a spatial range of a maximum of 100 km.

The model, which is an extension of the elementary Gaussian model, essentially is an Eulerian model, whose parametrizations are more closely related to the transport properties of the atmosphere. It has been shown (Eppel et al. 1991) that the model also performs a quite realistic dispersion for episodes with a temporally evolving boundary layer. In the model the transport equation

$$\frac{\partial}{\partial t}C + U(z,t)\frac{\partial}{\partial x}C + V(z,t)\frac{\partial}{\partial y}C - W(x,t)\frac{\partial}{\partial z}C$$
$$= K_y(x,z,t)\frac{\partial^2}{\partial y^2}C + \frac{\partial}{\partial z}K_z(x,z,t)\frac{\partial}{\partial z}C - RC + S,$$

which describes the concentrations $C = C(x,y,z,t)$ of a pollutant above a flat terrain, is solved numerically using a moment reduction in cross-wind direction in order to reduce a three-dimensional problem to a two-dimensional problem along the wind and vertical directions. This procedure preserves a high resolution of the plume structure especially for simulating the transport at distances far from the sources. As the numerical grid is oriented vertically in mean wind direction, the cross-wind resolution can be chosen very high for the entire domain of the model. This advantage contrasts to grid models, which either highly resolve only the source region, while distances far away from the sources are not covered by the grid, or the model domain is adequate to simulate transport far from the source, but the grid spacing is too crude to resolve the structure of the plume close to the emission source.

Besides the removal parameter R, which represents dry and wet deposition processes, a negative sink velocity W is defined in the model, which enables taking into account the effect of buoyancy of the plume centerline. Sources of pollutants are denoted by S. The vertically variable wind velocities U and V in the main- and cross-wind direction of the plume enable the simulation of a sheared wind field in the boundary layer. For stable and neutral atmospheres the diffusion coefficients K_y and K_z are determined from horizontal and vertical velocity variances (e_h and e_v) and the time scales t_y and t_z:

$$K_y = t_y \cdot e_h$$
$$K_z = t_z \cdot e_v.$$

The timescales t_y and t_z were determined by fitting the horizontal and vertical extent of calculated plume cross-sections to in-situ measurements. The velocity variances are calculated from dynamic equations for the time derivatives of e_h and e_v which contain the effects of diffusion, shear, buoyancy, redistribution and dissipation. These equations are solved by the input of vertical temperature profiles in addition to velocity profiles.

Convective conditions are parameterized by a phenomenological approach (Lenschow and Stephens 1982), which superimposes on the standard model a separation of the plume in an updraft and a downdraft branch. A detailed description of the procedure is given in Eppel et al. (1991).

4 Inverse Modelling of Source Strengths

A reliable estimation of source strengths of gaseous mercury exclusively from measurements of concentrations in the surrounding atmosphere near the sources is impossible without modelling. The strong variability of the meteorological parameters which determine the dispersion can only be taken into account using a numerical model. On the other hand, results of a dispersion model based on very uncertain and roughly estimated emission data most likely do not reflect the real atmospheric conditions. Therefore, we decided to combine both methods. The strategy was to use the measurements as constraints in fitting the model to the emission source strengths. This procedure can be described as an inverse modelling, because the model is applied oppositely to the usual way, in which the transmission is calculated from known emission sources. Applying inverse modelling, the problem can be formulated as follows:

In the surroundings of an ensemble of emission sources n measurements of the concentrations $c_i(x_i,y_i)$ of gaseous mercury at different points i exist. The location of the measurements in a model grid, here in a downwind and cross-wind direction of the plume, is given by x_i and y_i. Uncertainties of the measurements are denoted by δci. The dispersion model calculates the concentrations $C_i(x_i,y_i) = f(x_i,y_i,\alpha)$, where α is the vector of free parameters being fitted. In this case, α represents the source strengths of the emission sources at the BSL Werk Schkopau, which are treated variably in the procedure. In order to minimize the differences between in-situ measurements and model results, a multiparameter function is defined for a least-squares fit as a chi square:

$$\chi^2(\alpha) = \sum_{i=1}^{n} \frac{[f(x_i, y_i, \alpha) - c_i]^2}{\delta c_i^2} .$$

For the minimum, which is found at $\chi^2 = \chi^2(\alpha_{min})$, a statistical test for a χ^2-distribution with $[n - \dim(\alpha_{min})]$ can be made to check the performance of the fit. The parameter vector α can usually be estimated by applying different methods for minimizing χ^2. All investigations which are described in the following sections are based on the simplex method (Nelder and Mead 1965). This genuine multidimensional minimization procedure has the advantages that no accurate start parameters have to be estimated, different ranges of the parameters can be defined and it is even rather robust with respect to cross fluctuations in the function value. Figure 5 shows the flow chart of the inverse modelling procedure.

5 Theoretical Tests with the Inverse Modelling Procedure

The applicability of the inverse modelling procedure for emission estimates at the industrial site BSL Werk Schkopau was tested using characteristic vertical

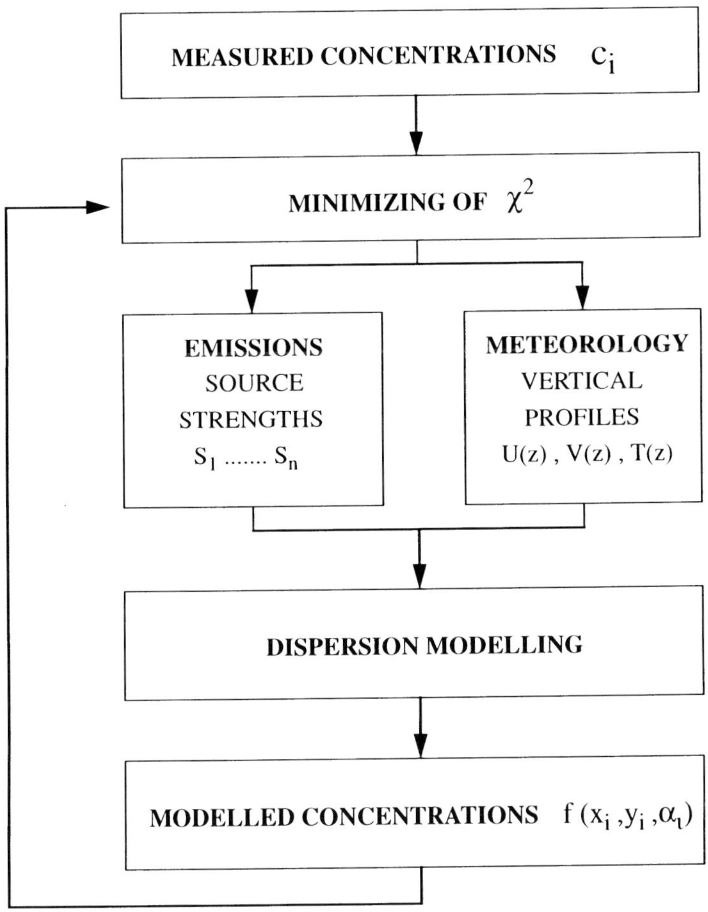

Fig. 5. Flow chart of the invers modelling procedure

profiles of wind velocities and temperatures. For these investigations meteorological data for the year 1994 were compiled as input for the dispersion model. The vertical profiles of the input data represent realistic meteorological conditions for different Grosswetterlagen (GWL) which occurred at the site BSL Werk Schkopau in the summer and winter of 1994 with atmospheric currents originating from north (N), south (S), east (E) and west (W) directions. These weather situations favour the atmospheric dispersion of gaseous mercury during stable conditions, low inversion layers or surface inversions usually occurring at night and in the early morning hours. An overview of the chosen cases for 1994, which was the period of time of several field measurement campaigns, is given in Table 2. The date, the sector of wind direction, the wind velocity at 1000 m height, the GWL following Hess/Brezowski, the type of the air mass and the representativeness of each profile are indicated. Except for the northerly current mean surface inversion layer heights of all profiles in 1994

Table 2. Characteristic weather situations, which favour the dispersion of air pollutants at the position BSL Werk Schkopau in 1994. All cases are valid for the early morning at 6 UTC

case	sector of wind direction	wind velocity in 1000 m height [m/s]	GWL	air mass	boundary layer	frequency of similar cases in 1994
January 17	N	5.1	BM	artic polar	no inversion, stable	1
July 1	N	3.1	SEa	warm continental	inversion at 100 m	7
February 4	E	8.7	SEz	warm continental and maritime	inversion at 250 m	10
July 12	E	7.6	HM	warm continental	inversion at 300 m	14
August 1	S	5.8	Sa	mediterranean tropic	inversion at 250 m	8
February 20	S	10.8	HFz	russian polar	no inversion, stable	4
December 10	W	9.8	Wz	maritime	inversion at 150 m	18
June 10	W	9.3	Wz	atlantic tropic	inversion at 100 m	12

corresponding to the individual sectors of wind are within 50 and 350 m. Totally, about 200 cases with surface inversions in the early morning occurred, from which 70 are represented by the cases depicted in Table 2. The representativeness of the individual profiles can be described as typical for monthly meteorological situations in 1994. Therefore, an adequate basis for the interpretation of the results of the mercury dispersion modelling was expected when using these profiles.

The performance of the inverse modelling procedure was investigated for the meteorological cases described in Table 2. An ensemble of three source areas, in the model domain exactly positioned as the source configuration at the BSL Werk Schkopau, was prescribed. Firstly, a reference run with the dispersion model was performed to determine concentrations of mercury at the ground in the surrounding of the sources. This was done in order to have some information about the values which should be measured. According to the distribution of the different sources at the factory premises at the BSL Werk Schkopau, which are specified above, source strengths of the three mercury emission source areas in the ratio 3:2:1 (the locations of L66, I54/H56 and F44 at BSL Werk Schkopau, respectively) were defined for the reference run. The approximate distance between neighbouring sources was 200 m. For testing the procedure, a number of results at individual grid points of the dispersion model were taken from each reference run (properties from this run are hereafter denoted theoretical) in order to reproduce this run when applying the inverse modelling technique. The location of the theoretical measurements inside the plume was chosen in a crosswind direction for different distances from the sources in a downwind direction. For each cross-section three or four data points were selected. The next step of the performance test was running the inverse modelling procedure for a different number of theoretical measurements. For the minimizing procedure maximum values of the source strengths were defined to be a factor of 3 of the theoretical source strengths. These maximum emissions were chosen as starting values. The results are depicted in Table 3.

In all model experiments the theoretical measurements originated from the region within a distance of 1000 and 3000 m from the sources. This area of the plume was preferred, because strong gradients of mercury concentration were expected to occur under different dispersion conditions. A number of statistical tests were performed to examine the accuracy of the results by increasing the number of measurements for a stable meteorological case on January 17. For these test runs a dry deposition velocity of 0.1 cm/s was considered. Wet deposition was neglected. The base run using the information of four measurements at a distance of 2000 m shows a quite satisfying result. The differences between modelled and theoretical source strengths for the individual sources are within a range of 14%. However, in this case, the chi square probability of 0.70 was relatively low. Instead, the statistical probability when

Table 3. Model experiments applying the inverse modelling technique with a maximum of 100 iterations. Emissions originating from an ensemble of three emission source areas (as L66, 154/H56 and F44 at BSL Werk Schkopau) in a distance of 200 m. The meteorological condition, January 17, 6 UTC, represents a stable case. The tests of the method have been performed using different numbers of measurements with different locations in the dispersion plume of gaseous mercury

number of measurements in distance from the sources	source 1 $\left\lvert \frac{S_{mod}}{S_{theo}} - 1 \right\rvert$	source 2 $\left\lvert \frac{S_{mod}}{S_{theo}} - 1 \right\rvert$	source 3 $\left\lvert \frac{S_{mod}}{S_{theo}} - 1 \right\rvert$	all sources $\left\lvert \frac{S_{mod}}{S_{theo}} - 1 \right\rvert$	χ^2	$P\langle\chi^2\rangle$
4 in 2000 m	0.06	0.14	0.09	0.01	0.1317	0.70
3 in 2000 m 3 in 3000 m	0.07	0.14	0.08	0.01	0.0574	0.99
3 in 2000 m 3 in 2500 m 3 in 3000 m	0.04	0.10	0.06	0.01	0.1532	0.99
3 in 2000 m 3 in 2200 m 3 in 2500 m 3 in 2700 m 3 in 3000 m	0.05	0.11	0.10	0.01	0.2776	0.99
4 in 1000 m	0.01	0.04	0.05	0.01	0.0035	0.95
4 in 1500 m	0.01	0.03	0.04	0.01	0.0002	0.99
4 in 3000 m	0.07	0.22	0.21	0.01	0.0469	0.80

Deviations of the modelled source strength $S_{1\ mod}$, $S_{2\ mod}$, $S_{3\ mod}$ from the theoretical values $S_{1\ theo}$, $S_{2\ theo}$, $S_{3\ theo}$, the corresponding minimin of chi square and the probability P for assumed measurement uncertainties of 10% are presented.

using six measurements, three at 2000 m and three at 3000 m, of 0.99 shows quite good agreement between model and measurements. Increasing the number of measurements to a density of 15, it can be seen that the accuracy of the individually determined source strengths does not significantly increase. The use of 15 measurements shows no significant further improvement in the results. A much stronger variability in the results can be seen by changing the location of the four measurements inside the plume. From all model experiments it can be summarized that the inverse modelling procedure reaches a minimum difference between modelled and theoretical source strengths of less than 5% by the using only four measurements. Therefore, relative to the source configuration at the BSL Werk Schkopau, four measurements of gaseous mercury at a distance of 1500 m appeared to be satisfactory. With further increasing distances, the accuracy of the results for individual emission sources decreases. An important result for all test runs was that the total source strength at BSL Werk Schkopau was almost reproduced by this procedure with uncertainties less than 1%.

Following these findings, a second series of test runs was performed in order to reproduce the results using four measurements at a distance of 1500 m for the other meteorological conditions indicated in Table 2. Now all cases were reproduced by the inverse modelling procedure within the same low uncertainties as shown in Table 2. The differences between modelled and theoretical concentrations in nearly all runs were lower than 1 ng/m^3. The results above confirmed the good performance of the inverse modelling technique and suggest an optimal distance of 1000 m for obtaining in situ measurements.

6 Emissions of Gaseous Mercury

For applying the inverse modelling procedure described above, data from several measurement campaigns in the period from December 1993 to April 1994 were used to estimate the range of source strengths in a first guess, which were used as start values for the minimizing procedure. These estimates are based on vertical wind- and temperature fields interpolated from neighbouring meteorological stations. The measurements which were used to determine source strengths were detected with peak values of 235 ng/m^3 within a distance of 1000 m from the sources. The total source strength of the BSL Werk Schkopau including the three closed chlor-alkali factories, the one chlor-alkali factory still in operation and the partly removed acetaldehyde production plant was roughly estimated to be lower than 10 kg/day. In this estimation of the source strength an error of a factor of 2 was assumed.

In June 1994 a 3-day measurement campaign was undertaken in order to estimate the source strengths at the BSL Werk Schkopau with higher accuracy. The meteorological data of this dry episode were simultaneously measured by radio-sonde ascents and theodolite observations to obtain more reliable data at the BSL Werk Schkopau for the dispersion modelling.

Within this period of time the highest concentrations of total gaseous mercury, some 100 ng/m^3, outside the factory premises were measured during the early morning hours of June 14. These concentrations at the BSL Werk Schkopau, which are relatively high compared to the background values, were exclusively measured under extremely stable meteorological conditions near the center of the dispersion plume. With regard to the application of the inverse modelling procedure, it was expected that a higher accuracy can be attained in the emission estimates if such peak concentrations are taken into account.

The dry deposition, in the model parameterized by a dry deposition velocity, was assumed following Petersen et al. (1995). Since gaseous mercury in the atmosphere mainly occurs as elemental mercury (Hg^o) and to a small portion as divalent inorganic mercury [Hg(II)(g)], an effective dry deposition velocity was taken into account. For this effective dry deposition velocity the contribution of Hg^o was assumed to be negligibly small compared to Hg(II)(g) with a value which was adopted to be 4 cm/s. Because the portion of divalent inorganic mercury was unknown, three different portions of 0, 15 and 30 of this species were considered for the inverse modelling runs. These portions correspond to dry deposition velocities of 0.0, 0.5 and 1.0 cm/s respectively. Table 4 shows the results for three source areas (denoted sources 1, 2, 3) including four inactive plants and one site still in production (source 4). The influence of different dry deposition velocities was also considered. All source strengths were calculated from concentrations which were measured on June 14 in the early morning hours between 05:00 and 06:00 h Universal Time Coordinated (UTC). Following the findings presented in Table 3, these measurements were obtained along a horizontal cross-section at a distance of approximately 1500 m. The resulting source strengths of the emission sources were within a range 0.2 and 1.7 kg/day. Compared to earlier estimates which were compiled by Hellwig and Neske (1990), the total source strength of all sources in 1995, after the closure of several factories, was a factor of 50 times lower. For the assumptions of different dry deposition velocities the results vary by a factor of 2. In both extreme cases of assuming that no dry deposition takes place or the dry deposition velocity amounts to a maximum value of 1 m/s, the total source strengths were estimated to be 1.5 and 3.5 kg/day respectively. Figure 6 shows the dispersion plume at 05:30 h UTC. It can be seen that three of the source areas, including all inactive plants, are located relatively close to each other, while the emission source 4, which still produces chlorine and sodium hydroxide, is 1000 m further away

Table 4. Results of invers modelled source strengths of four source areas at BSL Werk Schkopau (L66, I54/H56, F44 and P159) in kg/d dependent on different effective dry deposition velocities

source	v_d = 0 cm/s	v_d = 0.5 cm/s	v_d = 1.0 cm/s
source 1	0.7	1.0	1.7
source 2	0.4	0.6	0.8
source 3	0.2	0.3	0.5
source 4	0.2	0.3	0.5
all sources	1.5	2.2	3.5

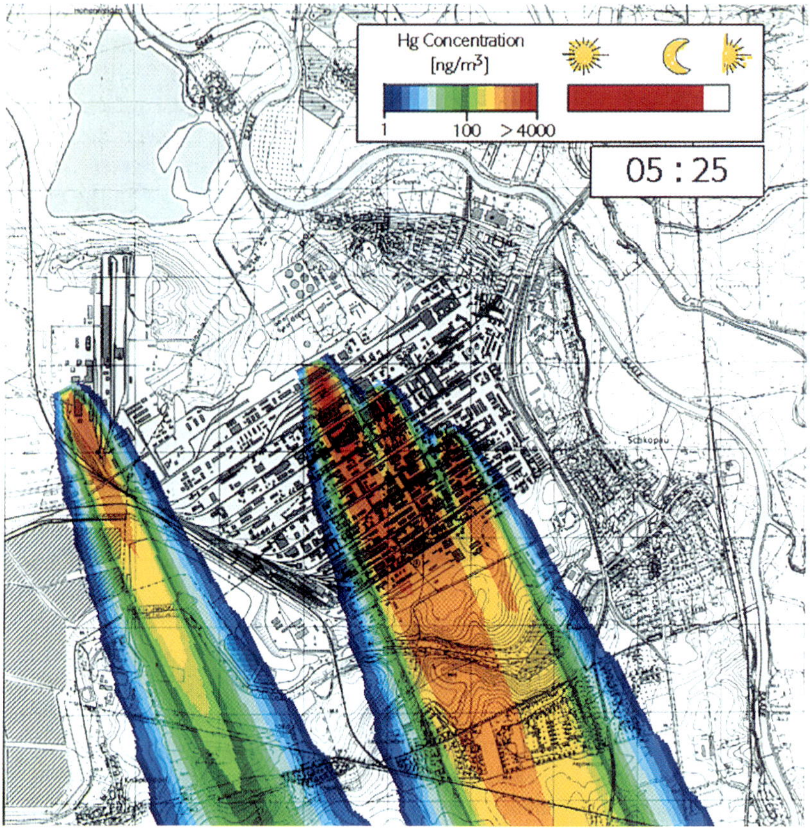

Fig. 6. Dispersion of gaseous mercury at BSL Werk Schkopau near Halle/Saale in the early morning of June 14, 1995, at 5:25 UTC

from the others. The source strength of this factory was calculated separately from measurements in a second dispersion plume. Since the source strength of this factory was independently measured directly at the low stack of the source by the same amount of 0.3 kg/day, which was estimated by inverse modelling, a high accuracy of all other source strength estimates can be expected.

During 10–12 UTC on June 14 the meteorological conditions changed to a convective situation and the dispersion plume of mercury was completely different. The concentrations measured in the center of the plume at a distance of 3000 m reached values of 80–90 ng/m^3. Using these measurements, the results shown in Table 4 were reproduced. The resulting source strengths correspond well with the result for the early morning hours.

However, in order to obtain a better estimate of the uncertainty, the accuracy of the measurements was varied in the model runs. Table 5 shows the corresponding uncertainties of the source strengths calculated by using measurements containing different errors with a maximum of 50%. Because the inverse modelling approach demands for the input mean concentrations in

Table 5. Accuracy of invers modelled source strengths of four source areas at BSL Werk Schkopau in kg/d dependent on different uncertainties of the used measurements. Results for June 14, 1994 5.30 UTC

source	$\delta c_i = 10\%$	$\delta c_i = 30\%$	$\delta c_i = 50\%$
source 1	10%	31%	52%
source 2	11%	30%	51%
source 3	7%	21%	35%
source 4	10%	31%	51%

the model grid, defined here as 50 by 20 m, the errors are assumed to be mainly due to the representation of a grid element by a few or only one measurement point inside such an area. The uncertainties of the emission source strengths are strongly dependent on the representation of a grid element by the measurements, which, in turn, is connected to the number and the location of the measurements.

7 Conclusions

Emissions of atmospheric mercury originating from closed production installations at the contaminated industrial site BSL Werk Schkopau were successfully determined by using the method of inverse modelling. The method, whose good performance was supported by several theoretical tests, has the advantage that neighboured individual source strengths can be specified using field measurements.

Based on measurements in 1993 and 1994 the source strengths at BSL Werk Schkopau of four inactive factories were calculated to be in the range between 0.2 and 1.7 kg/day. Since these emissions of atmospheric mercury compared to 1988 (Hellwig and Neske 1990) show a considerable reduction by a factor of 50, the results are expected to be useful for the actualization of the European emission inventory for mercury.

The significant change in the emission in this area after the reunion of Germany implies that the strong south-to-north gradients of atmospheric mercury concentrations over Europe found by Petersen et al. (1995) should be more smoother since the closure of the factories at the BSL Werk Schkopau. Further regional transport modelling using updated emission inventories with regard to specification of the south to north transmission is motivated by this study.

The method itself was quite useful for experimental planning, because the number of measurements which are needed for a successful source strength estimation, including the location of the measurements, can be simulated on the computer. This can be done before the field experiments are performed. Therefore the method is recommended for experimental planning of measurement campaigns near emission sources.

Acknowledgements The authors wish to acknowledge the financial support provided by the German BMBF, research contract no. 149 077 45. Thanks are due to Rolf-Dieter Wilken, Helmut Schiller, Gerhard Petersen and Dieter Eppel for computer codes and valuable suggestions during the research.

References

Axenfeld F, Münch J, Pacyna J (1991) Europäische Test-Emissionsdatenbasis von Quecksilberkomponenten fuer Modellrechnungen. In: Petersen G (ed) Belastung von Nord- und Ostsee durch ökologisch gefaehrliche Stoffe am Beispiel atmosphärischer Quecksilberverbindungen. Abschlussbericht des Forschungsvorhabens 10402726 des Umweltforschungsplanes des Bundesministers fuer Umwelt, Naturschutz und Reaktorsicherheit. Im Auftrag des Umweltbundesamtes, Berlin, External report GKSS 92/E/111. GKSS Research Centre, Max-Planck-Strasse, D-21502 Geesthacht, Germany

Ebinghaus R, Krüger O (1996) Emission and local deposition estimates of atmospheric mercury in north western and central Europe. In: Baeyens W, Ebinghaus R, Vassiliev O (eds) Global and regional mercury cycles: sources, fluxes and mass balances. NATO-ASI-Series. Kluwer, Dordrecht

Ebinghaus R, Kock HH, Jennings SG, McCartin P, Orren MJ (1995) Measurements of atmospheric mercury concentrations in northwestern and central Europe – comparison of experimental data and model results. Atmos Environ 29(22):3333–3344

Eppel DP, Petersen G, Misra PK, Bloxam R (1991) A numerical model for simulating pollutant transport from a single point source. Atmos Environ 25a (7):1391–1401

EPRI (1994) Mercury atmospheric processes: a synthesis report, prepared by expert panal on mercury atmospheric processes, 16–18 March 1994, Tampa, Florida. EPRI/TR-104214 Worksh Proc Sept 1994

Fitzgerald WF, Gill GA (1979) Subnanogram determination of mercury by two-stage gold amalgamation and gas phase detection applied to atmospheric analysis. Anal Chem 51:1714

Hellwig A, Neske P (1990) Zur komplexen Erfassung von Quecksilber in der Luft. In: Aktuelle Aufgaben der Messtechnik in der Luftreinhaltung, VDI Berichte 838, 457–466, VDI Verlag, Duesseldorf, Germany

Lenschow DH, Stephens PL (1982) The role of thermals in the convective boundary layer. Boundary Layer Meteorol 19:509–532

Nelder JA, Mead R (1965) Comput J 7:308

Petersen G, Iverfeldt A, Munthe J (1995) Atmospheric mercury species over central and northern Europe. Model calculations and comparison with observations from the nordic air and precipitation network for 1987 and 1988. Atmos Environ, 29 (1):47–67

Removal of Mercury from Gases by Modified Zeolites in a Stationary Bed

O. Petzoldt, H.J. Fell, and H.W. Bräuer

1 Introduction

Mercury pollution of the environment caused by humans is in the range of 8000 to 40 000 t/a with 6000 to 10 000 t/a of this originating from the mercury-processing industry, 2000 to 20 000 t/a from ore and mineral processing, and up to 10 000 t/a from combustion of fossil fuels.

For years, the quantities of mercury used industrially have been declining, not least of all due to the toxicologic classification of this heavy metal with respect to the environment and above all to human health. Nonetheless, there are still some areas of use, for instance electronic components, fluorescent lamps, batteries, thermometers, manometers, or dental fillings, where mercury remains indispensable – at least for the time being.

Contamination of gases with mercury must be anticipated in all thermal processes that gain application either directly or indirectly in the areas where mercury is used as mentioned above, for example when incinerating waste.

Several countries have meanwhile placed enacted laws limiting the emission of mercury in flue gases, the limits in Germany being between 50 and 200 $\mu g/m^3$ depending on the type of plant. Complying with these limits frequently requires treatment of the gases to remove mercury. This may be done by adsorption as in the Medisorbon process described here.

2 Problems Involved

Mercury is present in incineration plant flue gases primarily in elemental form as Hg° and in oxidized form as $HgCl_2$. Possessing a solubility of 73 g/l at 25 °C, $HgCl_2$ is well soluble in water and therefore can for the most part be extracted together with HCL and SO_2 in the flue gas scrubbers installed for this purpose. Removal is improved when mercury-combining chemicals are added to the scrubbing solution.

Environmental Science
Mercury Contaminated Sites (ed. by R. Ebinghaus et al.)

Elemental mercury is retained practically not at all in wet scrubbers and only to a small degree with flyash, meaning that a separate process step is necessary to remove Hg° from flue gas.

Adsorption methods employing activated carbon are particularly well suited for removal of mercury. The absorbent used may be impregnated with iodine, silver, sulphuric acid or elemental sulphur (Henning et al. 1988)

Sulphur- and sulphuric-acid-impregnated activated carbon has been used in fixed-bed adsorbers for several years with great success to clean natural gas (Daza 1993). In contrast with natural gas, however, precleaned flue gases generally contain a certain amount of SO_2 that is similarly adsorbed on the carbon and is catalytically oxidized by the residual oxygen in the flue gas to SO_3, which reacts with the moisture present in the gas to form sulphuric acid. This sulphuric acid absorbs additional moisture until the activated carbon finally becomes soaked its adsorptive capacity declines and corrosion problems set in.

As a means of overcoming these problems, activated carbon is practically no longer employed in fixed-bed adsorbers for cleaning flue gases. Instead, the adsorbent is continuously exchanged, as in the moving-bed or turbulent-contact method. These methods actually utilize the formation of sulphuric acid for additional removal of mercury. The spent adsorbents can either be incinerated in the same plant, if suitable facilities for reduction of pollutant levels are provided, or they can be sent for external disposal together with the flyash from the upstream dust collectors.

Another fundamental problem associated with use of carbonaceous adsorbents for cleaning gases containing oxygen is the risk of fire. Heat generated in the adsorber may not be completely removed, resulting in formation of hot spots within the bed, which may lead to spontaneous ignition. Licensing authorities therefore usually make the use of carbon-containing adsorbents contingent on the provision of appropriate safety measures, which in Germany are summarized in the LIS Report No.97 issued by the North Rhine-Westphalian State Institute for Source Emission Control (1991). Commonly, this covers carbon monoxide monitoring, automatic fire extinguishing and inertization facilities.

Owing to these problems, efforts have long been directed at developing non-flammable adsorbents that preferably do not retain either SO_2 or moisture and are catalytically inactive.

3 Process Development, Pilot Plants, Test Results

In the quest for a suitable alternative to flammable carbonaceous adsorbents, Lurgi first undertook preliminary trials in 1990 to 1992 on a laboratory scale with a wide variety of inorganic substances. These were subsequently continued in a focused direction and in close cooperation with Degussa using a modified zeolite specially developed by the latter.

Zeolites are crystalline alkaline aluminosilicates exhibiting a three-dimensional open lattice structure which forms defined micropores with a closely limited pore radius distribution (see Fig. 1). Before now, zeolites were excluded from use in adsorptive cleaning of the gases by the types commercially available being highly selective to water vapour adsorption, resulting in very quick exhaustion of their adsorptive capacity. By special synthesis or dealumination processes it is, however, possible to manufacture zeolites that are extremely hydrophobic, catalytically inactive, resistant to high temperatures, and stable to acids. Moreover, they do not adsorb SO_2 or HCl, thus ruling out corrosion problems due to acid formation. Non-flammability of zeolites prevents spontaneous ignition of the adsorbent. With appropriate binders, zeolites can be preformed in any desired geometry to achieve optimum pressure drop. DEGUSSA zeolites utilized in the Medisorbon process (chemisorptive removal of mercury from gases) are preferably extruded in cylindrical shape with particle sizes of a few millimetres.

Metallic mercury is removed by chemisorption. To achieve this, the zeolites are impregnated with elemental sulphur in a specially designed rotating kiln. The

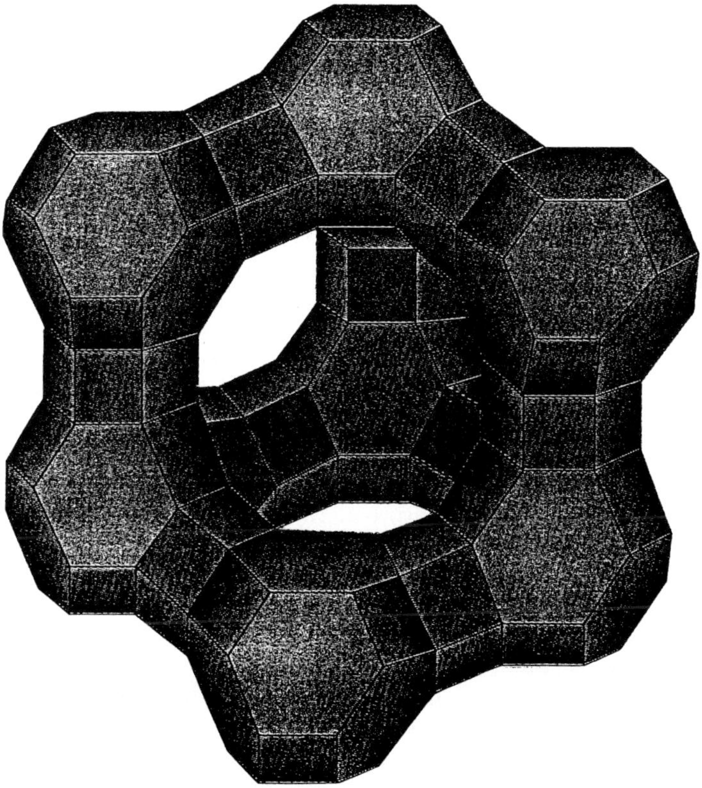

Fig. 1. Structure of dealuminated zeolite (Wessalith Day)

metallic mercury is adsorbed and then reacts with the sulphur to form mercuric sulphide, a non-volatile water-insoluble compound, thereby becoming fixed.

Long-term tests conducted at Lurgi's research laboratories have delivered important results in removal of metallic mercury. As an example, when treating raw gas containing 1 to 5 mg/m^3 of mercury, the concentration measured in the clean gas remained lower than 0.01 to 0.03 mg/m^3 even after 13 000 h of operation.

4 Description of Process and Plant

With the Medisorbon process, chemisorptive removal of mercury from the flue gas of incineration plants takes place at a temperature of about 60–85 °C. Higher temperature entails the risk of sulphur being lost from the adsorbent, while lower temperature results in slowing of the conversion reaction and moreover requires additional energy to reheat the flue gases before they are released through the stack.

In order to avoid increased retention of water by the adsorbent, relative humidity of the gases must not exceed 85%. This also rules out condensation of moisture in the adsorber. Optimum utilization of the theoretical adsorbent run time requires that the dust burden be kept as small as possible. Although finest dust particles mostly pass through the adsorbent bed, dust concentration in the gas to be cleaned should not be more than 10 mg/m^3 so as to reliably preclude excessive pressure drop across the bed and, with it, the necessity to replace the adsorbent prematurely.

The heart of a Medisorbon plant is the zeolite-packed adsorber. This is a horizontal-flow vertical adsorber (see Fig. 2). A particular advantage of the vertical type is the possibility of replacing only the first, possibly thin layer of the bed in the event of a high dust burden causing premature increased pressure drop, leaving the remaining layers in place. Furthermore, when using different adsorbents, it is possible to keep them separate, which simplifies their regeneration. Another obvious point is easier discharge of the separate layers from a vertical adsorber, which can be accomplished by gravity through a rotating-vane gate on a vertical adsorber. This type of adsorber is built as a very compact unit and does not contain any valves or similar controls needing to be actuated regularly in normal operation. The adsorbents employed for removal of mercury from gases by the Medisorbon process are now exclusively modified zeolites manufactured by Degussa.

The gas heater necessary for conditioning the gas is used together with fresh air intake during startup for heating the adsorber to operating temperature before admitting the moist gas as a means of preventing condensation in the adsorber. Depending on whether the gas blower is installed upstream or downstream of the Medisorbon system, auxiliary provision of a fresh air blower may be required for startup, but it may be of much smaller capacity than the gas blower. This fresh air blower is then used during the operation of filling the adsorbent into the adsorber to exhaust air containing a small amount of adsorbent dust and clean it in a bag filter installed in an adsorber bypass before it is released to the atmosphere.

Trials can be undertaken prior to designing the full-scale plant in order to ensure guarantee levels of clean gas purity or pollutant removal rates, especially when treating unusual gases such as those originating from thermal soil reclamation or metallurgical processes.

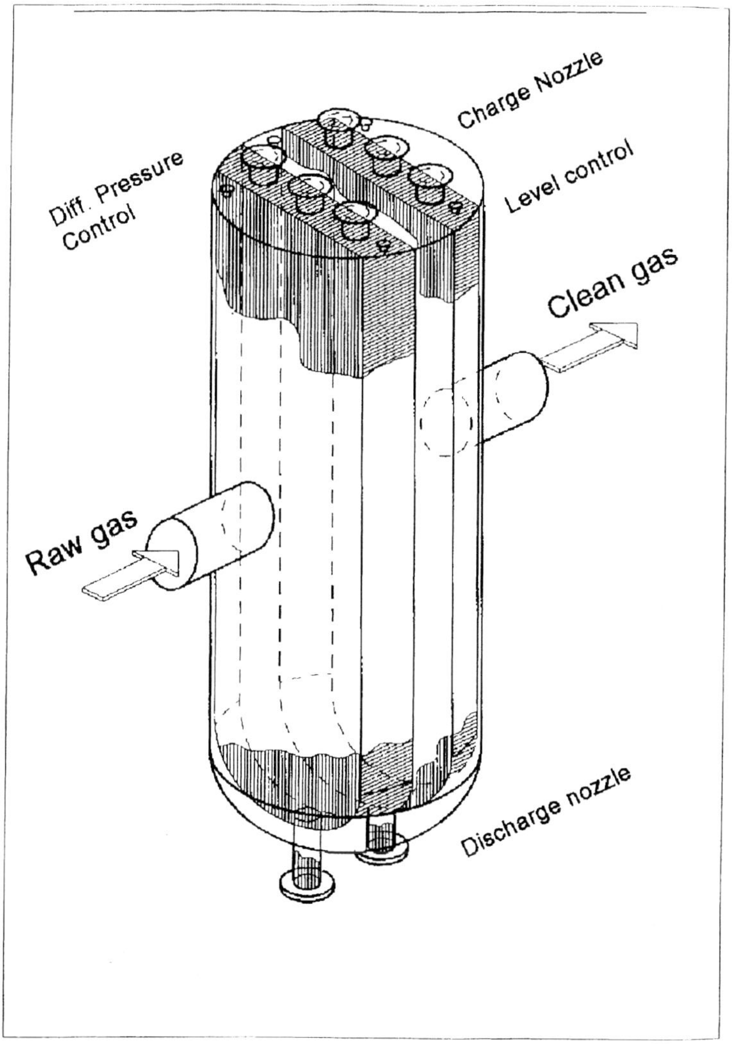

Fig. 2. Medisorbon fixed bed adsorber

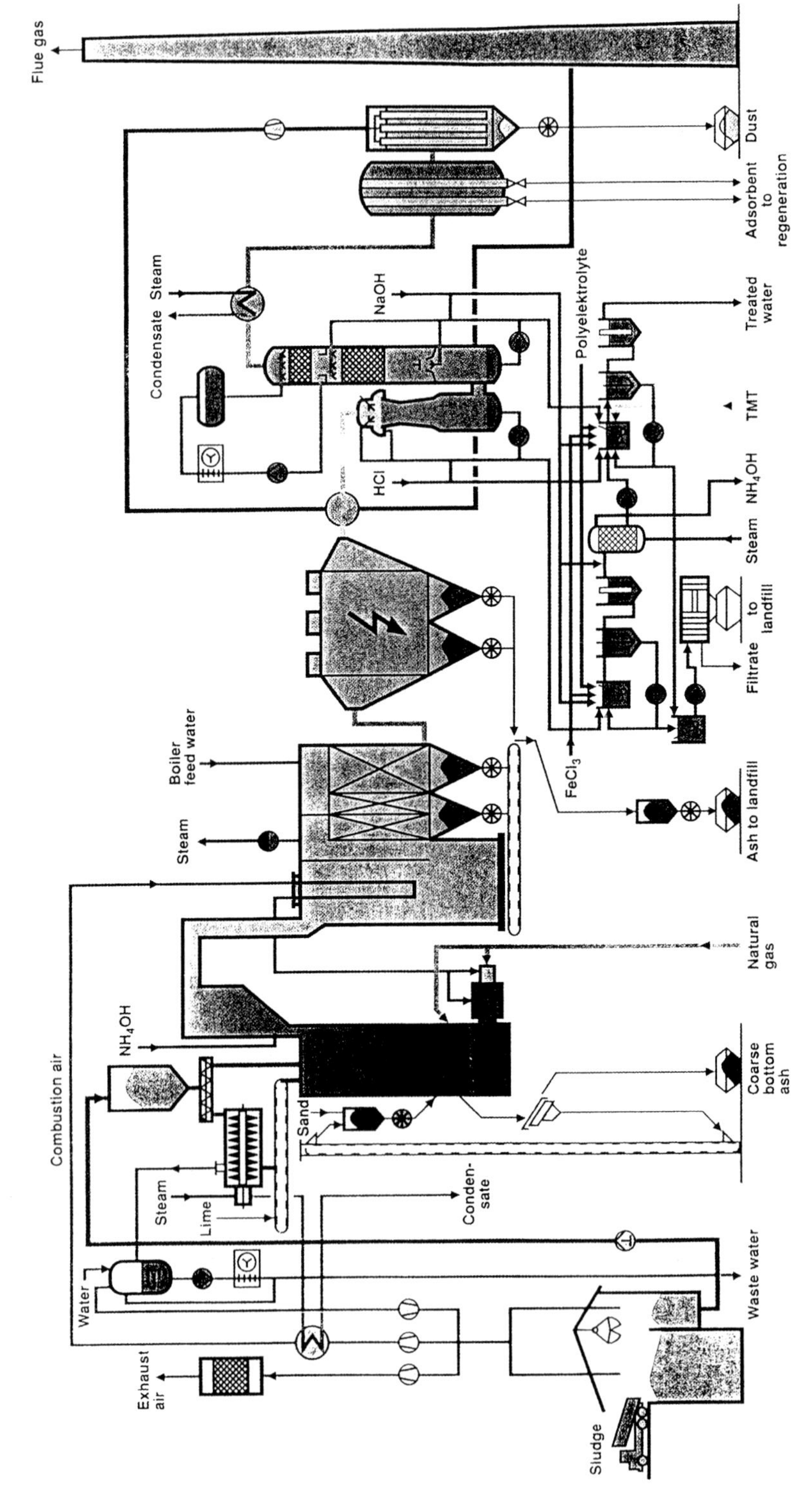

Fig. 3. Use of Medisorbon process in sewage sludge incineration plant

Table 1. Operating data of sewage sludge incineration plant

Capacity	Approx. 50 000	tDS[1]/a
Treatment streams	3	–
Flue gas flow	3 × 18 500	m^3/h (STP moist)
Adsorber inlet temperature	75	°C
Hg in raw gas	300–600	µg/m^3 (STP dry)
Maximum Hg level (guarantee)	<50	µg/m^3 (STP dry)
Attained Hg level	6–10	µg/m^3 (STP dry)

DS = dry substance.

In addition to laboratory test units, for which the gas is generated synthetically, several mobile pilot plants are available with gas treatment capacities between 10 and 130 m^3/h. These are self-contained units and are normally operated on a sidestream taken from the main flow of gas. At the present time, Lurgi has three such pilot units in operation, one at a hazardous waste incineration plant as previously mentioned and two at sewage sludge incineration plants.

Figure 3 illustrates integration of the Medisorbon process into the flue gas cleaning train of a three-stream industrial-scale sewage sludge incineration plant in the Netherlands.

Principal data of this system, which was commissioned in March 1993, are summarized in Table 1.

Operating experience to date confirms that Medisorbon working with Degussa zeolite constitutes a process low in operating and maintenance requirements. During operation to date, there has been no appreciable rise in pressure drop across the adsorbers.

5
Regeneration of Saturated Adsorbent

Degussa-modified zeolite as used in the Medisorbon process possesses high resistance to heat. Even when repeatedly heated to temperatures of 900–1000 °C, it does exhibits no measurable loss of mechanical strength or adsorptive capacity. Although the zeolite has a very long run time before the adsorptive capacity is exhausted – for example, with a mercury concentration of 1 mg/m^3 in the inlet gas and 95% removal, the theoretical run time of a 10-cm-thick layer is 1 year, that of a 50 cm layer 5 years – it may be necessary to reckon with a shorter practical run time owing to retention of dust and possible coadsorption of other contaminants. Under these conditions, thermal regenerability of the zeolites is a major factor in reducing operating costs.

Regeneration of mercury-saturated zeolite is performed, for example, in systems otherwise used for processing mercury-containing residuals such as batteries, fluorescent lamps and the like, in thermal soil reclamation units or activated carbon regeneration systems. As in processing of these residuals, the mercury is recovered in elemental form during regeneration of the zeolite and can be recycled. The regenerated zeolite is reimpregnated and similarly reutilized.

The zeolite can be regenerated and estimated four to six times before needing to be finally discarded, when it can be disposed of, for instance, by incineration or depositing in a controlled or secure dump. Definite figures on the number of reuses are not yet available.

6 Summary

Compared with other adsorption methods which mainly employ carbonaceous adsorbents, the main advantage of the Medisorbon process lies in the safety potential of the zeolite it uses. This means that it does not require the safety devices, fire-extinguishing facilities, extensive instrumentation and so on normally prescribed with other processes. There is no danger of spontaneous ignition.

Owing to the hydrophobic nature of the zeolite and its catalytic inactivity, soaking and corrosion problems caused by formation of sulphuric acid do not have to be contended with. These selective characteristics are naturally a drawback when other substances such as SO_2, HCl, HF or other mercury compounds like ionogenic form (Hg^{2+}) need to be removed as well. In these instances, additional process stages must be provided in the gas treatment plant, for instance customary wet scrubber systems.

The Medisorbon plant is practically free from need for maintenance and supervision. Removal, regeneration and refilling of the adsorbent is only necessary at intervals of a few years.

Relatively high cost of the adsorbent compared with carbonaceous materials can be offset by savings in running costs, long run time and useful life of the adsorbent, absence of regular maintenance procedures, and lower first cost of the equipment.

Naturally, it will be necessary to make proper comparison of all costs to determine whether the Medisorbon process is the most economical alternative for any actual project.

References

Daza L, Mendirioz S, Pajares JA (1993) Mercury elimination from gaseous streams. In: Applied catalysis B: Environmental, 2. Elsevier, Amsterdam, pp 277–287

Henning KD, Keldenich K, Knoblauch K, Degel J (1988) Impregnated activated carbon for mercury removal. Gas Sep Purif 2:20–22

Yan TS (1994) A novel process for Hg removal from gases. Ind Eng Chem Res 33:3010–3014

Mercury Speciation of MSWI Flue Gas on Pilot and Industrial Scales

J. Wang, Z. Xiao, and O. Lindqvist

An on-line system has been applied to measure mercury (Hg) speciation from a pilot combustion simulator and from an industrial scale Municipal Solid Waste Incineration (MSWI) plant in Göteborg, southwest Sweden. Under simulated waste incineration conditions, 60–80% Hg was found in its oxidized form Hg(II). At the MSWI plant, equipped with a modern flue gas cleaning system, Hg at concentrations about 10–15 $\mu g\ m^{-3}$ was emitted to the atmosphere, of which 15–20% was Hg(II).

1 Introduction

Mercury is a highly toxic non-essential element with no known biological function (Carty and Malone 1979). From the 1950s to the early 1970s, Hg poisoning events were mostly reported as an acute local problem. Since then, our understanding of the emission, transport and deposition of mercury in the environment has improved, especially in the last 10 years. The importance of mercury cycling in the atmospheric pool has been established, and Hg pollution has been described today as a regional and a chronic problem. The notion of mercury as a global pollutant has been recognized worldwide, and an elevated Hg concentration in freshwater fish is a current problem in many countries (Lindqvist et al. 1991; Lindqvist 1994).

Waste incineration and coal combustion are two major artificial sources of mercury emission into the atmosphere. Flue gases contain mercury in different physical and chemical forms, dependent on the concentrations of trace gases such as SO_2 and HCl, and on combustion conditions. The speciation and association of Hg to different ligands regulate the toxicity, transport pathways and residence time in different compartments of the ecosystem. The need to develop and apply a suitable method to measure industrial emission from different types of combustion processes has been emphasized since the first international Hg conference in 1990, and speciation information is still badly needed in modelling work describing Hg cycling in the environment and its final fate (Rodhe 1996).

Liquid solutions and solid absorbents have been used for selective collection of different Hg species from flue gases (Bergström 1986; Metzger and Braun 1987; Bloom et al. 1993; Larjava 1993). Recently, an on-line method for measuring both elemental and total mercury in flue gas has been developed in our laboratory, and applied to study Hg speciation in flue gas (Lindqvist and Schager 1990; Wang

Environmental Science
Mercury Contaminated Sites (ed. by R. Ebinghaus et al.)

et al. 1995). On-line measurement of Hg speciation may be a superior tool for understanding the reaction kinetics and mechanisms involved, especially in flue gas cleaning control. This chapter describes further work applying the on-line method to a pilot-scale combustion simulator (Lindqvist and Schager 1990; Hall 1992) under controlled conditions, and to the Sävenäs MSWI plant in Göteborg, providing a real industrial case study. Results obtained from these investigations are presented and discussed.

2
Experimental

2.1
The Flue Gas Simulator

Experiments were performed with a natural gas-fired flue gas generator consisting of a burner, a combustion chamber and a flue gas duct about 12 m in length (Fig. 1).

Desired concentrations of SO_2, HCl and Hg^0 were fed to the system via the burner. A Testo 33 Electrochemical Gas Analyzer was used for on-line monitoring of temperatures inside and outside the duct and concentrations of O_2, SO_2, NO and CO. The maximum output of the system is about 10 kW^{-1} h and the gas flow rate was maintained at 680 l min^{-1} by a flow meter through the 80 mm i.d. flue gas duct. The temperature in the combustion chamber was about 700 °C. Oxygen concentration of about 10% and water content of about 6% were used throughout

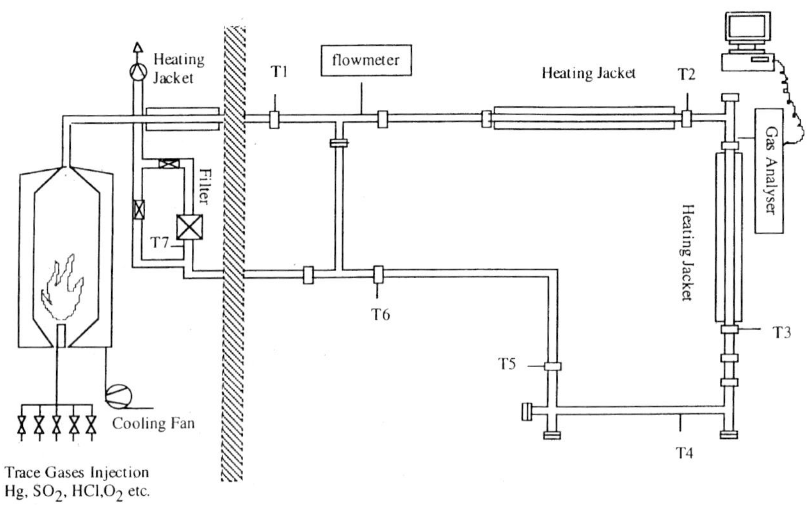

Fig. 1. Schematic diagram of the pilot scale flue gas simulator system

the investigation. Sampling of mercury was made at post T5, where the temperature was about 180 °C.

2.2
The Sävenäs MSWI Plant in Göteborg

Measurement was performed at the Sävenäs MSWI plant in Göteborg. The annual amount of waste treated at this plant is about 300 000 tons, of which about 70% is household waste and the rest is of industrial origin. The plant has been equipped with modern electrostatic filters for collection of fly ash, and two wet scrubbers which respectively wash out water-soluble compounds such as HCl, HF, dioxin, and heavy metals including Hg. Typical concentrations of different pollutants in flue gas before and after passing through the two wet scrubbers are summarized in Table 1.

Figure 2 is a schematic diagram of the installation at the Sävenäs MSWI plant and the sampling points chosen for this investigation. Point C is also where the plant monitors for other environmental pollutants.

Table 1. Typical concentration of pollutants in flue gas before and after wet scrubber, according to GRAAB (1993)

Pollutants	Before scrubbers ($mg\ m^{-3}$)	After scrubbers ($mg\ m^{-3}$)
HCl	700–1000	0–10
HF	1–10	0.1–0.3
Hg	0.05–0.1	0.015–0.025
Dioxin	$2\text{–}5 \times 10^{-6}$	$0.1\text{–}0.4 \times 10^{-6}$

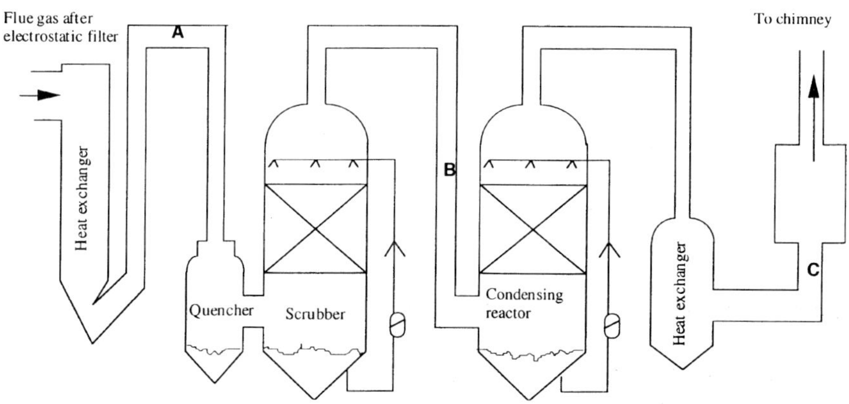

Fig. 2. Sampling points at Sävenäs MSWI Plant in Göteborg. A before the two scrubbers (160 °C), B between the two scrubbers (60 °C), and C after two the scrubbers (60 °C)

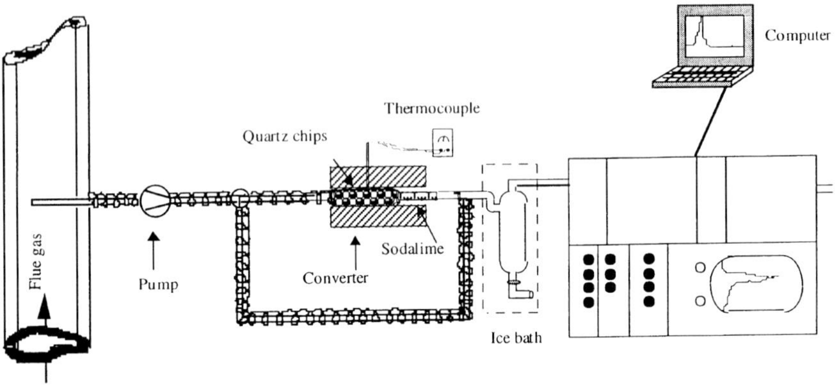

Fig. 3. The experimental set-up for measuring Hg^0 and Hg_{tot} by using thermal converter

2.3 The Sampling Setup

The commercial instrument Semtech, together with the thermal converter used in this investigation, have been described previously (Wang et al. 1995). Measurement results are collected every 10 s, and total measurement times for both Hg^0 and total Hg are at least half an hour for all results reported in this chapter.

The sampling scheme, applied both to the simulator and the MSWI plant, is illustrated in Fig. 3.

A Teflon-lined pump was used to drive the flue gas to the Semtech. Hg^0 can be detected directly, while total Hg was measured by converting the Hg(II) portion in the flue gas with the thermal converter. To avoid condensation of Hg(II) and water vapor on the sampling system, heating bands were used to heat the sampling Teflon tube and the pump to temperatures above 100 °C. The rate of flow of the flue gas through the thermal converter was about 1–1.5 l min^{-1}. An icebath was adapted to quickly cool the gas from several hundred °C just beyond the converter to roughly room temperature, before it passed to the detector of the AAS.

3 Results and Discussions

3.1 Results Obtained at the Combustion Simulator

Hall (1992) reported that typical raw gas concentrations in coal combustion are 4–10% O_2, 100–1000 ppm SO_2 and 1–100 ppm HCl, and correspondingly in waste incineration process 6–15%, 100–300 ppm and 400–1000 ppm. Similar concentrations of these gases were chosen in the simulator used in this investigation.

Table 2. Mercury speciation from simulated flue gas at different concentrations of SO_2 and HCl

$[O_2]$	Experimental conditions				Hg speciation				Percent
(%)	[HCl] (ppm)	$[SO_2]$ (ppm)	[CO] (ppm)	[NO] (ppm)	Hg_{tot} ($\mu g\ m^{-3}$)	Hg_{II*} ($\mu g\ m^{-3}$)	Hg^0 ($\mu g\ m^{-3}$)	$Hg_{tot\#}$ ($\mu g\ m^{-3}$)	Hg_{II}/Hg_{tot} (%)
9.7	100	137	50	39	93.90	43.35	50.56		46.17
9.5	100	305	52	57	189.34	80.98	108.36	179.05	42.77
9.7	100	1060	58	107	142.87	55.55	87.32	155.17	38.88
9.8	400	137	147	40	78.31	39.41	38.90		50.33
9.7	400	308	267	27	113.30	75.59	37.71		66.72
9.8	400	1015	280	80	142.49	97.25	45.24	153.01	68.25
9.7	1000	115	55	44	118.70	73.23	45.48		61.69
9.5	1000	313	71	54	278.96	198.70	80.26	263.70	71.23
9.2	1000	1055	160	76	99.12	77.76	21.36	110.93	78.45

Hg_{II*} was calculated from the difference of Hg_{tot} and Hg^0 measured by Semtech on-line system. $Hg_{tot\#}$ was collected using $KMnO_4$ solution absorption.

Experimental results showing the combined effect of Hg interaction with different concentrations of SO_2 and HCl are summarized in Table 2.

The performance of the on-line converter system for total mercury measurement has been calibrated against the Swedish Standard (1992) method of using $KMnO_4$ absorption solution for total mercury sampling from flue gas. The same flue gas stream was led to the $KMnO_4$ bubbler from the outlet of the Semtech system. The difference between the two methods is about 10%, as seen in Table 2.

It has previously been reported that addition of HCl increases the fraction of oxidized Hg(II) in flue gas, while addition of SO_2 reduces it (Lindqvist and Schager 1990; Hall 1992). The combined effect of both SO_2 and HCl can be seen from Table 2: the higher the concentration of HCl in the system, the larger the fraction of Hg(II) formed, irrespective of variation in the SO_2 concentration. The influence of the SO_2 concentration in reducing the fraction of Hg(II) in the system is not verified at high levels of HCl, implying that HCl is the key compound in determining the proportion of Hg(II) generated in these combustion processes. About 40% Hg(II) was formed under coal combustion conditions, and under typical waste incineration conditions the amount is 60–80%. No relation can be seen between Hg speciation and concentrations of CO, NO and O_2 in this investigation.

3.2 Results Measured at the Sävenäs MSWI Plant in Göteborg

The measurement results at the Sävenäs MSWI plant are summarized in Table 3.

Results obtained using the on-line converter system have also been compared with those using the Swedish Standard (1992) method. A difference of some 20% was noticed, especially at low Hg concentration levels.

It can be seen from Table 3 that at point A (cf. Fig. 2), before the two scrubbers, the concentration of total Hg is 30–80 $\mu g\ m^{-3}$ consists of less than 50% of Hg(II),

Table 3. Mercury concentration (μg m^{-3}) in flue gas measured at Sävnäs MSWI plant in Göteborg

Samples	Measured by Semtech		Sampled by $KMnO_4$	Hg_{II^*}/Hg_{tot} (%)
	Hg^o	Hg_{Tot}	Hg_{Tot}	
After the two scrubbers				
950428-2		14.99	16.88	
950428-3(I)		15.56	12.59	
950428-3(II)		10.61	10.20	
950508-2	7.06	8.28	10.43	14.73
950508-3(I)	5.51	6.73	6.14	18.13
950508-3(II)	7.56	9.43	11.88	19.83
Between the two scrubbers				
950509	41.98	52.32	51.62	18.67
950510		32.70	33.36	
950511-1(I)	23.88	37.88		36.96
950511-1(II)		25.87	23.09	
950511-2		30.28	24.06	
950511-3	21.22	31.42		32.46
Before the two scrubbers				
950516-1	16.60	29.32		43.38
950516-3		27.04	23.67	
950517-1(I)	18.34	36.50		49.75
950517-1(II)		87.51	83.72	
950517-2	21.38	38.37		44.28
950518-2(II)		57.53	52.09	
950518-5		77.23	74.38	

Hg_{II^*} was calculated as the difference between Hg^o and Hg_{tot} by Semtech.

which is lower than in the simulator. This is most probably due to the difference in combustion environments between the controlled simulator and the varied industrial incinerator; for example, the origin and composition of waste material may change from day to day or faster. At point B, about 30% Hg(II) was found. At point C, after the scrubbers, there was still about 15–20% of Hg(II) in the flue gas. Yet all the Hg(II), which is highly water-soluble, should be dissolved and rinsed out by the scrubber water. The Hg(II) found between the two scrubbers is most probably due to the rereduction of Hg(II) by SO_2 (Munthe et al. 1991) formed during combustion, or by other metal ions – such as Sn^{2+} – which can reduce Hg(II) under the operating conditions. Investigation of this is proceeding in our laboratory.

4 Conclusions

The on-line thermal converter has been successfully applied to measure Hg speciation of MSWI flue gas on both pilot and industrial scales. Uncertainties in the method, compared with the Swedish Standard method, vary from 10% under laboratory conditions to 20% under real industrial conditions.

The results obtained from the simulator indicate that about 40% and 60–80% Hg is found as Hg(II) in flue gas in typical coal combustion conditions and waste incineration conditions, respectively. In real combustion processes, however, the physical and chemical environments can differ from those of the simulated conditions in this study, so that the Hg(II) percentage may also vary. This portion of Hg(II) will be emitted directly into the atmosphere if no further cleaning measures are adopted.

At the Sävenäs MSWI plant in Göteborg, where two wet scrubbers have been installed, the total concentration of mercury emitted into the atmosphere was found to be 10–15 $\mu g\ m^{-3}$, of which about 15–20% was in Hg(II) form.

Acknowledgements The authors thank C. A. Pedersen, C. Lundgren and E. Larssen at Göteborg Municipal Waste Incineration Plant for their support and assistance during the measurements at Sävenäs, and Dr. M Abul-Milh at the Department of Inorganic Chemistry, Göteborgs University for running the flue gas simulator and technical assistance in sampling.

References

Bergström JGT (1986) Mercury behaviour in flue gases. Waste Manage Res 4:57–64

Bloom N, Presto E, Miklavcic VL (1993) Flue gas mercury emissions and speciation from fossil fuel combustion. Presented at: Conf on Managing air toxics: state of art, Capitol Hilton, Washington, DC, 13–15 July 1993

Carty AJ, Malone SF (1979) The chemistry of mercury in biological system. In: The biogeochemistry of mercury in the environment chap 20 Nriagu JO (ed) Elsevier North-Holland Biomedical Press, Amsterdam

GRAAB (1993) Environment report 94:1, Gothenburg Municipal Waste Incineration Plant, Box 95, 401 21 Göteborg, Sweden. Göteborgsregionens Avfalls AB, Raport 94:1, Miliörapport 1993 (in Swedish)

Hall B (1992) An experimental study of mercury reactions in combustion flue gases. Thesis, Department of Inorganic Chemistry, Chalmers University of Technology and University of Göteborg, Sweden

Larjava K (1993) On the measurement of volatile metal species at elevated temperature, thesis, Espoo, Technical Research Center of Finland, VTT publications 137

Lindqvist O (1994) Chemical transformation mechanisms, a synthesis report. Workshop Proc EPRI Rep No TR-104214. Convened 16–18 March 1994, Tampa, Florida

Lindqvist O, Schager P (1990) Continuous measurements of mercury in flue gases from waste incinerators and combustion plants. VDI Ber 838:401–421

Lindqvist O, Johansson K, Austrap M, Andersson A, Bringmark L, Hovsenias G, Håkansson L, Iverfeldt Å, Meili M, Timm R (1991) Mercury in the Swedish environment – recent research on cause, consequences and corrective methods. Water Air Soil Pollut 55:1–261

Metzger M, Braun H (1987) In-situ mercury speciation in flue gas by liquid and solid sorption systems. Chemosphere 16 (4):821–832

Munthe J, Xiao Z, Lindqvist O (1991) The aqueous reduction of divalent mercury by sulfite. Water Soil Air Pollut 55:621–630

Rodhe H (1996) The global mercury cycle – how large is the human impact? 4th Int Mercury Conf, 4-8 Aug 1996, Hamburg, Germany, Abstr p 2

Swedish Standard SS 02 84 23 (1992) Air quality – stationary source emissions – determination of total mass concentration of mercury – flameless atomic absorption methods, Box 3295, 103 66 Stockholm, Sweden (Standardiseringskommissionen i Sverige (SIS) (in Swedish)

Wang J, Xiao Z, Lindqvist O (1995) On-line measurement of mercury in simulated flue gas. Water Air Soil Pollut 80:1217–1226

Mercury Saturation Profile Across the Sulphur-Impregnated Activated Carbon Bed

Z. Špirić and M. Hraste

Mercury is very often removed as an impurity and environment pollutant from processing fluid gas streams by adsorption on a fixed bed. Considering the ecological and technological problems related to the presence of mercury in natural gas, INA-Naftaplin has erected at the gas treatment plant Molve (Croatia) processing units that insure complete mercury removal. This chapter deals with the results of 3 years' experience in practical field application, showing activated carbon mercury removal from natural gas.

1 Introduction

Even in small amounts, mercury and its compounds have an extremely harmful effect on human health (WHO 1989). Moreover, mercury corrosion dramatically endangers processing plants and facilities. Therefore, prevention of mercury entering into the environment and industrial processes is absolutely critical. A frequently applied procedure to remove mercury as an impurity and environment contaminant from processing fluid gas streams is adsorption on a fixed bed (Ruthven 1984). The adsorbent is primarily designed to extract elemental mercury, i.e. mercury in its vapour state. It is a special type of sulphur-impregnated activated carbon with a great adsorption capacity for mercury vapours (Hutchins 1979). This type of activated carbon contains around 15% elemental sulphur, distributed in a thin layer across an extremely large surface area of porous granules of activated carbon that creates insoluble mercury sulphide by chemical reaction with elemental mercury from natural gas.

Unfortunately, a major problem in the preparation and performance of this investigative task, as well as in the evaluation of the results, was the lack and/or inavailability of comparable analytical data from other authors, e.g. qualitative and updated references from this field of research.

This chapter deals with results of more than 3 years' experience in practical field application, showing activated carbon mercury removal from natural gas streams.

The purpose of our research was to define mercury loading and distribution, i.e. saturation profile in the activated carbon bed and variation in concentration of elemental sulphur through the bed under real processing conditions.

Environmental Science
Mercury Contaminated Sites (ed. by R. Ebinghaus et al.)

2
Experimental

In order to investigate and provide precise and accurate data (original processing results), and to establish crucial information and knowledge about the activated carbon bed saturation profile, we studied a mercury removal system for natural gas under actual processing conditions (Table 1b).

The mercury removal unit (MRU) consists of an adsorber (Fig. 1b) located in the process plant, treating water-washed natural gas (Table 1a). Besides hydrocarbons (more than 70% methane), natural gas contains CO_2, H_2S, and near saturated levels of water vapour. The MRU contains 64 m^3 commercial granular pelletised sulphur-impregnated activated carbon (Table 1c) and is designed to remove mercury to a level no greater than 5 $\mu g/m^3$.

Table 1. Typical process features of the Mercury Removal Unit (MRU)

a) Inlet stream composition (vol%)	
C_1	69.32
C_2	3.33
C_3	0.79
i-C_4	0.21
n-C_4	0.23
i-C_5	0.12
n-C_5	0.08
C_{5+}	0.18
CO_2	23.74
N_2	2.0
H_2S	150 ppm_v
RSH	30 mg/m^3
H_2O	0.11 mol%
Hg	1000 $\mu g/m^3$
b) Process design conditions	
Flow rate	5 000 000 Nm^3/day (218 564 kg/h)
Temperature	55–60 °C
Pressure	5200–6000 kPa (52–60 barg)
Molecular weight	23.5 kg/kmol
Density	58.5–62.8 kg/m^3
Viscosity	0.011 cP
Max. dP	57 kPa (0.57 bar)
c) Activated carbon	
Total sulphur content	15%
Apparent bulk density	560 kg/m^3
Moisture	11.8% m/m
Ash (gravimetric)	4.4% m/m
Sodium (AAS)	0.185% m/m
Potassium (AAS)	0.055% m/m
Average pore diameter	15.9×10^{-10} m (15.9 A)

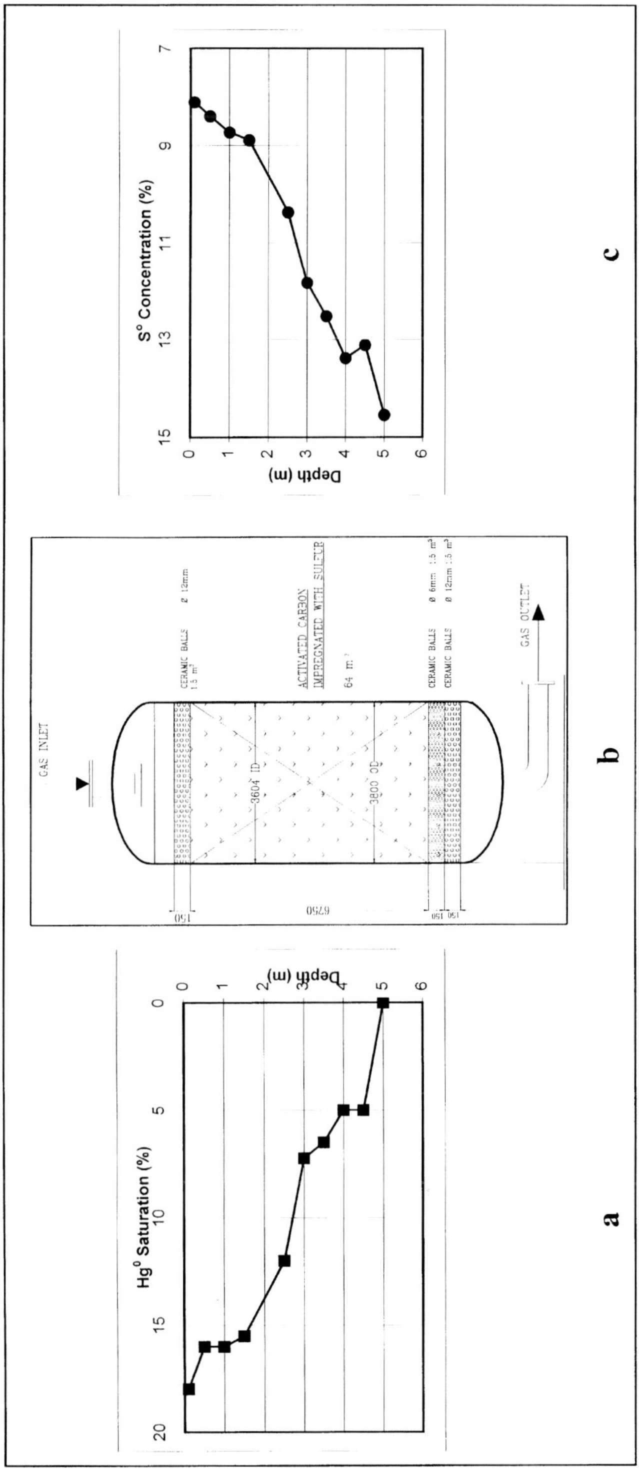

Fig. 1. Hg° saturation and the S° content profiles, after 830 days of MRU run

3
Methods

The levels of mercury in natural gas were detected and quantified at the selected sampling points (inlet to and outlet from the adsorber bed), applying the mercury analyzer based on fluorescence detection technique (Sir Galahad instrument, manufactured by PS Analytical) and the "gold film" mercury analyzer (Jerome 301 Gold Film Mercury Detector).

As to problems related to preservation and storage of field samples containing mercury, it is recommended to conduct the detection and analysis of activated carbon mercury content on site, immediately upon unloading, since there is no need for special preparation prior to sample treatment by means of suitable field analytical devices. Due to the high mercury content (1–20% w/w), sensitivity is of minor importance. Therefore, in order to complete the investigation, we have used the standard Eschka gravimetric procedure (Duschak and Schuette 1925) that adequately covers the required concentrations, resulting with reliable and reproducible data under the given circumstances.

However, the determinations of high mercury levels in saturated activated carbon become difficult due to desensitization of commercially available mercury analyzers. Namely, the analytical instruments developed recently are capable of detecting increasingly low levels, along with a high sensitivity for activated carbon contaminant overload (Bloom 1992). Therefore, for the determination of total mercury content in the saturated activated carbon, we used graphite furnace atomic absorption spectrometry (GFAAS), or cold vapour atomic absorption spectrometry (CVAAS). These procedures and methods are applicable for the determination of total mercury after the oxygen flask combustion (thermal decomposition). Such methods are quite suitable, but not for the activated carbons containing high levels of mercury (not only in the form of HgS), and in the presence of some aromatic hydrocarbons and readily inflammable compounds, this being the case with condensate present in treated natural gas.

Sulphur content in mercury-saturated activated carbon, as total sulphur, was detected and measured by the X-ray fluorescence spectroscopy method.

4
Results and Discussion

The sulphur-impregnated activated carbon bed was installed and utilized from April 1993 till September 1995. [approximately 830 working days, passing a total amount (Q) of 2.26×10^9 Nm^3 of the natural gas across the bed]. The mercury removal efficiency results obtained during this test period clearly indicated good performance of the sulphur-impregnated activated carbon bed. The natural gas mercury test analyses indicate typical reduction from inlet average 516 $\mu g/m^3$, to an average outlet mercury concentration of 0.12 $\mu g/m^3$ (for one of the test time

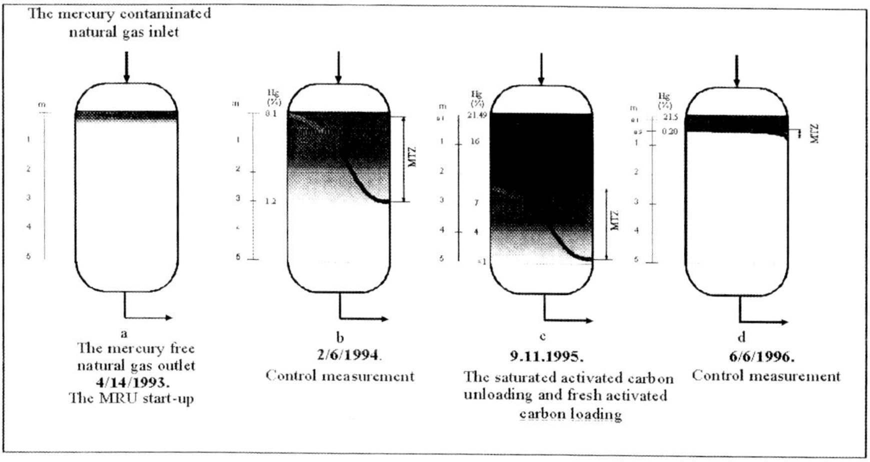

Fig. 2. MTZ adsorber's run review

periods), with excellent bed efficiency, more than 99.98% (Špirić and Vadunec 1996).

After the first year (Fig. 2b), and especially after 2 years of operation (Fig. 2c), analyses performed on activated carbon samples obtained from the mercury removal bed indicated (unexpectedly) high mercury loading, requiring a changeout of the activated carbon load. Following the carefully prepared unloading and replacement schedule, the level of saturation of the activated carbon withdrawn from the adsorber vessel was investigated. According to the programme, samples were taken from each 0.5 m of the bed depth. The spent activated carbon was replaced by the fresh load. Due to the presence and influence of numerous variables (Biscan and McNamara 1989; Bartel et al. 1993; McNamara and Wagner 1995), such as velocity (i.e. gas flow rate), contact time, temperature, pressure, composition of gas, as well as concentration of mercury vapour (frequently quite non-uniform), it is extremely hard to accurately predict the effective capacity, the "breakthrough curve", or the expected adsorber bed life. Using sample analysis data along with determination of fillup weight and sulphur content in replaced activated carbon, we determined the behaviour/progress of the mass transfer front wave, i.e. mass transfer zone (MTZ), as well as the dynamics and efficiency of the adsorption process. Thus, the primary goal of our research was establishment and interpretation of actual activated carbon saturation profile through the bed, based on defined properties of used activated carbon.

The results of analysis of the unloaded activated carbon (Table 2) revealed an expected, almost linear regularity of activated carbon mercury load distribution across the depth of the adsorber bed (Fig. 1a). Simultaneously, results showed an unexpectedly high mercury load, even above the one suggested (as the maximum value) by adsorbent manufacturers (Špirić 1996). As a result of the very high level of mercury content in the natural gas, an extremely high mercury saturation was observed in the surface layer. This was confirmed by analyses (more than 21% on the top of the layer), revealing also considerable "wetness" of the carbon. One of the possible explanations for this wetness is that, along with gaseous phase, there is also a liquid phase. Moreover, probable reasons for this phenomenon may be found in reactions and mechanisms present during the pretreatment stage, i.e. in the "water wash" column. This would involve not only the possibility of condensation of higher hydrocarbons within adsorbent pores, but also the possibility that water flowing from the wash water system is (over saturated) with mercury (liquid carryover). Occasionally, saturated steam, loaded with mercury

Table 2. Laboratory analysis data on saturated activated carbon, after approximately 830 days of MRU run

Sample no.	Depth (m)	Bulk Density (kg/m^3)	Carbon weight (g)	S° content		Hg content			Carbon loading
				(g)	(%)	Theoretical (g)	Measured (g)	Efficiency (%)	Measured (%)
1	0.1	750.2	0.120	0.010	8.11	0.061	0.026	42.40	21.49
2	0.1	750.2	0.100	0.008	8.11	0.051	0.015	30.30	15.36
3	0.5	744.2	0.172	0.017	9.86	0.106	0.027	25.80	15.90
4	1	728.3	0.133	0.012	8.73	0.073	0.020	28.06	15.31
5	1	728.3	0.152	0.013	8.73	0.083	0.026	31.07	16.95
6	1.5	718.8	1.039	0.092	8.89	0.577	0.180	31.14	17.30
7	1.5	718.8	1.051	0.093	8.89	0.584	0.166	28.41	15.80
8	1.5	718.8	0.085	0.008	8.89	0.047	0.014	30.00	15.96
9	1.5	718.8	0.072	0.006	8.89	0.040	0.011	27.70	15.39
10	1.5	718.8	0.979	0.087	8.89	0.544	0.144	26.38	14.70
11	2	677.8	0.101	0.012	11.47	0.072	0.010	13.69	9.81
12	2	677.8	0.136	0.016	11.47	0.097	0.012	12.84	9.21
13	2	677.8	0.131	0.015	11.47	0.094	0.012	12.78	9.16
14	0–2 m average	743.4	0.181	0.021	11.47	0.130	0.027	21.04	15.08
15	2.5	705.9	0.203	0.021	10.37	0.132	0.023	17.66	11.45
16	2.5	705.9	0.131	0.014	10.37	0.085	0.016	18.30	11.86
17	2.5	705.9	0.110	0.011	10.37	0.072	0.013	17.83	11.56
18	3	678.4	0.145	0.017	11.83	0.107	0.010	9.59	7.09
19	3	678.4	0.215	0.025	11.83	0.159	0.014	8.92	6.60
20	3	678.4	0.184	0.022	11.83	0.136	0.013	9.27	6.85
21	3.5 middle	667.8	0.182	0.023	12.52	0.143	0.012	8.57	6.71
22	3.5 middle	667.8	0.121	0.015	12.52	0.095	0.006	6.67	5.22
23	3.5 m wall	651.0	0.197	0.025	12.52	0.154	0.014	8.85	6.93
24	3.5 m wall	651.0	0.151	0.019	12.52	0.118	0.010	8.47	6.29
25	4.0 m middle	663.0	0.207	0.028	13.38	0.173	0.009	5.11	4.27
26	4.0 m middle	663.0	0.238	0.032	13.38	0.199	0.011	5.65	4.27
27	4.0 m wall	684.6	0.221	0.030	13.38	0.184	0.012	6.71	5.61
28	4.0 m wall	684.6	0.168	0.023	13.38	0.141	0.008	5.65	4.72
29	4.5	696.2	0.170	0.022	13.11	0.139	0.008	5.97	4.90
30	5.0 mixed	692.2	0.130	0.019	14.54	0.118	0.001	1.17	1.06
31	5.0 mixed	692.2	0.135	0.020	14.54	0.123	0.001	1.07	0.97
32	5.0 mixed	692.2	1.113	0.162	14.54	1.012	0.010	1.03	0.94

from the water wash column and entering the adsorber vessel, evaporate on activated carbon, depositing a considerable amount of mercury. It is assumed that such a large quantity of mercury cannot react instantaneously even with activated carbon, and even less so with impregnated sulphur contained in activated carbon pores. Furthermore, formation of stable and insoluble mercury sulphide is there by prevented. Mercury accumulates, therefore, and deposits on activated carbon particle's surface. As a possible confirmation to this assumption can be taken the fact that the samples of investigated saturated and replaced activated carbon show a "decrease" in mercury saturation after a certain utilization time. Evidently, this may be the mercury that has not been converted into stable and insoluble mercury sulphide upon chemisorption with elemental sulphur.

A reduced concentration of impregnated sulphur (8%) is evidenced at the top of the bed layer, when compared to original carbon. In the deeper layers of adsorber bed, sulphur content rises to values typical for the fresh carbon (15%). Regarding the above assumption, we were able to find out and to establish non-uniformity of (total) sulphur content across the adsorber bed (Fig. 1c). This means that during a certain stage of activated carbon utilization, sulphur becomes flushed and migrates (redistribution from upper towards lower layers), indicating possible problems and disturbances in the process, and/or even certain omissions in the general design. Considerable amounts of sulphur may become irrecoverably lost from activated carbon, since they may be dissolved and/or flushed by higher (liquid) hydrocarbons if they were not extracted in an appropriate manner during previous stages (separation, pretreatment). Therefore, after the first year of plant operation, when problems with reduction in sulphur content in activated carbon became evident, a process improvement was introduced, related to change of gas flow direction. Natural gas was sent towards the heat exchanger, and then to the adsorber and not, as originally designed, to the adsorber and then to the heat exchanger and warming-up. The new arrangement prevented adverse impact of liquid hydrocarbons on sulphur present in activated carbon. But, also elevated temperatures (100 °C and above) contribute to reduction of sulphur impregnated activated carbon efficiency, due to escape of certain quantity of sulphur from carbon granule caused by evaporation and/or formation of SO_2. Already at a temperature of >65 °C sulphur sublimes, i.e. passes from the solid to the gaseous state, and this aspect of behaviour should also be included into design and selection procedure.

Further, during the adsorption-chemisorption process, phenomenon whereby mercury adheres to impregnated activated carbon surface, the sulphide product is contained – fixed – within the activated carbon pores. Due to the physical properties, activated carbon is able to absorb also a whole series of organic compounds as well as water, thus considerably reducing its adsorption capacity for mercury. We also assumed, that some of the finest pores, which have greatest adsorption capabilities, may have been totally filled with sulphur during impregnation stage, whereas an amount of sulphur will be distributed in the deepest pores of activated carbon, in the section where any contact and reaction with mercury becomes impossible.

We can only assume that the observed reduction and redistribution of sulphur within the adsorber bed did not lead to dramatic decrease in adsorption efficiency, i.e. the mercury removal efficiency was not significantly affected. Even at sulphur contents of 8% the theoretical capacity for mercury, assuming formation of HgS, greatly exceeds the observed 21% Hg content. In the absence of available comparative investigations and data, it is quite difficult to reach a final conclusion. Nevertheless, the results and mentioned phenomenon lead us to the following conclusions:

- Our experience indicates that sulphur-impregnated activated carbon can be used for the purpose of natural gas mercury purification with excellent results.
- There is a certain boundary concentration of sulphur that becomes impregnated and is introduced into activated carbon pores. Beyond this limit, an increase of impregnate quantity becomes useless, since further addition of sulphur does not result in increased efficiency.
- The adsorption and chemisorption process applied to remove mercury from natural gas can be successfully controlled and optimized by correct distribution of sulphur within the activated carbon pore system.

The objective of this publication is to share our experimental and research results with the natural gas and petrochemical industry to help create fundamentals for design, production and utilization of an optimal adsorbent. Analysis of replaced/saturated activated carbon will provide a valuable and essential technological basis to support important decisions related to management, security and protection of process plants and the reduction of the hazards to our working and living environment caused by the presence of mercury in the treated natural gas.

5
Conclusions

We have defined mercury loading and distribution, i.e. saturation profile in the activated carbon bed, together with variation in concentration of elemental sulphur through the bed under actual process conditions. The results clearly indicate almost linear regularity of activated carbon mercury load distribution profile across the depth of adsorber bed. Results also showed an unexpectedly high mercury load, which was higher than suggested by adsorbent manufacturers and indicated by literature data.

References

Barthel Y, Cameron CJ, Gracco F, Orieux A, Sarazin P (1993) Mercury Removal from Wet Natural Gas, Gas Processors Association 10th Continental Meeting, 2–4 June 1993, Paris, France

Biscan DA, McNamara JD (1989) Mercury Detection and Removal: Field Experience in the U.S. and Abroad, Proceedings 14th LNG International Conference, May 1989, pp. 8–17
Bloom NS (1992) Considerations in sampling for and analysis of mercury at uncharacterised spill sites, in Charlton DS and Harju JA, eds., Workshop on Mercury Contamination at Natural Gas Industry Sites: Chicago, July 1992, Gas Research Institute, Environment and Safety Research Department, GRI-92/0214, pp 45–72
Duschak LH, Schnette CN (1925) The Metallurgy of Quicksilver, U.S. Bureau of Mines Bulletin 222, pp. 7–11
Hutchins RA (1979) Activated-Carbon Systems for Separations of Liquids, in Schweitzer PA (Ed.), Handbook of Separation Techniques for Chemical Engineers, pp. 1-415–1-447, McGraw-Hill, New York, 1979
McNamara JD, Wagner NJ (1995) Process Effects on Activated Carbon Performance and Analytical Methods Used for Low Level Mercury Removal in Natural Gas Applications, AIChE Spring National Meeting, Houston, February 1995
Ruthven DM (1984) Principles of Adsorption and Adsorption Processes, Wiley, New York
Špirić Z (1996) Mercury Removal from natural gas with sulphur impregnated activated carbon–our experience, 4th International Conference on "Mercury as a Global Pollutant", August 4–8, 1996, Hamburg
Špirić Z, Vadunec J (1996) Protection against Mercury Impact on Natural Gas Production Equipment and Environment, Energetic marketing, Zagreb, EGE 14/96, pp. 61–64 (in Croatian)
WHO (1989) Environmental Health Criteria 86: Mercury – Environmental aspects, Geneva

II.d Remediation

Basic Study on the Remediation Technology of Mercury Compound-Contaminated Soil by Low-Temperature Thermal Treatment

A. Matsuyama, H. Iwasaki, K. Higaki, H. Yabuta, T. Sano, and H. Akagi

1 Introduction

Japan has a long history of mercury contamination, the most tragic example being the organic mercury poisoning of Minamata Bay, and the subsequent outbreak of Minamata disease, which killed so many people. Today in Japan there are many cases of mercury contamination of soils in the vicinity of agricultural and industrial chemical plants etc. In other parts of the world, mercury contamination problems, such as those caused by the smelting of gold in the Amazon basin and other tropical regions, have become a worldwide concern.

In Japan, mercury-contaminated soils have generally been treated with sodium sulfide (Na_2S) or cement, which makes the mercury insoluble. One of the problems with this approach, however, is that the contaminant, mercury, remains in the soil after the treatment is completed, which limits the use of the soil. In the United States and Europe it is still common to treat mercury-contaminated soils with the method used in Japan, which is to make the mercury insoluble and stable. In recent years, however, other methods have been proposed. One is the soil-washing method, which involves washing mercury-contaminated soil with chelating agents or acids to remove the mercury. Another is the soil-heating method, which involves heating mercury contaminated soil to moderate to high temperatures (600–900 °C) to steam and remove the mercury (Larry and Jose 1990). These methods, however, present the following problems.

1. Treatment is expensive and thorough safety measures for workers must be implemented.
2. Heating of soil to such high temperatures may have the disadvantage of inducing a substantial change in the soil characteristics. Given these problems, a new soil treatment method is required to ensure the safe treatment of contaminated soils. Based on this understanding, we studied another method for removing mercury from contaminated soil in our laboratory.

The results of heating tests on minerals containing mercury indicate that mercury compounds tend to be unstable, and, even if in a stable form such as mercuric sulfide, start to evaporate when heated at a low temperature of 200–210 °C. The results particularly tend to occur in such as minerals as granite (Hamaguchi et al. 1976).

Environmental Science
Mercury Contaminated Sites (ed. by R. Ebinghaus et al.)

As a result of our studies, we confirmed that, even when mercury in the soil is in the form of mercuric sulfide, this starts to evaporate when the soil is heated at the relatively low temperatures of 200–300 °C.

Further, when the soil is heated with a catalyst, a sufficient amount of mercury is removed quickly. This chapter discusses the results of our basic laboratory study.

2 Materials and Methods

2.1 Equipment and Apparatus

Based on existing study reports (Ono et al. 1993), we established a system for testing the heating of mercury contaminated soils. Figure 1 shows the testing system used.

With our testing system, the heating temperature, gas flow rate, and pressure in the equipment can be adjusted. The heating temperature of the contaminated soils was automatically adjusted to preset temperatures with a temperature sensor inserted into the test soils and a temperature control device. The speed with which the temperature was raised to a preset level was uniform at 20–30 min. The pressure in the equipment was adjusted by a slyduck, which changed the voltage for operation of the aspirator for pressure adjustment. The pressure was read by a pressure meter. The gas flow rate was read and adjusted with a flow meter with a gas flow rate adjustment valve. The set values for the tests are outlined below.

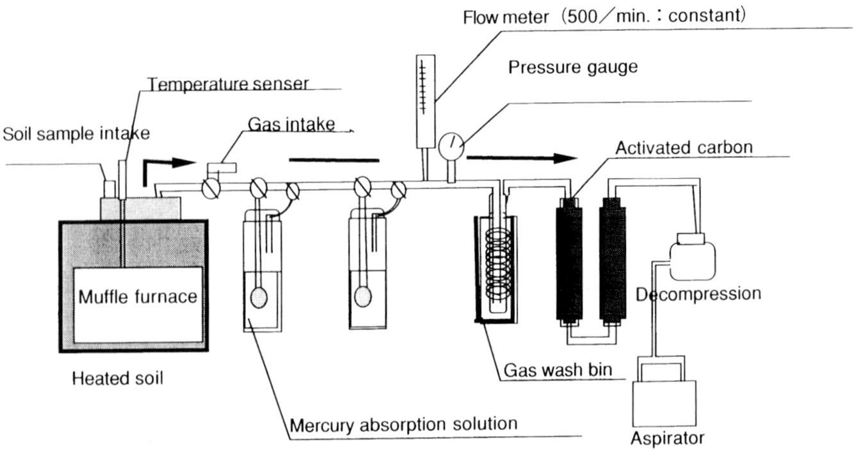

Fig. 1. System for heating tests

- Heating temperature: 100–200–300 °C.
- Pressure in the system: 99–96 kPa.
- Gas flow rate in the system: 700–800 ml/min.

The total mercury in the heated soil was measured with the semiautomated Hg analyzer Model Hg-3500 cold-vapor atomic absorption analysis device made by Sanso Seisakusho Co. Ltd., Tokyo, Japan.

Element analysis of each test soil was done with a JSX-3200type, an X-ray fluorescent element analysis device made by JEOL. Ltd., Tokyo, Japan. The crystal structures of the minerals in the soil were analyzed with a M18XHF22-type X-ray diffractometer made by Mac Science Co.Ltd., Yokohama, Japan.

2.2 Test Materials

2.2.1 *Soils*

- Volcanic ash soil (Kanto loam) from Ikuta, Kanagawa Prefecture.
- Standard sand from Toyoura, Yamaguchi Prefecture.
- Quartz sand reagent Kanto Chemical Co., in corporation. Lot No. 848TA494436.
- Contaminated soil from Tokyo, Japan.

After the volcanic ash soil and contaminated soil were collected, they were air-dried indoors and screened to select grains with a diameter of 1–2 mm for the tests. Table 1 shows the results of X-ray fluorescent element analysis and the amount of organic material in each soil sample.

2.2.2 *Reagents*

After the package of pure red mercuric sulfide (Kanto Chemical Co., in corporation Lot No. 305N1611 hereafter referred to as mercuric sulfide) was opened and washed several times with 1 N HCl. It was then washed again with pure water until no chlorine reaction with silver nitrate was detected in the solution. In preparation for testing, the red mercuric sulfide was then dried at 105 °C with a fan drier. All other reagents used in the tests were guaranteed reagents made by Kanto Chemical Co., and were not treated.

2.3 Systems for Reaction

The following four systems for reaction were used in the tests.

1. Artificially contaminated soil made for the tests. After 5 g of each specified soil was collected in beakers and 0.012 g of mercuric sulfide was weighed out with

Table 1. Characteristics of test soils

		Standard sand	Volcanic ash soil	Quartz sand	Contaminated soil
Components (%)	Na	1.03			
	Mg		4.04		2.26
	Al	4.46	27.20		21.63
	Si	85.52	33.06	Pure SiO_2	49.04
	S		0.38		0.33
	K	6.78	1.12		5.60
	Ca		0.65		1.65
	Ti	0.52	2.76		1.63
	V		0.16		
	Mn		0.51		0.39
	Fe	1.69	30.12		16.49
	Hg				0.98
	Organic substance		5.24		2.42

an automatic balance (four effective digits), the pure mercuric sulfide was added and mixed with each soil to create the artificially contaminated soil specially made for the tests. The concentration of the artificially contaminated soil was set at 2000 mg Hg/kg of dried soil, the highest concentration among mercury contaminated soils sampled in Japan.

2. Artificially contaminated soil made for the tests and the catalysts.

The catalysts listed in Table 2 were added to each artificially contaminated soil made in process (1). The amounts of catalyst added to the soil was ten times greater (5×10^{-4} mol) than the amount of mercury mol (5×10^{-5} mol) in the soil.

3. Artificially contaminated soil made for the tests and the iron chloride.

Table 2. Compounds added to the soils

Metal elements	Chemical formulas
Fe	$FeCl_2 \cdot 4H_2O$ FeOOH Fe_3O_4 Fe_2O_3 $FeSO_4$
Mn	$MnCl_2$ MnO_2 $MnSO_4$
Ni	$NiCl_2 \cdot 6H_2O$ NiO $NiSO_4 \cdot 7H_2O$
Zn	$ZnCl_2$ ZnO $ZnSO_4 \cdot 7H_2O$

As a catalyst, iron chloride was added to each artificially contaminated soil made for the tests. The amount of iron chloride added to the soil was equal to, 10 times greater than, and 100 times greater than the amount of mercury mol (5×10^{-5} mol) in the soil.

4. Contaminated soil from a site in Japan and the iron chloride.

As a catalyst, iron chloride was added to 5 g of contaminated soil from a site in Japan. The amount of iron chloride added to the soil was equal to and ten times greater than the amount of mercury mol in the contaminated soil.

2.4 Accuracy of the Tests by Reaction System

The above tests, in which various reaction systems were used, were repeated two or three times. The test results are average values. Repeat test values were determined as follows.

Before determining the number of samplings for each repeat test sample, it was necessary to understand the uniformity of the artificially contaminated soil made for the tests and the contaminated soil from a site in Japan. Therefore, mercury concentrations in the standard sand, volcanic ash soil, and contaminated soil from a site were measured to check for uniformity. Artificially contaminated soils were made with the standard sand and volcanic ash soil using the procedures outlined in process (1) above. In the analysis, eight 0.2–0.3 g samples were taken from each test sample. After the artificially contaminated soils were wet-decomposed with acid, total mercury concentration was measured. Table 3 shows deviations in the analysis. Therefore, the total mercury concentrations of each repeat test sample were determined by analyzing total mercury concentration with three samples from each repeat test sample and by calculating average values.

Table 3. Uniformity of the artifically contaminated soils made for the tests and contaminated soil

		Artificially contaminated Soil made with standard sand	Artificially contaminated soil made with volcanic ash soil	Contaminated soil
Amount of mercury added (mg)/5 g		9.3	11.8	
Analysis values (mg)	1	9.4	12.7	385.6/kg dired soil
	2	8.6	12.1	420.5
	3	9.0	12.0	321.0
	4	8.5	13.0	462.1
	5	10.6	12.0	400.3
	6	8.0	11.0	358.6
	7	8.7	13.0	392.6
	8	7.9	12.7	411.3
Average values		8.8	12.3	394.0
Standard deviation		±0.9	±0.8	±83.9

2.5
Method for Total Mercury Analysis in the Soils

After collecting a fixed amount (0.2–0.5 g) of sample material for analysis, 2 ml of a mixed solution (HNO_3 + $HClO_4$ = 1:1), 5 ml of sulfuric acid, and 1 ml of H_2O were added to the samples and digested at 230–250 °C on a hot plate for 20 min. The mixture was cooled to room temperature, and distilled water was added to the digested sample solution, which made the solution 50 ml. Then the mercury in the solution was gasified by reduction with stannous chloride and measured through cold-vapor atomic absorption analysis (Akagi and Nishimura 1991).

2.6
Method for Removing Iron Oxide from the Soil

The method used to remove iron oxide from the soil conforms with the Aguirera and Jackson method. (Aguilera et al. 1953; Oba 1966) For 1 g of sand, 1 g of sodium dithionite ($Na_2S_2O_4$), and 30 ml of sodium citrate ($Na_3C_6H_5O_7 \cdot 2H_2O$)solution of 0.3 mol w/v were added, and the mixture was heated and stirred in a water bath (70–80 °C) for 20 min. After the supernatant was removed with the gradient method, once again distilled water was added, the mixture was stirred, and the supernatant was removed. These washing procedures were repeated several times. Then the sand from which iron was removed was dried with a fan drier (105 °C) to create soil for the test.

2.7
Analysis of Crystal Structures in the Soils and Element Analysis of the Soils

Before heating, the required amounts of artificially contaminated soils made for the tests were collected and powdered with an agate mortar. After heating, the samples were taken out of the reaction furnace and cooled to room temperature in a desiccator. Then the required amounts of sample material were collected and powdered with an agate mortar. The powdered samples were analyzed with X-rays and X-ray fluorescent tests.

3
Results and Discussion

3.1
Heating of Artificially Contaminated Soils Made for the Tests

Based on the results of the heating tests conducted on the minerals described above, the standard sand (weathered granite), the quartz sand which was the primary component of granite, and volcanic ash soil, were used for testing and mercuric sulfide was added to create the specified mercury contaminated soils. The mercury contaminated soils were placed in heat-resistant glass vessels and heated using the test system in Fig. 1 under the conditions described above. Figures 2 and 3 show the results. As Fig. 2 shows, when the soils were heated to 200 °C, the amount of mercury removed changed very little. However, some mercuric sulfide in the standard sand and volcanic ash soil was removed by heat. As Fig. 3 shows, however, when the soils were heated to 300 °C, there was a significant change in the amount of mercury removed. Of the artificially contaminated soils, the highest percentage of mercury was removed from the standard sand, followed by the volcanic ash soil, and the quartz sand. Less mercury was removed from artificially contaminated soil made with volcanic ash soil than from artificially contaminated soil made with standard sand. This is thought to be due to the specific heavy metals absorption capacities of the organic substances and the soils. In other words, when mercuric sulfide was contained in a volcanic ash soil relatively rich in organic substances compared to standard sand, even when it was decomposed in the soil by heat, divalent mercury was absorbed by the organic acids in the soil, such as humin acid, or specifically absorbed by the clay in the soil (Cheam and Gamble 1974; Honda and Tachikawa 1978), therefore the mercuric sulfide may not have removed smoothly. These results show that more mercury was removed by heating when mercuric sulfide was mixed with soil materials or in soil than when heated as pure mercuric sulfide. However, we confirmed that the effect of mercury removal by heat-generated evaporation differed depending on the minerals contained in the artificially contaminated soil. The results match closely those of heating tests on minerals containing mercury (Hamaguchi et al. 1976).

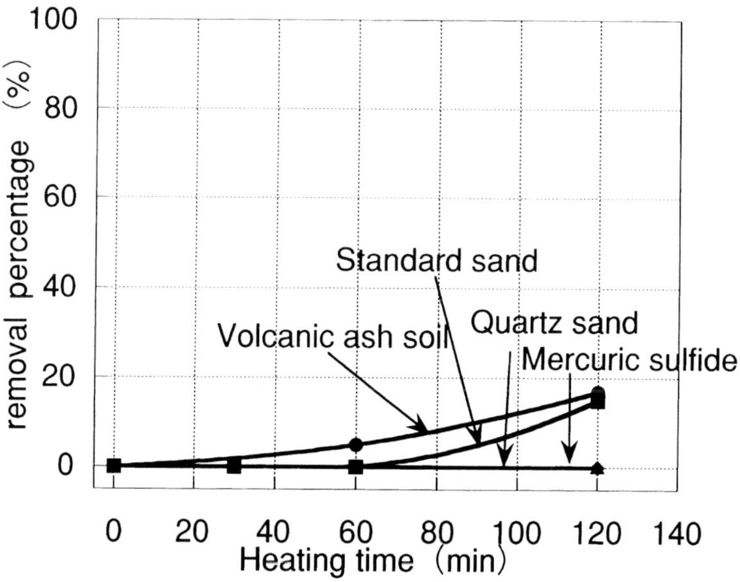

Fig. 2. Removal characteristics of contaminated soils heated to 200 °C

3.2 Differences in the Percentages of Mercuric Sulfide Removed by Heat from Mercury-Contaminated Soil Made with Standard Sand, Mercury-Contaminated Soil Made with Quartz Sand, and Pure Mercuric Sulfide

The most notable results of the tests described above were the differences in the percentages of mercury removed by heat between the quartz sands and pure mercuric sulfide, and the differences in those of mercury-contaminated soils made with standard sand and mercury-contaminated soils made with quartz sand. First, we will discuss the differences in the percentages of mercury removed over time by heat in the quartz sands and mercuric sulfide. As Table 1 showing the X-ray fluorescent test results indicates, the quartz sand used in this test did not contain metals other than silicon (Si). The heating test results show that the mercury removal percentages of mercury-contaminated soil made with quartz sand clearly improved over time compared to those of the mercuric sulfide. According to the results of X-ray powder analyses conducted before and after heating, no great differences in the peaks were observed, and no product was generated when the quartz sand and mercuric sulfide were heated. The results suggest that mercuric sulfide was not removed through evaporation caused by a heat-induced reaction between the quartz sand and the mrcuric sulfide, but by the physical and chemical characteristics of the quartz sand when it was mixed with mercuric sulfide and heated.

Next is a discussion on changes in heat-generated removal percentage over time in the mercury contaminated soils made with standard sand and mercury-contaminated soils made with quartz sand. As Fig. 2 shows, when the soils were heated to 200 °C, the amount of mercuric sulfide removed changed very little for 2 straight h. However, as Fig. 3 shows, when the soils were heated to 300 °C, there was a significant change in the amount of mercuric sulfide removed. After 2 h of heating, the percentage of mercuric sulfide removed in the standard sand was

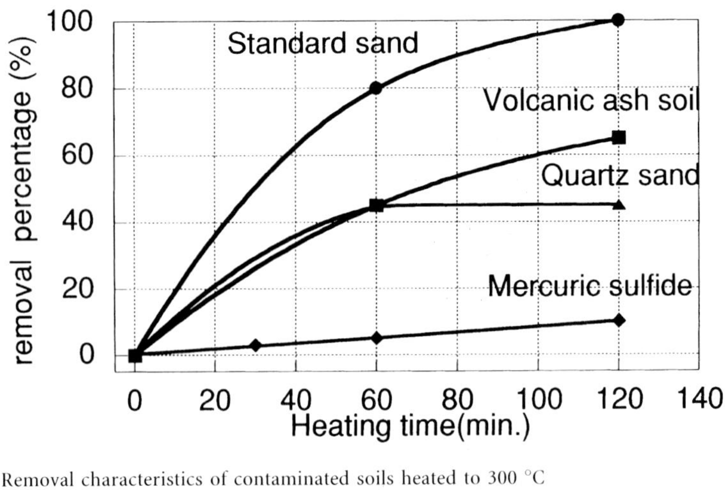

Fig. 3. Removal characteristics of contaminated soils heated to 300 °C

double that of the quartz sand. While heating the artificially contaminated soils in our tests, the colors of the quartz and standard sands did not change when they were heated to 200 °C, but the standard sand became red (clearly observable with the naked eye) when heated to 300 °C.

As Fig. 4 showing the results of powder analysis with X-rays demonstrates, the color change in the artificially contaminated soil made with heated standard sand is thought to be a function of the generation of Fe_2O_3 (hereinafter referred to as

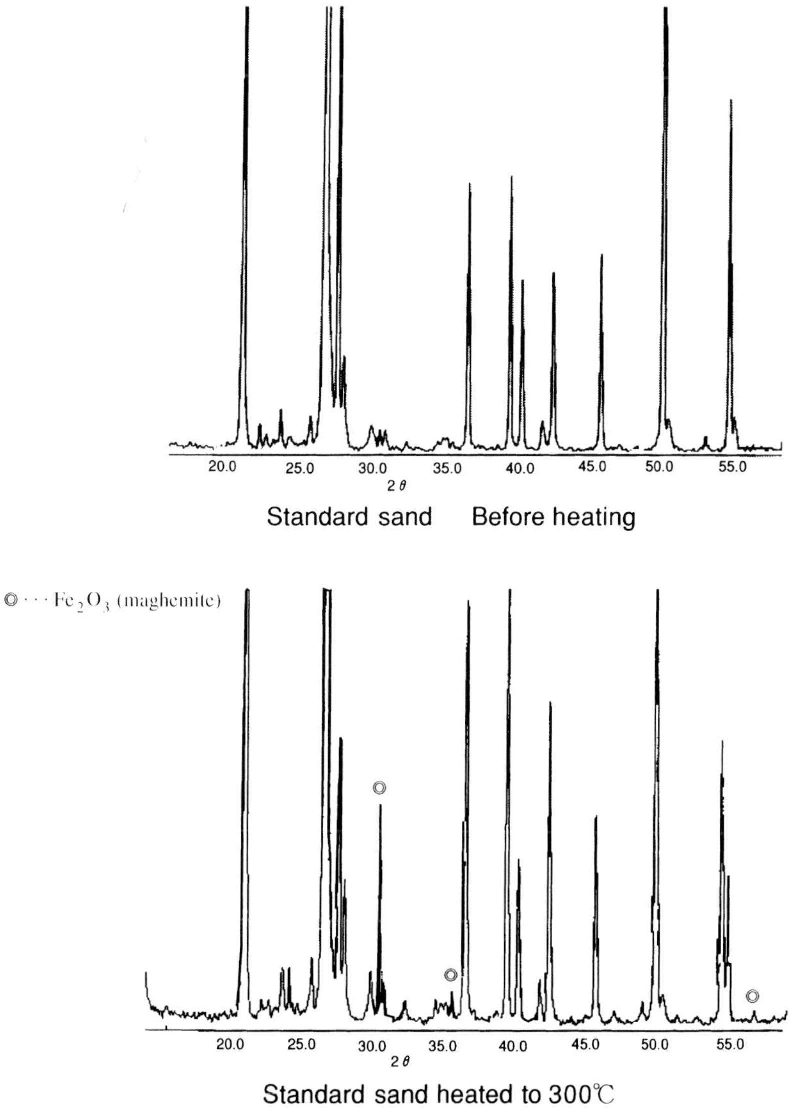

Fig. 4. Results of powder analysis with X-rays

iron oxide). Iron oxide was not generated in the artificially contaminated soil made with standard sand, which was not heated. It is known that iron, which is a weathering product of granite, adheres to the surface of standard sand and exists in the form of amorphous iron hydroxide and oxidized iron hydroxide (FeOOH). These iron components change to iron oxide after heating, although the temperatures that effects this change differ according to form (Nakajima 1994). Based on the test results and existing reports, the differences in mercuric sulfide removal percentage between the heated artificially contaminated soil made with quartz sand and artificially contaminated soil made with standard sand were thought to be a function of chemical changes in the iron component and so on of the standard sand and soils. Therefore, if the iron component and so on of the standard sand were removed, the removal percentage of mercuric sulfide when the soil is heated should drop. Therefore, we removed iron from the standard sand with the method described above, and the sand from which the iron was removed was heated to 300 °C. Figure 5 shows the test results and Table 4 shows the results of X-ray fluorescent analysis of the standard sand before and after iron removal. As Fig. 5 shows, compared to the standard sand, less mercuric sulfide was removed by heating from sand from which the iron component was removed. In the heating tests, unlike standard sand, the color of the sand from which iron was removed did not change to red after being heated for straight 2 h. Further, as Table 4 shows, iron and titanium underwent the greatest chemical changes when the iron component was removed from the standard sand. Given this, it is likely that the differences in the percentage of mercuric sulfide removed by heating between the artificially contaminated soil made with standard sand

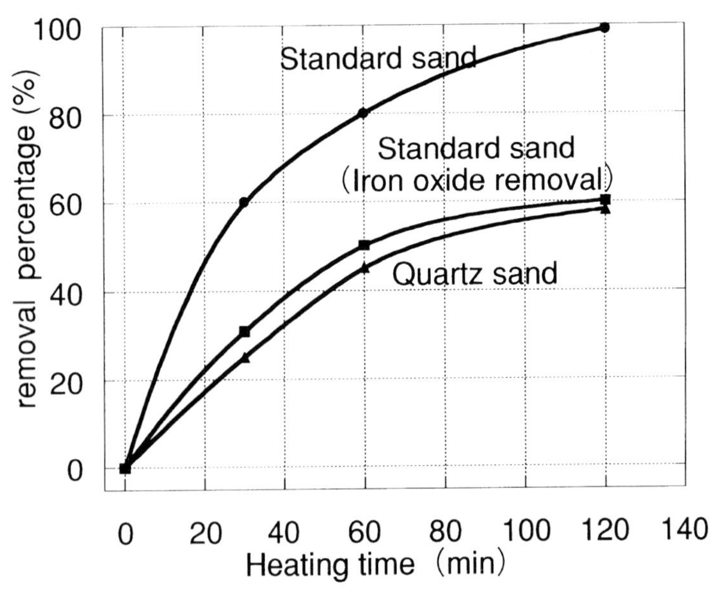

Fig. 5. Removal characteristics of sand (from which iron was removed) heated to 300 °C

Table 4. Results of X-ray fluorescent analysis of sand

		Standard sand	Standard sand (iron oxide removal)
Components (%)	Al	4.28	4.24
	Si	86.51	87.11
	K	4.04	7.33
	Ti	0.50	0.24
	Fe	1.51	0.95

and that made with quartz sand was attributable to the iron component, which was extracted with the iron-removal process.

We confirmed that mercuric sulfide can be removed when it is mixed in soil and heated below 583.5 °C, the temperature at which mercuric sulfide sublimates. It is considered that the evaporation of heated mercuric sulfide was effected by the forms and amounts of the silicas, such as quartz, contained in the soils, and the existence of metal components such as iron, and their amounts.

3.3 Catalyst Selection

The heating tests on the artificially contaminated soils made for the tests, which were outlined above, show that, of the artificially contaminated soils, the highest percentage of mercury was removed from standard sand heated to 300 °C. It took at least 2 h, however, to remove at least 99% of the mercury in the soil. With the volcanic ash soil, only 65% of the mercury was removed after at least 2 h of heating at 300 °C. Therefore, to ensure the efficient removal of mercury from all types of soils, it is necessary to use a catalyst that facilitates quick reaction. Based on previous test results, we decided to select a catalyst from among transition metal compounds, such as iron. As Table 2 shows, we added typical types of oxides, sulfides, and chlorides of each transition element to the artificially contaminated soil made with quartz sand. To check reactions, the amount of each compound added was made the same as the amount of the transition metal elements, an amount that was ten times greater than that of the mercury mol contained in the artificially contaminated soil. Samples of each artificially contaminated soil containing each compound were heated to 100, 200, and 300 °C. Figure 6 shows the test results.

According to the test results, of the catalysts used to facilitate the removal of mercuric sulfide, $FeCl_2{\cdot}4H_2O$ (hereinafter referred to as iron chloride) proved to be particularly effective at removing mercury. Only about 50% of the mercuric sulfide was removed from artificially contaminated soil with an initial value of 10 mg-Hg/5g quartz sand when the soil, which contained iron chloride, was heated to 200 °C for 1 h. However, over 99.5% of it was evaporated, leaving total mercury concentration at 0.1 mg or less, when the soil was heated to 300 °C for 1 h. According to the test results, compounds other than iron chloride demonstrated some effectiveness, but iron chloride and $NiCl_2$ (hereinafter

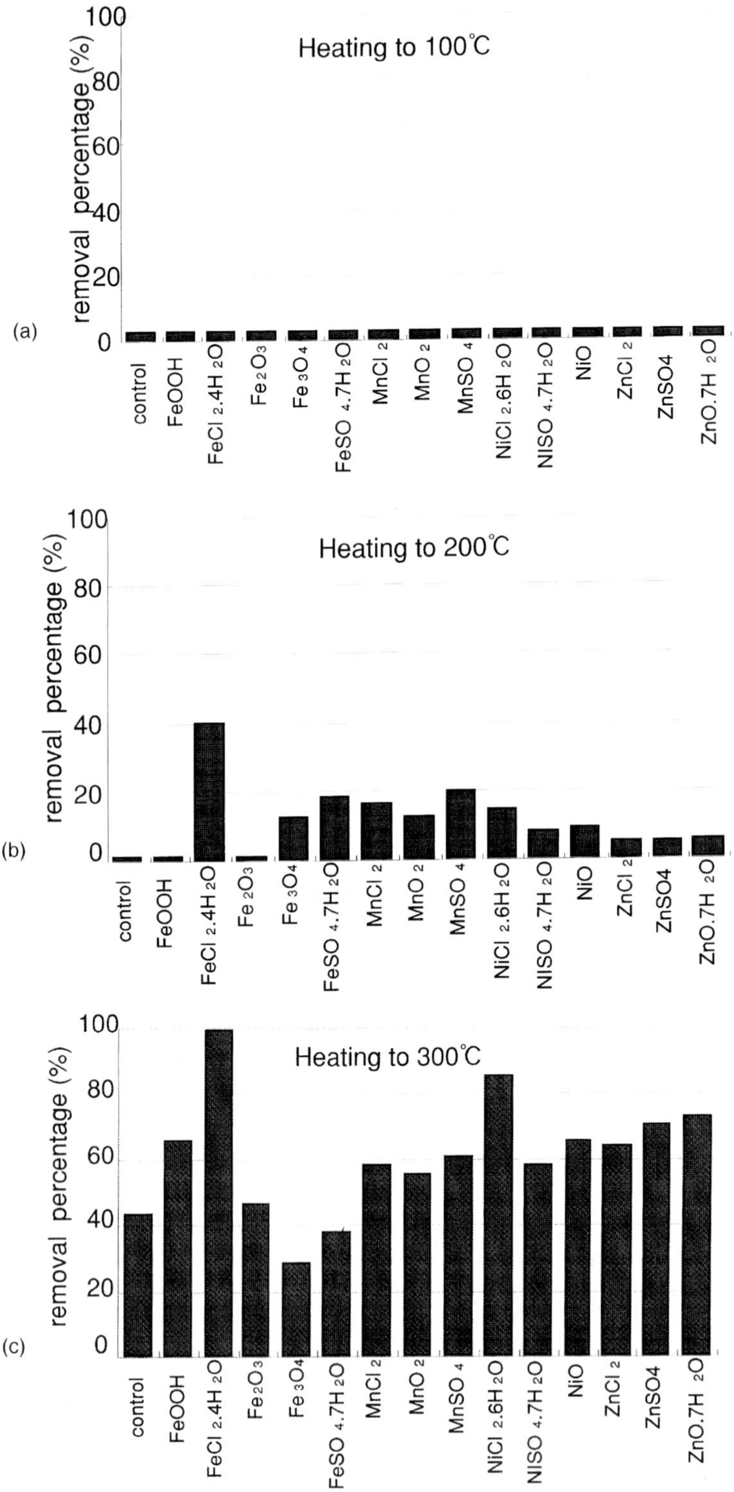

Fig. 6. Effect of reaction accelerators

referred to as nickel chloride) were clearly the most effective at accelerating the removal of mercuric sulfide. Both iron and nickel are in the group of eight transition elements and thus have similar chemical characteristics. None of them, for example, forms an amalgam with mercury.

As Fig. 7 showing the results of the powder analysis with X-ray demonstrates, when artificially contaminated soil was heated to 100 °C, no peak in the generation of iron oxide was observed, but as the heating temperature rose to 200 and 300 °C, the peak in the generation of iron oxide became more prominent, the iron chloride was denatured, and iron oxide was generated. Some nickel chloride was also denatured and nickel oxide was generated. This shows that the denaturation of transition metal compounds such as iron and nickel when they are heated has significant effect in the evaporation and removal of mercuric sulfide. Further, it is thought that the influence of chloride in transition metal compounds when they are heated cannot be ignored because the denatured products do not contain chloride components.

As a result, iron chloride was clearly the most effective compound at accelerating evaporation, and in the form of iron, it is abundant in the environment and is thus considered to have little environmental impact. We have concluded that iron chloride is the most appropriate compound to serve as a catalyst.

3.4
Effect of the Addition of Iron Chloride

We added iron chloride to the artificially contaminated soils made with quartz sand and volcanic ash soil and examined the effects of heating temperature and the amount of iron chloride added on the percentage of mercury removed by heating. We heated the artificially contaminated soils to 200 and 300 °C to examine the relationship between heating temperature and mercury removal percentage over time. We examined the relationship between the amount of mercury removed and the amount of iron chloride used over time by changing the amounts of iron chloride added. The amounts of iron chloride added to the soils were equal to, 10 times greater than, and 100 times greater than the amounts of mercury mol contained in the artificially contaminated soils made for the tests. Figures 8 to 11 show the test results.

As Figs. 8 and 9 show, as the amount of iron chloride was increased, the percentage of mercuric sulfide removed when the artificially contaminated soils were heated to 200 °C increased. After 2 h of heating, when the amount of iron chloride added to the artificially contaminated soils was 100 times greater than the amount of mercury mol contained in the artificially contaminated soil made with quartz sand, almost 95% of the mercuric sulfide was removed. However, even when the amount of iron chloride added to the artificially contaminated soil made with volcanic ash soil was the same as that added to artificially contaminated soil made with quartz sand, only about 60% of the mercuric sulfide was removed. As Figs. 10 and 11 show, when the artificially contaminated soils were heated to 300 °C, regardless of the amount of iron chloride added, at

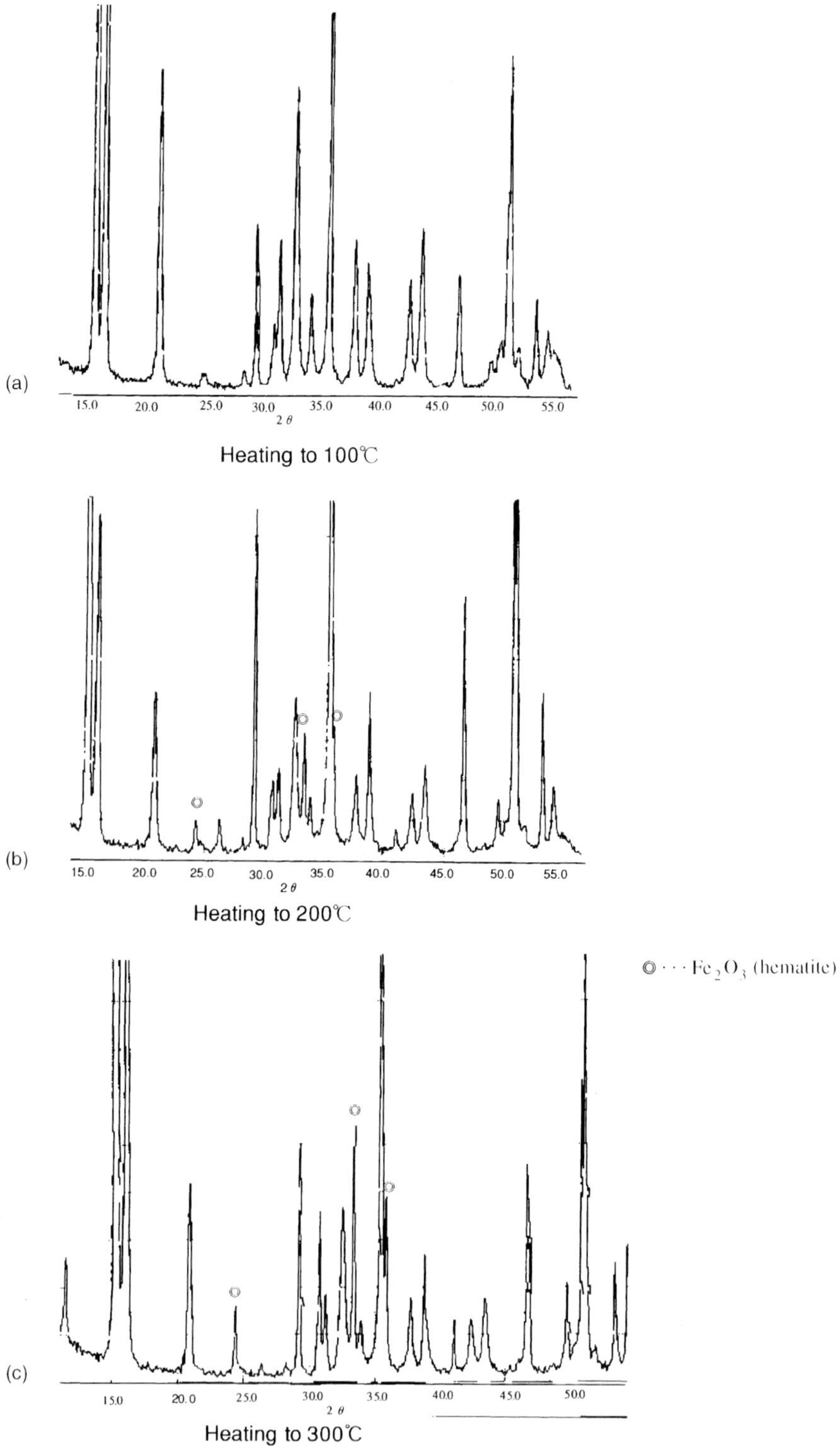

Fig. 7. Results of X-ray analysis of the denaturation of iron chlolide according to heating temperature

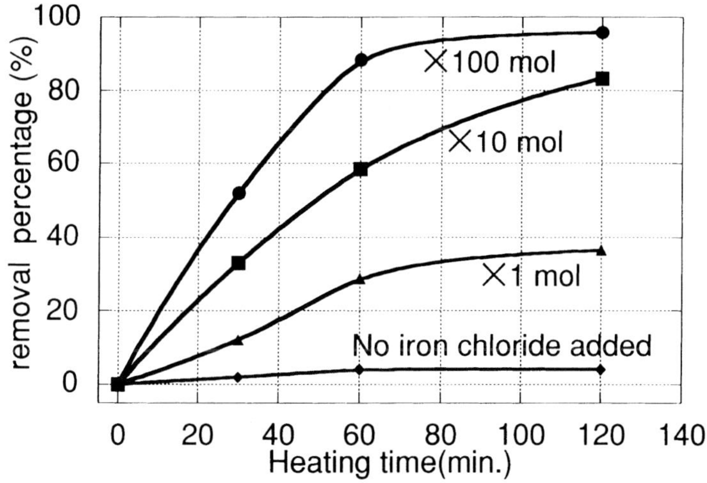

Fig. 8. Variable amounts of iron chloride added to quartz sand (heated to 200 °C)

least 95% of the mercury was removed from both artificially contaminated soils. The removal characteristics clearly changed over time. When the artificially contaminated soils were heated to 300 °C, almost 100% of the mercuric sulfide was removed in about 30 min to 1 h after the start of heating. However, when the artificially contaminated soils were heated to 200 °C, the mercuric sulfide did not remove quickly, but steadily and gradually over time. Given these results, it is evident that the effective heating temperature for the mercuric sulfide removal recation is greatly affected by the addition of iron chloride and the amount of iron chloride to be added to the soils is greatly affected by heating temperature.

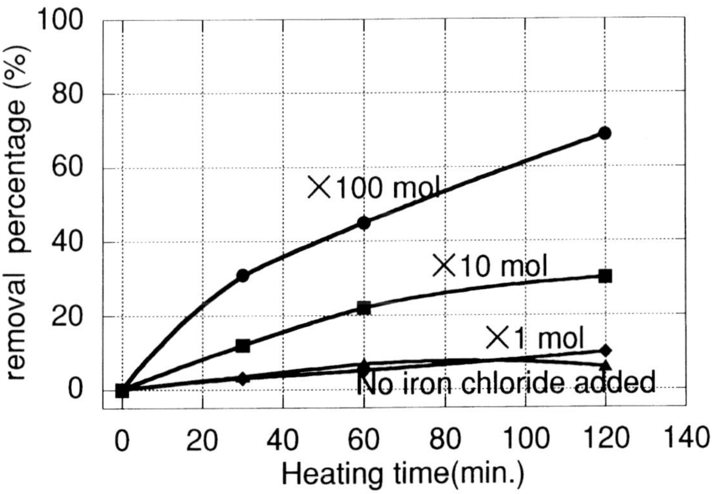

Fig. 9. Variable amounts of iron chloride added to volcanic ash soil (heated to 200 °C)

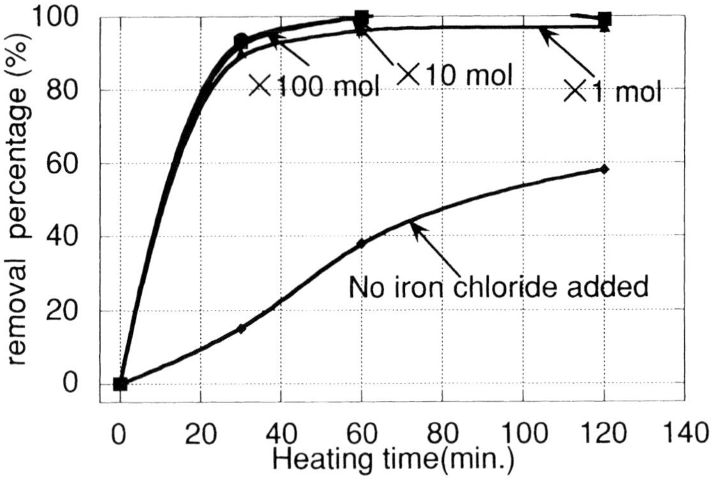

Fig. 10. Variable amounts of iron chloride added to quartz sand (heated to 300 °C)

When the artificially contaminated soils made with standard sand and volcanic ash soil were heated to 300 °C, the iron chloride added to the artificially contaminated soil made with volcanic ash soil was slower at removing mercury than the iron chloride added to the artificially contaminated soil made with standard sand. The difference, however, was minor. From our testing, we have concluded that although it is difficult to determine the effect of iron chloride added to artificially contaminated soils (except for sand or sandy soil) when the soils are heated to 200 °C, when they are heated to 300 °C, regardless of soil type, the amount of iron chloride that should be added is equal to or slightly more than the amount of mercury mol contained in the artificially contaminated soils.

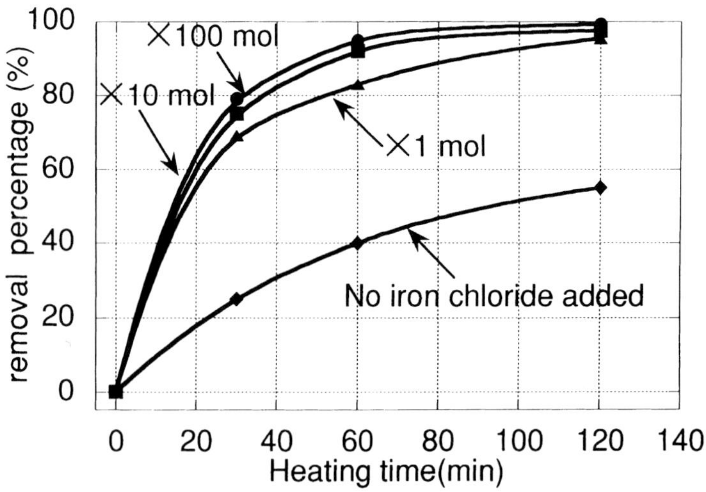

Fig. 11. Variable amounts of iron chloride added to volcanic ash soil (heated to 300 °C)

3.5 Examination of the Mechanisms of Heat-Generated Removal Reactions

Based on our test results, we examined the heat-generated mercuric sulfide removal mechanisms when iron chloride was added. With our testing system, since no water was added before the heat was applied and the heating temperature exceeded 200 °C, there was hardly any moisture in the artificially contaminated soils when the reactions occurred. Therefore, it can be considered that all substances causing the reactions were bonded solids when the reactions occurred.

Iron chloride, which was used in our tests as a catalyst, is known to denature quickly when heated in air (Mizushima 1964). Denaturing starts when it is heated to around 200 °C, but when it is heated to 250 °C it changes and finally converts to iron oxide and hydrogen chloride gas. Therefore, if iron chloride is heated to 250 °C or more in air, bivalent iron chloride becomes trivalent iron oxide and the electrons that are discharged when iron chloride is denatured by heating can be used to forcefully form a reducing atmosphere between the solids. This is obvious from the test results dicussed above, from the fact that the percentage of mercuric sulfide removed depended on the heating temperature, and the fact that the generation of iron oxide peaked as the heating temperature rose, as shown in Fig. 7 of the X-ray analysis results. When iron chloride and mercuric sulfide are simply mixed in the atmosphere at room temperature, nothing happens. It is thought, however, that when they are mixed and heated, an oxidation-reduction reaction atmosphere is forcefully generated, and that an oxidation reduction occurs between it and the mercuric sulfide. We think that if some mercuric sulfide is decomposed by an oxidation reduction reaction caused by the denaturation of iron and the characteristics of the silica, an oxidation reduction atmosphere can be formed in the soil, which leads to an oxidation reduction reaction.

There is a test report in which mercuric sulfide was dispersed in water. In the report, when trivalent iron ions also existed in the water, they were reduced to bivalent mercury ions, and mercuric sulfide was solubilized at the same time. It is also reported that mercuric sulfide may solubilize when bivalent iron ions are oxidized to trivalent iron ions, and that there is a strong correlation between solubilization and the oxidation-reduction potential of metal ions (Akagi et al. 1975). Further, as we mentioned earlier, the effect of chloride in transition metal compounds when artificially contaminated soils are heated cannot be ignored because the denatured products do not contain chloride components.

A report on mercury behavior in the discharge of waste treatment facilities explains that when hydrogen chloride is heated in air to a low temperature of around 300 °C, it oxidizes abruptly and chlorine gas is generated (Urabe and Yokokawa 1990). Given this, it is likely that some of the hydrogen chloride gas, which is generated when iron chloride is denatured by heat, is oxidized and chlorine gas is generated. As mercury reacts furiously with chlorine even at room temperature, it is necessary to take the oxidation-reduction reaction caused by chlorine into consideration.

Given these results, we conclude that the reactions we observed in our tests were oxidation-reduction reactions in the soil, reactions in which the characteristics of the silicas, such as quartz, and the denaturation of iron chloride by heating, were utilized.

3.6 The Effect of Iron Chloride Added to Samples of Mercury-Contaminated Soils from a Site in Japan

Next, we conducted tests using samples of contaminated soil from a site in Japan. In the tests, we added iron chloride to soil samples and then heated the samples to examine mercury removal performance. The soil used in our tests was obtained from the site of a former chemical plant. The soil was volcanic ash soil with a mercury concentration of about 400 mg/kg of dried soil. We assumed that the mercury was in the form of mercuric sulfide to calculate the amount of mercury mol contained in the soil. We added iron chloride in amounts equal to and ten times greater than that of mercury mol contained in the soil, and then heated the soil-iron chloride mixture to 300 °C for 1 h. As Fig. 12 shows, even when the catalyst was not added, about 60% of the mercury was removed when the soil was heated to 300 °C. By adding the catalyst, however, the percentage of mercury removed clearly increased. Further, when the amount of iron chloride added was ten times the mercury mol, more than 99% of the mercury was removed.

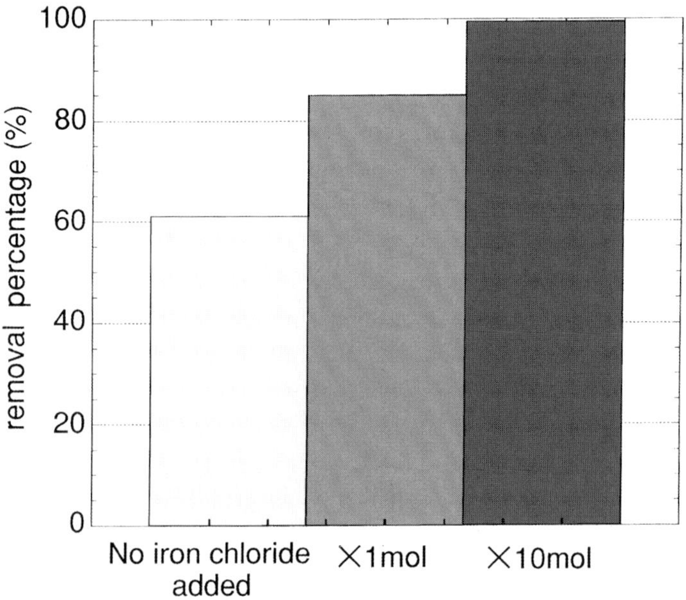

Fig. 12. Effect of iron chloride added to samples of mercury-contaminated soil in Japan

4 Conclusion

We examined a method for purifying mercury contaminated soils by heating the soils at low temperatures.

1. Mercuric sulfide evaporates at temperatures below 583.5 °C (the physical and chemical sublimation point of mercury) when it is heated in minerals, such as standard sand.
2. When mercuric sulfide is dispersed in minerals, such as in standard sand, and is heated, it evaporates at 200–300 °C. The same phenomenon was observed when pure quartz sand was used. Therefore, we conclude that the evaporation of heated mercuric sulfide was induced by the physical and chemical charcateristics of the quartz.
3. The removal of mercuric sulfide from mercury-contaminated soils is facilitated by adding iron chloride ($FeCl_2.4H_2O$) to and then heating the soils.
4. The efficiency of the mercuric sulfide evaporation reaction generated by heat and the additon of iron chloride to contaminated soils and the amount of iron chloride to be added to the soils is greatly affected by heating temperature. As our test results show, the percentage of mercury removed increased with an increase in heating temperature. At a heating temperature of 300 °C, regardless of soil type, at least 99% of the mercuric sulfide can be removed; the amount of iron chloride that should be added is equal to or slightly greater than the amount of mercury mol contained in the soil. We conclude that this is due to the dramatic physical change that occurs to iron chloride when it is heated to 250 °C.
5. According to references, when iron chloride is heated to 200–250 °C in air, it becomes denatured and converts to iron oxide and hydrogen chloride gas. Therefore, when iron chloride is heated in air, bivalent iron can be converted to trivalent iron, allowing an oxidation-reduction atmosphere to be forcefully made and a reaction between it and the mercuric sulfide to be generated.

Our test results also suggest that coexisting silicas have an important function in the removal of mercuric sulfide through heating. Given these two points, we consider the reactions we observed in our tests to be oxidation-reduction reactions in the soil, reactions in which the characteristics of the silica, such as quartz, and the denaturation of iron chloride by heating, were utilized.

References

Aguilera NH, Jackson ML (1953) Iron oxide removal from soils and clays. Soil Science Society of America Proceedings, vol 17. The Soil Science Society of America, Wisconsin, pp 359–364

Akagi H, Nishimura H (1991) Speciation of mercury in the environment. In: Suzuki T et al. (eds) Advances in mercury toxicology. Plenum Press, New York, pp 53–76

Akagi H, Fujita Y, Takabatake E (1975) Solid-photosensitized methylation of inorganic mercury in aqueous acetic acid solutions. The Chemical Society of Japan, Tokyo, pp 1273–1279

Cheam V, Gamble DS (1974) Metal-fulvic acid chelation equilibrium in aqueous $NaNO_2$ solution. Hg(II), Cd(II), and Cu(II) fluvate complexes. Can J Soil Sci 54:413–417

Hamaguchi H, Umezaki Y, Kanno S, Kitamura S, Sugino T, Fujii M, Wakita H (1976) Mercury. In: Mercury sub-committee (eds) Chemistry of earth. The Chemical Society of Japan, Tokyo p 10 (in Japanese)

Honda K, Tachikawa R (1978) Adsorption strength of mercury in soil relationships among particle diameter, the composition of organic substances, and the amount of mercury adsorbed. Jpn Soil Sci Plant Nutr 49: 378–382

Larry R, Jose L (1990) Final best demonstrated available technology (BDAT) background document for mercury-containing wastes D009, K106, P065, P092, and U151. U S Environmental Protection Agency, Washington, p 5–1

Oba H (1966) Method for identification of the primary minerals in soils. Jpn Soil Sci Plant Nutr 37th edn, vol 1. Japanese Society of Soil science and Plant Nutrition, p 1 (in Japanese)

Ono A, Kato H, Mitsuhashi I, Nakashio M, Nozaki M (1993) Heat processing of mercury contaminated soils. In: Hiraoka M (ed) Proc 4th Annu conf of the Japan Society of Waste Management Experts, The Japan Society of Waste Management Experts, Tokyo, pp 843–844

Mizushima M (1964) Encyclopaedia chimica. In: Encyclopaedia chimica editing committee (eds) Iron chloride. Kyoritsu Press, Tokyo, p 1050 (in Japanese)

Nakajima S (1994) Changes in earth color. In: Fukagawa M (ed) Color of stones and soils. Kinmiraisha, Tokyo, pp 178–205 (in Japanese)

Urabe T, Yokokawa H (1990) Thermodynamic study of the forms of mercury in combustible gas from waste. In: Hirayama N (ed) Waste management research, Vol 1. The Japan Society of Waste Management Experts, Tokyo, pp 10–18

Electroleaching: a Mobile Cleanup Process for Mercury Contaminated Materials

J. Thöming, L. Sobral, and R.D. Wilken

To date, there is no satisfactory technique known for simultaneous remediation of materials contaminated with mercury and other metals. In the case of mercury-polluted tailings from Brazilian gold-mining areas, there is the opportunity to extract mercury and traces of gold hydrometallurgically, thus making the remediation process cost-effective. Hydrometallurgical techniques are widely used in ore treatment, whereas soil remediation applications are rare. This chapter outlines two variants of an electroleaching process and gives an overview of its thermodynamic fundamentals. Model calculations are presented as well as results from laboratory experiments.

The leaching agent, hypochlorous acid, can be produced by electrolytic oxidation of a sodium chloride solution. After leaching, the leachate is treated electrolytically to deposit the metals in the elemental form and to regenerate the leaching agent for recirculation. The leaching mechanism is a mercury oxidation and a complexation of mercuric ions by chloride ions, forming, for example, a tetra-chloro-mercurate-complex. The main process parameters, such as pH, Eh and element concentrations, control the speciation of mercury. It is illustrated how leaching efficiency and electrolytical mercury deposition are both dominated by the mercury species in solution.

The electroleaching could be conducted in two different reaction systems: (1) as a 1-reactor process with an in-suspension production of the leaching agent (EL-IS) and (2) as a 2-reactor process with an in-line production (EL-IL). While the more costly EL-IL is capable of recovering mercury sufficiently (>99% removal rate) even from highly contaminated German industrial samples, the EL-IS is the best solution for less contaminated Brazilian gold-mining residues. After treatment, the mercury concentration is lower than 1 mg/kg and more than 82% of the residual gold can be recovered.

1 Introduction

A major source of mercury issues to the environment has traditionally been the chemical industry such as chlor-alkali plants. Another source is the gold mine industry that uses elemental mercury to amalgamate gold in gravity concentrates. Amalgamation has been used as an effective means of separating gold from host rock. Amalgamation of run-of-mine ores has gradually diminished for health and environmental reasons and the process is now virtually obsolete, but it is still commonly employed in the treatment of gravity concentrates and has the advantages of being a simple and cost-effective process.

The conventional technique to treat mercury-bearing residues deals with the thermal decomposition. During this process, the mercury is volatilized by increasing the temperature and immediately condensates. The use of this route entails some drawbacks, such as: the need of using large-capacity equipment for treating great amounts of residues, and the difficult task of dealing with the large

Environmental Science
Mercury Contaminated Sites (ed. by R. Ebinghaus et al.)

volume of mercury vapours, which requires sophisticated condensation systems to assure no leak to the atmosphere.

Avoiding these disadvantages of thermal remediation, wet extraction processes are becoming more and more interesting. This work presents a contribution to the extraction and recovery of mercury from mercury-bearing materials. The main goal of the present work is to provide an improved and economical process for the hydrometallurgical extraction and recovery of mercury that can be applied to low, and to some extent, to high mercury content residues.

There is not enough information in the literature regarding hydrometallurgical recovery of mercury from residues (US PTO 1993). There are, however, extensive works on the environmental problems which are caused by mercury. The lack of information on recovery processes is due largely to the chemistry of the mercury-water system, and the lack of very low solubility compounds which could be used to reduce mercury concentrations in solution. In addition, there is the problem that mercury forms strong complexes with many ligands, but this is, at least, an advantage when considering the possibilities of chemicals for leaching mercury from residues.

2 The Electroleaching Process

The process described here is called electroleaching because electrolysis is employed for generating the leaching agent.

The leaching process takes place by the action of hypochlorous acid or hypochlorite ions, generated during the electrolysis of:

1. A suspension made up of residue in an aqueous sodium chloride solution (in-suspension production of leaching agent: EL-IS, Fig. 1).
2. An aqueous sodium chloride solution in a separate electrolysis cell (in-line production of leaching agent: EL-IL, Fig. 2).

The leaching agent generation is the result of the oxidation of chloride ions to chlorine gas, which in contact with the aqueous solution produces hypochlorous acid and/or hypochlorite ions, depending on the solution pH. In the situation where sodium chloride is used as the electrolyte, the reaction occurring at the anode is

$$2Cl \Leftrightarrow Cl_2 + 2e \qquad (1)$$

and the reaction taking place at the cathode is

$$2H_2O + 2e \Leftrightarrow 2OH + H_2. \qquad (2)$$

In a cell in which the anode and cathode products are not separated, there is a measuring process which can result in

$$Cl_2 + 2OH \Leftrightarrow ClO + H_2O + Cl \qquad (3)$$

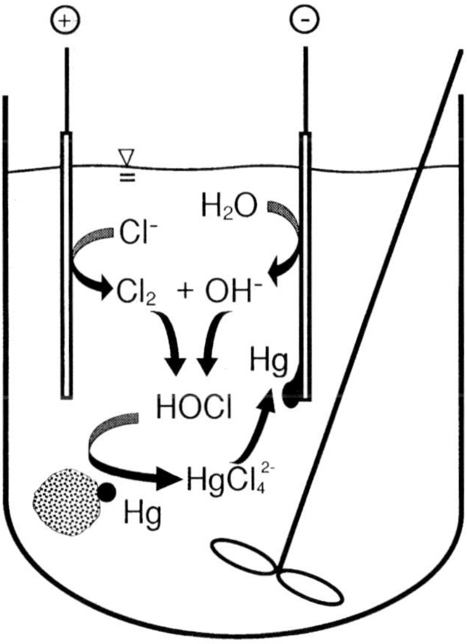

Fig. 1. Electroleaching with in-suspension production of leaching agent (EL-IS)

or, in the case of lower pH values, in

$$Cl_2 + H_2O \Leftrightarrow HClO + H^+ + Cl^- \, . \tag{4}$$

Both the free chlorine and the hypochlorite or the hypochlorous acid, respectively, act as oxidizing agents and are also chemically reactive on their own.

The success of this process will depend, to some extent, on how effective the chlorine reaction is with the aqueous solution, generating the oxidizing agents ($HClO$ and ClO^-), which promote the mercury dissolution, according to the following reaction:

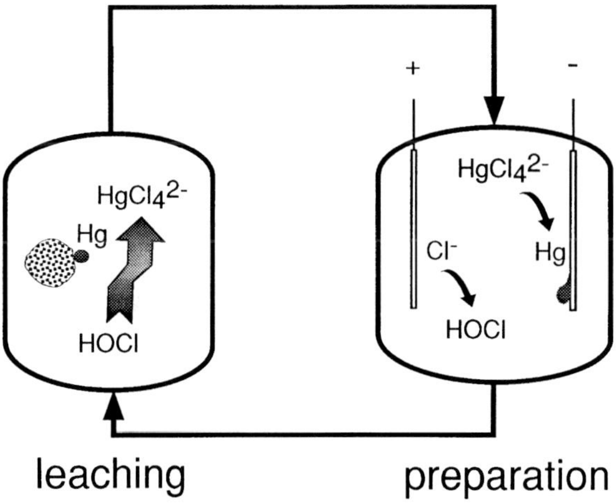

Fig. 2. Electroleaching with in-line production of leaching agent (EL-IL)

$$Hg + HClO + 3Cl \Leftrightarrow HgCl_4^2\ ) + OH \tag{5}$$

The dissolved mercury, mainly as $HgCl_4^2$, is electrolytically deposited on the surface of the cathode:

$$HgCl_4^2 + 2e \Leftrightarrow Hg + 4Cl\ . \tag{6}$$

3
Theoretical Considerations

Thermodynamic equilibrium calculations, which provide knowledge of species stability, are well known for their accuracy. Nevertheless, they can give engineers only a rough idea about the state of the solid/liquid extraction system. This is due to the low temperature of these processes – usually 20–80°C. In this temperature range kinetic factors often cause metastable states differing from predicted stabilities. This fact can be taken into account by the right choice of species, that may occur before calculations are done.

The behaviour of mercury during leaching can be controlled by pH, redox potential and the concentration of complexing agents, and it depends also on the dilution of mercury. This can be illustrated by thermodynamic stability diagrams, or more specifically Eh-pH diagrams which show predominant stability areas of the mercury species. These areas are calculated by solving equilibrium equations and also constraint equations, in which case the method proposed by Haung and Cuentas (1988) was used.

In Fig. 3 the mercury/water system is compared for different concentrations of chloride, the complexing agent. This diagram is calculated for the total amount of mercury which could occur during leaching of sample G (described below). The data for calculation were taken from Bailey et al. (1982), Barin (1993) and Knacke et al. (1991).

For the calculation procedure, the activity coefficients were taken into account but have only little influence on the distribution of the predominant species. The activity coefficients can be calculated after Debye/Hückel only if the ionic strength (I) is low ($I < 0.005$ mol/kg). For highly concentrated electrolytes ($I < 6$ mol/kg) the equation of Bromley (1973; Rastogi and Tassios 1980) provides good approximations.

In the pure system, without any complexing agent present, the pH has to be below 3, at a very high redox potential, to dissolve the main part of the mercury. Whereas if chloride is added to the system, mercury becomes soluble even at neutral pH. Nevertheless, a high redox potential is needed for dissolution to occur which is provided by the hypochlorite. During the electroleaching process, the solution pH should not be above 8 so as to avoid precipitation of mercury oxide (HgO), as insoluble as mercurous chloride, which will remain in the residue as such.

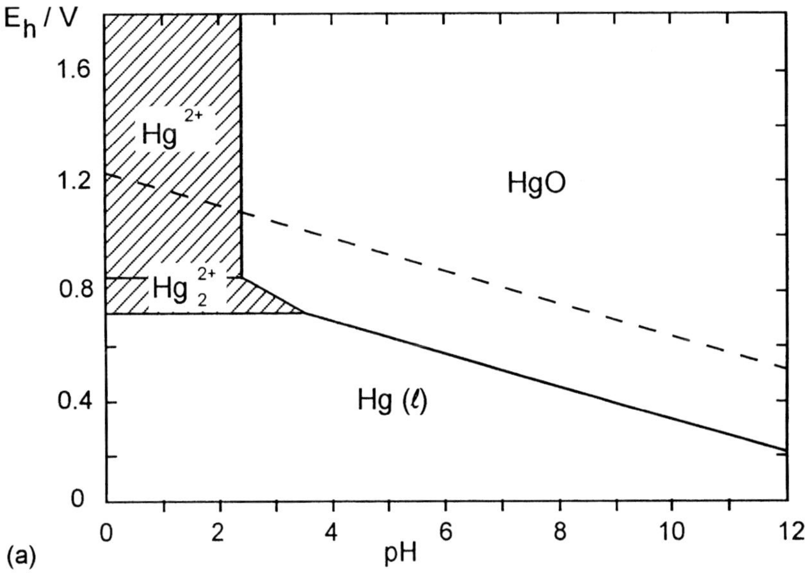

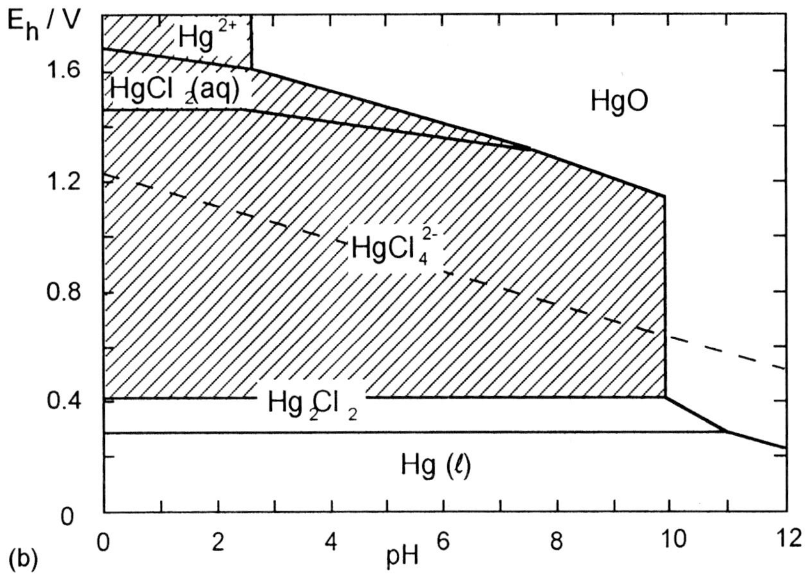

Fig. 3a,b. Potential pH diagrams comparing leaching conditions for mercury: 5 mmol/l in **a** pure aqueous solution and **b** with addition of complexing agent: 1 mol/l NaCl. *Hatching* indicates areas where soluble species predominate

There is the same effect if we calculate for gold instead of mercury, which means that traces of gold, in the case of gold-mining residues, can be extracted also:

$$2Au + 3Cl_2 \Leftrightarrow AuCl_4 + 2Cl \quad . \tag{7}$$

According to Fig. 3, the sodium chloride concentration plays a very important role in mercury dissolution for producing the most soluble species ($HgCl_{42-}$), which means that the NaCl concentration of 1.0 mol/l is more than suitable to assure the safe production of that species.

The dissolved mercury can be extracted from the solution by cathodic reduction, Eq. (6) being assumed as the predominant reaction during the deposition of mercury, from a solution containing more chloride than mercury. This half-cell reaction has a reversible electrode potential, under standard conditions, of 0.4033 V(SHE). This potential was calculated from standard Gibbs energy data (Balej 1985).

Predictions can be made also for electrolytic deposition. This process step is sensitive to chloride concentration. If the soil is highly contaminated, the concentration of dissolved mercury is also consequently high. This enhances calomel formation according to Eq. (8). The equilibrium constant is calculated after data taken from Lindsay (1979).

$$HgCl_4^2 + Hg^0 \Leftrightarrow Hg_2Cl_2 + 2Cl \;\; ; \; \log K = 5.32. \tag{8}$$

This equation describes the disproportion equilibrium between mercuric ions and elemental mercury on the one hand and calomel on the other. If too much calomel is present, it forms a non-conducting thin layer on the cathode, which can bring the process to a stop.

As electrolysis continues, gold is also dissolved and codeposited with mercury, generating an amalgam. The phase diagram of the system Au-Hg (Hansen 1958) indicates that at ambient temperature, alloys with more than 5% gold are solid and single phase.

4 Bench-Scale Experiments

4.1 Materials and Methods

The soil samples from Germany (G) were taken from an area of an inactive chlor-alkali plant, representing an average of 11 samples described in detail by Windmöller et al. (1996). The Brazilian sample (B) is a residue from small-scale gold production of a primary gold ore in Mato Grosso state. In this artisanal production the fine-grained gold particles are separated by amalgamation from a gravity concentrate produced by sluice boxes. The residual material from the amalgamation step, by using pans or rotating drums, is kept in tanks covered with water (Table 1).

Table 1. Total mercury, detected species after Windmöller et al. (1996), amount of sand (>63 μm) and loss of ignition (LOI) of samples from Germany (G) and Brazil (B)

Sample	Total Hg (mg/kg)	Species	Particle size >63 μm (%)	LOI (%)
G	220	Hg^0/Hg^{2+}	84	2.14
B	6	$Hg^0/Hg_2^{2+}/Hg^{2+}$	92	0

The mercury concentrations in all solid material probes were analyzed after ultrasonic digestion in concentrated nitric acid at 70°C by FIAS-AAS (flow-injection cold vapour atomic absorption spectrometer) by using a Perkin-Elmer FIAS 400 flow injection unit, a Perkin-Elmer 1100B spectrometer and an AS 90 autosampler.

To carry out EL-IL, the leaching agent was generated in a 500-ml electrolysis cell by using graphite electrodes (0.21 dm^2 each). Experiments were carried out galvanostatically by using two different current densities: 2.4 A/dm^2 and 4.8 A/dm^2. The pH was changed between 4 and 8, at chloride concentrations of 0.2 and 1 mol/l. Those solutions were taken for the leaching of the soil, which was done at room temperature in a closed vessel. A mass of 25 g of the soil was suspended in 250 ml of solution (S/L = 1/10). The pH was kept constant automatically.

During the leaching process, samples were taken and filtered (0.45 μm). The concentrations of HOCl and NaCl were measured immediately by titration with As_2O_3, while dissolved mercury was analyzed, after acidification, by AAS. Additionally, emissions of chlorine and mercury vapours were monitored by absorption in sodium hydroxide solution. The absorbed chlorine dismutates into hypochlorite and chloride ions, which were monitored by a chloride-sensitive electrode.

For mercury depletion, the leachate was separated from the solid by vacuum filtration and pumped back into the electrolytic cell. A potentiostat (Wenking STP 84) and a lugging capillary were used to keep the cathodic potential constant during experiments.

The EL-IS experiments were also run in a closed vessel but with two graphite electrodes dipping into the pulp. In this case, kinetic measurements were carried out by taking well-defined samples of the suspension with constant solid concentration. Both, solid (>0.45 μm) and filtrate were analyzed separately, and the amount of deposited mercury could be calculated as being the difference between the total mercury concentration in the residue, before leaching, and the sum of the remaining mercury concentration in the leaching cake and in the electrolyte (aqueous phase).

The voltammetric studies were carried out by using a glassy carbon rotating disc electrode. A conventional Pyrex cell assembly was used incorporating a reference electrode separated from the bulk solution by a Luggin capillary.

All solutions used for voltammetric studies contained NaCl (AnalaR) as supporting electrolyte and mercury as mercuric chloride ($HgCl_2$). Solutions were made up using deionized distilled water and were de-aerated with oxygen-free nitrogen before steady-state polarization curves were recorded. A nitrogen flow was maintained over the solution, during the experiments, so as to avoid the oxygen getting into the cell.

The influence of hypochlorite concentration in the current efficiency for mercury deposition was calculated evaluating cyclic voltammograms. For this purpose, the charge of the cathodic branch was measured at varying hypochlorite concentrations.

4.2 Results

4.2.1 *Voltammetric Studies*

The limiting current density measured at a potential of −0.90 V(SCE) is illustrated in the Levich plot (Fig. 4) for mercury concentration of 5×10^{-4} mol/l. From the slope of the Levich plot and using the literature value for the kinematic viscosity of 8×10^{-7} m^2 s^{-1} (Washburn 1929), the calculated diffusion coefficient is 4.3×10^{-9} m^2 s^{-1}. This result is in close agreement with the reported value in the literature (Matlosz and Newman 1986).

Cyclic voltammograms of solutions containing 0.4 mmol/l dissolved mercury, 1 mol/l chloride and 0–15 mmol/l HOCl indicate that there is no influence of HOCl on mercury deposition but on the current efficiency, which decreases exponentially as the HOCl concentration increases (Fig. 5).

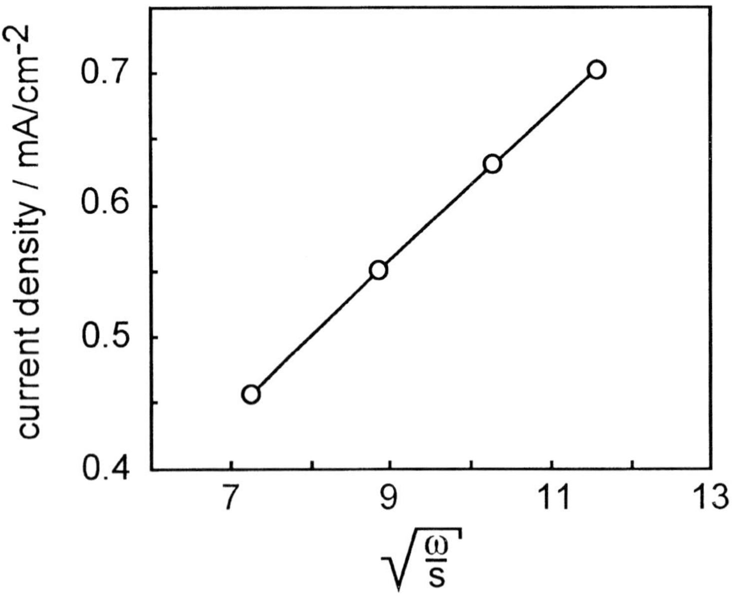

Fig. 4. Limiting current density against square root of the electrode angular velocity. (Hg^{2+}) = 5×10^{-4} mol dm^{-3}, (NaCl) = 0.83 mol/l

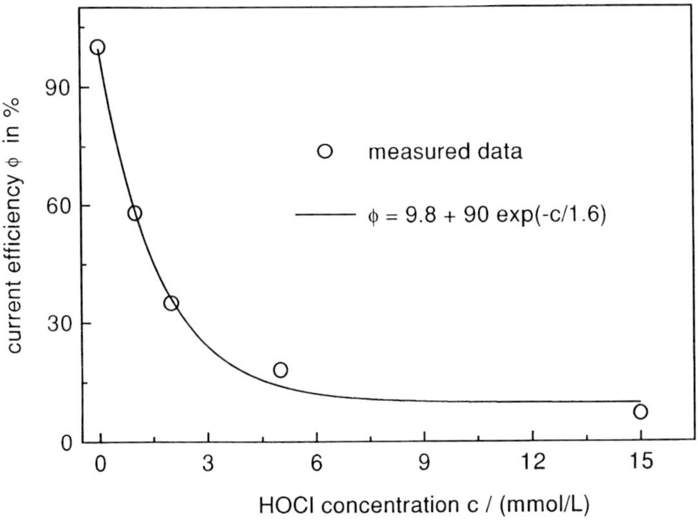

Fig. 5. Influence of varying (HOCl) on mercury deposition

4.2.2
Electroleaching with In-Line Production of Leaching Agent (EL-IL)

Agitated Leaching. The leaching agent could be produced electrolytically following a first order reaction. The yield of hypochlorous acid depends slightly on the pH with an optimum at pH = 6 and increases with increasing chloride concentration. In a 1 mol/l sodium chloride solution 40 mmol/l of hypochlorous acid were produced within 3 h.

The leaching kinetics by using a German sample (G) is shown in Fig. 6. The treatment of this soil is very fast and removes sufficient mercury. Although all mercury is dissolved after few minutes, there is evidence that hypochlorite keeps on reacting. This is due to the oxidation of some soil constituents producing non-volatile products, which is indicated by a slight LOI increase of 0.1%. This could be due to oxidation of organic matter, and possibly generating halogenated hydrocarbons (TOX). Figure 7 shows the influence of pH on the reaction rate. At pH values above 6 the reaction rate decreases. The consumption of hypochlorite is independent of pH.

The Brazilian soil sample, which has no LOI, shows the same hypochlorite effect (Fig. 8). The slight increase in residual mercury over time, from 0.7 to 1 mg/kg (about 80% removal rate), indicates that in this range of mercury concentration the readsorption of dissolved mercury has to be taken into account. Such a readsorption can be avoided by increasing the chloride concentration in solution. If the chloride concentration is decreased from 3 to 1 mol/l the final mercury concentration also decreases, and becomes more distinct, and only 70% of the mercury could be kept in solution after 1 h. This

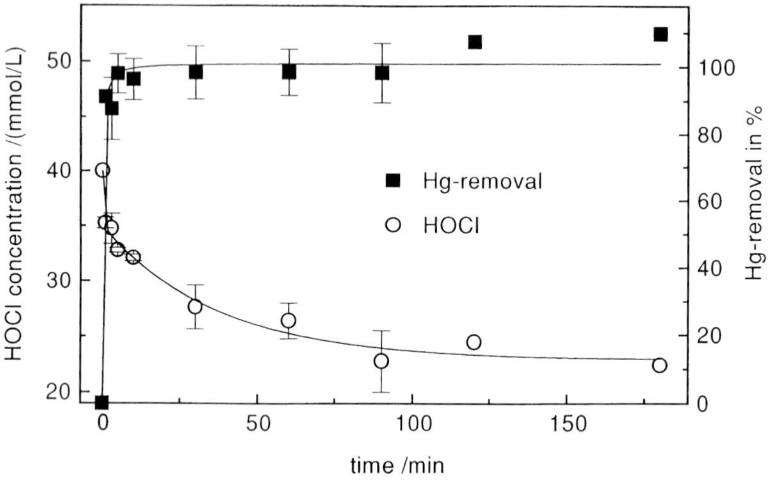

Fig. 6. Leaching kinetics for the German sample (G) during EL-IL

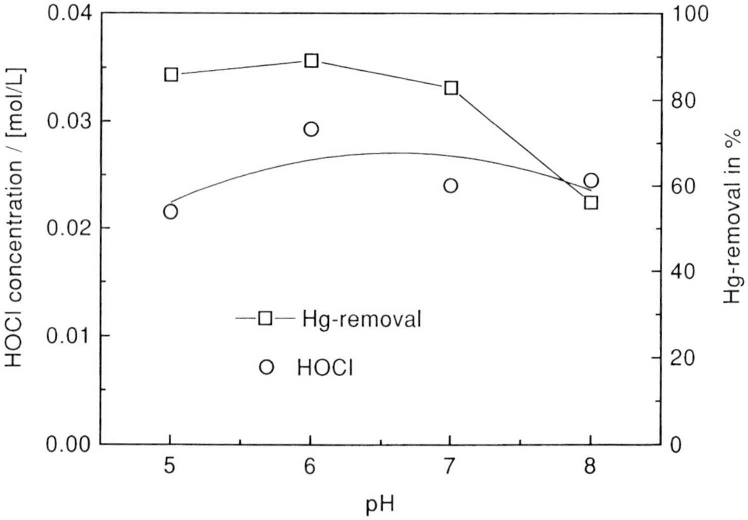

Fig. 7. Influence of pH on the reaction rate and consumption of hypochlorite

readsorption effect is assumed to be responsible for the remaining 1 mg/kg of mercury in the soil.

Electrolytic Purification of the Leachate

Results of electrodeposition of mercury, from the resulting leachates, are shown for the worse case which is a highly concentrated leachate. Figure 9 shows how the current efficiency varies with time during the potentiostatic electrodeposition

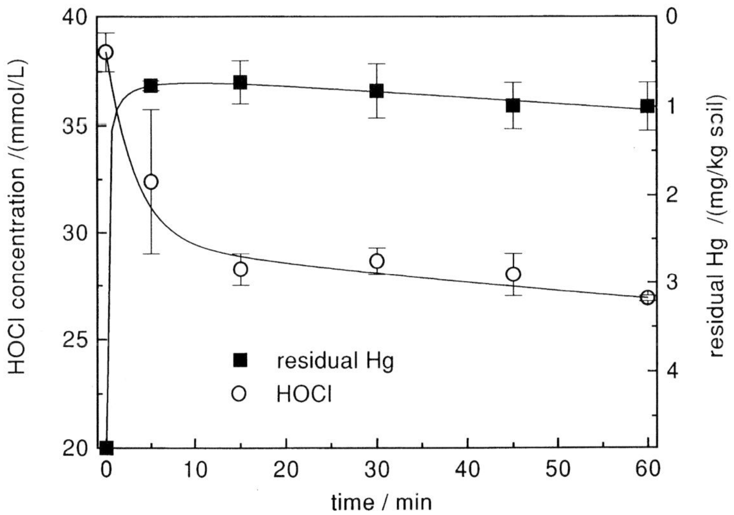

Fig. 8. Leaching kinetics for the Brazilian sample (B) during EL-IL

of mercury considering two different cathode potentials. The speed of depletion, indicated by the concentration values, is somehow faster at the higher potential, and so is the current efficiency. Actually, further cathodically increasing of the potential would stimulate the decomposition of water (hydrogen evolution) and so decreasing the current efficiency. Although the best potential for mercury depletion is −1 V (Ag/AgCl, 3 mol/l KCl), some other potentials could also be safely used. So this electrodeposition process step can easily be run galvanostatically.

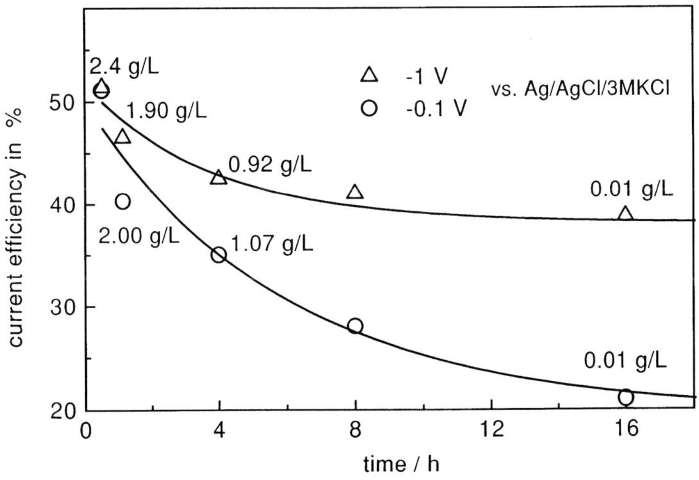

Fig. 9. Current efficiency for potentiostatic electrodeposition of mercury

4.2.3
Electroleaching with In-Suspension Production of Leaching Agent (EL-IS)

During EL-IS, both the concentration of hypochlorite and dissolved mercury stay low. This is due to consecutive reactions which are hypochlorite consumption by oxidation and cathodic deposition of the dissolved mercury. Figure 10 shows the decrease in mercury and gold concentrations in suspended Brazilian soil (sample B). For the German sample (G) 87% of the mercury could be removed. This removal rate seems to be high actually the remaining concentration of 28 mg/kg is still too high. Abrasion of deposited mercury by soil particles is assumed to cause the high equilibrium level of about 50 mg/l (Fig. 11).

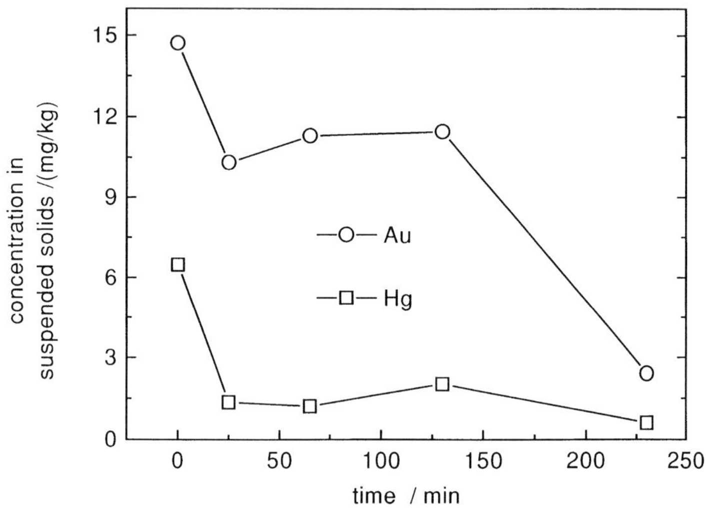

Fig. 10. Decrease in mercury and gold concentrations in suspended Brazilian soil (sample B) during EL-IS; pH = 6, I = 1A

5
Conclusions

The experimental results permit the conclusion that, by choosing suitable operational conditions, it is possible to reduce mercury concentration to low values with high extraction efficiency (>99%). For those residues with low mercury content (from the gold-prospecting areas), we recommend the use of the so-called electroleaching with in-situ production of the leaching agent (EL-IS). In this case, it was possible to reduce the mercury concentration to less than 1 ppm, which is not always low enough to be considered suitable for discharging. This indicates the necessity of extending the electrolysis time and of washing the residue after finishing the leaching process.

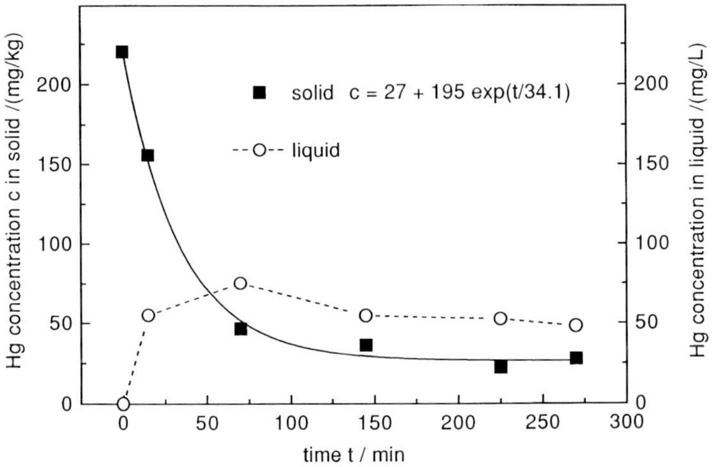

Fig. 11. Mercury concentrations in suspended German soil (sample G) during EL-IS; pH = 6, I = 1A

In the case of dealing with high mercury content residues and using the EL-IS, it is imperative to use special devices to extract the elemental mercury from the reaction system, accumulated on the cathode surface, otherwise the solid particles, in suspension, drag the elemental mercury away from the cathode surface providing the reaction between elemental mercury droplets and Hg^{2+} ions, generating the quite insoluble mercurous chloride (Hg_2Cl_2) which would remain in the residue. In fact, dealing with highly contaminated mercury-bearing residues, it is recommended, as mentioned, to use in-line production of the leaching agent in a completely different reaction system. According to the results from these leaching tests, without electrodeposition (Fig. 6), using only the leaching agent (HClO), it was realized that, despite the mercury removal rate being nearly 100%, after analyzing the residues, there were still some few ppm of mercury left. The liquid remaining in the residue, after solid/liquid separation, still contains soluble mercury species which are readsorbed during washing with pure water. Therefore, the aqueous solution used for the washing procedure should contain, as a first step, chloride ions, as a complex agent, to keep the remaining mercury as soluble as possible, as mentioned above ($HgCl_4^{2-}$). Following this previous treatment, a secondary washing step, using only water, has to be carried out in order to avoid high salinity waste being released to the environment.

According to the mercury extraction yield results from the leaching tests, varying the pH, it was realized that by performing the electroleaching process in an acidic range of pH (between 4 and 6), the mercury dissolution is enhanced due to the generation of hypochlorous acid rather than hypochlorite ions. There are residues from gold-prospecting areas that still contain, apart from elemental mercury, some small amount of gold , which in this pH range ends up by being dissolved and co-deposited with mercury, producing an amalgam. Consequently, this is definitely a suitable argument for convincing the gold prospectors to treat their residues.

In order to improve the space-time yield of the electrolysis cell, it is necessary to increase the surface area of the cathode where the deposition takes place. Therefore, several configurations of electrochemical reactors have been proposed (Robertson 1978). Technical feasibility of removing traces of mercury by using extended surface area electrodes was shown. Robertson et al. (1978) reduced the level of mercury to below the accepted concentration of 0.01 ppm in waste water by employing the Swiss-roll cell with a cadmium-coated stainless steel cathode. The use of graphite felt as an efficient porous electrode for the removal of traces of mercury ions from aqueous electrolyte solutions was investigated by Oren and Soffer (1983). Matlosz and Newman (1986) investigated a flow-through porous electrode, made of reticulated vitreous carbon, for removal of mercury from contaminated brine.

The final leaching solution should not be discharged, since it contains not only high salinity but also some remaining soluble mercury species in concentration not suitable for discharging. The recycling of this solution is advisable, using the residue treatment process.

Acknowledgements This work was supported by the bilateral Brazilian-German cooperation program, the International Bureau at GKSS and the Federal Ministry of Education, Scientific Research and Technology (BMBF).

References

Bailey SM, Churney KL, Nuttall RL (1982) The NBS tables of chemical thermodynamic properties, selected values for inorganic and C1 and C2 organic substances in SI units. J Phys Chem Ref Data 11, Suppl 2

Balej J (1985) Mercury. In: Bard AJ, Parsons R, Jordan J (eds) Standard Potentials in aqueous solutions. Marcel Dekker, New York, pp 265–285

Barin I (1993) Thermochemical data of pure substances. VCH, Weinheim

Bromley LA (1973) Thermodynamic properties of strong electrolytes in aqueous solutions. AIChE J 19: 313–320

Hansen H (1958) Constitution of binary alloys. McGraw-Hill, London

Haung HH, Cuentas L (1989) Construction of Eh-pH and other stability diagrams of uranium in a multicomponent system with a microcomputer. I. Domains of predominance diagrams. Can Metallurg Q 28. (3) 225–234

Knacke O, Kubaschewski O, Hesselmann K (1991) Thermochemical properties of inorganic substances. Springer, Berlin Heidelberg New York

Lindsay WL (1979) Chemical equilibria in soils. John Wiley, New York

Matlosz M, Newman J (1986) Experimental investigation of a porous carbon electrode for the removal of mercury from contaminated brine. J Electrochem Soc 133:1850–1859

Oren Y, Soffer A (1983) Graphite felt as an efficient porous electrode for impurity removal and recovery of metals. Electrochim Acta 28:1649–1654

Rastogi A, Tassios D (1980) Estimation of thermodynamic properties of binary aqueous electrolytic solutions in the range 25–100° C. Ind Eng Chem Process Des Dev 19:477–482

Robertson PM, Scholder B, Theis G, Ibl N (1978) Construction and properties of the Swiss-Roll electrolysis cell and its application to waste water treatment. Chem Ind (Lond), 1 July, pp 459–465

US PTO (1993) Bibliographic information file, CASSIS/BIB, US Department of Commerce, Patent and Trademark Office, Washington

Washburn EW (1929) International critical tables of numerical data, physics, chemistry and technology. McGraw-Hill, New York, vol III, pp 79 and vol V, pp 15

Windmöller CC, Wilken RD, Jardim WF (1996) Mercury speciation in contaminated soils by thermal release analysis. Water Air Soil Pollut 89:399–416

Mercury Decontamination of a Chloralkali Plant in Alexandria, Egypt

T. Meschede and R. Vogelsberger

A description is given of the decontamination of a chlor alkali plant which became contaminated with mercury during more than 30 years of production. The project comprises the clean-up of the contaminated areas, the dismantling and demolition of equipment and buildings, and the disposal of contaminated materials in a specially designed landfill. Extensive high-level health and safety measures have been applied in order to protect workers and the environment.

1 Introduction

The Misr Chemical Industry (MCI) chlor alkali facility, located in an industrial suburb west of Alexandria, Egypt, is the nation's primary source of caustic soda and chlorine, with the latter being used for the disinfection of drinking water in cities and villages throughout Egypt and the surrounding countries. The facility was originally built in the 1950's and comprised two production lines utilizing the mercury-amalgam electrolysis process. As a result of over 30 years of operation, several portions of the production and by-product units have become contaminated with mercury to various degrees.

MCI has initialized a major modernization effort comprising two major undertakings: the mercury decontamination and partial demolition of the existing electrolysis facilities, and the construction of a new electrolysis unit using mercury-free membrane technology.

The bank which finances the project specified that the environmental decontamination activities must be an integral part of the overall project, and that cleanup procedures and standards typical for Germany and the USA should be utilized where possible. In particular, exposure of workers and the surrounding neighborhood to mercury emissions during decontamination activities must be minimized. In addition, the timing of the decontamination of the existing facilities and the construction of the new unit had to be coordinated such that there was no interruption in the chlorine production.

This project represents one of the few major decontamination efforts at a mercury-amalgam facility, and is one of the most sophisticated environmental remediation projects ever undertaken in this part of the world.

Environmental Science
Mercury Contaminated Sites (ed. by R. Ebinghaus et al.)

2 Extent of Contamination

An extensive sampling campaign was conducted in August and September 1992, involving real-time air monitoring and several hundred samples of wastewater, sediments, soil, concrete, and interior and exterior building surfaces. The sampling results indicated that mercury concentrations exceeded acceptable levels (based on German and USA standards and guidelines) at numerous locations, with particular hot-spots in and around the two cell houses, related workshops, and at a mercury distillation/recovery unit.

Peak mercury vapor concentrations were found above 800 $\mu g/m^3$ in the interiors of the two electrolysis cell houses, compared to the maximum allowable work-place concentration of 50 $\mu g/m^3$ in the USA. Mercury concentrations up to 7.5 mg/l were found in the wastewater trenches, compared to the standard of 0.05 mg/l mercury under the European Community Regulations (EWG-82a) for Waste Water Discharges from Chlor Alkali Plants. Contaminated soil and dirt were found in many areas of the plant, with particularly high concentrations near the distillation unit, up to over 200 000 mg/kg mercury. Peak levels in concrete surfaces of 120 000 mg/kg were measured. Core borings indicate that mercury contamination extends several centimeters into the walls, floors, and ceilings of the cell houses.

Typically, mercury concentrations in soil or concrete above 10 mg/kg are considered to be contaminated.

3 Selection of Feasible Remediation Technologies

An extensive feasibility assessment was performed of various remediation processes and technologies that have been applied in Europe and North America for mercury-contaminated sites, including mechanical/hydraulic soil washing, chemical extraction, thermal treatment, landfill deposition, and others. A comparison was made of the advantages and disadvantages of each technology with respect to performance, reliability, risks to humans and the environment, costs, and other factors, with specific consideration of the particular site conditions at the MCI facility.

The selected decontamination method comprises a combination of techniques including:

- high-pressure water cleaning of all surfaces,
- in-situ chemical fixation for lightly contaminated surfaces,
- controlled removal/dismantling of highly-contaminated structures (e.g. the two cell houses)
- deposition of all solid wastes in a secure landfill.

The review of available and feasible decontamination technologies revealed that, regardless of which technology was chosen, the target value of 10 mg/kg Hg in the treated materials can barely be reached. At the time of writing, no large scale-treatment plants which could cope with the detected high mercury levels were in operation or available at all. The treatment would cause very high costs, and an additional landfill would be needed for the deposition of treated or untreated materials with mercury concentrations above the target value. Based on the findings, it was decided to refrain from the treatment of the wastes and, instead – as there are no existing hazardous waste landfills in the country – to utilize the available budget for the construction of a suitable state-of-the-art and secure landfill for the long-term deposition of all solid wastes resulting from this decontamination project. In this case "long-term" means at least 100 years, by which time new technologies may be available to completely detoxify the residues.

An appropriate landfill site was selected in an uninhabited desert area approximately 80 km southwest of Alexandria. The siting process, landfill design, and the entire MCI decontamination strategy was reviewed and approved by the Egyptian environmental and other involved authorities in Cairo.

4 Decontamination Concept

The overall strategy was to conduct the decontamination and demolition activities in such a manner as to minimize the exposure of workers and the environment to contaminated dust and mercury vapors. To accomplish this, the use of manual labor for demolition was minimized, and a series of dust suppression measures were implemented, such as the use of water sprays and vacuums at points of potential dust emission.

A comprehensive work plan was prepared for all of the facility decontamination activities. A detailed inventory was made of all facility surface areas, structures, and process units, with a description of the degree of contamination and the recommended decontamination method for each item.

The decontamination concept for the MCI facility includes three general phases of activity. Phase A involved facility cleanup, site preparation, and the construction of treatment and storage facilities. Several facilities were constructed on site, including a cleaning pad for high-pressure washing of contaminated items, a laboratory for analysis of solid and aqueous samples, a waste-water treatment plant for demercurization of cleaning wastewater (from general site cleanup and decontamination activities), a vehicle-wash for trucks and cars leaving the site, and a personnel decontamination unit for the work crews. A general cleanup of the entire plant was performed to remove debris, cleanup the dust and dirt, and then pressure-clean the concrete surfaces, followed by the initial pressure-cleaning of all mercury-contaminated process units and buildings. An interim storage facility was installed on site to store contaminated materials until the off-site landfill was completed.

Phase B consisted of the construction of the new electrolysis membrane plant, of the shutdown of the old electrolysis units and the rehabilitation of those portions of the facility that were reused in the new membrane plant.

Phase C involved the decontamination and, where appropriate, the demolition of contaminated equipment, the electrolysis units, and related structures. The two existing cell houses were dismantled and demolished, and all contaminated soil and concrete surfaces from the plant (above 10 mg Hg/kg) were excavated and replaced with clean materials. Metals and equipment with no or low contamination were reused or scrapped. The contaminated waste materials are placed in the off-site landfill. An extensive monitoring program for the classification of dismantled and demolished materials, the quality control, and the final verification of site cleanup had been implemented.

A particular feature of the decontamination project was the use of cutters and cranes to dismantle the heavily contaminated cell houses in a piece-by-piece manner, along with water sprays for dust suppresion. This technique was selected over conventional methods of demolition via explosives or wrecking ball because of the reduced potential for the spread of mercury contaminated dust and fumes. The excess spray water was collected and – together with the cleaning water from the cleaning pad and vehicle wash – treated before discharge.

All loose soil and rubble was solidified in concrete blocs to minimize mercury losses during storage, transport, and filling into the landfill. The handling of the concreted materials was easier than that of loose rubble. For example, conventional cranes and trucks could be used for transportation and packing. The measurement and quantity control of processed contaminated rubble and soil could be done by simply counting the blocs.

Also, the release of mercury vapors from contaminated materials could be reduced considerably by concreting. However, during the hot summer season it was found that mercury vapor levels in the close vicinity of the concreted blocs were still too high in comparision with maximum allowable work-place concentrations. Therefore, work was discontinued during daytime hours and the transportation and landfilling of the concrete blocs is executed only during the cooler night shifts.

5
Health and Safety

Throughout the project, extensive measures were foreseen to protect the health and safety of the workers, especially with respect to potential exposures to mercury. Personal protective equipment was available and routine monitoring of working conditions is conducted, as well as periodic medical examinations of the work crews. In addition, all of the personnel received an extensive introductory training on health and safety and frequent updates and refresher courses. The specific procedures for avoiding and minimizing potential mercury hazards were detailed in a comprehensive Health and Safety Plan developed for this project.

However, the best safety of workers was provided by the extensive use of demolition machines instead of manual work.

The main features of the health and safety plan had been developed based on the worldwide remediation experience of Dames & Moore and relevant European and US regulations and technical standards, including German regulations (Gefahrstoffverordnung; guidelines of the Employment Accident Insurance Fund of the Industry) and US Occupational Safety and Health Administration standards (29 Code of Federal Regulations 1910 and 1926).

While focusing on mercury hazards, the health and safety plan also addressed hazards due to general activities at demolition and construction sites (electric shock, burns, falls, motor vehicle accidents, asbestos fibers, entry into confined spaces, etc.) and specific hazards related to past and ongoing chemical production at the MCI site, such as inhalation of chlorine gas and physical contact with strong acids and caustics.

6 Description of Landfill

In the design of the landfill, special consideration was given to the use of construction methods and materials which are familiar to local contractors and which are suitable to the harsh climatic conditions. Therefore, asphaltic concrete, which is a common construction material in Egypt and has excellent long-term physical qualities, was selected as the liner material. Synthetic liners were rejected because of questionable long-term integrity, operational difficulties in

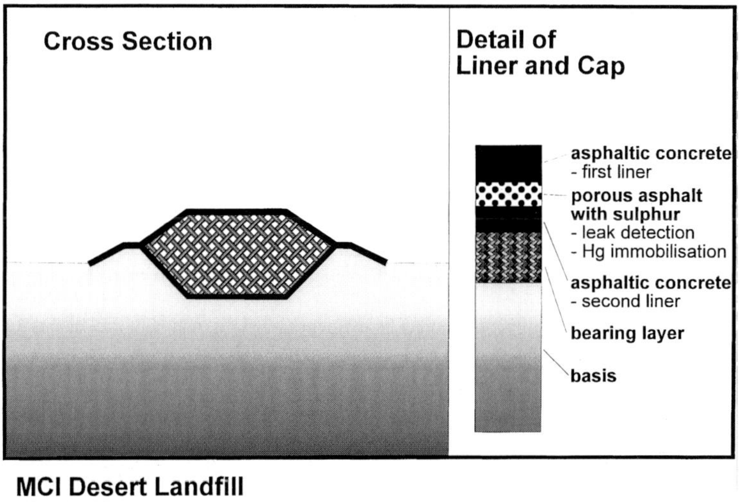

Fig. 1. Schenatic view of landfill

the desert environment, and the expense associated with installation and repairs by a specialized expatriate crew.

A schematic view of the landfill is shown in Fig. 1. The landfill contains an impermeable base and cover comprised of two layers of asphalt concrete, a vapor-monitoring system for leak detection, and a layer of sulfhur between the liners to act as a mercury immobilization barrier. In addition, both cap and base are designed to allow for the injection of sealants to provide additional containment should leaks be detected in the future. The total volume of contaminated material which had to be disposed of in the landfill is 15 000 m^3. Due to the placement of concrete blocs, only few voids were left, which were filled with sulfhur and sand. Since the disposed material is cast into concrete, no settlement can occur which could cause ruptures in the landfill cover.

The groundwater is located about 60 m below the ground surface, which provides a high level of protection in the unlikely case of mercury leakages. Monitoring of surrounding soil and vegetation was performed before and during the expected 2 years of landfill operations to determine any off-site contamination.

7 References

Burkhard G, Egloffstein T (eds) (1995) Asphaltdichtungen im Deponiebau. Expert Verlag, Renningen-Malmsheim

Decree of the Prime Minister No. 338/95 promulgating the executive statutes of the law on environment as enacted by law No. 4/1994. In: The executive regulations of the law on environment, Serial No 25, The Middle East Library for Economic Services, Cairo, Egypt

DVWK (1996) Deponieabdichtungen in Asphaltbauweise, Sirtschafts-und verlagsgesellschaft Gas und wasser, Bonn, Heft 237

EEAA (1995) Guidelines for full environmental impact assessment of industrial establishments (DRAFT-undated, handed over to Dames & Moore by Dr. El Zarka, EEAA, CAIRD on 14 October 1995; Unpublished)

Fleischer, Schramm, Brückner (1991) Die Quecksilberverseuchung in Bayern, Interpellation, Ausschuß für Landesentwicklung und Umweltfragen, Bayerischer Landtag, 12. Wahlperiode, Drucksache 12/1608, München

Grubitzsch J (1994) BUNA: Modellprojekt zur Sanierung einer Queckslilber-Altlast. TerraTech Jahrg 3, Nr 1, pp 38–39

Law No. 4/1994 promulgation the law concerning environment. In: Set of laws and decrees on environment and pollution. The Middle East Library for Economic Services, Cairo, Egypt

Lobenhofer M (1995) Quecksilberschaden durch Bodenaustausch saniert. TerraTech, Jahrg 4, Nr 5, S 48

Petts J, Eduljee G (1994) Environmental impact assessment for waste treatment and disposal facilities Wiely & Sons, Chichester

Renner J (1991) Großtechnische Anwendung eines kombinierten chemisch-physikalischen und thermischen Verfahrens zur Behandlung kontaminierter Böden. In: Franzius V (ed) Sanierung kontaminierter Standorte 1991. Erich Schmidt Verlag, Berlin

Renner J (1993) Sanierung einer quecksilberhaltigen Altlast. TerraTech, Jahrg 2, Nr 1, S 56–58

Schuldt M (1990) Prüfwerte für Bodenverunreinigungen. In: Franzius V (ed) Sanierung kontaminierter Standorte 1990. Erich Schmidt Verlag, Berlin

v Schellendorff GB, Geyti A (1991) Methoden der Sanierung quecksilberkontaminierter Böden.In: Franzius V, Wolf K, Brahdt E (eds) Handbuch der Altlastensanierung, Müller Verlag, Heidelverg
World Bank (1991) Technical paper number 139 – environmental assessment sourcebook: vol I Policies, procedures, and cross-sectoral issues. World Bank, Washington, DC

II.e Mercury Contamination in Aquatic Systems

Mercury Subsurface Maxima in Sediments: a Diagnostic for Anthropogenic Origins

G.E. Glass, J.A. Sorensen, G.R. Rapp Jr., M. Balcer, and L. Schwarzkopf

Multimedia environmental surveys and monitoring studies have shown the major sources of mercury contamination in the Lake Superior watershed to be from former industrial and municipal direct discharges and present airborne sources. Walleye from the St. Louis River, the largest US tributary to Lake Superior, and siscowet trout from the lake continue to show elevated mercury tissue levels, limiting safe human and wildlife consumption. The main routes of mercury bioaccumulation in fish appear to be from airborne deposition and recent contaminated sediment sources and through the benthic and planktonic food chains. No evidence for significant background from geologic sources is indicated in upstream river sediments or from Lake Superior sediment core samples dating from about 2000 to 8000 years. B. P.

The identification of possible mercury contamination sources from individual reservoirs and river reaches focused on measurements of mercury concentrations in biota. The indigenous biota are an integrator of mercury exposure, with the residue magnitudes as a function of location used to narrow the range of possible contamination sources or source areas. Mercury concentrations in fish populations showed a pattern consistent with localized exposure when upstream samples were compared with those confined within a downstream series of reservoirs. The linkage between mercury in fish and the food chain was also confirmed by increased mercury levels measured in plankton and benthos. Sources of mercury contamination to the foodchain are indicated by the significant correlation of mercury concentrations between surface sediments and mayfly larvae.

Documentation of sediment contamination was accomplished by extensive core sampling of river and reservoir sites. Eighty sediment core samples from the six reservoirs on the lower St. Louis River show some sites with mercury concentration maxima of 2 to 4 μg/g at 1 to 1.5 m sediment depth in deep water, while other sites exhibit mercury concentration maxima of 0.8 to 2.0 μg/g at various sediment depths from the surface to 1 m. The greatest surface (0–20 cm) sediment mercury concentrations occurred in shallow water and wetland areas of a reservoir where water levels fluctuate daily to produce hydropower for peak demands.

1 Introduction

The continuing uptake of mercury into the aquatic food chain has contaminated fish in rivers and lakes of the Upper Midwest and elsewhere in North America. Fish Consumption Advisories issued by the Minnesota Dept of Health (MDH 1996) establish meal limits based mostly on this contaminant. Some studies indicate that environmental mercury levels are increasing. A 3 to 5% increase per year for fish mercury levels was noted by Swain and Helwig (1989) when they compared data from two time periods for a small number of Minnesota lakes. An

Environmental Science
Mercury Contaminated Sites (ed. by R. Ebinghaus et al.)

increase has also been documented in atmospheric mercury concentrations (Slemr and Langer 1992). This is consistent with elevated mercury levels observed in soil (Nater and Grigal 1992) and sediment (Swain et al. 1992), which may be explained by recent atmospheric mercury deposition (Sorensen et al. 1994, 1990). Mercury deposition from precipitation measured in the past 3 years was significantly higher than for the previous period 1990–1992 (Glass et al. 1996). However, some studies of mercury in wetlands show maximum values in mercury corresponding to accumulations during the period 1935–1980 (Benoit et al. 1994).

Our assessment studies have contributed to quantifying the extent of mercury contamination throughout the St. Louis River, the largest US tributary to Lake Superior, and its watershed, and have identified some contaminated hot-spots (Glass et al. 1990, 1992).

A survey of mercury in water, plankton, sediments (Glass et al. 1990), and plankton and fish (Sorensen et al. 1992) along the lower St. Louis River (located between the cities of Cloquet and Duluth; Fig. 1), demonstrated the likelihood that sediments in the lower reservoirs may influence mercury residues. In the 1992 study it was found that mercury concentrations in standard-sized walleye appear greater near the lower reservoirs than they do upstream or in the lower estuary (Fig. 2).

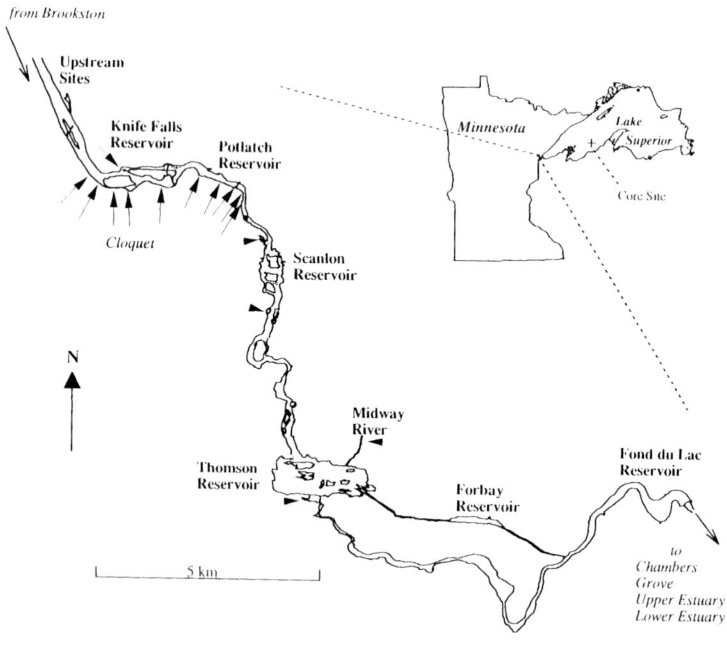

Fig. 1. The lower portion of the Saint Louis River, between Cloquet and Duluth, Minnesota, showing the six study reservoirs. *Bold arrows* show approximate points of former discharge of industrial and municipal wastes (WLSSD 1973) along the river before a regional collection system and comprehensive industrial and municipal waste treatment facility began operations in Duluth, Minnesota, in 1978

The results of a preliminary 1992 survey of sediment mercury concentrations as a function of depth in three cores taken from the Thomson, Forbay, and Fond du Lac Reservoirs show significant maxima in mercury levels (1 to 2 μg/g Hg) in strata corresponding to major industrial discharges from wood-fiber processing and paper-making (in Cloquet, Minnesota) between the 1940s and early 1960s, see Fig. 3 (MPCA 1996). Other historical uses of mercury in the watershed may have also contributed to the accumulated levels of mercury in the river sediments (Table 1).

Although significant mercury concentration maxima in the three cores were found relatively deep in the sediment strata (1–2, 0.5–0.75, and 2–3 m for Thomson, Forbay, and Fond du Lac reservoirs, respectively), they were all taken in the deeper water portions of the reservoirs. Other areas of these reservoirs, such as areas of higher currents or shallower water, may not have sediments with the same characteristics. That is, the highly contaminated portion of the sediment may be much nearer to the sediment-water interface as a result of varying sedimentation rates and sediment redistribution during periods of high flow, and thus may make an important contribution to mercury bioaccumulation and fishery contamination.

The goal of this study was to identify mercury-contaminated sediment areas in the lower St. Louis River that may be contributing to the existing fish and food chain contamination. The contaminated sediment areas with potentially high rates of mercury bioaccumulation in the fish food chain may become a priority for developing mitigation efforts.

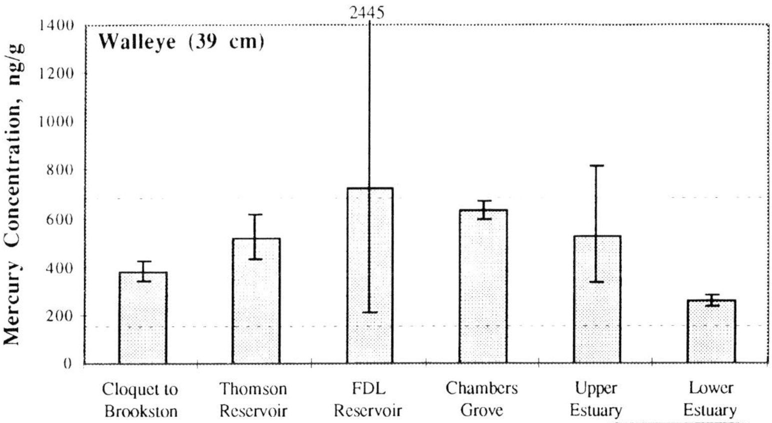

Fig. 2. Fish mercury concentrations (calculated for standard sized 39-cm walleye) in muscle tissue caught at lower reservoirs and reaches on the lower St. Louis River and estuary of Lake Superior (Sorensen et al. 1992). *Error bars* are 95% confidence intervals. *Dotted lines* indicate health consumption advisories: one fish meal per week (*lower*) and one fish meal per month (*upper*). (MDH 1996)

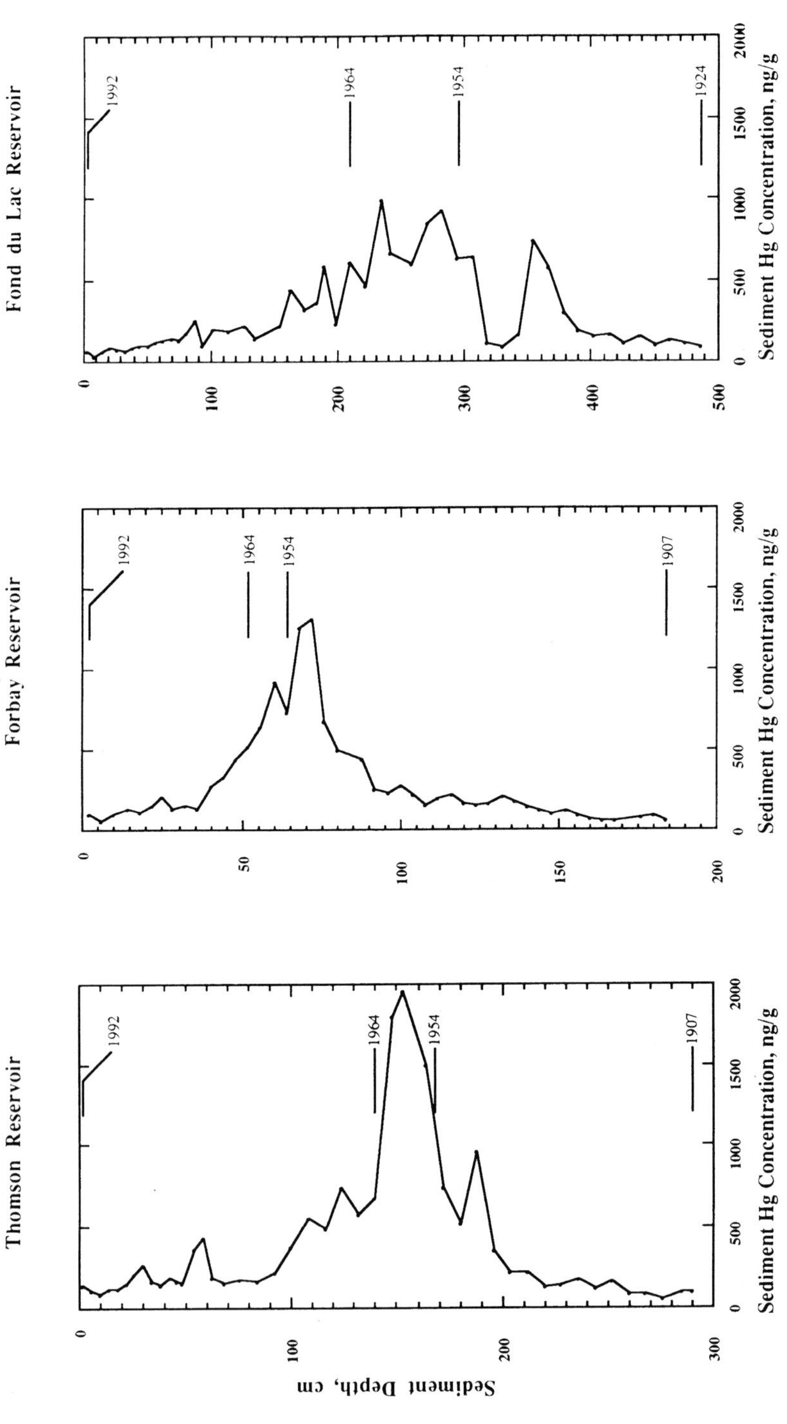
Thomson Reservoir
Forbay Reservoir
Fond du Lac Reservoir
1992
1964
1954
1907
1924
Sediment Hg Concentration, ng/g
Sediment Depth, cm
0
500
1000
1500
2000
100
200
300
400
50
150

2 Methods and Measurements

2.1 Study Area

The study area consisted of the lower six reservoirs (Knife Falls; Potlatch; Scanlon; each 30–50 acres; Thomson, 339 acres; Forbay, 36 acres; and Fond du Lac, 217 acres) of the St. Louis River, shown in Fig. 1. Specific sampling sites within each reservoir were chosen to represent a variety of sediment depositional zones and morphology. A Lake Superior sediment core was also analyzed to confirm background sediment mercury levels.

2.2 Sampling Methods

2.2.1 *Sediment Core Collection*

Sediment core samples were collected in the summers of 1995–1996 using a 5-cm diameter by 1-m-long modified Livingstone piston corer (Wright 1967) operated from the center of an anchored pontoon using methods described in Faegri and Iverson (1965). Eighty cores (penetration $\leq$1 m) of reservoir and river sediments were taken to provide more complete coverage of the reservoir basin. Replicated cores were taken at selected sites within 5 m of each other in an effort to estimate sediment mercury variability over short distances.

The Lake Superior sediment cores were obtained in 1980 using box and gravity coring deep-water techniques (Johnson and Fields 1984; T. Johnson pers. comm. 1997).

2.2.2 *Sediment Core Sectioning*

Immediately after obtaining a St. Louis River core sample, the open end was temporarily covered while the outside of the core tube was rinsed with ambient water to remove clinging sediment. The cover was then removed and a ziploc bag (labeled with site and sediment strata depth information) was placed against the tube end while the core was extruded. After extruding 10 cm of core (distances marked on the plunger), the sediment was sliced cleanly from the end of the core

Fig. 3. Sediment mercury concentrations in the three lower reservoirs of the Saint Louis River. Strata dating at 1954–64 were determined by Cs 137 measurements by D.J. Steck, Schaefer Environ. Radiation Lab., St. John's University, Collegeville, Minnesota, pers. comm. (Glass et al. 1996; MPCA 1996) tube with a stainless steel knife. This process was repeated for the entire length of the core sample. Notable characteristics regarding odor, color, texture, and layering of the sediments were noted and recorded. Subsamples were transported by cooler and kept at 4 °C until analyzed.

Water depths at each sampling site were determined by lowering a weight attached to a cord with calibrated length gradations. Values were recorded to the nearest centimeter.

The Lake Superior sediment core was kept in a plastic core tube at room temperature, split laterally, and subsampled (just prior to analysis) at 1-cm widths every 5 cm using a stainless steel spatula. Portions for analysis were kept refrigerated in individual ziploc plastic bags until analyzed.

The width of each subsample determines the resolution of the resulting measurement. Too great a width of subsamples could obscure a possible elevated mercury concentration layer by dilution with sediments from nonelevated years. Too narrow a subsample (and limited analyses resources) would leave too much of the core underrepresented by analyses. We chose a 1-cm subsampling width because it offered sufficient resolution to resolve possible high deposition episodes while giving reasonable coverage of the entire sediment core.

2.2.3 *Sediment Dating*

During the preliminary 1992 sediment survey of the three lower reservoirs, replicate cores were taken for dating analyses. Those cores were sectioned in the

Table 1. Usages and practices involving mercury in the St. Louis River watershed region of western L. Superior: prospective sediment mercury contributors

Known mercury sources (usage/mechanism)	Lake Superior environment affected
Hide/fur processing (tanning and treatment)	Air and estuary
Lighthouses (lens weight bearing)	Shores and islands; lake sediments
Gold prospecting (ore assay and extraction)	Air- and upper watershed
Explosives and primers (ore extraction, ammunition, and fireworks)	Air- and upper watershed, estuary
Sintering ore [iron pellet (and cement) production]	Air- and watershed, estuary
Iron/steel mills (iron content analysis, switches)	Estuary
Pulp/paper mills (slimicides in fiber production, switches)	Air- and watershed, estuary
Ships [load leveling (clinometer), ballast, and trim lights]	Sites of wrecks and salvage
Farming and silviculture (insecticides and seed treatments)	Air and runoff
Reservoir formation/operation (mobilization and transport)	Reservoirs and runoff
Medical and dental usage and treatments [blood pressure gages, bactericides (Mercurochrome), amalgams]	Air, sewage, and landfills
Laboratory tests (preservatives, reagents, toxicity tests, and pressure and volume measuring instruments)	Air, sewage, and landfills
Stream flow measurements (pressure gauges)	Various rivers and streams
Household paints and electrical (slimicides, switches, thermostats, batteries)	Air, sewage, and landfills

field into 4–8-cm increments and subsamples were sent to Daniel J. Steck, Schaefer Environmental Radiation Lab., St. John's University, Collegeville, Minnesota, for counting Cs-137 and other radioisotopes. Sediment strata measuring the highest and second highest counts were assigned 1954 and 1964 dates, respectively, based on nuclear weapons testing activities (MPCA 1996). To the bottom of the core was assigned the date of reservoir construction, and to the top, the year of sampling.

The Lake Superior core sediments were dated using paleomagnetic (Johnson and Fields 1984) and carbon-13 (Huber 1993) methodology.

2.2.4
Benthos and Associated Surface Sediments

A benthic population survey was conducted in the spring of 1995 (May 31 to June 6) near the sediment study sites. Benthic organisms were qualitatively sampled using Ekman dredges and processed on site by screening and washing using a standard bottom sieve bucket (Wildco, Wildlife Supply Co., Saginaw, Michigan). Two types of 15 × 15 × 15 cm Ekman dredges were used. For water depths less than 1.5 m, a handle-mounted stainless steel Ekman dredge was used; for deeper water, a brass Ekman dredge on a nylon cord was employed. Both were operated such that the top 10 cm (approx.) of sediment was sampled. Between 5 and 22 dredge samples were collected per site depending on the abundance of benthic biomass. A small portion of each dredged sediment sample was subsampled near the middle of the dredge (using a stainless steel spoon), from the top 0–4 cm, and transferred to a labeled ziploc bag for storage at 4 °C until analyzed for mercury content. The remaining sediment sample was sieved for organisms on site using the general procedures used for benthos screening and separation by taxa as found in USEPA (1990).

Benthic samples were placed in labeled ziploc bags and kept on ice until reaching the lab, where they were kept at 4 °C until identification, final screening, and washing were completed no later than 2 days after sampling. Processing included depuration of live organisms in water for 24 h to remove stomach contents. For each sampling site, organisms were separated by taxa, weighed, and stored frozen in plastic weighing boats until analyzed. The primary taxa of interest for mercury content in this study was *Hexagenia*. Selected taxa from samples with sufficient biomass were analyzed for mercury to determine uptake for that trophic level.

2.2.5
Contamination Prevention Measures

All reservoir sampling was done using clean sampling techniques. This involved the use of clean, mercury-free latex gloves by personnel handling core and benthos samples, mercury-free sample containers protected using ziploc storage bags, and the precleaning of all sampling equipment. Both the Ekman dredge and the piston corer were rinsed with on-site water several times before and after each sample was taken. Clean techniques were also used for subsampling and analysis of the Lake Superior core.

All other materials such as gloves, larger ziploc bags, measuring utensils, and glassware were tested for mercury contamination by soaking for 1 to 4 days in a solution made of a dichromate/nitric acid preservative and distilled water. Mercury concentrations in the soak solutions were found to be below the total mercury detection limit (<2 ng/l) in water.

2.3 Analyses Methods

2.3.1 *Mercury in Sediment*

Analyses for mercury concentrations in sediment were conducted using cold vapor atomic absorption spectrometry (CVAA) according to USEPA (1991) method 245.5 with minor modifications (Glass et al. 1990). Samples were analyzed wet and in some cases as wet and dried pairs. All results are reported on a dry weight basis. Instrument calibration was accomplished using three laboratory working standards (60, 180, and 540 ng Hg/l) and two reagent blanks at the beginning and end of each analytical run.

Sediment samples were thoroughly mixed by kneading the ziploc bag for up to 5 min followed by stirring the contents with a spatula before removing an aliquot from the bag for analysis. Aliquots were placed directly into analysis flasks and weighed. In addition, a separate aliquot was transferred to a crucible for percent moisture determination.

2.3.2 *Mercury in Benthos*

Analyses for mercury concentrations in benthos were conducted using USEPA (1991) method 245.6 with minor modifications (Glass et al. 1990). Between 7 and 75 mg of dried (105 °C overnight) benthos tissue was used per analysis. Instrument calibration was performed at the beginning and end of each analytical run using three laboratory working standards (15, 30, and 60 ng Hg/l) and two reagent blanks. Analytical precision was approximated by analyzing similar organisms (species and size) from the same sampling site. All results are reported on a dry weight basis.

2.3.3
Statistical Analysis

Statistical analysis were performed using Systat and Excell software and Apple Macintosh computers.

2.4
Quality Assurance

2.4.1
Mercury Measurements

Analytical precision was determined by analyzing every tenth sample in duplicate. At least 1 out of every 15 samples was also analyzed for spike recovery. The standard addition was added as mercuric chloride in a dichromate/nitric acid preservative matrix. Analytical method blanks (four per run) were prepared by adding all analytical reagents to a flask that contained no sample. Two certified sediment (NBS 1645) or tissue samples (NRC DORM-1) were included in each sediment and benthos analyses run, respectively.

Acceptance criteria for quality assurance checks were ± 20% for accuracy and ± 30% for precision. Samples which failed to meet acceptance criteria for an analytical run were reanalyzed.

Results of quality assurance measurements and statistics are summarized in Table 2. Because of the variability in sediment compaction, strata comparisons between replicate cores can become slightly misaligned. Misaligned strata and sharply varying mercury concentrations with strata depth can account for most of the mercury concentration variability observed for comparisons of replicate core samples.

Table 2. Statistical summary of various quality assurance measurements[a]

Parameter	Relative percent deviation[b]		Certified reference recov.[c] (%)	Standard addition recovery (%)	
	Lab	Field		of samples	of certified std.[c]
Hg in sediment[d]	16.9 (152)	42 (67)†	103 ± 1 (85)	109 ± 9 (110)	105 ± 2 (42)
	8.9 (128)		102 ± 1 (81)	100 ± 1 (81)	104 ± 2 (40)
Hg in benthos		26 (28)	99 ± 2 (23)	107 ± 6 (8)	98 ± 4 (12)
			97 ± 2 (22)	102 ± 7 (6)	98 ± 4 (12)

[a] Values in parentheses are the number of sample or sample pair analyses for entry. For double row entries, the first refers to results for all possible cases while the second refers to cases passing acceptance criteria.

[b] Sample pairs are generated by duplicate aliquots (Lab column) or replicate samples (Field column). Calculated as 100% × (/value 1/-value 2/)/(value1 + value2)/2 averaged over all data pairs for each parameter. †Entry is 31 (8) if only surface strata are considered.

[c] Based on standard reference materials: sediment – NBS #1645; tissue – NRC #DORM-1.

[d] Regression slope for analyzed dry vs analyzed wet (both as dry weight) was 0.94 ± 0.01 (n = 55).

3 Results and Discussion

3.1 Mercury in Surface Sediments (0–10 and 10–20 cm)

The surface-sediment mercury concentration values for the 0-10 cm and 10-20 cm strata range from 15 and 28 ng Hg/g at the upstream site to 740 and 820 ng Hg/g at the Scanlon Reservoir (core SC-2) and to 2100 and 1800 ng Hg/g at the Thomson Reservoir (cores TH-36, -12), respectively. The average values across sites within each reservoir are plotted in Fig. 4 and show that the levels increase from the surface to depth in every case. The average mercury concentration ratio (0-10 cm to 10–20 cm) across reservoirs is 0.64 (one average per reservoir); this ratio is different from 1.0 at >0.99 confidence. This is in contrast with that observed in Minnesota lakes, where sedimentation rates are much lower and mercury levels typically decrease with strata depth (Swain et al. 1992). Because the surface sediment strata represent the bioturbation layer, the layer in which the benthos inhabit, then the degree of mercury exposure from sediments would appear higher at the Thomson, Scanlon, and Forbay Reservoirs as compared with the other reservoirs.

3.2 Mercury Bioaccumulation Levels in Benthos

Mercury concentrations in *Hexagenia* ranged from 36 to 168 ng Hg/g dry weight and showed no significant relationship with animal size (weight or length). Box plots of mercury concentrations for each site are shown in Fig. 5.

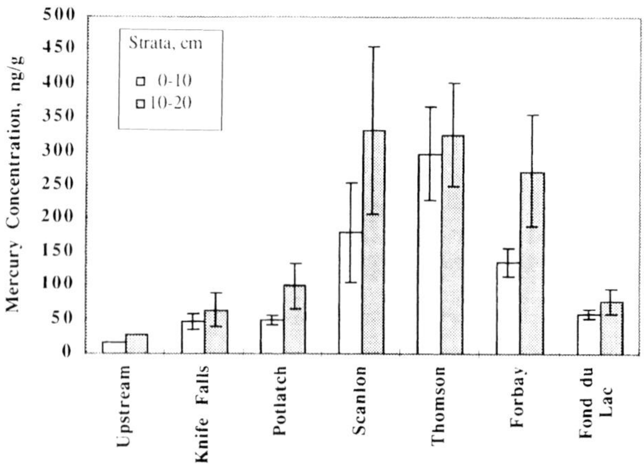

Fig. 4. Average mercury concentrations in surface sediments of the Saint Louis River upstream, and from the six lower reservoirs (Fig. 1). *Bars* indicate standard errors. The upstream area has one sampling site and no standard error was calculated

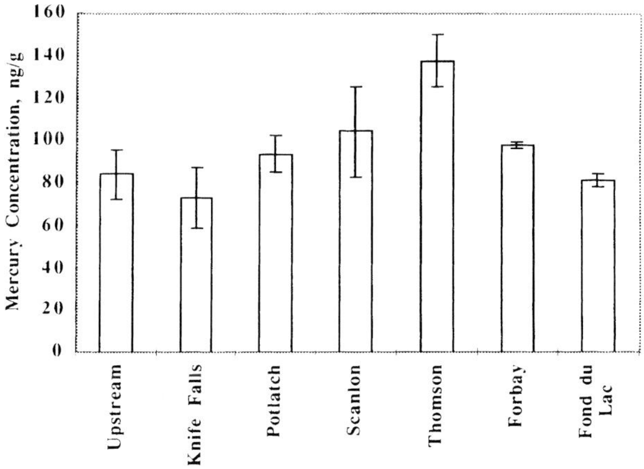

Fig. 5. Mercury concentrations in *Hexagenia*, dry weight basis, first averaged within each site then averaged across sites for each of the lower reservoirs on the Saint Louis River. *Bars* indicate standard errors

A statistically significant correlation (0.94, $p = 0.002$) between mercury in *Hexagenia* and surface sediment emerges if the average values of mercury in *Hexagenia* and surface sediments (0-10 cm, core samples) are compared across all reservoirs. A positive correlation was also observed for a similar comparison of mercury in *Hexagenia* and associated Ekman grab sediment, though not statistically significant. These results suggest that contaminant levels in *Hexagenia* are more closely related to average sediment conditions over a large-scale (reservoir) than to the site-specific environment from which they were collected.

Both numbers of organisms and biomass for the lower reservoirs (Thomson, Forbay, Fond du Lac) are significantly less than those for the upstream sites and upper reservoirs with the exception of numbers of organisms for the Forbay Reservoir (M. Balcer, pers. comm.).

3.3
Sediment Mercury Background Concentration Levels

Sediment mercury concentrations are plotted for each 10-cm section of the top 1 m of core samples obtained from the upstream river site and the six lower reservoirs, see Fig. 6. The upstream site showed an average sediment concentration of 16.9 ± 7.8 (SD) ng Hg/g for nine of the ten strata sections. The 20–30-cm section showed a concentration of 80 ng Hg/g, indicating a relatively small, upstream intermittent source, possibly from upstream discharges or other anthropogenic activity in the watershed.

The low sediment concentrations in this core can be compared with the sediment samples from cores taken from Lake Superior which provide a more extensive record of historic mercury background concentrations (Fig. 7). The mean value from 128 subsamples, every 5 cm across the length of the 635 cm Lake Superior core sample, was 25.1 ± 8.3 (SD) ng Hg/g. These low-level concentrations represent lake sediment accumulations over a period of about 8000 to 10 000 years. If there were episodic inputs of mercury from geologic sources into the Lake Superior watershed or airshed during the period 2000–8000 B. P., then the resulting increases in water concentrations and accumulations in algae and diatoms and their subsequent deposition and incorporation into the sediments should have been reflected in measured sediment levels as subsurface maxima. No significant maxima were observed for the time spans represented by the subsamples analyzed from the Lake Superior core, except for elevated surface layers, as previously reported by Thomas (1974). The interpretation of changes in global mercury cycling as reflected in Lake Superior sediments must be carefully evaluated in light of the multiplicity of anthropogenic (Table 1) and natural sources (Rasmussen 1994) making contributions to the total mercury deposition to the watershed.

3.4 Sediment Mercury Subsurface Maxima

Subsurface maxima in mercury sediment concentrations were observed in most cases where anthropogenic mercury sources were previously known to exist, Fig. 6. The sediment mercury core profiles obtained from five sampling sites within the Knife Falls (KF) Reservoir showed two of the five sampling sites with subsurface maxima at 30–40 and 80–90 cm, respectively, with Hg concentrations of 1200 and 300 ng Hg/g, respectively (cores KF-6, -5). The other three sampling sites were all below 70 ng Hg/g (Fig. 6a). The major known sources of mercury were industrial discharges and landfill seepage located on both sides of the river. The industrial sources were from wood fiber products, which began in the early 1900s (MPCA/WDNR 1992), and ceased discharging to the river on or before 1978 when they became part of the Western Lake Superior Sanitary District.

The sediment mercury core profiles obtained from four sampling sites within the Potlatch (PH) Reservoir showed subsurface maxima in all cores at different sediment depths with Hg concentrations ranging up to 500 ng Hg/g (Fig. 6a; Core PH-2). This reservoir received loadings from upstream sources as well as those from other direct discharge sources.

The sediment mercury core profiles obtained from eight sampling sites within the Scanlon (SC) Reservoir all showed subsurface maxima in cores at different sediment depths and Hg concentrations (Fig. 6b). One core (SC-4) showed two extremely high concentrations of 2480 and 3560 ng Hg/g at 70–80 and 90–100 cm sediment depth, respectively.

This reservoir received loadings from upstream sources as well as those from paper manufacturing and other industries and municipal discharges up to 1978. Mercury-containing compounds were used as slimicides in the period from about 1950–1970s and substantial quantities were discharged with industrial process wastewater directly to the river. The most commonly used mercury compound was PMA (phenyl mercuric acetate). If the approximate dates are

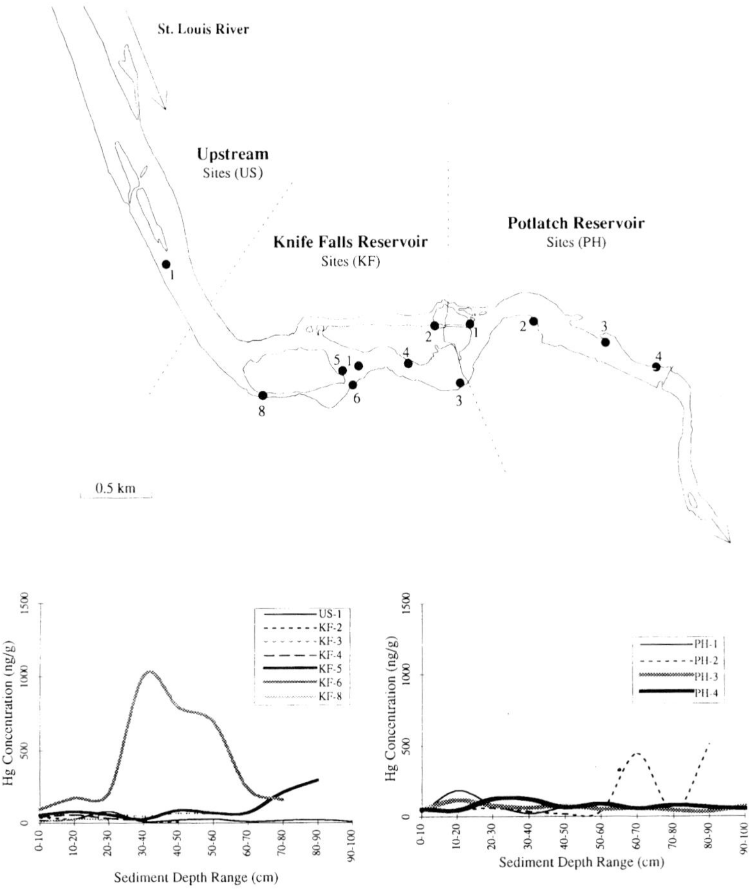
St. Louis River
Upstream
Sites (US)
Knife Falls Reservoir
Sites (KF)
Potlatch Reservoir
Sites (PH)
0.5 km
Hg Concentration (ng/g)
US-1
KF-2
KF-3
KF-4
KF-5
KF-6
KF-8
PH-1
PH-2
PH-3
PH-4
Sediment Depth Range (cm)

(a)

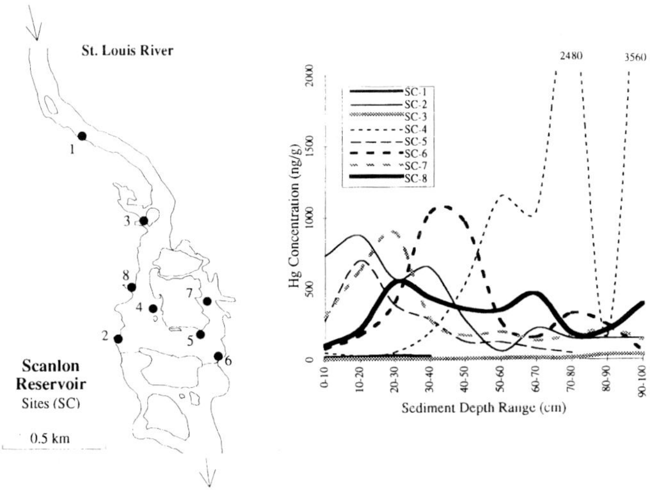
St. Louis River
Scanlon
Reservoir
Sites (SC)
0.5 km
2480
3560
Hg Concentration (ng/g)
SC-1
SC-2
SC-3
SC-4
SC-5
SC-6
SC-7
SC-8
Sediment Depth Range (cm)

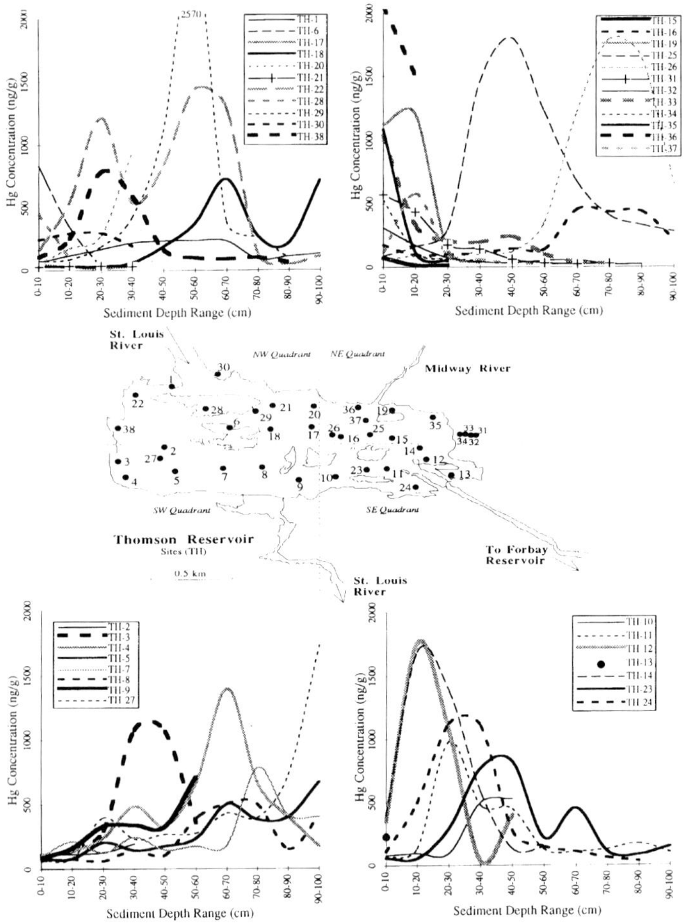
Hg Concentration (ng/g)
Sediment Depth Range (cm)
St. Louis River
Midway River
NW Quadrant
NE Quadrant
SW Quadrant
SE Quadrant
Thomson Reservoir
Sites (TH)
0.5 km
To Forbay Reservoir
St. Louis River

(c)

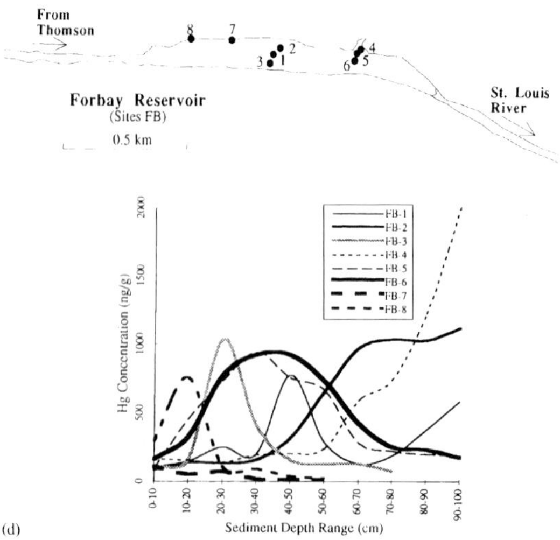
From Thomson
Forbay Reservoir
(Sites FB)
0.5 km
St. Louis River
Hg Concentration (ng/g)
Sediment Depth Range (cm)

(d)

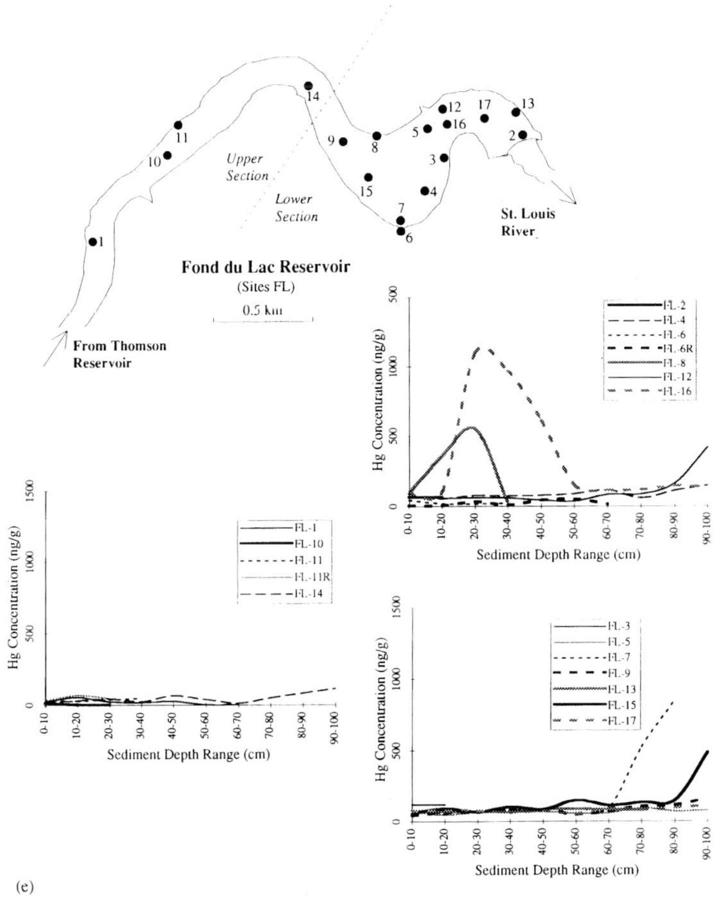

Fig. 6a–e. Summary of mercury concentrations in sediment cores from various sampling sites upstream (*US*, **a**) and within the six lower reservoirs of the Saint Louis River. The upper two reservoirs of this study (Knife Falls: *KF*, **a** and Potlatch: *PH*, **a**) exhibited fewer contaminated sites and lower overall mercury concentrations in sediment than those downstream (Scanlon: *SC*, **b**, Thomson: *TH*, **c**, Forbay: *FB*, **d**), and Fond du Lac (*FL*, **e**)

correct in Fig. 3, then most of the mercury loadings to the river and river sediments were made during the 1950s and 1960s.

The two Scanlon Reservoir cores showing the lowest concentrations with no subsurface maxima were taken adjacent to an industrial landfill and may reflect more recently deposited solid materials related to the solid waste disposal site.

The sediment mercury core profiles obtained from 38 sampling sites within the Thomson (TH) Reservoir showed most sites with subsurface maxima and secondary maxima in cores at different sediment depths and Hg concentrations (Fig. 6c). Thirteen of the cores had maximum mercury concentrations between 1500 and 2570 ng Hg/g occurring at a variety of sediment depths.

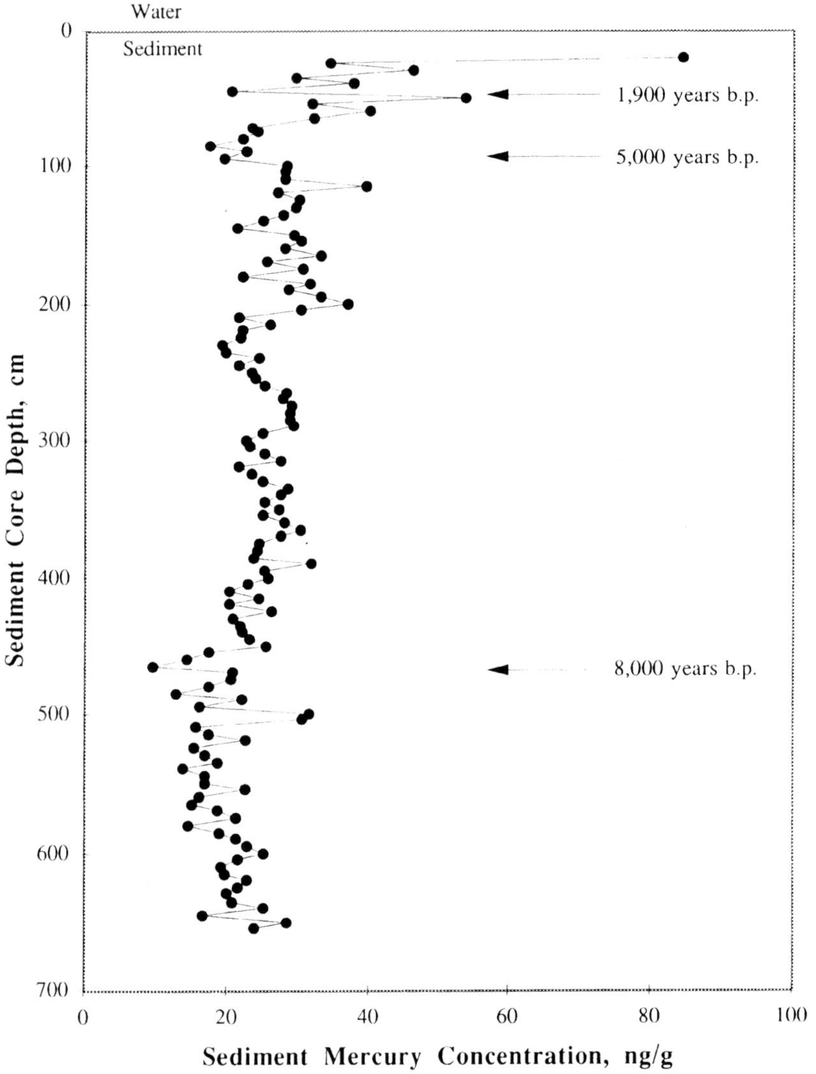

Fig. 7. Mercury concentration (ng/g) as a function of depth into the sediment for a Lake Superior sediment core, collected in 1980 (latitude 47° 11′ 04″ N; longitude 89° 21′ 06″ W; water depth of 180 m, Fig. 1). Most surface sediments were not represented, but show significantly higher mercury concentrations (Thomas 1974) which decrease rapidly with sediment depth (in the first 30 cm). Paleomagnetic (Johnson and Fields 1984) and carbon-13 (Huber 1993) dating indicated sediments were about 1900 to 8000 years B. P., as indicated

Thomson, the largest reservoir of the six studied, has a dozen small islands and an additional inlet, the Midway River. The complexity of the flow patterns coupled with the size, depth, and reservoir operational aspects make the interpretation of depositional patterns difficult. A number of additional parameters (including particle size, organic carbon, and total weight loss on ignition) are in the process of being measured in order to interpret the controlling variables that are determinant in establishing the depositional profiles.

The sediment mercury core profiles obtained from seven of eight sampling sites within the Forbay (FB) Reservoir showed significant subsurface maxima at a variety of different sediment depths. Notable Hg concentration maxima ranged from 750 ng Hg/g at 20–30 cm to 2000 ng Hg/g sediment at 90–100 cm sediment depth (Fig. 6d). This reservoir is primarily a deep channel and is used to divert the flow from the Thomson Reservoir to hydroelectric generators at the level just above the Fond du Lac Reservoir, the last and deepest reservoir on the St. Louis River.

Nine of the 17 cores from the sampling sites on the Fond du Lac (FL) Reservoir had mercury levels that did not exceed 250 ng Hg/g sediment (Fig. 6e). Subsurface maximum concentrations of 500 and 1220 ng Hg/g sediment occurred at a sediment depth of 20–30 cm for two (core FL-6R, -16) of the remaining eight. The other six cores had maxima below 1 m of sediment depth. For the eight sites exhibiting mercury concentrations greater than 250 ng/g, the mercury stratigraphy was complex, especially for sediment depths >1 m, and most likely represents sediment mercury loadings modified by sediment scouring in upstream reservoirs during major flood events.

3.5 Surface Maxima of Sediment Mercury Levels

Nine of the 38 sediment cores from the Thomson Reservoir had mercury concentration maxima at the *surface* (0–10 cm section) with values of 200, 300, 400, 500, 600, 800, 1100, 1100, and 2000 ng Hg/g (Fig. 6c). All of these cores were obtained from areas of relatively low total sediment accumulations, 20 to 40 cm, and most from an area of the reservoir which consists of shallow wetlands. The wetlands are subject to substantial water-level fluctuation as the reservoir is used for hydro power peak load generation and cycles on a 24-h schedule. It has not been determined whether or not these cores represent the total mercury loading captured by low, limited rates of sedimentation and remain near the sediment-water interface, or whether they represent some special circumstance such as a new loading source, or both. These core sites may represent areas of highest potential exposure for mercury bioaccumulation in today's aquatic food chain.

Additional studies are underway to (1) determine if possible new mercury sources have contributed to the high surface sediment concentrations, and (2) identify factors controlling reservoir sedimentation, mercury accumulation, and bioturbation influences in mercury sediment deposition.

4 Summary and Conclusions

Mercury analyses of a Lake Superior sediment core revealed no subsurface mercury maxima deposition episodes between about 2000 and 8000 years B. P., where dated values were available. This indicates that natural sources of mercury deposition and sediment accumulation were relatively low within the watershed, in contrast with recent times.

The lowest mercury concentrations, mean values of 17 and 25 ng Hg/g sediment (dry wt.), occurred at the furthest upstream site on the St Louis River and in Lake Superior deeper strata sediments, respectively.

The upstream reservoirs of this study (Knife Falls and Potlatch) exhibited fewer contaminated sediment strata and lower overall mercury concentrations than those downstream (Scanlon, Thomson, Forbay, and Fond du Lac). This indicates that significant loadings of mercury had occurred from localized anthropogenic sources.

The highest mercury concentrations (3500 ng Hg/g sediment, dry wt.) occurred in a sediment depositional zone of the Scanlon Reservoir at 90–95 cm depth into the sediments.

The occurrence of elevated mercury levels in sediment strata appear as subsurface maxima in concentration profiles as a function of depth into the sediments. These maxima indicate a beginning and ending of significant anthropogenic mercury loadings to the river system over a specified time period, and show dilution by low mercury-containing sediments at both the start and end of the loading period.

Several sampling sites exhibited high mercury concentrations (300 to 2000 ng Hg/g sediment, d. w.) in surface strata (0–10 cm). These observations are contrary to the expectation that because point source mercury discharges ceased during the 1970s, all sediment mercury contamination would be buried by relatively clean sediments. This confirms the hypothesis that varying sedimentation rates, high flow episodes, and possible new sources have caused the highest mercury concentrations to exist in the surface strata for some sediment-sampling sites.

Average mercury concentrations (per reservoir) in *Hexagenia* are correlated with average mercury in surface sediment levels, with the highest averages of both observed in the Thomson Reservoir.

The observations of elevated mercury in surface sediments of certain reservoir areas, and the significant positive correlation between mercury in sediment and *Hexagenia*, indicate that sediment mercury can act as a mercury contamination source to fish food organisms, and therefore, indirectly influence fish residue levels within those reservoirs.

The areas exhibiting high mercury levels in surface sediment on the Thomson Reservoir are the focus of further studies on testing mitigation approaches and identifying bioaccumulation mechanisms.

Acknowledgements and Disclaimer We acknowledge and thank the Great Lakes National Program Office, USEPA for major financial support (Grant No. GL995478) and the Duluth EPA lab for salary (G.E.G.), space, and equipment support for this project. For their parts in sampling and analysis we thank Dan Engstrom (UM LRC), James Huber, Randy Rude, Kent Schmidt, Vadim Melnichuk, Ed Swain (MPCA), and Mark Zacker (FDL). We also thank Jane Reed, Charles Lippert, and Kim Nieman for assistance in data presentation.

The mention of trade names and sources of commercial products is for identification and does not constitute endorsement or recommendation for use. The views expressed in this chapter are those of the authors and do not necessarily reflect the policies of their respective employers or of the USEPA.

References

Benoit JM, Fitzgerald WF, Damman AWH (1994) Historical atmospheric mercury deposition in the mid-continental United States as recorded in an ombrotrophic peat bog. In: Watras CJ, Huckabee JW (eds) Mercury pollution: integration and synthesis. Lewis, CRC Press, Boca Raton, pp 187–202

Faegri K, Iverson J (1965) Field techniques. In: Kummel B, Raup D (eds) Handbook of paleontological techniques. WH Freeman San Francisco, pp 482–494

Glass GE, Sorensen JA, Schmidt KW, Rapp GR Jr (1990) New source identification of mercury contamination in the Great Lakes. Environ Sci Technol 24:1059–1069

Glass GE, Sorensen JA, Schmidt KW, Huber JK, Rapp GR Jr (1992) Mercury sources and distribution in Minnesota's aquatic resources. MPCA, 520 Lafayette Rd, St Paul, Minnesota, April, 120 pp

Glass GE, Sorensen JA, Austin JJ, Schmidt KW, Kallemeyn LW, Hedman SC, Rapp GR Jr (1996) Mitigating mercury in Minnesota lakes and streams. Report to the Legislative Commission on Minnesota Resources. Minn Pollut Ctrl Agency, 520 Lafayette Rd, St. Paul, Minnesota, 64 pp

Huber J K (1993) A late glacial pollen sequence from Lake Superior. Curr Res Pleistocene 10:89–90

Johnson TC, Fields J (1984) Paleomagnetic dating of postglacial sediment, offshore Lake Superior, Minnesota-Wisconsin, USA. Chemical Geol 44:253–265

MDH (1996) Fish consumption advisory. Minnesota Dept of Health Minneapolis, Minnesota, 68 pp

MPCA (1996) Preliminary contaminant assessment of the Thomson, Forbay, and Fond du Lac reservoirs. Report to USEPA, Grant No X995766o1. Water Quality Division, 520 Lafayette Rd, St Paul, Minnesota, April, 262 pp

MPCA/WDNR (1992) The St. Louis River system remedial action plan. St Paul, Minnesota, Madison, Wisconsin

Nater EA, Grigal DF (1992) Regional trends in mercury distribution across the Great Lakes States, north central USA. Nature 358:139–141

Rasmussen PE (1994) Current methods of estimating atmospheric mercury fluxes in remote areas. Environ Sci Technol 28:2233–2241

Slemr F, Langer E (1992) Increase in global atmospheric concentrations of hg inferred from measurements over the Atlantic Ocean. Nature 355:434–437.

Sorensen JA, Glass GE, Schmidt KW, Huber JK, Rapp GR Jr (1990) Airborne mercury deposition and watershed characteristics in relation to mercury concentrations in water, sediments, plankton, and fish in Minnesota lakes. Environ Sci Technol 24:1716–1727

Sorensen JA, Glass GE, Schmidt KW (1992) Mercury concentrations in fish, plankton, and water from the St. Louis River watershed including reservoir lakes. Report to the Fond du Lac Band, Natural Resources Dept, Cloquet, Minnesota

Sorensen JA, Glass GE, Schmidt KW (1994) Regional patterns of wet mercury deposition Environ Sci Technol 28:2025–2032

Swain EB, Helwig DD (1989) Mercury in fish from Northeastern Minnesota lakes: historical trends environmental correlates and potential sources. J Minn Acad Sci 55:103–109

Swain EB, Engstrom DR, Brigham ME, Henning TA, Brezonik PL (1992) Increasing rates of atmospheric mercury deposition in Midcontinental North America. Science 257:784–787

Thomas RL (1974) cited in Matheson DH, Munawar M (1978) Limnology of Lake Superior. J Great. Lakes Res 4:249–275

USEPA (1990) Macro invertebrate field and lab methods. Secs: 5.2, 5.5, 6.1, 6.3, 6.4. EPA/600/4-90/030, Nov. Office of Research and Development, Washington, DC

USEPA (1991) Methods for the determination of metals in environmental samples. EPA/600/4-91/010, June. Office of Research and Development, Washington, DC

WLSSD (1973) Water quality management plan: operational analysis of existing treatment facilities. June 14. Appendex B. Existing facilities management inventory. Western Lake Superior Sanitary District, Duluth, Minnesota, 54 pp

Wright HE Jr (1967) A square-rod piston sampler for lake sediments. J Sediment Petrol. Sept:975–976

Concentration of Mercury and Other Heavy Metals in Central India

K.S. Patel, R.M. Patel, A.N. Tripathi, C.K. Chandrawanshi, P.K. Pandey, S. Chikhalikar, A. Kamavisdar, and S.G. Aggarwal

The contamination sites for the pollution of Hg and other heavy metals, i.e. Fe, Mn, Cr, Co, Ni, Cu, Zn, Cd, Pb and Tl, have been detected in the central part of India in the mineral- and coal-rich areas which lie between 18° and 23° N latitude and 80° and 84° E longitude. The heavy metals are found to be associated with the natural resource materials, i.e. minerals and coal, and are emitted into the environment by both natural and human activities. Thirty one sediments samples derived from 20 different ponds and 10 different river basins were collected and analyzed by AAS. The pollution level of Hg and ten other heavy metals in water and sediments, distribution of Hg in water and sediments, and their sources are described.

1 Introduction

The heavy metals are very toxic and found to be associated in nature with the sulphide and sulpho-salt ores of the ferrous and non-ferrous alloying elements, and with fossil fuel (Oehme 1978). Of these, Hg is one of the most toxic and highly volatile elements. The environmental chemistry of Hg and other heavy metals, i.e. Fe, Mn, Cr, Co, Ni, Cu, Zn, Cd, Pb, Tl etc., in water and sediments has been extensively investigated in temperate climates, but largely neglected in the tropical climate of Asia (Rask and Metsaelae 1991; Wilken and Hintelmann 1991; Belkin and Sparcke 1993; Stigliiani et al. 1993; Iskander and Vega Carrillo 1994; Colina et al. 1996; Fabbri et al. 1996; Glass et al. 1996; Ikingura and Mutakyahwa 1996; Wiener et al. 1996). In central India, the area lying between 18°and 23° N latitude and 80°and 84° E longitude is most sensitive to heavy metal pollution due to the occurrence of minerals and coal and their utilization. Several steel, cement and aluminium industries, and thermal power plants mainly using coal as a source of energy are full at work in this region. These minerals and coal are found to be associated with Hg and other heavy metals at trace levels. Due to natural and human activities (e.g. exploitation of minerals and coal, smelting of minerals, burning of coals, combustion of gasoline, atmospheric deposition of the heavy metals, weathering of rocks and erosion of soils) the metal pollution in the environment of this region is increasing. Sediment is an efficient sink or source for the release of the metals to the aquatic or other living organisms. Therefore, the distribution of heavy elements, i.e. Fe, Mn, Cr, Co, Ni, Cu, Zn, Cd, Pb, Tl and Hg, in the sediments of reservoirs situated in different geographical environments, i.e. industrial belt, thermal power plant, coal mines and the iron

Environmental Science
Mercury Contaminated Sites (ed. by R. Ebinghaus et al.)

mineral belt of central India, has been investigated to understand their concentration level, chemical forms, dispersion and sources.

2 Materials and Methods

2.1 Equipment

A GBC model 932 Atomic Absorption Spectrophotometer (AAS) equipped with GF-3000 graphite furnace and a model 3000 HG cold vapour accessory was used to monitor the elements. The details of the employed analytical conditions are shown in Table 1.

2.2 Reagents

All chemicals used were analytical grade reagents (Merck). The standard solution of the elements supplied by Merck was used. Ultrapure concentrated H_2SO_4 (70.0%, v/v), HNO_3 (96.0%, v/v) and $HClO_4$(70.0%, v/v) acids were employed to leach the heavy metals. A 0.1% w/v $KMnO_4$ solution in 1.0 M H_2SO_4 acid was used to release non-labile Hg in water.

2.3 Sampling

The water and sediment samples were collected in summer, June 1996, at the 31 sites. They were collected from sampling stations belonging to 20 different

Table 1. Operating conditions and detection limit of the GBC 932 AAS equipped with a GF-3000 graphite furnace

AAS	Element	Detection limit (μg/ml)	Wave length (nm)	Slit width (nm)
CV-AAS	Hg	0.0004	253.6	0.5
Flame AAS	Fe	0.05	248.3	0.2
	Co	0.05	240.7	0.2
	Ni	0.04	232.0	0.2
	Cu	0.025	324.7	0.5
	Zn	0.008	213.9	0.5
	Mn	0.02	279.5	0.2
	Pb	0.06	217.0	1.0
	Cr	0.05	357.9	0.2
	Cd	0.009	228.8	0.5
Graphite furnace AAS	Tl	0.001	276.8	0.5

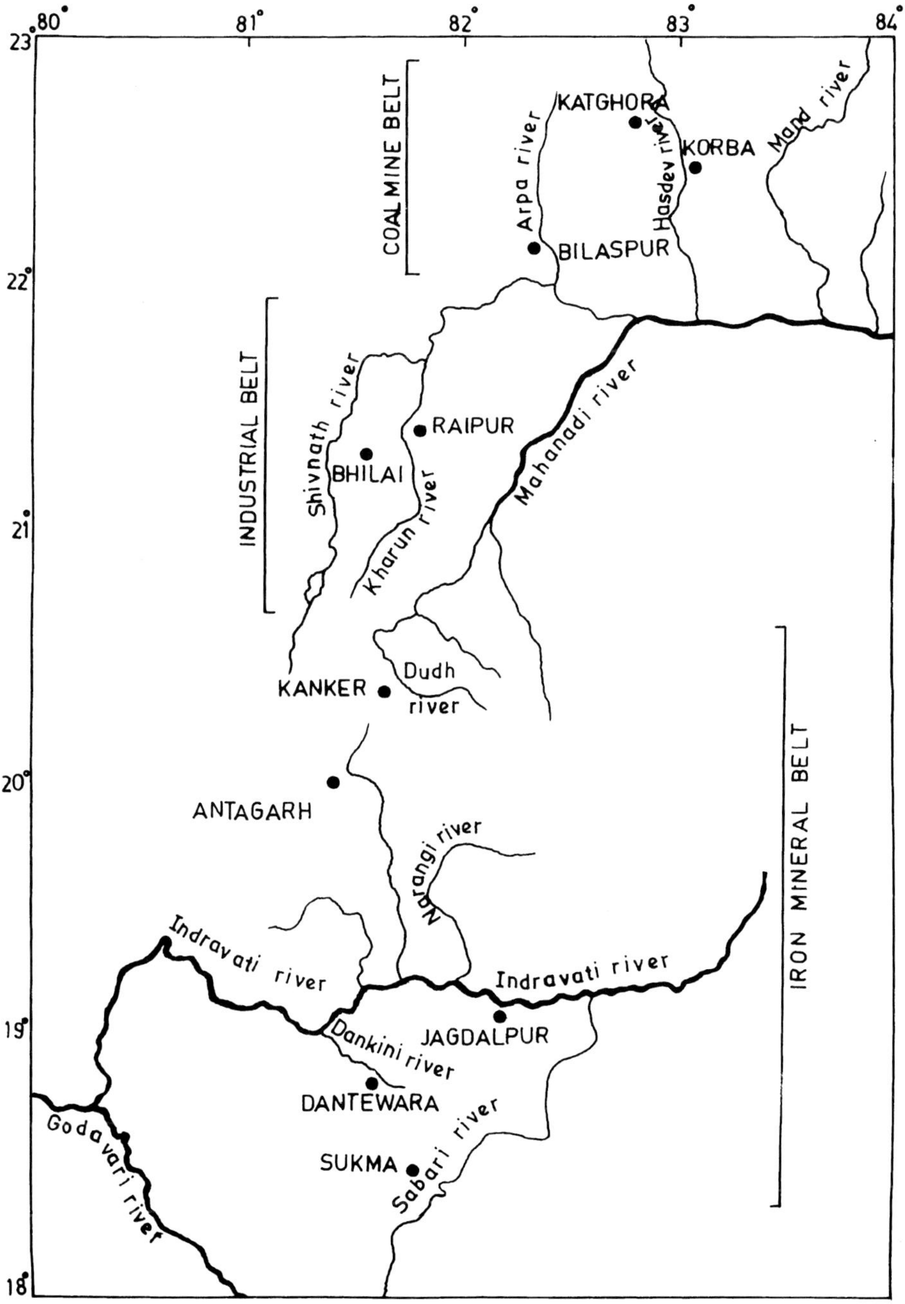

Fig. 1. Geographical map of central India

Table 2. Concentration of Hg in water and sediments of central India

Area	Location	Site	Surface water				Sediments		
			Colour	pH	Conc. (ppb)	Av. conc. (ppb) in pond (in river)	Colour	Conc. (ppm)	Av. conc. (ppm) in pond (in river)
IB	Rajim	Pond	Y	8.3	2.0		Y	0.25	
		Mahanadi river	CL	7.6	1.4		Y	0.18	
	Bilaspur	Pond	YG	8.2	3.2		Y	0.29	
		Arpa River	CL	7.7	2.5		Y	0.22	
	Raipur	Urla pond-I	Y	8.6	4.1	5.6	Y	0.51	0.83
		Urla pond-II	Y	8.3	6.0	(2.7)	Y	0.80	(0.26)
		Urla pond-III	YG	8.2	11.3		Y	1.33	
		Dangania pond	YG	8.4	5.4		GY	0.85	
		Nayapara pond	YG	8.5	6.0		GY	0.74	
		Kharun river	Y	8.3	2.3		Y	0.30	
	Bhilai	Hudco pond	Y	8.5	4.5		YBr	0.28	
		Kutela bhata pond	G	8.6	2.7		Br	0.19	
		Luchaki pond	GY	9.0	11.3		GB	3.01	
		Shivnath river	CL	7.8	4.5		Y	0.34	
TPP + CMB	Korba	Pond-I	YG	7.8	5.1		G	0.36	
		Pond-II	YG	7.9	6.3		G	0.34	
		Pond-III	YG	8.0	7.0		G	0.62	
		Pond-IV	YG	7.9	4.5	5.5	G	0.56	0.46
		Husdev river	CL	7.6	2.8	(2.5)	Y	0.36	(0.32)
	Katghora	Pond	Y	7.0	5.7		Y	0.41	
		Husdev river	CL	7.7	5.9		Y	0.27	

Table 2. (contd.)

Area	Location	Site	Surface water				Sediments		
			Colour	pH	Conc. (ppb)	Av. conc. (ppb) in pond (in river)	Colour	Conc. (ppm)	Av. conc. (ppm) in pond (in river)
IMB	Kanker	Pond	G	7.2	7.0		RY	2.18	
		Dudh river	YR	7.3	5.8		R	1.36	
	Jagdalpur	Pond	G	7.1	13.3		Br	4.58	
		Indravati river	YR	7.7	11.7	12.8	Y	3.41	3.35
	Sukma	Pond	G	7.2	16.6	(10.4)	Br	6.72	(2.50)
		Gorani river	YR	7.6	13.6		Y	4.27	
	Antagrah	Pond	G	6.9	10.0		YBr	2.70	
		Antagarh river	YR	6.1	9.3		Y	1.59	
	Dantewada	Pond	G	7.3	17.1		YR	2.98	
		Dankani river	YR	7.3	11.8		Y	1.85	

IB = industrial belt; TPP + CMB = thermal power plant + coal mines belt; IMB = iron mineral belt; Y = yellow, CL = colourless, YG = yellowish green, G = Green, GY = greenish yellow, YR = yellowish red, Br = Brown, YBr = yellowish brown, GBr = greenish brown, RY = reddish yellow.

ponds, and 10 different river basins of central India (Fig. 1). The colour of the surface water and sediments was recorded (Table 2). The pH of the water at the site was measured. At all sampling sites the pond sediments were clayey, whereas the river sediments were sandy. The water samples were filtered through Whatman filter paper no. 42. The wet sediment samples were sieved through a 500-μm mesh sieve and stored in a polyethylene bag. The sediment samples were oven-dried at 60 °C for 48 h.

3 Procedure for Determination of Heavy Metals

Five grams of dried and weighed sediment samples were transferred into a 250-ml conical flask and treated with 3 ml each of concentrated H_2SO_4 and HNO_3 acids and 1 ml of concentrated $HClO_4$ acid at 4 °C. The conical flask was stored for 1 week after closing the mouth with a stopper. The extract was filtered with Whatman filter paper no. 42 by washing with the HNO_3 acid (0.1 M) and diluted to 50 ml with deionized double distilled water. The heavy metals in the extract were monitored by AAS techniques, i.e. acetylene flame, graphite furnace, cold vapour AAS using calibration curves prepared under similar conditions (Table 1).

4 Procedure for Determination of Hg in Water Samples

Ten ml of the filtered water sample was taken in a 50-ml beaker and the labile Hg (the inorganic species reducible with borohydride) was determined by cold vapour AAS. Total Hg (the inorganic plus the organic) was determined by treating 10 ml of the filtered water sample with 0.5 ml $KMnO_4$ solution and waiting for at least 6 h at room temperature (20 ± 2 °C) for complete liberation of the non-labile Hg (organic mercury). The total Hg content was determined by cold vapour AAS as above.

5 Results and Discussion

5.1 Mercury in Water

The maximum temperature of surface water during day in the 1st week of June was 40 ± 3 °C. All surface waters except some river waters were coloured. The pH of the surface water ranged between 6.1 and 9. The concentrations of the labile (the inorganic), non-labile (the organic) and total (the inorganic plus the organic) mercury in the surface water of 30 reservoirs are given in Table 2. The concentration of total mercury in 20 ponds and 10 river basins was found to be in the range of 2.0–17.1 and 1.4–13.6 ppb, respectively. The pH of water affects the Hg level and a very high Hg concentration was seen in the surface water, whose pH was >6.9. The stagnant surface water of ponds contained a very high Hg concentration (<17.1 ppb), which may be due to contamination with greenish algal bloom or iron compounds at a high level. However, the Hg concentration was also very high in mobile (river) water of the iron belt area, possibly due to contamination with iron compounds at a high level (Table 2). The concentration of organic as well as total Hg was higher in the stagnant than in the mobile water and the ratio of organic to inorganic Hg was in the range of 9.0–24.5, and 1.9–3.6 in 20 ponds and 10 river basins, respectively (Table 3). The dispersion of mercury in the groundwater was examined in water near the sampling sites. The groundwater samples were collected using water-sampler equipment (Pandey et al. 1979). Hg-contamination of the groundwater was determined and found in the range of 0.4–3.5 ppb. The effect of the depth profile of the groundwater aquifers on the Hg level was seen to be significant in the dolomite rock-bearing areas i.e. Raipur and Bhilai (Table 4). The highest level of mercury pollution in the water of the iron mineral belt was 17.1 ppb, mostly due to natural phenomena, i.e. weathering of iron ore, soil erosion or atmospheric deposition (Kamavisdar and Patel 1997; Patel and Tripathi 1997).

5.2 Mercury in Sediments

Sediment samples of ponds and rivers from differently located geographical environments of central India, the industrial towns (Raipur, Bhilai, Bilaspur), thermal power plant (Korba), coal mine area (Katghora), and remote iron mineral areas (Kanker, Jagadalpur, Sukuma, Antagargh, Dantewada) were

Table 3. Speciation of Hg in surface water

Location	Site	Speciation of Hg in surface water (ppb)			Ratio Non-labile:labile
		Labile	Non-labile	Total	
Raipur	Urla pond-III	0.9	10.2	11.3	11.3
	Kharun river	0.8	1.5	2.3	1.9
Bilaspur	Pond	0.3	2.9	3.2	9.6
	Arpa river	0.6	1.9	2.5	3.2
Korba	Pond-I	0.2	4.9	5.1	24.5
Antagarh	Pond	0.7	9.3	10.0	13.3
	River	2.0	7.3	9.3	3.6
Kanker	Pond	0.7	6.3	7.0	9.0
	River	1.6	4.2	5.8	2.6

Table 4. Distribution of Hg between surface and groundwater

Location	No. of sites	Av. conc. in surface (ppb)	Rock type	Conc. of Hg in tube well and ground water	
				Tube well depth (m)	Hg concentration (ppb)
Raipur	5	6.5	Dolomite	10	2.0
				20	1.2
				30	0.9
				60	0.4
Bhilai	3	6.2	Dolomite	20	3.1
				60	1.1
Bilaspur	1	3.2	Granite	10	2.0
Korba	4	5.5	Granite	50	1.3
Kanker	1	7.0	Iron rock	50	3.5
Anthagarh	1	10.0	Iron rock	70	2.9

collected as described earlier. The metal contents in the minerals and coal are of two types: (1) those which volatilize at furnace temperature, i.e. Hg and, to some extent, Pb, Tl, Zn and Cd, and (2) those which may be non-volatile and may remain in the waste or ash as oxides and silicates. The chemical database and statistics of the heavy metals in the sediments obtained are summarized in Table 5. The migration processes of these heavy metals in the sediments are connected with oxides and hydroxides of Fe and Mn which adsorb these heavy metals from lakes or rivers because of a high negative charge and large surface area. In both pond and river basin sediments, the occurrence of heavy metals has a positive correlation with Fe and Mn contents (Figs. 2–4). In the iron-mineral area, the occurrence of the heavy elements, i.e. Hg, Tl, Pb, Cr, Co, Ni, Zn and Cu, has excellent correlation with the concentration of Fe and Mn in the sediments but in the industrialized areas, only the non-volatile elements, i.e. Cr, Co, Ni, Cu and Zn, have an excellent positive correlation with the Fe and Mn concentrations unlike the volatile elements, i.e. Hg, Tl, Pb and Cd (Figs. 2–4). All heavy metals except Cd had a higher concentration in pond sediments. The concentration of Cu and Zn was higher in the sediments derived from coal mines and coal-burning areas. The highest concentration of Pb was detected in pond sediments of the industrial belt, due to emission by various sources, i.e. iron-melting plants, thermal power plants, vehicles using leaded gasoline, use of leaded pipes, atmospheric deposition, etc.

5.3 Effect of Depth Profile

The concentration of Hg and other elements in different depth profiles of sediments of the Urla pond, Raipur and Shivnath river sediments was determined. The concentration of elements, i.e. Mn, Fe, Cd, Hg and Tl, increased, as the depth profile of sediment increased whereas the concentration

Table 5. Average, range and std. dev. of concentration Hg and other elements in sediments in central India

Area	Type of sediment	No. of sediment		Concentration of elements (ppm)										
				Fe	Mn	Cr	Co	Ni	Zn	Cu	Cd	Pb	T1	Mg
IB	R	4	a	10 500	210.6	28.2	4.8	36.4	34.7	12.4	0.31	4.34	3.83	0.26
			b	9600–11 400	180–241	25.4–31.6	4.6–4.9	32.6–40.6	24.4–47.3	11.7–13.0	0.28–0.35	4.28–4.41	3.41–4.36	0.21–0.31
			c	770	25.0	2.6	0.09	3.35	9.7	0.65	0.03	0.06	0.39	0.04
	P	10	a	21 000	400.7	35.6	10.1	50.8	45.4	17.6	0.14	13.51	5.67	0.82
			b	8900–23 360	296–486	29.0–42.8	8.1–12.1	45.3–56.5	36.5–57.2	10.6–22.5	0.07–0.18	10.8–15.52	3.34–7.89	0.73–0.89
			c	1542	69.8	5.0	1.35	3.76	6.5	3.9	0.04	1.5	1.5	0.05
TPB + CMB	R	2	a	9360	45.0	11.4	1.5	12.5	32.1	10.6	0.35	0.53	1.80	0.32
			b	8980–9420	42–48	10.8–12.0	1.5–1.6	12.2–12.8	30.4–33.8	10.5–10.7	0.35–0.36	0.51–0.54	1.72–1.87	0.32–0.33
			c	3.11	4.24	0.85	0.07	0.42	2.4	0.14	0.007	0.02	0.11	0.007
	P	5	a	16 500	72.6	12.6	1.8	16.8	52.8	14.4	0.15	0.60	3.21	0.46
			b	13 000–18 500	64.2–86.4	9.8–15.5	1.1–2.5	12.2–19.3	43.4–61.7	9.8–17.4	0.11–0.19	0.54–0.65	3.12–3.34	0.44–0.49
			c	2179.4	9.3	2.4	0.51	2.82	6.85	3.0	0.03	0.04	0.09	0.02
IMB	R	5	a	11 600	85.7	17.7	2.7	18.6	4.8	3.4	0.20	4.81	4.52	2.50
			b	10 700–12 800	71.7–99.3	10.1–26.2	2.2–3.0	11.8–27.7	4.6–5.0	3.0–3.7	0.14–0.24	4.35–5.20	3.71–5.03	2.32–2.64
			c	924.6	11.6	7.4	0.3	6.1	0.2	0.3	0.04	0.34	0.51	0.12
	P	5	c	21 000	92.4	25.0	4.9	20.5	5.2	4.1	0.13	4.60	4.83	3.35
			b	19 500–23 800	64.7–124.3	21.0–30.2	4.2–5.3	14.2–29.7	4.5–5.6	3.6–4.6	0.10–0.17	4.30–4.92	4.52–5.28	3.26–3.43
			c	1856	21.9	3.6	0.43	5.9	0.45	0.4	0.03	0.3	0.3	0.07

a = average concentration; b = range; c = standard deviation <±>; IB = industrial belt; TPB + CMB = thermal power plant + coal mine belt, IMB = iron mineral belt; R = river; P = pond.

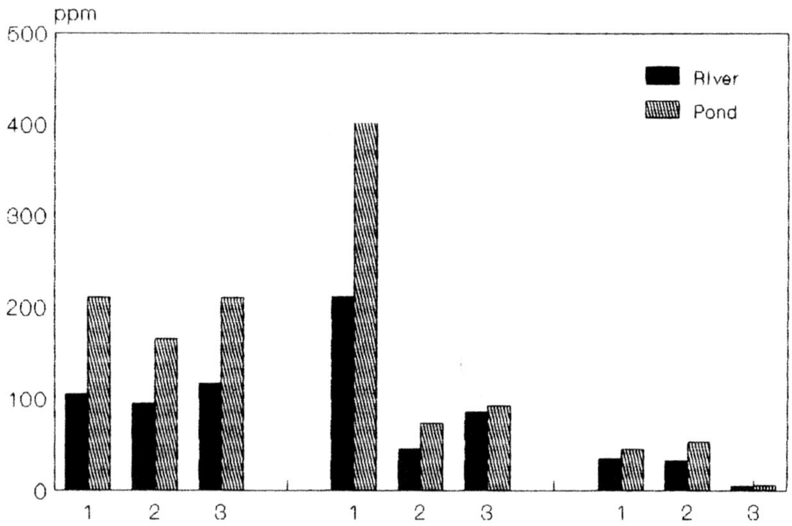

Fig. 2. Average concentration of Fe, Mn, Zn in sediments. 1 = IB, 2 = TPP + CMB, 3 = IMB, [Fe] = C × 10

of other elements, i.e. Co, Ni, Cu and Pb, showed a reverse trend (Fig. 5). The concentration of Zn increased in the river basin as the depth profile of sediment increased, but a reverse trend occurred in pond sediment. Among them, Cd had no correlation with the depth profile. The concentration of pb decreased significantly as the depth profile of sediment increased, and mostly concentrated at the surface of the sediment.

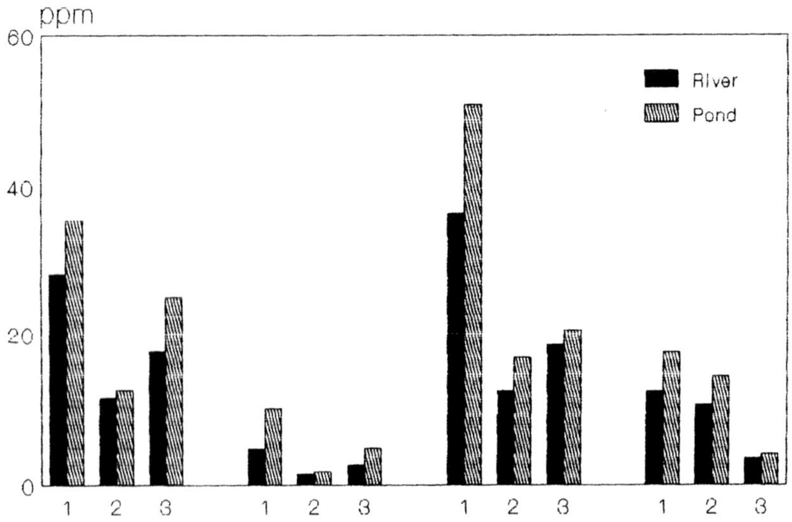

Fig. 3. Average concentration of Cr, Co, Ni, Cu in sediments. 1 = IB, 2 = TPP + CMB, 3 = IMB

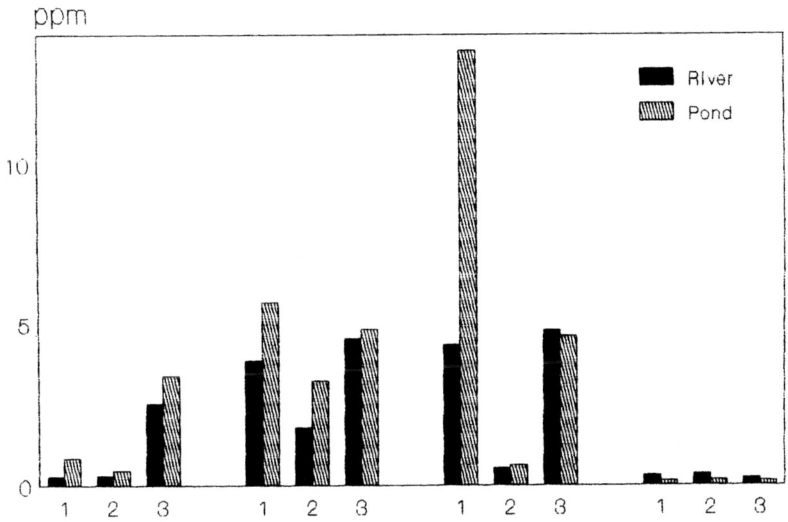

Fig. 4. Average concentration of Hg, Tl, Pb, Cd in sediments. 1 = IB, 2 = TPP + CMB, 3 = IMB

5.4 Sources

Central India is very rich in minerals (i.e. ore of iron, calcium, aluminium, copper, tin, etc.) and coal. These materials, i.e. minerals and coal, are found to be associated with heavy metals at trace levels and their burning, smelting, exploitation, weathering and atmospheric deposition are the main sources of contamination of the environment in this region (Table 6).

5.5 Comparison of Mercury Pollution

A very high concentration of Hg in river and sea basin sediments in industrialized countries with a temperate climate like the USA or Europe was reported (Rask and Metsaelae 1991; Wilken and Hintelmann 1991; Belkin and Sparcke 1993; Stigliiani et al. 1993; Iskander and Vega Carrillo 1994; Colina et al. 1996; Downs 1996; Fabbri et al. 1996; Glass et al. 1996; Horvat et al. 1996; Ikingura and Mutakyahwa 1996; Wiener et al. 1996). However, a very low concentration of Hg at lower ppb levels in the surface water was detected. In tropical countries like India, the reverse pollution trend was observed. The lake and river sediments were found to be contaminated with a moderate amount of Hg (0.26–3.3 ppm), however, the surface water had a relatively high concentration of Hg in the range of 1.4–17.1 ppb as in the southern part of the country (Ayyadurai et al. 1983).

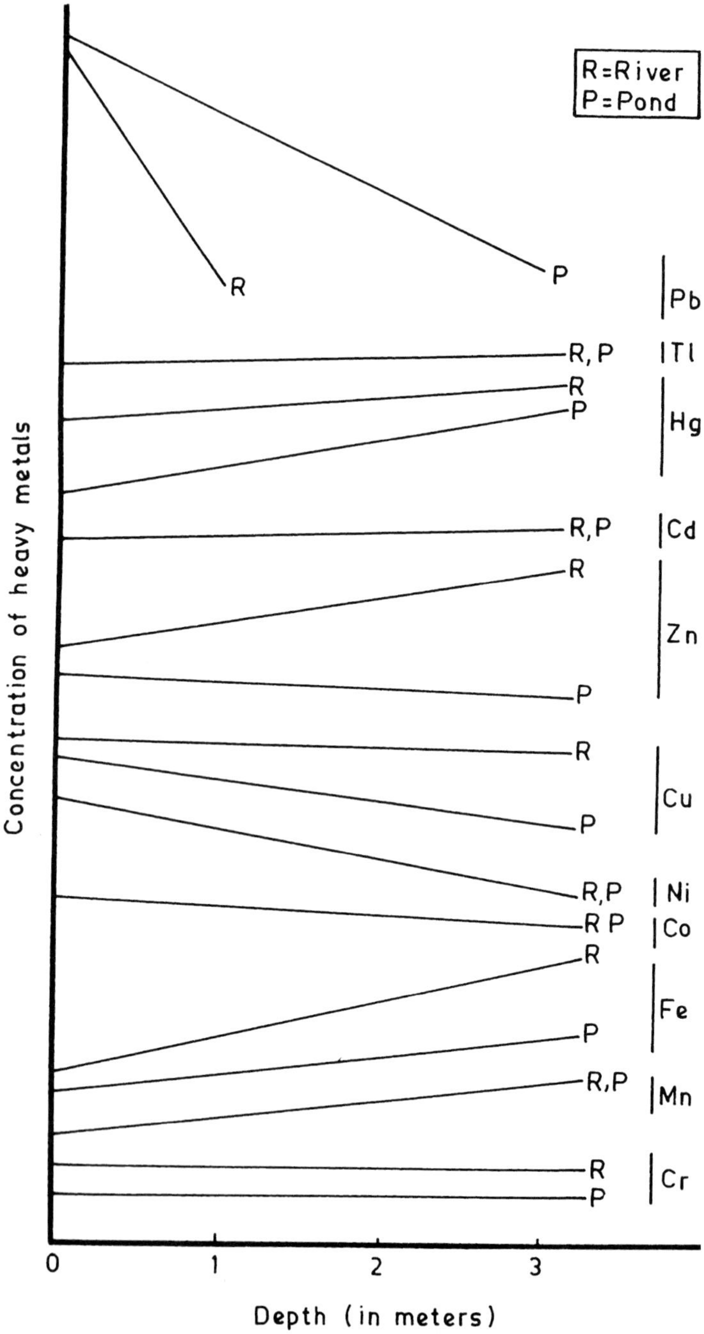

Fig. 5. Downward dispresion trend of heavy metals in sediment

Table 6. Possible pollution sources for the contamination of the sediments in central India

Elements	Main sources
Fe, Mn, Cr, Co, Ni, Cu, Zn, Cd, Hg	Iron and dolomite melting, coal burning furnace, rock weathering, atmospheric deposition, etc.
Pb	Iron and dolomite melting, coal burning furnace, rock weathering, combustion of leaded gasoline, atmospheric deposition, use of leaded pipes

6 Conclusion

The natural background of elements, i.e. Fe, Mn, Cr, Ni, Cu, Co, Zn, Pb, Tl and Hg, determined in different sediments of the unpolluted area, was 500, 11.0, 1.0, 1.2, 0.7, 0.3, 4.0, 0.07, 0.06 and 0.05 ppm, respectively. Most of the sediments in central India, extending over an area of $\approx 2.0 \times 10^5$ km^2, are highly polluted (<18-fold) with metal, i.e. Fe. In addition, the iron belt region covering an area of $\approx 6.0 \times 10^4$ km^2 and the industrial belts, i.e. Raipur and Bhilai, are highly polluted with other heavy metals, i.e. Mn, Cr, Ni, Pb, Tl and Hg. The Hg level in the surface water is very high (17.1 ppb) and mostly present in organic form. The main possible source of Hg and other heavy metals in the water and sediments of central India is through weathering of the natural geological resource materials, exploitation and smelting of iron ore and dolomite, burning of fossil fuel, i.e. coal, gasoline, atmospheric deposition of heavy metals, use of leaded pipes, etc. Extensive research will be needed for the exact mapping of the level of heavy metal pollution in the environment of central India to determine the sources, mobility, sink, receptors, bioaccumulation, transformation and toxicities.

Acknowledgements We are thankful to the Department of Environment and Forests, New Delhi, for financial support of this work through project No. 19/51/92-RE. The Alexander von Humboldt Foundation, Bonn, Germany, is gratefully acknowledged for the donation of the GBC graphite furnace, and for financial support to attend the mercury conference held in Hamburg, August 1996. We wish to express our sincere thanks to Dr. Ralf Ebinghaus, GKKS, Geesthacht, Germany, for encouragement and financial support. We are also thankful to Prof. V. Kuban, Univ. of Brno, Czech Republic, for his kind help in presenting this work at the mercury conference held in Hamburg.

References

Ayyadurai K, Kamalam N, Rajagopal CK (1983) Mercury pollution in water in Madras city. Indian J Environ Health 25:15–20

Belkin HE, Sparcke HM (1993) Hg, As, Sb and Se contents of sediments from the Kuskokwim River, Bethel, Atlanta, USA. Environ Geol 22:106–110

Colina M, Oliveros S, Marquez R, Ledo H (1996) Mercury level in water and sediments from Sinameica Lagoon and its inflows, Venezuela. Mercury as a Global Pollutant Conf, August 4–8, 1996, Hamburg, p 363

Downs I (1996) Mercury and methylmercury in rivers Yare, Norfolk, U.K. Mercury as Global Pollutant Conf, August 4–8, 1996, Hamburg, p 367

Fabbri D, Felisatti O, Lombardo M, Trombini C, Vassura L (1996) The lagoon of Ravennk: characterization of mercury-contaminated sediments. Mercury as a Global Pollutant Conf, August 4–8, 1996, Hamburg, p 307

Glass GE, Sorensen JA, Rapp GR Jr, Schwarzkopf L (1996) Mercury sources, deposition, and mitigation in the L. Superior watershed. Mercury as a Global Pollutant Conf, August 4–8, 1996, Hamburg, p 381

Horvat M. Faganeli J, Planinc R, Covelli S, et al. (1996) Mercury pollution in Trieste Bay. Mercury as a Global Pollutant Conf, August 4–8, 1996, Hamburg, p 491

Ikingura JR, Mutakyahwa MKD (1996) Source of mercury contamination and exposure in Tanzania, Mercury as a Global Pollutant Conf, August 4–8, 1996, Hamburg, 316

Iskander FY, Vega Carrillo HR, Acuna EM (1994) Determination of mercury and other elements in La Zacatecana Dam sediment in Mexico. Sci Total Environ 48:45–49

Kamavisdar A, Patel KS (1997) Iron in drinking water of central India. Indian J Environ Health (in press).

Mueller HW, Schwaighofen B, Kalman W (1994) Heavy metal contents in river sediments. Water Air Soil Pollut 72:191–203

Oehme FW (1978) Toxicity of heavy metals in the environment part 1–2. Marcel Dekker, New York

Pandey SP, Narayanswamy VS, Hasan MZ (1979) Quality of well waters of Nagpur with regard to nitrates and nitrites (phase I), Indian J Environ Health 21:35–46

Patel KS, Tripathi AN (1997) Precipitation of mercury and other heavy metals in central India J Environ Monit Assess (submitted)

Rask M, Metsaelae TR (1991) Mercury concentration in northern pike, *Esox Lucius* L. in small lakes of EVO area, southern Finland. Water Air Soil Pollut 56:369–378

Stigliiani WM, Jaffe PR, Ariderberg S (1993) Heavy metal pollution in the Rhine basin. Environ Sci Technol 27:786–793

Wiener JG, Rada RG, Frazier BE, Engstrom DR (1996) Stratography of mercury in depositional sediments in the contaminated Sudbury River, Massachusetts, USA. Mercury as a Global Pollutant Conf, August 4–8, 1996, Hamburg, p 309

Wilken RD, Hintelmann H (1991) Mercury and methylmercury in sediment and suspended particles from the river Elbe, North Germany. Water Air Soil Pollut 56:427–437

Fluvial Transports of Mercury Pollution in the Ill River Basin (Northeastern France): Partitioning into Aqueous Phases, Suspended Matter and Bottom Sediments

J.L. Probst, A. Messaïtfa, G. Krempp, and P. Behra

1 Introduction

The increase and dispersion in aquatic systems of mercury, which is not very abundant in the global environment, are mainly due to the industrial development and intensification of human activities (halogen chemicals, dental and pharmaceutic industries, agriculture, gold mining, etc.). In the Alsace region (northeastern France), the Ill river and one of its tributaries, the Thur, are the most polluted because of industrial wastewaters from the chlorine and soda industry which uses mercury as a cathode in electrolysis.

The Ill river, which is a tributary of the Rhine, drains the eastern side of the Vosges Massif and the Alsatian plain (Fig. 1). In this region, groundwater plays an important role in the economy and in hydrosystem equilibrium. In the upper part of the Ill drainage basin, the river greatly contributes to the groundwater supply (Esteves 1989; Roeck 1992), but the upper part of the Ill river presents strong positive anomalies in mercury concentrations which can be attributed to the industrial plants devoted to chlorine products in Vieux Thann (Dinh 1982).

Within the framework of French scientific programmes (i.e. GIP Hydrosystemes and Programme Environnement, Vie et Sociétés of CNRS), we have been studying the fate of mercury inputs into the Ill river from the pollution source in the upper part of the basin (Vieux Thann) to the downstream part of the basin, before the confluence with the Rhine, near Strasbourg (Probst 1996) since 1994.

The main objective of this study is to follow the fate of mercury in the river system, particularly in the different phases, i.e. aqueous phase, suspended matter and bottom sediments. It is also to determine the influence of physicochemical processes on the mercury fluvial transports in the different phases. This study is focused on the influence of the chloride concentration on mercury content in the solution, in a region where the chemical composition of the river water is highly influenced by saline industrial activity from potash mines, and also on the influence of sediment particle size on the distribution of mercury in river-suspended matter and in river bottom sediments.

Environmental Science
Mercury Contaminated Sites (ed. by R. Ebinghaus et al.)

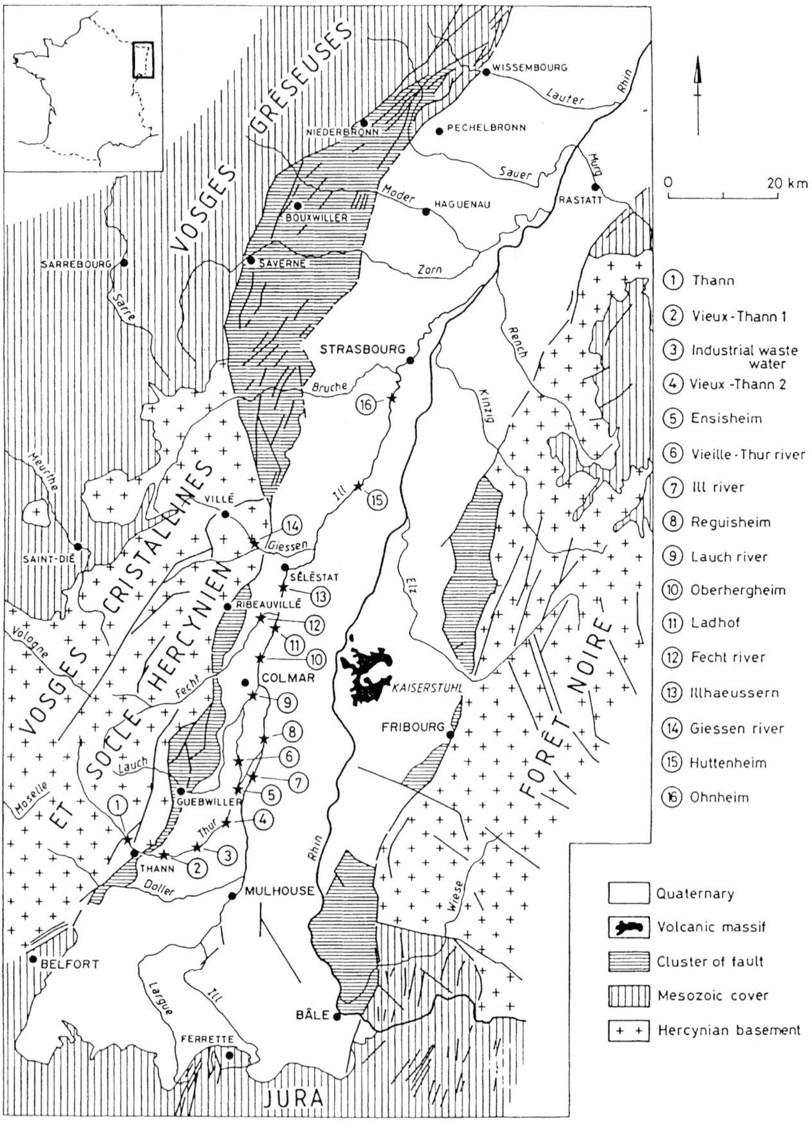

Fig. 1. Location map of the sampling stations in the Thur and Ill river basins

2
Material and Methods

Samples were collected during two low-water periods (October 1994 and July 1996) and two high-water periods (May 1995 and February 1996) at 16 sampling stations distributed along the Thur and the Ill rivers and at the outlet of the main tributaries (Vieille Thur, Lauch, Fecht and Giessen; Fig. 1). For each water period, the samples were collected the same day for all stations; the hydrological conditions were thus the same for all the stations.

River water samples were immediately filtered in the field under nitrogen pressure using 0.45-µm Millipore filters. For mercury analysis in the solute phase, the filtrate was put in a 500 ml Pyrex glass bottle with 20 ml of HNO_3 (35%) and 20 ml of H_2SO_4 (50%). To avoid mercury losses when the bottle is opened just before analyzing the filtrate, the bottle is closed with Parafilm and a cap. Some drops of $KMnO_4$ (6%) are added to the filtrate just before analyzing to oxidize mercury. It is important to note that $KMnO_4$ is added at the last moment rather than in the field after filtration because it could be observed that by adding $KMnO_4$ in the field, a manganese precipitate is formed and traps mercury between the field and the laboratory.

Part of the filtrate is treated with HNO_3 to analyze trace elements by ICP-AES (Samuel and Rouault 1983) or ICP-MS using a quick scanning method (± 10% accuracy). The same techniques were used to analyze trace elements present in sediments. Major elements were determined for part of the filtrate without treatment in the field using the following classical techniques (Krempp 1988; Probst et al. 1992): atomic absorption for Ca^{2+}, Mg^{2+}, Na^+ and K^+ (accuracy 1 µmol/l), ionic chromatography for Cl^-, SO_4^{2-}, NO_2^-, NO_3^- and PO_4^{3-} (accuracy 1 µmol/l), Gran titration with HCl for alkalinity (accuracy 1 µmol/l) and colorimetric method for H_4SiO_4 (accuracy 1 µmol/l).

The suspended matters and the bottom sediments were dried at room temperature to avoid mercury loss. As one can see in Fig. 2, when the drying temperature of the sediments increases, the mercury loss increases and the

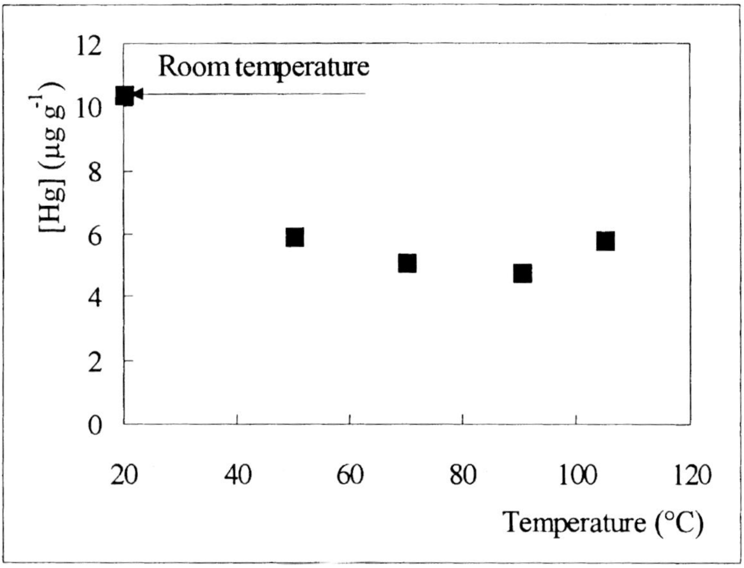

Fig. 2. Effect of the drying temperature of sediment on loss of mercury

mercury contents in the sediment decrease by about 50%, going from 20 to 80–100 °C drying temperature. For mercury analyses, the sediments are chemically digested with nitric and sulphuric acids in the presence of $KMnO_4$. Mercury is analyzed by cold vapour flameless atomic absorption spectrometry (CVFAAS) on a Perkin Elmer 430 with a preconcentration of gold powder if necessary (Krempp 1996). To avoid water vapour condensation on windows, the detection cell is heated at a temperature higher than the dewpoint. The absolute detection limit is 5 ng of mercury in the initial analyzed solution.

Mercury contents were analyzed in the different phases (aqueous phase, suspended matter and bottom sediments) and in the different granulometric fractions of the sediments. The bottom sediments were riddled to separate the different sand fractions (fine 50–200 μm and coarse >200 μm). The clay and silt fractions (<50 μm) were separated by syphoning the sediments in suspension according to Stokes law [clays (<2 μm), fine silts (2–20 μm) and coarse silts (20–50 μm)]. This granulometric separation exercise was made only for the bottom sediments collected in February 1996.

3 Spatial and Seasonal Variations in Mercury Content in the Ill River Basin

Mercury concentrations measured at the different stations upstream to downstream are low in the solutions, as already shown for other rivers (Beim and Grosheva 1996; Grosheva and Hayashi 1996; Ikingura and Mutakyahwa 1996). These concentrations are sometimes even below the detection limit, even in industrial wastewater. All aqueous concentrations of mercury remain under the standard limit allowed for drinking water (1 $\mu g\ l^{-1}$, FAO/OMS 1972; EC 1980).

Nevertheless, the concentrations present a generally decreasing trend upstream to downstream in the basin, which can be attributed to river water dilution by the different tributary inputs and to sorption processes onto colloids (Fig. 3a).

It is interesting to note that the mercury concentrations in the aqueous phase are higher during the low-water period of July 1996 than during the high flows of February 1996. This difference between low and high-water periods must be related to the difference in solute mercury concentration in industrial wastewater (169 $ng\ l^{-1}$ in July and under the detection limit in February) and to the methylation processes of bottom sediment mercury. Contrary to the solutions, the suspended matter (Fig. 3b) and the bottom sediment (Fig. 3c) show high concentrations of mercury, as already shown for other rivers (Dauvalter et al. 1996; Dauvalter and Rognerud 1996; Ikingura and Mutakyahwa 1996). This is due to sorption processes onto sediments because of the high total dissolved solid concentrations in the river water (see next section) and of the high affinity of mercury for solid phase (Chapman et al. 1982; Dinh 1982; Behra 1987; Tiffreau et al. 1995; Tavares et al. 1996). However, as with the solutions, mercury

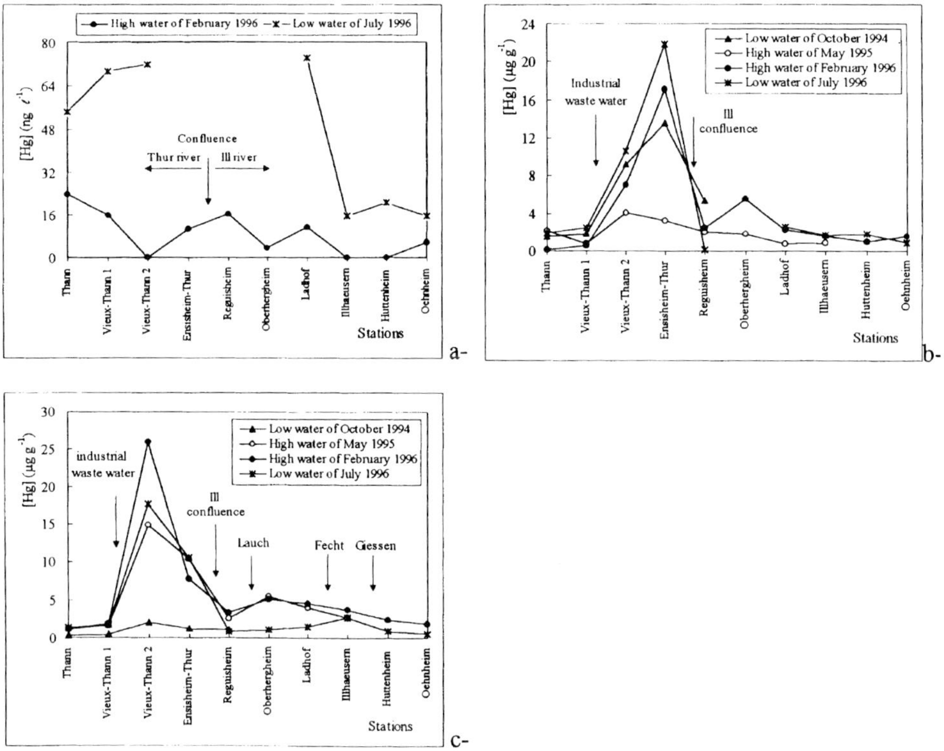

Fig. 3. Longitudinal profiles of mercury contents in the Ill river basin going from the Thur river (upstream) to the Ill river (downstream). **a**- In solution **b**- In suspended matters **c**- In bottom sediments

concentrations decrease upstream to downstream in the Ill river basin. The suspended sediments supplied by industrial wastewater to the Ill river in Vieux Thann station (between Vieux Thann 1 and 2) contain 13 to 438 µg g^{-1} mercury depending on the sampling period. Then, these mercury-polluted sediments are progressively diluted in the river water by the input of unpolluted suspended matter coming from the physical erosion of the drainage basin.

In the river bottom sediments, concentrations of mercury are always greater than the regional geochemical background, which could be determined in this study by using previous measurements in river bottom sediments of several Ill tributaries which are not a priori polluted by mercury. As we can see in Fig. 4, the regional geochemical background value could be estimated at 0.3 µg g^{-1} mercury in the bottom sediments by taking the average of values between 0.05 and 0.5 µg g^{-1} mercury, which correspond to the world natural background limits (Chesterikoff et al. 1973; AFEE 1975; Martin and Scolari 1977; Lindqvist et al. 1984; Lindqvist 1986) and which are the most frequent values in the unpolluted areas of the Ill river basin. Nevertheless, it can also be seen in Fig. 4 that some bottom sediments present a relatively high mercury concentration, even in a

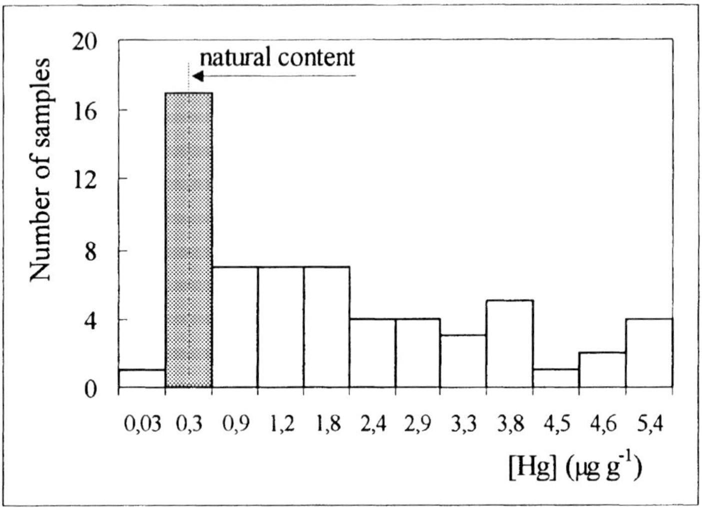

Fig. 4. Regional natural background of mercury content in river bottom sediments in the Ill basin (Data from Collin and Gendrin 1976 and Carbiener 1978)

priori unpolluted areas; but we are not totally sure that these areas have never been contaminated by mercury pollution in the past.

The differences that can be observed in Fig. 3 between longitudinal mercury concentration profiles corresponding to different sampling periods could be attributed to two main factors:

1. The amount of mercury discharged into the Thur river by industrial point pollution.
2. The climatic and hydrologic conditions of the sampling period, including the total river suspended sediment concentrations (TSS).

During the high- water period of May 1995, mercury concentrations in the suspended matter were generally lower than during the high-water periods of February 1996, in May 1995 the mercury concentration (110 µg g^{-1}) in the suspended sediments of the industrial wastewater being four times lower than in February 1996 (438 µg g^{-1}). In the Thur river part (until the Ill confluence), this difference could not be attributed to a dilution effect of the polluted suspended sediments by suspended matters coming from slope erosion because the total suspended sediment concentration is more or less the same during the two periods (Fig. 5) whereas in the Ill river part of the system, the dilution effect contributes to the decrease in mercury content in the sediment because the TSS concentrations are higher during February 1996 than during May 1995 (Fig. 5).

If one compares mercury concentrations in bottom sediments and in suspended matter, it is important to note that during the high-water periods of May 1995 and February 1996 the concentrations are higher in the bottom sediments than in the suspended matter, whereas during the low-water period of October 1994, on the contrary, the suspended matter is more concentrated. One

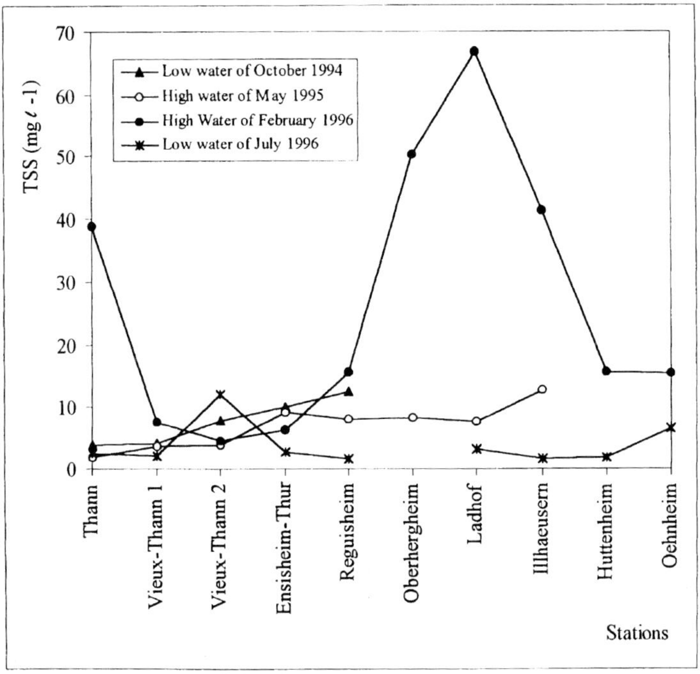

Fig. 5. Longitudinal profiles of total suspended sediment concentration (TSS) in the Thur and Ill rivers during the different sampling periods. a- High flows of February 1996 b- Low flows of July 1996

could explain these patterns by higher suspended sediment concentrations and higher water column depth, which, respectively, dilute the mercury concentration in the suspended matters and probably decreases the methylation processes of mercury in the bottom sediments. It is also important to note that the methylation processes which increase the losses of mercury in the bottom sediments are also limited by low temperature during the winter high-water periods. For example, it is interesting to note that the mercury content in the bottom sediment collected in the industrial channel is five times higher during the high-water period of February 1996 (winter time) than during the low flows of July 1996 (summer). Such methylation processes have already been observed on river basin or lake catchment scales, and the mercury fluxes which have been measured indicate relatively high specific values: 0.08–0.17 g methylmercury (MeHg)/km^2/a for the Gardsjön catchment in Sweden (Lee et al. 1995), 0.07–0.30 g MeHg/km^2/a for tributaries of Onondaga Lake in New York state, USA (Henry et al. 1995), 0.8–5.9 g MeHg/km^2/a for Swedish catchments receiving mercury atmospheric deposition (Johanson et al. 1991), 0.3–2.3 g MeHg/km^2 a for an unperturbed drainage basin in a Canadian boreal forest (St. Louis et al. 1994). If one considers that 2% (Mason and Morel 1993) to 89% (Gill and Bruland 1990) of total dissolved mercury originates from methylation processes, it is possible to estimate the methylation fluxes in the Ill river water (0.2–0.6 to 7–26 g MeHg/

km^2 per year). Even if these values are only rough estimations, they are in the same order of magnitude as the literature values quoted above.

The longitudinal profiles of July 1996 are rather exceptional because the high mercury concentrations which can be observed both in the suspended matter and in the bottom sediments must be attributed to the highest solute mercury concentrations (169 ng l^{-1}) measured in the industrial wastewater in Vieux Thann and to accidental pollution of the Thur from the wastewater treatment station in Cernay.

The mercury concentrations measured in the different tributaries of the Ill river after the confluence with the Thur river, i.e. the Lauch upstream Colmar, the Fecht between Colmar and Selestat and the Giessen downstream from Selestat are very low in the different phases and within the ranges of natural values.

4 Mercury River Fluxes in Particulate and Dissolved Phases

Using mercury content in the dissolved and particulate phases and river discharge measurements at different gauging stations, it has been possible to calculate the amount of mercury transported in solution and in suspension by the Ill river and its tributaries. For each sampling period, one can follow the spatial evolution of mercury fluxes going from the upstream part of the basin to downstream (Fig. 6a, b). Globally for the whole basin, the particulate river fluxes of mercury appear to be higher during the high-water period of February 1996 than during the low flows of July 1996 (Fig. 6a). This is mainly due to higher industrial mercury fluxes, higher river-suspended sediment transports and important remobilization of river bottom sediments. In contrast to particulate phases, the solute mercury river fluxes are higher during low flows than during high-water periods (Fig. 6b). This can be attributed to a higher solute mercury content in the industrial wastewater and to methylation processes of bottom sediment mercury which increase the mercury concentration in the solution. It is interesting to note in Fig. 6b that solute mercury fluxes are rather important downstream from Oberhergheim station, even if at this station the river discharge decreases to zero because of river water infiltration to groundwater. This means that downstream from this station, mercury in dissolved phases mainly originates from methylation processes of mercury previously adsorbed on bottom sediments.

If one compares the shape of the spatial evolution of particulate mercury fluxes during the two sampling periods (February and July 1996), it becomes evident that for the high- water period the fluxes increase until Oberhergheim and decrease in the lower part of the basin, whereas for the low flows it is the contrary. In the upper part of the river, the increase during high flows is due to remobilization of mercury-enriched bottom sediments while the decrease during low flows can be attributed to sedimentation processes. In the downstream part, after Oberhergheim station, the slope of the main channel decreases and the sedimentation processes become active even during high-water periods, as could

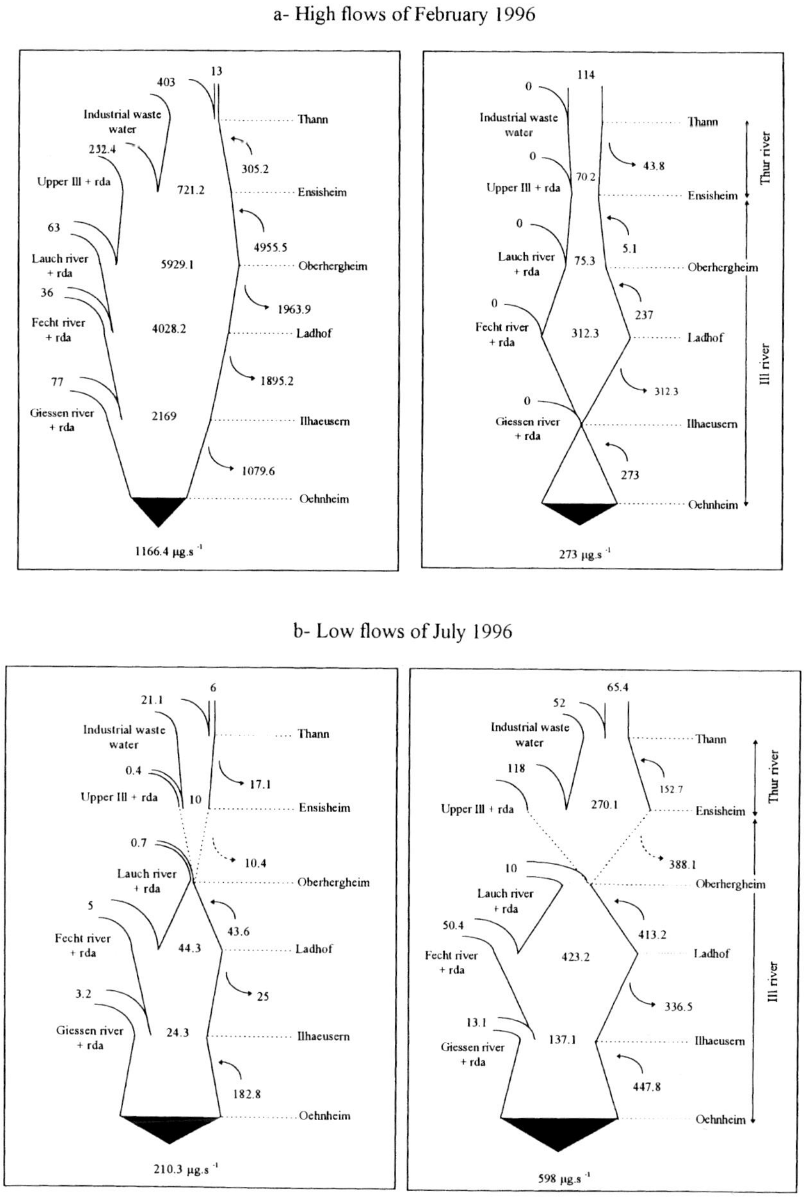

Fig. 6. Spatial evolution of mercury fluxes (in $\mu g.s^{-1}$) in particulate (left side) and dissolved (right side) phases in the upper Ill river basin during high (a) and low (b) water periods. rda: Residual drainage area

be observed in the spatial behaviour of total suspended sediment transports (Messaitfa 1997). On the contrary, during the low flows, the particulate fluxes of mercury increase in the downstream part, from Oberhergheim to Oehnheim, mainly because of methylation processes and biological autochthonous production, which increase the total suspended matter concentrations and the mercury content in suspended matter.

As one can see in Fig. 6a, b, it was possible, using input-output budgets between two sampling stations, to calculate the flux of particulate mercury which was lost by sedimentation processes or supplied by remobilization of bottom sediments. These mercury losses or gains during the river-suspended sediment transport could be compared in Fig. 7 with mercury content behaviour in bottom sediments for each river channel section. As seen in Fig. 7, there is a good relationship for each sample between the mercury concentration calculated for the remobilized or deposited part of the suspended sediment fluxes and the mercury content in the downstream bottom sediments.

5 Influence of Particle Size on Mercury Content in the Sediments

To determine the influence of particle size on mercury content, we decided to work on bottom sediments because the suspended matters are always fine fractions (clays and silts), particularly during the low-water flows. As can be seen in Fig. 8, the granulometry of the bottom sediments has a great influence on the distribution of mercury content. Fine particles (clays and silts, <50 μm) are more concentrated in mercury (in $\mu g\ g^{-1}$) than the coarse fractions (sand >50 μm).

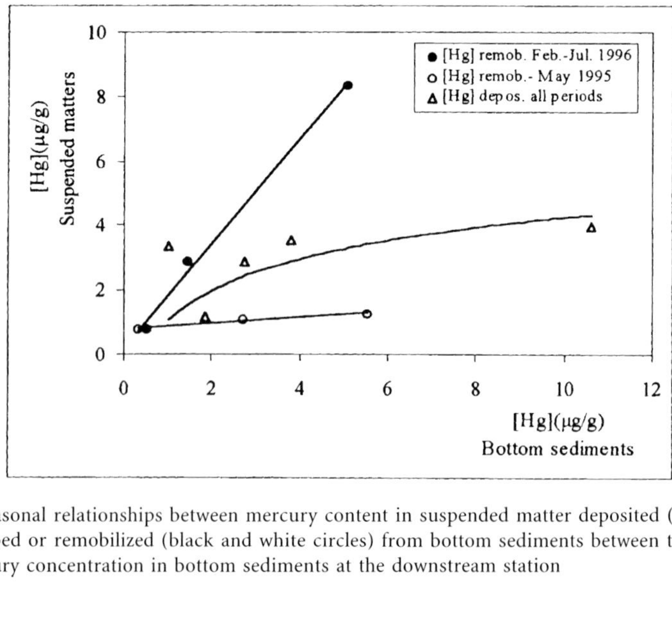

Fig. 7. Seasonal relationships between mercury content in suspended matter deposited (triangle) in the river bed or remobilized (black and white circles) from bottom sediments between two stations and mercury concentration in bottom sediments at the downstream station

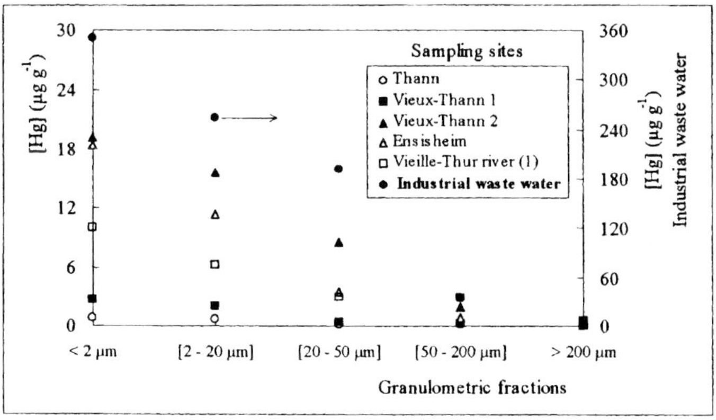

a

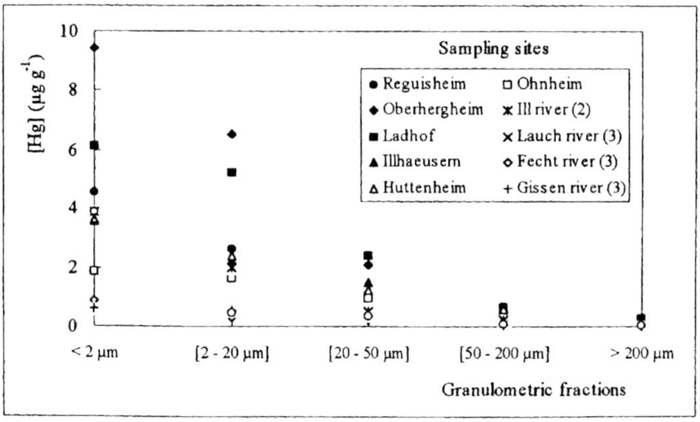

b

Fig. 8a,b. Mercury content in the different granulometric fractions of the bottom sediments collected during the sampling mission of February 1996 at the different stations of the Thur river (**a**) and of the Ill river basin (**b**)

These patterns are in agreement with previous results obtained by different authors (De Groot et al. 1971; Oliver 1973; Smith et al. 1973; Jaffe and Walters 1977; Robbe 1984). Thus, it is evident that mercury sorption is more important onto the finest particles (Reimers and Krenkel 1974; Reimers et al. 1974) which have the highest specific surface area, particularly onto the so-called clay fractions as is also the case for other metals (Kudo 1977; Boust 1981; Boust et al. 1981). Nevertheless, the abundance of mercury in the different fractions (Fig. 9) depends also on the relative abundance of these fractions. Indeed, even if the mercury concentrations are lower in the coarse fractions, the total mercury contained in these fractions is not negligible, if these fractions are abundant. This mercury sorption onto sandy sediments must be taken into consideration in such a study. This is the case, for example, for the stations of Thann and Vieux Thann 2 on the Thur river and Illhausern on the Ill river and for two other stations on the Vieille-Thur and Fecht rivers (Fig. 9). Of course, the sorption of

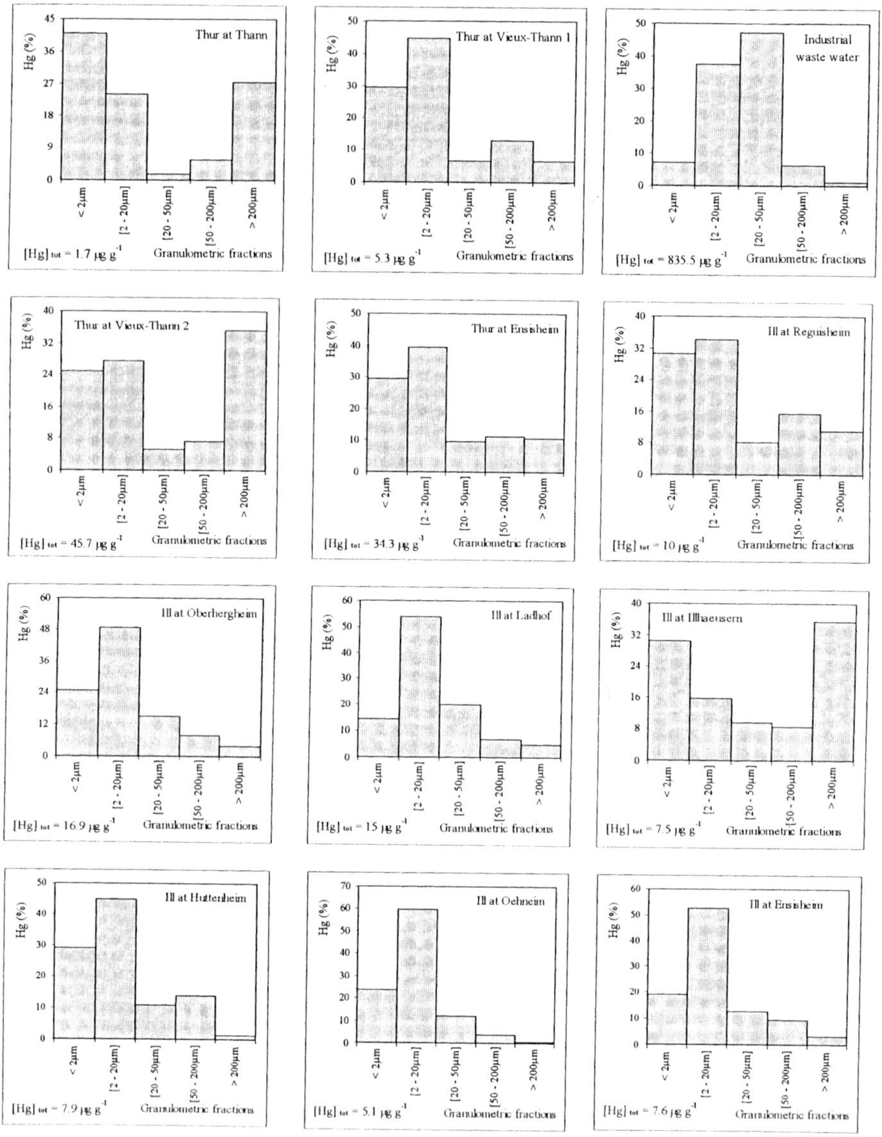

Fig. 9. Distribution of total mercury in different granulometric fractions of the river bottom sediments collected during the sampling mission of February 1996 at the different stations. The mercury content in each fraction is expressed as a percentage of total mercury content in the whole sediment

mercury onto sediments depends not only on the particle size distribution but also on the physicochemical, mineralogical and surface composition of these particles. ICP-AES and ICP-MS analyses show that mercury pollution in the sediments is associated with high concentrations of Al, Fe, Mn and Ti (Messaïtfa 1997).

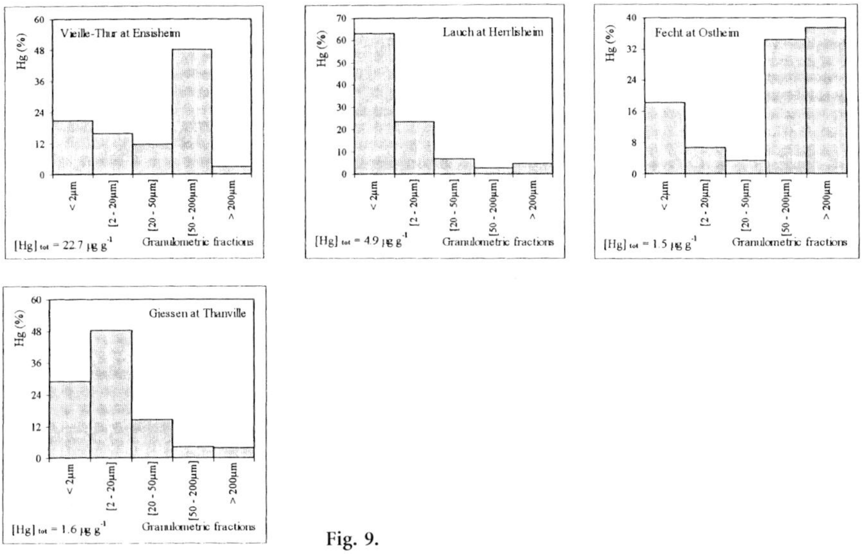

Fig. 9.

6
Influence of Chloride Concentration on the Mercury Content in the Different Phases

6.1
Field Measurements

The upstream part of the Ill river basin is also polluted by saline industrial activity of potash mines which supply mainly sodium chloride indirectly into the river. Consequently, chloride concentrations are very high in the Thur and Ill rivers. Contrary to what has already been reported by different authors (Beijer et al. quoted by Fagerström and Jernelöv 1972; Robin and Ottmann 1976; Frenet 1979; Behra 1987; Melamed and Villas Bôas 1996), mercury content in suspended matter and in bottom sediments of the Thur-Ill river system increases with chloride concentration (Fig. 10b, c). At the same time, mercury concentration in the solution decreases (Fig. 10a). Thus it is possible that increasing chloride concentration in the solution could play a role in mercury sorption onto metal (hydr)oxides present in the sediments.

6.2
Laboratory Experiments

In order to test the influence of chloride concentration on mercury sorption, we carried out two laboratory experiments during the same time: the first with bottom sediments of the Thur river collected at Ensisheim and the second with a mercury solution prepared in our laboratory. In the first experiment, we wanted to test the reactivity of a mercury-polluted sediment containing 10.4 μg/g of mercury to increasing NaCl concentration solutions by measuring the mercury concentration in the sediment and in the solution at each step of the experiment. The second experiment is not directly related to the field observations. Nevertheless, it is important to test the competition between mercury ($HgCl_4^{2-}$) and HPO_4^{2-} (which is well known to have a great affinity for particule surfaces; Sposito 1989) present in the solution in order to be sure that mercury is really trapped by manganese precipitate when the NaCl concentration of the solution increases. This hypothesis could be verified by measuring the evolution of mercury concentration when the NaCl concentration increases in the presence or absence of Na_2HPO_4.

The results obtained during the first experiment (Fig. 11a, b) show that the mercury desorbed from the sediment into the saline solution decreases when the salinity increases. Nevertheless, it can be observed that for the silt fractions (2–50 μm), the total mercury desorbed increases during the first three steps of salinity increase, until 3.4 mmol/l of NaCl, and after decrease, as for the clay fractions (<2 μm). It is surprising to note that when the NaCl concentration increases, the mercury contents decrease, not only in the solution but also in the sediments, and could not even be detected in the two phases when NaCl concentration reaches 0.1 mol/l. This behaviour could be explained by the sorption of mercury onto metal (hydr)oxides when salinity increases; but mercury trapped in this complex could not be detected. Indeed, the classical chemical digestion with nitric and sulphuric acids did not allow the release of mercury to be analyzed because at pH1 and high chloride concentration, the

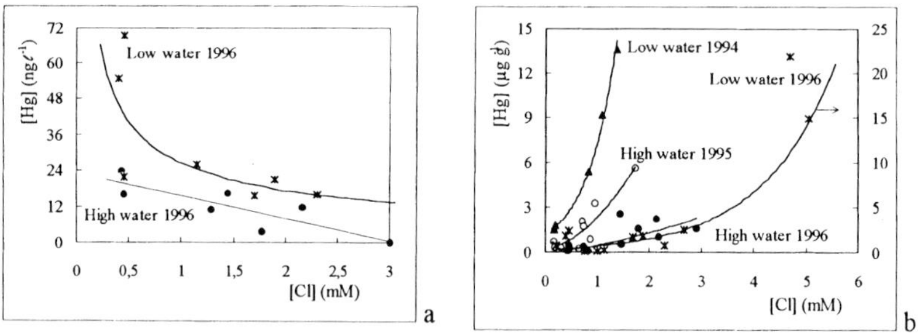

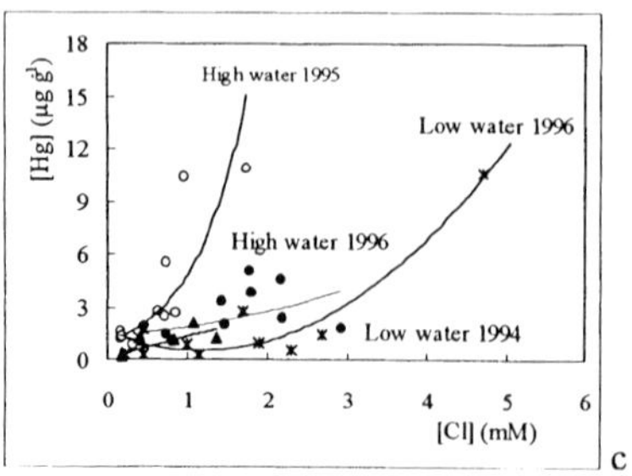

Fig. 10a-c. Variations in mercury concentrations in the different phases as a function of chloride concentrations in the solution for the different sampling periods in the Thur-Ill river system. a- Solution b- Suspended matters c- Bottom sediments

Table 1. Experimental conditions of the second experiment. Competition between mercury and HPO_4^{2-}

	Step 1	Step 2	Step 3
Experimental conditions	$-KMnO_4$ = 5 drops $-HNO_3$ (35%) = 4 ml $-H_2SO_4$ (50%) = 4 ml –[NaCl] = 0.001 – 0.4 M	$Na_2HPO_4.12H_2O$ (4.2 g l^{-1})	[Hg] = 5 μg l^{-1}
Reaction times (h)	24	24	24
Observations	Manganese precipitate	Addition of phosphate	Addition of mercury

stable mercury species ($HgCl_4^{2-}$) (Michard 1989) is strongly adsorbed onto metal oxides due to the oxide protonation surface ($\equiv$S-OH^+).

The second laboratory experiment exemplified the effect of manganese precipitation which could be observed during the first sampling because $KMnO_4$ was added to the solution in the field (see Material and Methods). The experimental conditions are described in Table 1; pH values remain acidic during the experiment.

As during the first experiment, the mercury concentration in the solution decreases to zero when the sodium chloride concentration reaches 0.1 mol/l in the absence of Na_2HPO_4 (Fig. 12a).

It is important to note that below 0.1 M NaCl, the solution remains coloured purple by the $KMnO_4$. Whereas when the NaCl concentration exceeds 0.1 M, a manganese precipitate appears at the bottom of the bottle and the solution becomes colourless. Coprecipitation with or sorption of mercury onto manganese precipitate could explain why the concentration decreased to zero in the solution.

In the presence of Na_2HPO_4, mercury and HPO_4^{2-} appear to compete to be adsorbed onto the manganese precipitate; but mercury presents a greater affinity on manganese precipitate than HPO_4^{2-}. Consequently, when the NaCl concen-

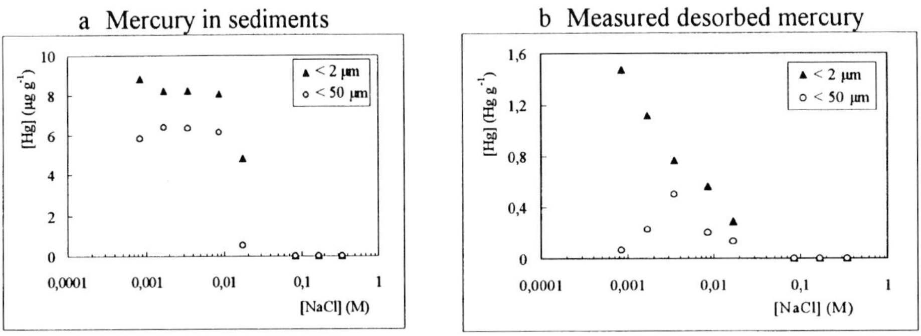

Fig. 11a,b. Results of laboratory experiments with the fine fractions (<50 μm) of the Thur river bottom sediments collected at Ensisheim. a Evolution of mercury content in the sediments (clay and silt fractions) when NaCl concentration in the solution increases. b Evolution of desorbed mercury concentration in the solution when the NaCl concentration of the solution increases.

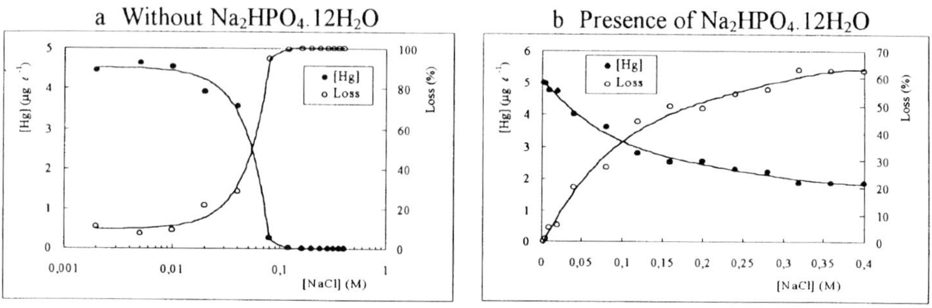

Fig. 12. Evolution of mercury concentration in the solution when chloride concentration increases. a- Without Na_2HPO_4. b- With Na_2HPO_4 (see experimental conditions in table 1). The losses of mercury are expressed in % of the initial mercury content in the solution (5 µg l^{-1}).

tration increases (Fig. 12b), the mercury content in the solution decreases also, but less than in the absence of Na_2HPO_4. For NaCl concentration greater than 0.1 M, the losses of mercury in the solution are only 30–60% of the initial mercury content (5 µg l^{-1}), instead of 100% in the absence of Na_2HPO_4. As quoted by Tiffreau (1996), Mn oxides are very good sorbents for mercury even at very low pH, due to their low point of zero charge.

7 Conclusion

The Ill river and one of its tributaries, the Thur river, show evidence of mercury contamination which could be attributed to industrial wastewater input in the upstream part of the Thur river. The study of mercury pollution behaviour in the Thur-Ill river system during four sampling periods shows that mercury is highly concentrated in suspended matter and in bottom sediments, whereas in solution, the mercury content is very low and often below the detection limit. Mercury concentrations in sediments are largely superior to the regional natural background of mercury content in the river bottom sediments, which could be determined in this study as 0.3 µg/g in the unpolluted areas of the different Ill tributaries. The longitudinal profiles of mercury content established from 16 measurement stations present more or less the same pattern in the different fractions (solution, suspended matter, bottom sediments): the mercury concentrations progressively decrease from the pollution source to the downstream part of the Ill basin by natural dilution processes due to water and suspended sediment inputs by the different tributaries. The differences observed between the sampling periods could be due to the initial mercury concentration in the pollution source and to hydrological and climatological conditions. High mercury concentrations in the industrial wastewater induce high mercury content in the river system. Low-water periods corresponding generally to high air temperature, low-water column depth and low-river suspended sediment concentration could be favourable to methylation processes in the river itself, decreasing the mercury content, particularly in the river bottom sediments. The mercury methylation fluxes which could be roughly estimated for the Ill river are in good agreement with the values literature. On the contrary, during high flows, these processes are relatively limited and mercury contents in the river system could be higher even if the dilution effect is more important.

Input-output budgets of mercury river fluxes from upstream to downstream in the basin showed the different processes which control the particulate and solute river transports of mercury pollution, such as mercury methylation in bottom sediments particularly during low-water periods, deposition or remobilization of suspended matter according to the hydrological season and the channel section.

A granulometric study of the bottom sediments shows that mercury concentrations are higher in the finest particles (clays and silts <50 μm) for all the stations, but the abundance of the different granulometric fractions must be taken into account in the global mercury budget of the sediment. That means that if coarse fractions (sand >50 μm) are abundant, the contribution of these fractions to total mercury content in the sediment must be considered.

It could be shown also that river water salinity could play an important role in the mercury sorption onto river sediments, particularly in the Ill river basin, where industrial saline activity supplies high sodium chloride wastewater to the Thur and Ill rivers. Our field measurements show that when the chloride concentrations increase in the river water, the decrease in mercury content in the solution is associated with an increase in mercury concentration in the suspended matter and in river bottom sediments. Laboratory experiments showed that when the salinity increases, mercury can be trapped, for example, by manganese precipitates. This example illustrates the possibility of solute mercury being adsorbed onto colloidal phases when salinity increases, as could be observed in the field.

New laboratory experiments and other field sampling missions are planned to confirm the different assumptions. It is also very important now to study mercury speciation in the different phases and to determine the impact of this mercury contamination on the groundwater and on the Ill river ecosystem by studying, for example, mercury content in mosses and in fishes.

Acknowledgements This study was supported by the GIP HYDROSYSTEMES (Groupement d'Intérêt Public) of BRGM-CEMAGREF-CNRS-IFREMER-INRA-ORSTOM-OIEau and the Programme Environnement, Vie et Sociétés (SEAH, Systèmes Ecologiques et Actions de l'Homme) of CNRS. The authors are grateful to the reviewers for their helpful comments on the first version of the manuscript.

References

AFEE (Association Française pour l'Etude des Eaux) (1975) Les micropolluants minéraux dans les eaux superficielles continentales. no 1. Le mercure. Synthèse bibliographique du Bureau National de l'Information Scientifique et Technique (BNIST), Paris, p 103

Behra P (1987) Etude du comportement d'un micropolluant métallique – le mercure – au cours de sa migration à travers un milieu poreux saturé: Identification expérimentale des mécanismes d'échanges et modélisation des phénomènes. Thèse 3e cycle, Université Louis Pasteur, Strasbourg, p 191

Beim A, Grosheva E (1996) Transport and bioaccumulation of mercury in the North Dvina river basin. Proc IV Int Conf on Mercury as Global pollutant, 4–8, Aug 1996, Hamburg, p 314

Boust D, Jouanneau JM, Latouche C (1981) Méthodologies d'interprétation des teneurs totales en métaux traces contenues dans les sédiments estuariens et littoraux Bull Inst Géol Bassin d'Aquitaine, Bordeaux 30:71–86

Boust D (1981) Les métaux traces dans l'estuaire de la Seine et ses abords. Thèse 3e cycle, University de Caen, p 186, + annexes

Carbiener R (1978) Etude écologique de la pollution par le mercure du bassin du Rhin en Alsace et de son évolution 1973–1977. UER de Pharmacie, Université Louis Pasteur, Strasbourg, p 149

Chapman PM, Romberg GP, Vigers GA (1982) Design of monitoring studies for priority pollutants. J Water Pollut Control Fed 54, 3:292–297

Chesterikoff A, Carru AM, Garban B, Ollivon D, Chesterikoff C (1973) La pollution de la Basse-Seine par le mercure (du Pecq à Tancarville). Centre Belge d'Etude et de Documentation des Eaux, Liège, no. 355–356, pp 269–276

Collin P, Gendrin P (1976) Bassin du Rhin. Pollution par les métaux lourds: mercure-plomb-cadmium. Campagne de surveillance 1975, Comité Régional de l'Amenagement des eaux Alsace, Colmar, p 42

Dauvalter V, Rognerud S (1996) Mercury in lake sediments in the border areas between Russia and Norway. Proc 4th Int Conf Mercury as a global pollutant, 4–8 Aug 1996, Hamburg, p 306

Dauvalter V, Moiseenko T, Rodyushkin I, Öhlander B, Ponter C, Kudryavtseva L, Sharov A, Ödmann F, Peinerud E (1996) Mercury in the Imandra lake, Murmansk region, Russia. Proc IV Int Conf on Mercury as Global pollutant, 4–8 August, 1996, Hamburg, pp 152

De Groot A J, De Goeij JJM, Zegers C (1971) Contents and behaviour of mercury as compared with other heavy metals in sediments from the rivers Rhine and Ems. Geol Mijnbouw, Gravehgue 50:pp 393–398

Dinh TD (1982) Contribution à l'étude des mécanismes de la pollution des eaux souterraines par le mercure. Examen du transfert en milieu naturel et approche fondamentale en laboratoire. Thèse de Docteur Ingénieur, Université Louis Pasteur, Strasbourg, p 124 + annexe

EC (European Communities) (1980) Directive du conseil du 15 juillet 1980 relative à la qualité des eaux destinées à la consommation humaine (80/778/CEE). J Off Commun Europ: L 229, 30 July 1980:11–29

Esteves M (1989) Etude et modélisation des relations aquifère-rivière dans le Ried de Colmar (Haut Rhin, France). Thèse de Doctorat de l'Université Louis Pasteur, CEREG, Strasbourg, p 233

Fagerström T, Jernelöv A (1972) Some aspects of the quantitative ecology of mercury. Water Res 6: 1193–1202

FAO/OMS (1972) Evaluation de certains additifs alimentaires et de contaminants: mercure, plomb, cadmium. 16e rapport du comité mixte FAO/OMS d'experts des additifs alimentaires. Genève, 4–12 Avril 1972

Frenet M (1979) Phénomène de fixation-désorption du mercure sur les argiles dans les eaux à salinité variable. Application à l'estuaire de la Loire. Thèse d'état, Universite. de Nantes, p 227

Gill GA, Bruland KW (1990) Mercury speciation in surface freshwater systems in California an other areas. Environ Sci Technol 24 (9):1392–1399

Grosheva E, Hayashi S (1996) Mercury transport and bioaccumulation in the Selenga river basin and estuary. Proc IV Int Conf on Mercury as Global pollutant, 4–8, Aug 1996, Hambourg, p 68

Henry EA, Dodge-Murphy LJ, Bigham GN, Klein SM Gilmour CC (1995) Total mercury and methylmercury mass balance in an alkaline, hypereutrophic urban lake (Onondaga lake, NY). Water Air Soil Pollut, 80:509–518

Ikingura JR, Mutakyahwa MKD (1996) Sources of mercury contamination and exposure in Tanzania. Proc IV Int Conf on Mercury as Global pollutant, 4–8, Aug 1996, Hamburg, p 316

Jaffe D, Walters J K (1977) Intertidal trace metal concentrations in some sediments of the Humber estuary. Sci Total Environ, 7:1–15

Johanson K, Aastrup M, Anderson A, Bringmark L, Iverfeldt A (1991) Mercury in Swedish forest soils and waters. Assessment of critical load. Water Air Soil Pollut 56:267–281

Krempp G (1988) Techniques de prélèvements des eaux natureles et des gaz associés. Méthodes d'analyse des eaux et des roches. Notes techniques de l'Institut de Géologie (nouvelle édition), Universiti. Louis Pasteur, Strasbourg, 19, p 79

Krempp G (1996) Méthode d'analyse du mercure à l'état de trace dans les eaux et les sédiments. Note technique du Centre de Géochimie de la Surface CNRS, Strasbourg, p 5

Kudo A, Townsend RD, Miller DR (1977) Prediction of mercury distribution in river sediments, J Environ Eng Div 4:605–614

Lee YH, Bishop K, Pettersson C, Iverfeldt A, Allard B (1995) Subcatchment output of mercury and methylmercury at Svartberget in northern Sweden. Water Air Soil Pollut, 80:455–465

Lindqvist O (1986) Fluxes of mercury in the Swedish environment: contributions from waste incineration. Waste Manage Res 4:35–44

Lindqvist O, Jernelöv A, Johansson K, Rodhe H (1984) Mercury in the Swedish environment; global and local sources. Report SNV PM 1816, distributed by National Swedish Environmental Protection Board, Box 1302, S-171 25 Solna, Sweden

Martin P, Scolari G (1977) Etude de l'évolution historique de la teneur en métaux lourds de certain sédiments du littoral français, de la méditerranée et du bassin rhodanien. Rev Int Océanogr Méd XLVII:85–89

Mason RP, Morel FMM (1993) An assessment of the principal pathways for oxidation of elemental mercury and the production of methylmercury in Brazilian waters affected by goldmining activities. In: Proc Int Symp Perspect Environ Geochem Trop Countries, Niteroi, pp 409–412

Melamed R, Villas Bôas RC (1996) Phosphate-background electrolyte interaction affecting the transport of mercury through a Brazilian oxisol. Proc IV Int Conf on Mercury as Global pollutant, 4–8, Aug 1996, Hamburg, 365

Messaïtfa A (1997) Transferts d'eau, de sédiments et de polluants associés sur le bassin de l'Ill: cas du mercure. Thèse de doctorat, Université Louis Pasteur, Strasbourg, 210 pp

Michard G (1989) Equilibres chimiques dans les eaux naturelles Publisud, Paris, 357 pp

Oliver BG (1973) Heavy metal levels of Ottawa and Rideau River sediments. Environ Sci Technol 7:135–137

Probst JL, (1996) Transports fluviaux de mercure sur le bassin amont de l'Ill. Premiers résultats. Rapport GIP Hydrosystèmes et Programme Environnement, Vie et Sociétés (SEAH) du CNRS, Strasbourg, p 16

Probst JL Nkounkou RR, Krempp G, Bricquet JP, Thiebaux JP, Olivry JC (1992) Dissolved major elements exported by the Congo and the Ubangui rivers during the period 1987–1989. J Hydrol 135: 237–257

Reimers RS, Krenkel PA (1974) Kinetics of mercury adsorption and desorption in sediments. J Water Pollut Control Fed 46, 2:352–365

Reimers RS, Krenkel PA, Sastry AK (1974) Capsulation of mercury sorption in natural and synthetic media. In: Jenkins stt (ed) Proc 7th Int Conf on Water pollution research, 9–13 Sept 1974, Paris. Pergamon Press, Oxford

Robbe D (1984) Interprétation des teneurs en éléments métalliques associés aux sédiments, Rapports des laboratoires, Série: Envir et Génie Urbain, Ministère de l'Urbanisme et du Logement-Ministère des Transports, Laboratoire Central des ports et chaussées, Paris, p 149

Robin M, Ottmann F (1976) Etude comparative de la fixation du mercure sur les sédiments estuariens et sur la kaolinite. Rev Int Océanogr Méd XLI-XLII:105–125

Roeck U (1992) Le transfert de mercure utilisé comme descripteur du fonctionnement hydrologique (échanges cours d'eau-nappe) dans la plaine du Rhin supérieur en Alsace. Thèse de doctorat de l'Université Louis Pasteur, Strasbourg

Samuel J, Rouault R (1983) Les méthodes d'analyses des matériaux géologiques pratiquées au laboratoire d'analyses spectrochimiques. Notes Techniques de l'Institut de Géologie, n° 16, Université Louis Pasteur, Strasbourg, p 46

Smith JD, Nicholson RA, Moore PJ (1973) Mercury in sediments from the Thames estuary. Environ. Pollut. 4:153–157

Sposito G (1989) The chemistry of soils. Oxford University Press, New York, 227 pp

St Louis VL, Rudd JWM, Kelly CA, Roulet NT, Heyes A, Beaty KG, Bloom NS, Flett RJ (1994) Mass-balance budgets of methyl mercury in two unperturbed boreal forest wetlands. In: Int Conf on Mercury as a global pollutant, 10-14 July 1994, Whistler, British Columbia, Abstr

Tavares CF, Pereira ME, Duarte AC, Millward GE (1996) Mercury in soils and plants from fields surrounding Estarreja channel (Aveiro Lagoon-Portugal) contaminated by a chlor-alkali plant effluent. Proc IV Int Conf on Mercury as Global pollutant, 4–8, August, 1996, Hamburg, 301

Tiffreau C (1996) Sorption du mercure (II) à l'interface eau/solide: étude expérimentale et modélisation. Thèse de doctorat, Université Louis Pasteur, Strasbourg, p 189

Tiffreau C, Lützenkirchen J, Behra P (1995) Modeling the adsorption of mercury (II) on (hydr)oxides I. Amorphous iron oxide and α-quartz, J Colloid Interface Sci 172:82–93

Distribution of Total and Methyl Mercury in a Creek Ecosystem Near Bombay

G.G. Pandit, R.M. Tripathi, S.K. Jha, and T.M. Krishnamoorthy

The Trans Thane Creek Industrial Area (TTCIA) in New Bombay, developed as a chemical zone by the Maharashtra Industrial Development Corporation (MIDC), houses a large number of major and small-scale industries which manufacture and store a number of petrochemical, pharmaceuticals, fine chemicals and pesticide formulations. The creek ecosystem has been monitored three times for total and methylmercury; once during the pre-and postmonsoon periods of 1994, and once during the premonsoon period of 1995. The concentration of total mercury was determined by CVAAS using $SnCl_2$ methodology for liberating vapour phase mercury. Reversed phase high performance liquid chromatography (RP-HPLC) with a UV detector was used for the separation and determination of methylmercury. The efficacy of the methodology was assessed through analysis of standard reference material obtained from IAEA (tuna fish, IAEA-350).

The spatial profile of total mercury in creek water depicted a distinct gradient with low concentration (0.03-0.3 μg/l) near the southern side and high concentrations (0.5–4.3 μg/l) towards north where industries are situated. The postmonsoon values showed only a marginal a decrease in concentration levels during 1994; however, a significant decrease in the concentrations during premonsoon in 1995 was observed. The concentrations of total mercury in sediment and biological species (fish and crabs) varied between (0.3–2.6 μg/g) and (0.03–2.5 μg/g) dry weight, respectively. The methylmercury component in sediment was distinctly lower, 5.9–65.5 ng/g (3–8% total Hg), compared to that in biological samples, 20.4–344.5 ng/g dry weight (33–97% of total Hg).

1 Introduction

Mercury is one of the heavy metals well known for its toxicity. It is widely dispersed and usually occurs at trace levels in the natural environment (Adriano 1986). These natural trace levels can only be detected in unpolluted open oceans (Bloom and Crecelius 1983) and in ambient air in remote areas like the polar region (Ebinghaus et al. 1994). The background levels are usually exceeded in coastal regions as a result of industrial inputs (Ebinghaus et al. 1994). Worldwide natural emissions of mercury are higher and were estimated to be 3500 t/a (Pacyma and Munch 1991). Mercury exists in several forms in the environment. Geochemical and environmental studies have shown that ionic mercury can be converted into highly toxic organomercury compounds (Callister and Winfrey 1986). The Trans Thane Creek Industrial Area (TTCIA) in New Bombay, developed as a chemical zone by the Maharashtra Industrial Development Corporation (MIDC), houses a large number of major and small-scale industries which manufacture and store a number of petrochemical, pharmaceutical, fine

Environmental Science
Mercury Contaminated Sites (ed. by R. Ebinghaus et al.)

chemical and pesticide formulations. In order to project long-term risks and ecological impact associated with the distribution of mercury in the environment, a study to determine total and methylmercury has been carried out in a creek ecosystem near Bombay. This chapter describes the spatial profile of total and methylmercury in the creek ecosystem based on the samples collected during premonsoon (May) and postmonsoon (October) periods.

2
Experimental

2.1
Study Area

The Trans Thane creek is a part of Bombay Harbour Bay, which is connected with the Arabian Sea in the south and the Ulhas river through Thane creek in the north. The bay is subjected to diurnal tides, the range of tidal rise observed being 4.8 and 0.8 m. The Thane Belapur stretch, which is about 20 km long and 2 km wide, encompasses a total area of about 2560 ha. The volume of tidal water on an average is 4.8×10^8 m^3 (Pillai et al. 1958). Along the east and west sides of the Thane creek, many industrial units have been established. The Thane creek, which is the ultimate recipient of all the liquid discharges from the industries, was monitored three times for total and methylmercury once during the pre- and postmonsoon periods of 1994 and once during the premonsoon period in 1995. The major industries located in the TTC industrial area are shown in Fig. 1, where the area has been divided into four sectors: (A) Airoli; (B) Ghansoli; (C) Koparkhairane and (D) Sanpada. The creek sampling locations such as confluence points of canals (A, B, C, D) with the creek upstream (A1, B1, C1, D1), downstream (A2, B2, C2, D2) and midway points of the creek (A3, B3, C3, D3) are also presented in Fig. 1.

2.2
Sample Collection

The creek water samples were collected about 0.2 km from the Airoli coast (A) as well as 0.2 km on the other side (A1 and A2). In addition, samples were also collected at the midpoint of the section (A3). Similarly, water samples were also collected at different locations of the coast such as Ghansoli, Sanpada and Koparkhairane. Water samples were filtered immediately with 0.45-μm membrane filter paper in acid-cleaned polythene bottles containing concentrated nitric acid with concentration adjusted to 0.5% nitric medium. Prior to collection of field samples, the acid-cleaned polythene bottles were tested for blank mercury contents. The bottles were filled with 0.5% HNO_3 overnight and known aliquots were used for blank evaluation. The 0.5% HNO_3 solutions showed no mercury

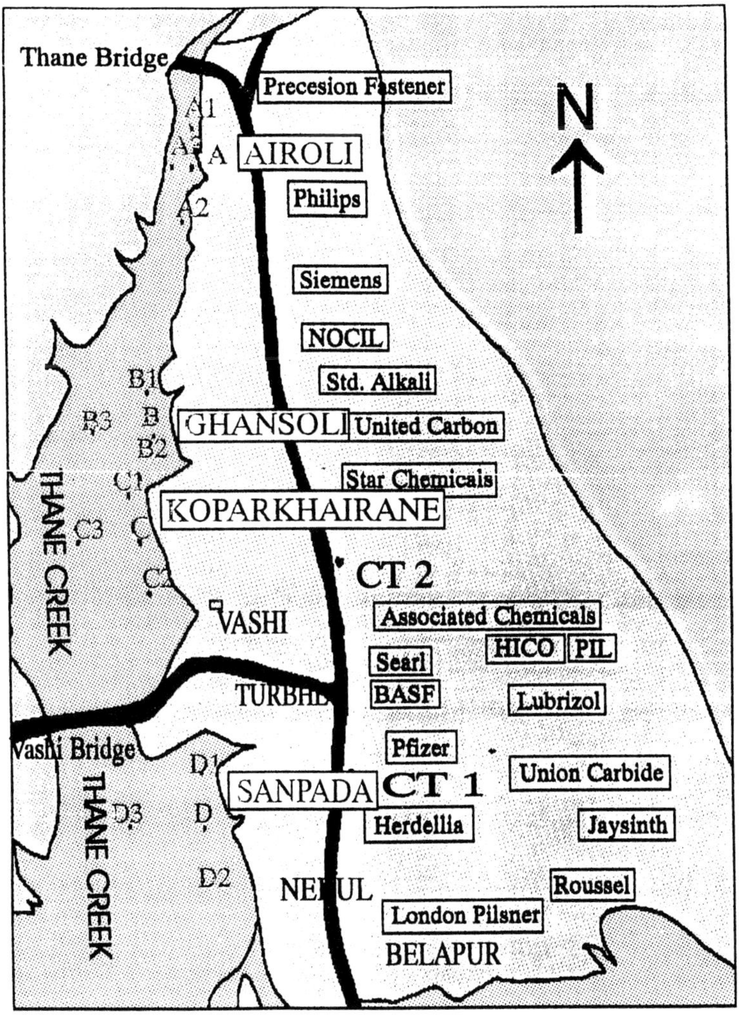

Fig. 1. Major industries in the TTC area and sampling locations for three cruises in the Trans-Thane Creek

concentrations, indicating the need for equilibrating the polythene bottles with concentrated HNO_3 prior to field use for collection of samples.

Sediment core samples were collected from the three locations, namely Airoli, Ghansoli and Koparkhairane, with a gravity corer of 5 cm diameter. Cores were frozen immediately after collection. Upon arrival at the laboratory, all the samples except those to be used for the percent moisture content experiment were freeze-dried.

Biological samples (fish and crabs) were collected during the May 1995 cruise by the fishermen in TTC area where water and sediment samples were collected.

Samples were washed to remove adhering sediment particles, dissected and the edible portions were separated, homogenized and freeze-dried.

2.3 Determination of Total Mercury

Acidified water samples were evaporated to reduce the volume by five to six times prior to analysis. A known aliquot is used for the determination of total mercury by cold vapour atomic absorption spectrometry (CVAAS, model EC,MA-5800 E) using $SnCl_2$ reduction methodology. The loss of mercury during evaporation was studied using radiotracer ^{203}Hg experiments and found to be < 5%. When the concentration of mercury in samples was very low, a coprecipitation technique with copper sulphide followed by chemical separation through ion exchange was employed for its determination (Doshi and Joshi 1987). This has facilitated improving the minimum detection limit by an order of magnitude.

The freeze-dried sediment and biological samples (1 g) were refluxed with 13 ml digestion mixture (HNO_3:H_2SO_4:$HClO_4$ = 10:2:1) in a Soxhlet apparatus for 6 h to extract the total mercury content. The concentrations were determined by CVAAS technique, as mentioned earlier.

2.4 Determination of Methylmercury

Separate aliquots of freeze-dried sediment and biological samples were homogenized, rinsed with acetone (Hight and Corcoran 1987) and extracted in 6 M HCl using a mechanical shaker for 12 h. The methylmercury component was extracted in toluene, followed by backextraction in sodium thiosulphate. Reversed phase high performance liquid chromatography (HPLC) with ultraviolet detection (UV) was used for separation and quantification of methylmercury. An ODS column was used for the separation of methylmercury using methanol-water (1:1) containing 2-mercaptoethanol buffered to pH 5 as the mobile phase (Pandit et al. 1996). The details of analytical parameters are given in Table 1.

Table 1. Details of analytical parameters for the estimation of methylmercury using RP-HPLC

Serial number	Parameter
1. Column	Octa decyl silane (ODS)
2. Mobile phase	Methanol-water (50:50) with 2-mercaptoethanol
3. pH	5 (adjusted with ammonium acetate and acetic acid)
4. Detector	UV-visible
5. Wave length	230 nm
6. Injection volume	20 µl
7. Sample medium	0.1 mM sodium thiosulphate

3
Limit of Detection and Validation

The minimum detection limit for total mercury is 2 ng (relative: 0.2 ng/ml), and this was improved by an order of magnitude using the CuS coprecipitation technique. The minimum detection limit for methylmercury is 0.5 ng (absolute). The efficacy of the methodology was assessed by analysis of standard reference material obtained from IAEA (*t*una fish, IAEA-350).

4
Results and Discussion

The spatial profile of the concentrations of total Hg in creek waters during the three cruises (May 1994, October 1994 and May 1995) is shown in Fig. 2. The water samples collected on first two cruises showed very high concentrations (0.5–

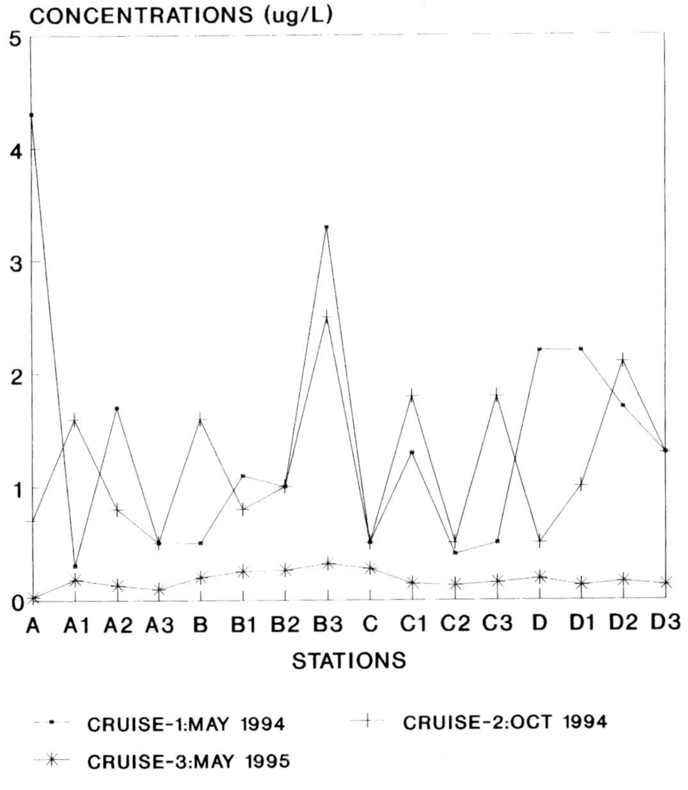

Fig. 2. Concentration of Hg in Thane Creek during three cruises (high tide)

4.3 μg/l) towards the north where industries are situated. During the premonsoon, the spatial profile of mercury showed a prominent peak (4.3 μg/l) at Airoli, followed by a small peak (1 μg/l) near Ghansoli and another at Sanpada (2 μg/l). The spatial profile indicated considerable reduction in concentrations during the postmonsoon period. The prominent peaks at the Airoli and Sanpada regions decreased considerably but maintained higher concentrations (1.3 μg/l) in the Ghansoli-Koparkhairane region. The spatial profile of total mercury in creek water during the third cruise (May 1995) indicates generally very low concentration (0.03–0.3 μg/l) due to the control measures adopted by the industries based on our earlier measurements. Comparison of discharge point and midway point concentrations of total mercury in water samples in the Thane creek are given in Fig. 3. Discharge points at Airoli and Sanpada showed higher levels of Hg compared to midway points, indicating a dilution effect. However, concentrations of total mercury at the discharge points at Ghansoli and Koparkhairane are less than those observed at the midway points at Ghansoli and Koparkhairane.

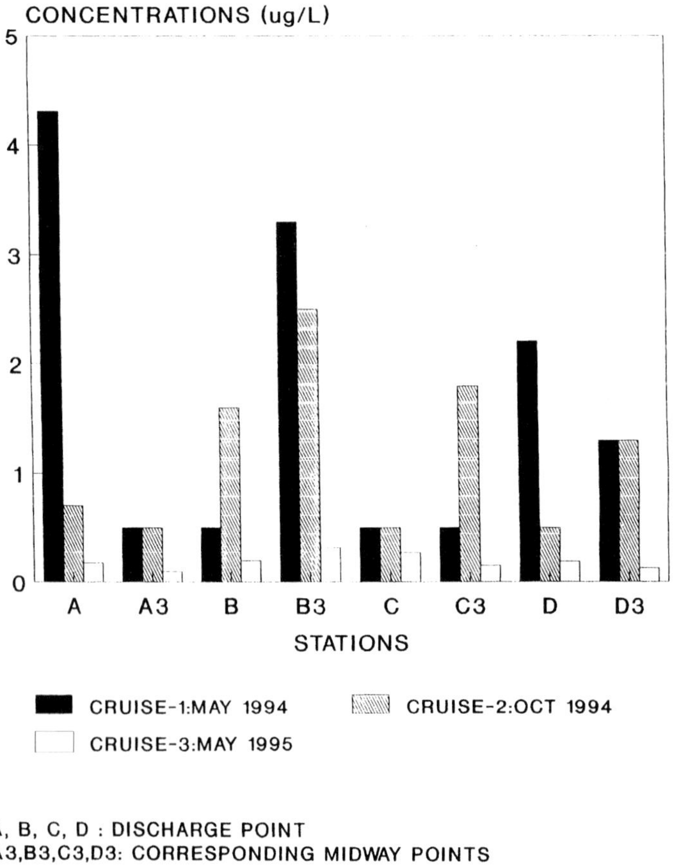

Fig. 3. Comparison of discharge point and midway point conc. of Hg in Thane Creek

Table 2. Concentration of total mercury and methylmercury in sediment samples collected at TTCA in may 1995

Serial number	Location	Concentration of mercury (ng/g d.w.)[a]		Methyl mercury (%)
		Total	Methyl	
1.	Airoli			
	Top layer	1600	46.8	2.8
	15 cm	811	n.d.	
2.	Ghansoli			
	Top layer	2140	65.5	3.1
	15 cm	21.5	n.d.	
3.	Koperkhairane			
	Top layer	75	5.9	7.8
	15 cm	64.7	n.d.	

n.d: not detected.
[a]Percentage of moisture ranged from 58–62.

Table 2 gives concentrations of total mercury and methylmercury in sediment samples collected from the TTC area during May 1995. The sediments from Ghansoli and Airoli showed high concentrations of Hg in the topmost segment, indicating the possibility of a mercury source in the nearby region. The level of total mercury and methylmercury at Koparkhairane showed very low concentration. Percentage of methylmercury to total mercury ranged from 2.8% (Airoli) to 7.8% (Koparkhairane) indicating more biological activity at Koparkhairane (Jha et al. 1996). The high values of total mercury in sediment samples may be attributed to the presence of clay material (48–65%) with a mean grain size of 2–9 µm, and the presence of montmorillonite with a minor amount of *k*aolinite- and *i*llite-type minerals.

Several fish used for consumption by the local inhabitants were analyzed for total and methylmercury. Table 3 summarizes the concentrations of total

Table 3. Concentrations of total mercury and methylmercury in various types of fish samples collected May 1995 in the Trans Thane Creek area

Serial number	Species	Biological name	Concentration of mercury (ng/g d.w.)[a]		Methyl mercury (%)
			Total	Methyl	
1.	Boi	*Mungil dusumieri*	62.5	20.4	32.6
2.	Black fish	*Tilapia mossambique*	137.5	80.2	50.3
3.	Red prawn	*Metapaenpios affinis*	67.8	42.3	62.4
4.	Cat fish	*Macrones vitattus*	176.1	123.0	70.0
5.	Prawn	–	548.1	310.1	56.5
6.	Salpa	–	154.4	94.2	61.0
7.	Dhoma	*Sciaena*	84.1	60.5	72.0
8.	Mudskipper	*Boleophthalmas*	109.2	50.9	46.6
9.	Crab	–	352.4	344.4	97.0

[a]Percentage of moisture ranged from 60–70.

mercury and methylmercury in a variety of biota. The concentrations of total mercury and methylmercury were found to vary from 62.5–548 and 20.4–344.4 ng/g (d.w.), respectively. The ratio of methylmercury to total mercury varied from 0.33 to 0.97 in different species. Species such as crab, prawn and cat fish, which are bottom feeders, showed a higher ratio of methylmercury to total mercury as compared to mid-column feeder species (dhoma, boi and black fish).

5 Conclusions

Thane creek is one of the most mercury-polluted creeks in the Bombay region. The spatial profile of total mercury in Thane creek water on the third cruise which was arranged in May 1995, showed lower concentrations (0.03–0.3 μg/l) due to control measures adopted by the industries based on our earlier measurements during May 1994 and October 1994. The percentage of methylmercury to total mercury in sediments varied from 2.8 to 7.8%, while in biological samples it varied from 33 to 97%. Bottom-feeder species (crabs, prawns and cat fish) have a higher methylmercury to total mercury ratio than mid-column feeder species (*b*oi, *d*homa and *b*lack fish).

Acknowledgement The authors wish to thank Dr. K.S.V. Nambi, Head Environmental Assessment Division, BARC for his interest and kind encouragement.

References

Adriano DC (1986) Trace elements in the terrestrial environment, chap 8 Springer, Berlin Heidelberg New York

Bloom NS, Crecelius EA (1983) On the chemical forms of mercury in edible fish and marine invertibrate tissues. Mar Chem 14:49–59

Callister, SM, Winfrey MR (1986) Microbial methylation of mercury in Upper Wisconsin river sediment. Water Air Soil Pollut 29:1873–1887

Doshi GR, Joshi SN (1987) Determination of mercury in natural waters. Ind J Technol 25:371–372

Ebinghaus R, Hintelmann H, Wilken RD (1994) Mercury cycling in surface waters and in the atmospheric species analysis for the investigation of transformation and transport properties of mercury. Fresenius J Anal Chem 350:21–29

Hight SC, Corcoran MT (1987) Rapid determination of methylmercury in fish and shellfish, J Assoc Off Anal Chem 70:22–28

Jha SK, Krishnamoorthy TM, Sastry VN, Mahadevan TN, Tripathi N, Anilkumar S, Kulkarni VV (1996) Preliminary investigations on sedimentation rates and mercury distribution in core sediments of Thane Creek. Proc 5th Natl Symp on Environment Saha Institute of Nuclear Physics, calcutta 28 Feb–1 March 1996, Bhabha Atomic Research Centre, Bombay pp 108–113

Pacyna JM, Munch J (1991) Anthropogenic mercury emissions in Europe. Water Air Soil Pollut 56:51–61

Pandit GG, Tripathi RM, Jha SK, Krishnamoorthy TM, Nambi KSV (1996) Standardisation of methylmercury estimation in fish and sediment samples by the HPLC-UV technique. Proc 5th Natl

Symp on Environment Saha Institute of Nuclear Physics, Calcutta 28 Feb–1 March 1996 Bhabha Atomic Research Centre, Bombay. pp 118–121

Pillai KC, Subaratnam T, Ganguly AK (1958) Tidal movement and water renewal rates in Bombay Harbour Bay Atomic Energy Establishment Trombay (AEET) India, REP. AEET/HP/RWD-2

Subject Index

Springer and the environment

At Springer we firmly believe that an international science publisher has a special obligation to the environment, and our corporate policies consistently reflect this conviction.
We also expect our business partners – paper mills, printers, packaging manufacturers, etc. – to commit themselves to using materials and production processes that do not harm the environment. The paper in this book is made from low- or no-chlorine pulp and is acid free, in conformance with international standards for paper permanency.

Springer

Printing: Saladruck, Berlin
Binding: Buchbinderei Lüderitz & Bauer, Berlin